Gulf Studies

Volume 21

This book series, indexed in Scopus, serves the growing scholarly interest in the dynamic, complex, and strategically important Arab Gulf region by offering an academic publication platform to scholars in the region and beyond. The series takes an interdisciplinary approach to documenting the changes occurring in Gulf societies. It aims to advance the understanding of Gulf societies and their interactions with the rest of the world from a non-Western perspective. The series explicitly gives scholars from the Arab Gulf region more voice and visibility in academic publications. It includes studies of the member states of the Gulf Cooperation Council (GCC), namely: Bahrain, Kuwait, Oman, Qatar, Saudi Arabia, and the United Arab Emirates, as well as Iran, Iraq, and Yemen. It accepts proposals for research monographs, edited volumes, and handbooks that explain how the Gulf societies' social, economic, and political landscapes have evolved over the past few decades. The series covers a wide range of topics within four broad themes, as follows:

(1) Social and Cultural Issues in the Gulf: Demography, migration, citizenship, gender, culture and identity, urbanization, education, new media, unemployment, youth, family, women empowerment, leadership, aging, human rights, sports, Islamic ethics, and Islam and society

(2) Politics and Security in the Gulf: Gulf defence and security, the GCC integration, Gulf politics and political systems, and the GCC's foreign policy

(3) Energy and Economics in the Gulf: The political economy of the Gulf, economic diversification, international investments, international trade, knowledge economy, energy security and geopolitics of energy, renewable energy development, environmental regulations, urban planning, and Islamic finance

(4) The GCC States and the World: The GCC and Iran, the GCC and Africa, the GCC and South Asia, the GCC and China, the GCC and East Asia, the GCC and Southeast Asia, and the GCC and Western powers.

Submission and Peer Review

All proposals will undergo peer review by the editorial board members. If accepted, the final manuscript will be peer reviewed internally by the editorial board as well as externally (single/double blind) by Springer ahead of acceptance and publication. Please submit your book proposal to the Series Editor, Md Mizanur Rahman: mizan@qu.edu.qa, and to Springer Publishing Editor, Alex Westcott Campbell: alexandra.campbell@springernature.com

Esmat Zaidan · Evren Tok · Asmaa Al-Fadala ·
Logan Cochrane
Editors

The Future of Education Policy in the State of Qatar

Springer

Editors
Esmat Zaidan
College of Public Policy
Hamad Bin Khalifa University
Qatar Foundation
Doha, Qatar

Evren Tok
College of Public Policy
Hamad Bin Khalifa University
Qatar Foundation
Doha, Qatar

Asmaa Al-Fadala
College of Public Policy
Hamad Bin Khalifa University
Qatar Foundation
Doha, Qatar

Logan Cochrane
College of Public Policy
Hamad Bin Khalifa University
Qatar Foundation
Doha, Qatar

ISSN 2662-4494 ISSN 2662-4508 (electronic)
Gulf Studies
ISBN 978-981-97-9666-3 ISBN 978-981-97-9667-0 (eBook)
https://doi.org/10.1007/978-981-97-9667-0

This Springer imprint is published by the registered company Springer Nature Singapore Pte Ltd.
The registered company address is: 152 Beach Road, #21-01/04 Gateway East, Singapore 189721, Singapore

Foreword

The landscape of education is constantly evolving, shaped by global trends, technological advancements, and the unique cultural contexts of each nation. In Qatar, a country that has long prioritized education as a cornerstone of its development, the future holds immense opportunities and challenges. As we stand on the threshold of a new era, weaving together tradition and modernity, local values, and global perspectives, it is more crucial than ever to reflect on where we are and envision where we need to go.

This book is a comprehensive exploration of the future of education in Qatar, offering insights from educators, policymakers, and thought leaders who are at the forefront of this critical field. It delves into the strategies, innovations, and policies that will shape the educational outlook in the years to come, ensuring that Qatar continues to build a knowledge-based society that is resilient, innovative, and globally competitive.

The chapters within this book examine current practices, forward-looking policy options, and critical recommendations. The subject of the *Future of Education in Qatar* is treated with the breadth and depth it deserves, highlighting critical gaps for ongoing exploration in topics such as the evolving roles of educators in a digital age, the impact of globalization on curriculum development, special education needs, vocational and technical training, gender disparities in education, and the challenges of modernizing education while preserving cultural identity. It is hoped that this book stimulates further research and discussion, contributing to a more nuanced and comprehensive understanding of educational development not only in Qatar but globally.

This book aspires to be a valuable resource for practitioners, policymakers, students, and researchers seeking to gain insight into the educational dynamics of Qatar. It offers a blueprint and inspiration for other nations aiming to reform their education systems by integrating modern advancements without discarding cultural values. The narrative highlights the critical role of adaptability, innovation, and global integration in education, painting a vivid picture of an education system that not only shapes the future of a nation but also defines its place in the global arena.

Through the lens of Qatar's National Vision 2030, this book explores the multidimensional and dynamic nature of Qatar's education policies and reforms. It illustrates how these reforms align with the country's broader goals of economic diversification and the delicate balance between globalized educational trends and local cultural values. The chapters do not follow a chronological sequence but rather investigate the deeper implications of these transformative changes on Qatar's social fabric and its trajectory into the future.

As we embark on this journey of exploration, it is important to remember that education is not just about imparting knowledge, but about empowering individuals to think critically, adapt to change, and contribute meaningfully to society. The future of education in Qatar will require us to reimagine traditional models, embrace new technologies, and foster a culture of lifelong learning.

This book serves as a guide for educators, policymakers, and all those who are passionate about the future of education in Qatar. Each chapter contributes to a larger mosaic, providing a comprehensive understanding of how education in Qatar is being rethought and reshaped to meet the challenges and opportunities of the twenty-first century. It is an invitation to engage in a dialogue on the opportunities that lie ahead and the steps we must take to ensure that our educational system not only meets the demands of the future but also sets new standards of excellence.

I hope that the insights and ideas presented in these pages will inspire you to think creatively about the future of education in Qatar, and to contribute to the ongoing transformation of our educational landscape.

Prof. Ibrahim Bin Saleh Al-Naimi
Undersecretary
Ministry of Education and Higher Education
Doha, Qatar

Acknowledgements

This book would not have been possible without the support and contribution of many individuals and institutions to enumerate.

First and foremost, we would like to express our deepest gratitude to our contributing authors, whose commitment and expertise have been the bedrock of this book. Their insightful work and knowledge have greatly enriched this book with innovative ideas and solutions, thereby placing education policies, reforms, and practices in Qatar within a comprehensive and detailed perspective. Their contribution gives a rich compilation of knowledge addressing the complexities and challenges of educational transformation, offering valuable lessons and inspiring future initiatives within Qatar and beyond.

We are immensely grateful to the Qatar Foundation, Hamad Bin Khalifa University, Qatar University, and other educational institutions in Qatar that helped us accomplish our research and provided us with valuable resources. We are also thankful to the Qatar National Research Fund (QNRF) for their support *under NPRP grant #12C-0804-190009,* entitled SDG Education and Global Citizenship in Qatar: Enhancing Qatar's Nested Power in the Global Arena, from the Qatar National Research Fund (a member of the Qatar Foundation). We also acknowledge their funding towards this book's Open Access publication, making its valuable content accessible to a wider audience.

This book is dedicated to the educators and students of Qatar, whose endless search for knowledge and excellence remains at the very core of the nation's educational advancement. It is their effort, we strongly believe, which is going to create an educational framework for the future in Qatar—resilient and vibrant, prepared to answer all the challenges of a world in rapid transformation.

Contents

Editors and Contributors

About the Editors

Esmat Zaidan is Associate Dean for Academic Affairs and Associate Professor at the College of Public Policy, Hamad Bin Khalifa University, Qatar Foundation. She previously held positions at Qatar University, United Arab Emirates University, and the University of Waterloo in Canada, where she earned her master's and Ph.D. degrees. Her previous professional appointments include the World Bank, the Norwegian Agency for Development Cooperation (NORAD), and the United Nations (UNDP). Dr. Zaidan's research focuses on sustainable development strategies and policy planning, with an emphasis on human-centered sustainability policies and people-centric cities. She explores the dynamics between technological deployment and human interaction. Her research also focuses on education and development, education policies and reforms, as well as the localization of SDGs in higher education.

Evren Tok is Associate Dean of Community Engagement and Associate Professor at the College of Public Policy, Hamad Bin Khalifa University. Dr. Tok is also the recipient of the Best Research Project Award in HBKU in 2022 and the Service Excellence Award in 2024.

Asmaa Al-Fadala Assistant Professor at HBKU's College of Public Policy, has 28 years in K-12 and higher education and policy development. She has been a governmental policy writer, professor, and author on leadership for learning, education transformation, teacher development, innovation, and SDGs. She holds a Ph.D. and an M.Phil. from the University of Cambridge in Educational Leadership and Policy.

Logan Cochrane is Associate Professor at the College of Public Policy at HBKU. For the last 15 years, he has worked in non-governmental organizations internationally, including in Afghanistan, Benin, Burundi, the Democratic Republic of Congo, Egypt, Ethiopia, Qatar, Saudi Arabia, South Africa, South Sudan, Tanzania and Uganda.

Contributors

Meshal M. Abdullah Department of Geography, College of Arts and Social Science, Sultan Qaboos University, Muscat, Oman;
Department of Ecology and Conservation Biology, Texas A&M University, College Station, TX, USA

Ammar Abulibdeh Applied Geography and GIS Program, Department of Humanities, College of Arts and Sciences, Qatar University, Doha, Qatar

Khalid A. Al-Ali Director of Higher Education Institutional Affairs, Ministry of Education and Higher Education, Doha, Qatar

Talal Al-Awadhi Department of Geography, College of Arts and Social Science, Sultan Qaboos University, Muscat, Oman

Amira D. Al-Balushi Department of Geography, College of Arts and Social Science, Sultan Qaboos University, Muscat, Oman

Asmaa Al-Fadala College of Public Policy, Hamad Bin Khalifa University, Qatar Foundation, Doha, Qatar

Reem Al-Hababi Gulf Studies, Qatar University, Doha, Qatar

Najlaa Al-Maadid Director of Planning, Quality, and Innovation, Civil Service and Government Development Bureau, Doha, Qatar

Hessa Al-Thani College of Education, Qatar University, Doha, Qatar

Amani Alabed College of Business, University of Doha for Science and Technology, Doha, Qatar

Kholoud Aldous Qatar Computing Research Institute, Hamad Bin Khalifa University, Doha, Qatar

Nouf Alkaabi Director of Planning, Quality, and Innovation, Civil Service and Government Development Bureau, Doha, Qatar

Mohammad S. Alkuwari College of Education, Qatar University, Doha, Qatar

Tasneem Amatullah Emirates College for Advanced Education, Abu Dhabi, UAE

Hira Amin College of Public Policy, Hamad Bin Khalifa University, Doha, Qatar

Loreta Armstrong Strategy & Development Office & College of Business & Economics, Qatar University, Doha, Qatar;
Research Assistant, Research Assistant, Qatar University, Doha, Qatar

Jennifer Bruder Carnegie Mellon University in Qatar, Doha, Qatar

Youmen Chaaban Educational Research Center, College of Education, Qatar University, Doha, Qatar

Abdul Rahman Mahmoud Chamseddine Georgetown University in Qatar, Ar-Rayyan, Qatar

Rosie Clayton WISE Research Fellows, Qatar Foundation, Doha, Qatar

Logan Cochrane College of Public Policy, Hamad Bin Khalifa University, Qatar Foundation, Doha, Qatar

Mustafa Colak Department of Business Administration, Faculty of Political Science, Social Sciences University of Ankara Ulus, Ankara, Turkey

Martina Dickson Emirates College for Advanced Education, Abu Dhabi, UAE

Xiangyun Du Department of Planning, UNESCO Center for PBL, Aalborg University, AalborgAalborg, Denmark

Muhammad Mubashir Ehsan Education Above All (EAA), Qatar Foundation, Doha, Qatar

Mohamed Abdelaziz Eissa Associte Professor of Finance, Department of Finance and Economics, College of Business and Economics, Qatar University, Doha, Qatar

Said Elbanna Research Professor of Strategic Management & Director of Center for Entrepreneurship & Organizational Excellence, College of Business and Economics, Qatar University, Doha, Qatar

Ali Farooq Qatar Computing Research Institute, Hamad Bin Khalifa University, Doha, Qatar

Hany Fazza Georgetown University in Qatar, Ar-Rayyan, Qatar

Noor Hamwy Carnegie Mellon University in Qatar, Doha, Qatar

Sara Hillman Texas A&M University at Qatar, Doha, Qatar;
Hamad Bin Khalifa University, College of Humanities and Social Sciences, Doha, Qatar

Bernard J. Jansen Qatar Computing Research Institute, Hamad Bin Khalifa University, Doha, Qatar

Soon-gyo Jung Qatar Computing Research Institute, Hamad Bin Khalifa University, Doha, Qatar

Mohamed Keshta Core Curriculum Program, Qatar University, Doha, Qatar

Salar M. Khan Academyati, Qatar Foundation, Doha, Qatar

Mohamed Mahgoub Pre-University Education, Ar-Rayyan, Qatar

Igor Jacky Dimitri Michaleczek Educational Research Center, College of Education, Qatar University, Doha, Qatar

Rula Momani Gulf Studies Program, Qatar University, Doha, Qatar

Pilira Stella Msefula Higher Education, Qatar Foundation, Doha, Qatar

Mary L. Newsome Assistant Dean, Office of Graduate Studies, Qatar University, Doha, Qatar

Richard O'Kennedy Research, Hamad Bin Khalifa University Research, Doha, Qatar;
Dublin City University, Dublin, Ireland

Ozcan Ozturk College of Public Policy, Hamad Bin Khalifa University, Doha, Qatar

Michael H. Romanowski Education Research Center, College of Education, Qatar University, Doha, Qatar

Joni Salminen School of Marketing and Communication, University of Vaasa, Vaasa, Finland

Abdellatif Sellami Education Research Center, College of Education, Qatar University, Doha, Qatar

S. Duygu Sever Arts & Sciences Department, Carnegie Mellon University in Qatar, Doha, Qatar

Atif Shafique WISE Research Fellows, Qatar Foundation, Doha, Qatar

Fawaz Hazza Abu Sitta UNESCO Chair in Anticipatory, Foresight Research Director at the Center for Futures Studies, University of Dubai, Dubai, United Arab Emirates

Rana Sobh College of Business and Economics, Qatar University, Doha, Qatar

Mohamoud Soufi Senior Learning and Development Specialist, Center for Entrepreneurship & Organizational Excellence, College of Business and Economics, Qatar University, Doha, Qatar

Evren Tok College of Public Policy, Hamad Bin Khalifa University, Qatar Foundation, Doha, Qatar

Cesar Wazen Director of International Affairs Office, Qatar University, Doha, Qatar

Heba Younis LSBU Business School, London South Bank University, London, UK;
Senior Lecturer of Enterprise and Entrepreneurship London South Bank University, London, UK

Esmat Zaidan College of Public Policy, Hamad Bin Khalifa University, Qatar Foundation, Doha, Qatar

Mounia Zidani The Office of Graduate Studies, Qatar University, Doha, Qatar

List of Figures

List of Tables

Chapter 1
Introduction: Future of Education in Qatar

Esmat Zaidan, Logan Cochrane, Evren Tok, and Asmaa Al-Fadala

Abstract This introduction delves into the multi-dimensional and dynamic nature of Qatar's education policy and reforms, highlighting their connection to the country's broader goals of economic diversification. It examines how Qatar navigates the delicate balance between globalized educational trends and local cultural values. This exploration provides insights into various dimensions of these reforms, shedding light on their impact, the challenges they pose, and the future directions they chart within a global context for Qatar's education system.

1.1 Introduction

This introduction delves into the multi-dimensional and dynamic nature of Qatar's education policy and reforms, highlighting their connection to the country's broader goals of economic diversification. It examines how Qatar navigates the delicate balance between globalized educational trends and local cultural values. This exploration provides insights into various dimensions of these reforms, shedding light on their impact, the challenges they pose, and the future directions they chart within a global context for Qatar's education system.

Its education system—tradition fusing with modernity, and local values intercepting with the global vistas at the crucible—is the heart of Qatar transformed into a pioneering nation. This chapter will take the readers through the transformational landscapes of educational reforms in Qatar, a narrative that traverses beyond the traditional folds of academic discourses. As the Qatar National Vision 2030 should also resonate, it is a story of a country in search of balancing its strong tradition with the needs of a changing global economy.

The chapter elucidates the diversified dimensions of the education system in Qatar, edifying how under the influence of the complex interlinking of cultural, social, and economic factors, these dimensions get shaped. Finally, the chapters of the book do not describe these changes or events that occurred in a chronological pattern but

E. Zaidan (✉) · L. Cochrane · E. Tok · A. Al-Fadala
College of Public Policy, Hamad Bin Khalifa University, Qatar Foundation, Doha, Qatar
e-mail: ezaidan@hbku.edu.qa

E. Zaidan et al. (eds.), *The Future of Education Policy in the State of Qatar*,
Gulf Studies 21, https://doi.org/10.1007/978-981-97-9667-0_1

instead investigate the much deeper implication of these transformational changes over the Qatari social fabric and its trajectory into the future. It is a canvas dotted with global educational trends, knowledge-based economy imperatives, and the old quest to find national identity in the face of the onslaughts of globalization. This narrative is not just about the educational institutions and policies but also about the people—students, educators, policymakers, and the wider community—who are both the architects and the beneficiaries of this transformative journey.

Each chapter, therefore, is intended to be precisely that: an individual piece of the larger mosaic that gives a comprehensive understanding of how education in Qatar is being rethought and reshaped to meet the challenges and opportunities of the twenty-first century. The book focuses on subtleties in new teaching methods, strategic plans for the development of higher education, and the embracement of technology or even more urgent necessity for the ethos of responsive cultures.

1.2 Transformative Shifts in the Context of the Qatari Education System: Balancing Globalization and Cultural Identity

In a country like Qatar, where there is a rapid shift from a traditional society to an increasingly knowledge-based economy, the education system has undoubtedly become a central domain for transformative policy shifts. These changes target dynamic locals armed with the ability to take part in the international higher learning education market while gaining mobility in obtaining access to higher learning and access to jobs, all in the protection of values, tradition, and language of the country. Amin and Cochrane (2023) provide a comprehensive overview of educational reform in Qatar and describe the intended and unintended consequences of changes in policy; they highlight the voucher system brought in after 2011 to further liberalize the education sector.

Education in Qatar has recorded a great expansion on its landscape through the years. The country experienced a growth from 15 schools in the late nineteenth century to a whopping 529 primary and secondary schools, which served approximately 327,841 students as of the last academic year 2019/2020 (Amin & Cochrane, 2023). The process reflects the depth and pace of the socio-cultural changes and institutional modifications, which have accompanied the increase in adult literacy levels. Over the next five decades, the government aims to transform its largely expatriate-dependent population, reliant on oil and gas, into an empowered local workforce well-versed in an economy driven by knowledge and globalization (Amin & Cochrane, 2023).

Ten years have, indeed, brought great growth to the number of private schools in the country. However, they were largely the product of increasing expatriate numbers, and to a lesser extent, Qatari citizens who were using government vouchers to subsidize private school fees. However, the data displayed that this is not the trend of privatization only; the government school system also increased, though slowly. Even when the state introduced the vouchers, most of the Qatari citizens were still found attending government schools. But if that trend were to continue further between government and private schools, it would likely lead to significant demographic changes emerging in these schools. Over time, these changes could result in various social transformations and divisions based on identity, language, gender, cultural values, status, higher education choices, and available jobs or life positions (Amin & Cochrane, 2023).

Cultural identity issues, the status of the Arabic language, global and local job markets, and rapid social change are all part of an intricate set of factors that indirectly challenge traditional gender roles; hence, the trend toward privatization is anything but straightforward. The education system in Qatar has undergone a remarkable transformation, aligning with the country's broader aspirations for growth and development. The country has been investing in preparing a generation of skilled, adaptable, and enterprising professionals through substantial investments, an emphasis on quality education, and openness to worldwide partnerships. As the country progresses toward the ultimate realization of its national vision, education clearly emerges as the most crucial contributor in steering Qatar toward an envisioned future enshrined with affluence, knowledge, and sustainability. The experience of an evolving Qatari education system is highly promising for the future of the nation and exemplary for the depth and range of educational reform it embraces within the region and, arguably, beyond (Amin & Cochrane, 2023).

1.3 Modernizing Education in Qatar—Keeping a Balance Between Global Excellence and Local Identity in the Knowledge-Based Economy

The last decade has seen radical reforms in Qatar's education sector, with modernizing systems set and achieved as standards of excellence like those in any part of the world. These reforms have been broadly outlined to shape the new MEHE landscape of education in Qatar. It has been through the implementation of professional standards and teachers' licenses, being a keyway of strengthening the quality in education within the country. The next step of developing comprehensive curriculum standards that also encompass non-academic areas of learning was a major step toward holistic education. There has been a remarkable transformation in teaching approaches, which incline more toward student-centered strategies, using modern information communication technology and focusing on scientific research in line with global educational trends (Education Sector in Qatar, 2020). This approach is progressive and falls in line with Qatar National Vision 2030, which aims to transit the country to a knowledge-based economy with sustainable development across all the sectors. A testimonial of such concerted efforts is the contribution of the education sector to the GDP of Qatar, which grew remarkably to 3.3% in 2018–19 from 1.3% in 2012–13 (Education Sector in Qatar, 2020). This has led to the growth of the expatriate population in Qatar, particularly in urban centers such as Doha, Al Wakrah, and Al Rayyan, resulting in an increasing demand for educational institutions. In response, there has been significant growth in private education, with many international institutions being established, offering diverse curricula and attracting more students, particularly from the expatriate community (Education Sector in Qatar, 2020).

Reflecting on its historical development, the education system in Qatar has made significant progress, since its inception. Starting with the early Quranic schools, it was based upon this foundation, and by the 1980s, the world of education was molded well in the country, due to the needs of Qatar's massive oil economy. The first primary school opened in 1952, the creation of the Ministry of Education came to pass in 1956, and in 1977, Qatar University was inaugurated. An important juncture on this long evolutionary path was the implementation of K-12 education reform transition in 2002, emphasizing an outcome-based curriculum that had clearly defined standards for teaching and student performance (Education Sector in Qatar, 2020).

Though there is a clear classification of Qatar's education sector into five distinct levels—Early Childhood, Pre-primary, Primary, Preparatory, and Secondary education—it reflects the country's commitment to deliver structured and stage-appropriate education. The Early Childhood schools, under the control of the Ministry of Administrative Development Labour and Social Affairs (MADLSA), are concerned about the foundational years for children up to the age of four. MEHE is committed to exerting its authority regarding the supervision of educational actions from pre-primary to secondary level, with a continuous focus to improve the educational setup along with a rise in research facilities. The education sector in the country is focused on raising the level of education among all citizens and residents, with special attention to the fact of merging students with special needs into the system of mass education (Education sector in Qatar, 2020).

But probably the most impressive part about Qatar's education diversity is that the guidelines of MEHE allow several expatriate communities to have their schools. The number of public and private schools in the country was 199 and 282, respectively, from pre-primary to tertiary levels. Most of these schools are in the densely populated municipalities of Doha and Al Rayyan, where both the public and private sectors witness high demand for education. The total enrollment in students across all schools increased substantially from 232,345 in 2012–13 to 325,510 in 2018–19. Projections suggest that the growth trend could be maintained, and the total students' strength might rise further to 353,555 by 2023–24. Accordingly, the number of schools is also expected to increase to 524 by 2023–24. This is because there will be an increase in the number of expatriates, the entry of private players into the system, and the induction of globally ranked educational institutions (Education Sector in Qatar, 2020).

The Educational Voucher Scheme of 2012 is another example of how the country is grappling to increase the quality of education in its institutions. Under this scheme, Qatari students receive government support to attend private schools. It fosters the nature of competitiveness of an educational environment and emphasizes the principle of diversity and choice of schools. This provides financial assistance that covers tuition fees of up to 28,000 Riyals per academic year for all stages. In 2017, 24,592 Qatari students availed themselves of the benefit under this scheme across 78 schools. As of the report, it stands at 99 schools benefiting approximately 20,000 students (Education Sector in Qatar, 2020).

Education in Qatar ranges from elementary to the tertiary level, with a global range of educational courses hosted in the country. Additionally, Education City hosts numerous local branches of international colleges and universities, featuring internationally accredited programs and institutions. This campus and the spread of international schools across the country have notably upgraded the presentation of the country in terms of private institutions for education. However, the schools' fees differ according to some reasons, such as location, additional services presented at the schools, and the investment (Education Sector in Qatar, 2020). Students who studied in the Qatar education system include a mix of Qatari and non-Qatari students. In 2018–19, the percentage of Qatari students was higher within public schools (54%), while the number of non-Qatari students studying in private schools was way higher (81%). This demographic, therefore, indicates varied needs and preferences in relation to the education being offered in different setups within the country. More Qatari females are enrolled in public schools, while non-Qatari males dominate the enrollment in private schools (Education Sector in Qatar, 2020).

Education reform and investment in Qatar are highly committed to bringing about a modern, world-class educational system (Zaidan et al., 2024). It is through this system that caters for the various needs of the population, including expatriates, that is playing an instrumental role in the transition of the nation toward a knowledge-based economy. The education sector has experienced great transformation and growth in recent years. The paper presents a brief description of the present and future analysis of this vital sector in the country.

The push toward educational privatization in GCC countries, including Qatar, reflects a complex interplay of neoliberal policy prescriptions, economic imperatives, and value shifts within societies. According to Amin and Cochrane (2023), privatized education has some benefits and raises questions of inclusiveness, equality, and the perpetuity of local culture and language. The direction of education policy in Qatar, therefore, to a large extent determined by government objectives is central to enabling the sector to deliver quality, inclusive education.

1.4 The Evolution Landscape of Education: Navigating Local Traditions amidst Global Trends in the Quest for an Economic and Knowledge-Based Transition

Regionally, and internationally, the Arabian Gulf, including Qatar, is prominently positioned at the forefront of every transition plan aimed at economic diversification and the development of knowledge-based economies. This aligns with new aspirations inspired by Western models, emphasizing education, internationalization, and transnational education, particularly through English-medium policies and the adoption of Western higher education frameworks.

Education reform in Qatar is an inseparable part of the Qatar National Vision 2030, which is based on sustainable development and the construction of an advanced society. The country has to transform into a knowledge-based economy, which necessitates dynamism and skill of its local workforce on the front of global competencies and, at the same time, puts a lot of emphasis on preserving traditional cultural values. The substantial government investments in education, represented by the creation of the Qatar Foundation for Education, Science, and Community Development, have set the country on an interesting journey of transformation (Amin & Cochrane, 2023). From modest beginnings with meager educational institutions in 1891 to 2020, Qatar's basic education system has transformed into an all-inclusive network of 529 schools, serving over 327,000 students, demonstrating the nation's unwavering commitment to education. This growth has taken the adult literacy rate above 93% now, reflecting how deeply educational improvements have come in the country (Amin & Cochrane, 2023).

Twinning local institutions with prestigious international universities marks the twin strategy adopted by Qatar in its higher education provision. This is an effort, most pronounced in Education City, that signals approach on the one hand, one of commitment to global educational standards, and on the other, signals a strategic foray toward portraying itself as a center of international education (Amin & Cochrane, 2023). However, this was a difficult task for Qatar to achieve because balancing the educational models of the world and saving the local culture from dilution was certainly a tough goal. The process of educational reforms, especially since 2002, has been aiming at the elevation of the quality of education in pursuance of Qatari values, heritage, and culture. Further attempts at improvement through the voucher system of 2012 and changes in the language of instruction point toward that (Amin & Cochrane, 2023).

The education system has observed that the trend toward privatization is on an increase in Qatar, and this trend is because of some factors, such as cultural identity, job market needs, and social norms that are changing. The government voucher system has enhanced the increasing tendencies of private schooling (Amin & Cochrane, 2023). The education system characterizes neoliberalism through its market-driven perspectives that make education one of the commodities much affected by consumer demand and profit maximization. This has transformed education from a social good to a market good, where the private education sector often offers better quality and more responsive alternatives compared to those present in government schools. One of the biggest beneficiaries of the internationalization of higher education could be one category: women, who can obtain high-status degrees from foreign countries without moving outside of their own countries (Ibrahim & Barnawi, 2022). Scholars argue that as beneficial as these policies may turn out to be, they inherently create systems that favor the private accumulation of wealth and a wider income disparity along with established differences. This occurs at the corporate level, and even at the individual level, since access to quality education becomes an issue of financial class privilege, thereby widening societal gaps (Ibrahim & Barnawi, 2022). On its own part, inclusive and equitable quality education is precisely needed for quality learning, not specifically mentioned in the Sustainable Development Goals (SDGs). SDG 4 precisely addresses inclusive and equitable quality education for all. However, the quality-based nature of private education somehow bypasses inclusiveness, more so regarding students with diversified learning abilities, either through high operational costs or at the expense of quality. Most private schools opt not to admit children with disabilities, hence limiting the inclusivity of education offered (Alpen Capital, 2018).

The future of Qatar's education system lies in adopting global education trends while preserving its local cultural values and ethics (Zaidan et al., 2023). This general trend of centralization in the government schools, perhaps most symbolically presented with the replacement of the Supreme Education Council by the Ministry of Education and Higher Education in 2016, clearly represents a strategy to deal with just such challenges. This is critical in the forming of an education system that will see to the preparation of citizens for both the needs of local society and those felt by the global market. Summarily, the education system of Qatar is at a crossroads, emanating from the wider aspirations of the country toward development and integration in the global scenario. However, with large investments and attention toward quality education systems that actively engage with international best practices, Qatar can ideally place itself in such a way that outstanding reform and innovation in education can be brought about in the coming years.

1.4.1 Education Reforms and the Transformation of the Education System in Qatar: Navigating Change and Tradition

In the MENA region, including Qatar, there has been a general trend toward neoliberal education reforms since the 1980s, even in formerly socialist Arab states. These reforms reflect integration into the capitalist world system and include the development of private educational sectors (Alayan et al., 2012). The UNDP Human Development Report 2016 indicates that Arab societies perform below the world average in educational attainment, achievement, and equitable access, often cited as justification for substantial educational reform (UNDP, 2016).

These reform efforts have focused on raising educational quality through improved teacher training, pedagogies that inspire critical thinking, and upgrading curricula to prepare students for the labor market demands and Western-inspired, English-language programs (Gelvin, 2021; Kirkpatrick, 2017). However, there are significant voices opposing these reforms, cautioning against the dilution of local culture and language and the implicit subjugation accompanying foreign educational tools and services (Alayan et al., 2012). One point to consider is that the reform, like any change, is difficult because it requires changes in human behavior. People often resist new ideas.

The nature of educational reforms in Qatar is complex and multifaceted, underscoring the challenges of aligning educational outcomes with the broader goals of economic diversification and the development of a knowledge-based economy. For example, the QCEA, later renamed "National Assessments," was introduced in 2004 as a standardized testing method to evaluate student performance in state-run independent public schools. Initially, the assessment revealed poor performance levels among students, particularly in math, science, and English. There was a significant improvement over the years, although it was unclear whether this was due to actual performance improvement or changes in the assessment levels (Mommani et al., forthcoming).

School-Based Reforms in the Early 2000s: Aimed at addressing curricular deficiencies, the reforms led to the establishment of independent schools with a focus on English and technical/scientific subjects. This shift, however, unintentionally de-emphasized Arabic and Islamic studies. Then Qatar collaborated with RAND for these reforms, which were based on principles of autonomy, accountability, variety, and choice. This initiative attempted to decentralize governance in schools and improve learning outcomes. However, the collaboration with RAND was terminated in 2012 due to partnership challenges.

Established in 1995, the Qatar Foundation played a significant role in higher education reforms. Education City, the headquarters of QF, hosts several branch campuses of prestigious universities from around the world. These reforms were not aimed at generating revenue but were intended to support Qatar's transition to a KBE by developing talent and promoting economic diversification (Knight, 2015). The QF's activities included academic programs, research initiatives, and science and research institutions. The Qatar National Research Fund, part of QF, encouraged research in fields of national interest and promoted international collaborations (Williams, 2007).

Furthermore, the reforms in Qatar faced challenges due to the nation's cultural and political context. The mentioned "zigzagging" of school reforms (Barnawi, 2017) in Qatar reflects the nation's ongoing struggle to balance modern and conservative approaches in education, particularly concerning language policy, the emphasis on STEM subjects, and maintaining cultural identity. The reforms also had cultural implications, particularly regarding women's participation in the labor market. Despite legal advancements promoting women's involvement in the workforce, traditional beliefs continued to influence their perceived roles.

Additionally, the adoption of Western educational models and a shift to English as the medium of instruction led to concerns about the potential negative impact on Qatari culture, values, and language. Eventually, there was a reversal in some of these reforms, with a renewed emphasis on Arabic studies. However, there is a need for reforms that are not only imported but also adapted to fit the local context and culture.

Regarding the diversity and scope of tertiary education offerings, Qatar University, which is notably the largest in the region, has progressively broadened and enhanced its academic programs. It is now distinguished as being among the top 200 universities globally and ranks in the top 3 within the Arab region. The university has introduced a range of new academic programs, notably in the fields of medicine and nursing. Other significant educational institutions in Qatar include the College of the North Atlantic, now known as Doha University of Sciences and Technology, and the University of Calgary, both of which offer various postsecondary educational programs. Education City, serving as the central hub for the Qatar Foundation, hosts international branch campuses from prestigious universities such as Texas A&M (specializing in engineering), Carnegie Mellon (offering business and computer science programs), HEC Paris (focused on business management), Virginia Commonwealth School of the Arts, Georgetown (concentrating on foreign service), Weill Cornell Medical College (focusing on medicine), and Northwestern (known for journalism and communication). Furthermore, modifications have been implemented in the scholarship awarding process to maximize benefits for students pursuing studies abroad. Additionally, there has been an emphasis on expanding opportunities in trades, technical, and vocational training to cater to students not inclined toward pursuing advanced academic degrees (Mommani et al., forthcoming) forthcoming. These initiatives, encompassing areas like healthcare, education, and government, are part of Qatar's strategic efforts to enhance its Information and Communications Technology (ICT) sector as a step toward establishing a Knowledge-Based Economy (KBE).

The education reforms at school and higher education levels in Qatar certainly have contributed toward the improvement of the system but not up to the major requirements for a real Knowledge-Based Economy (KBE). These are the specific challenges leading to these challenges in the transition process—part of the relatively short formal education history of Qatar and difficulties introduced as part of its political and economic framework. RAND Corporation was invited back after the reforms to assess their impact on the country. Among key findings in RAND's assessment was that, despite all the intensive efforts, the primary nature of teaching was still orientated toward an instructor whose teaching lacked student participation. Reforms at both the teaching methodology and the level of linguistic instruction were being done so that they were in synchronization with international market requirements, which directly or indirectly meant that less focus was on local history, culture, and identity. Although substantial changes have taken place, most of them were more of an adoption of foreign policies and practices rather than adaptation to local contexts (Romanowski & Du, 2022).

Understanding and dealing with these highly embedded and cultural aspects of resistance is most necessary. For example, according to Hofstede's Cultural Dimensions Theory, power distance is high in Qatar, which means in such societies, power imbalance is generally considered acceptable. That necessarily means that, with respect to the relationship involving parents and children, employers and employees, the government and its citizens, and education, among others, a teacher is considered a leader and knowledge giver, while students are considered learners and knowledge receivers (Mommani et al., forthcoming) forthcoming. Normally, people from Qatar perceive teachers to be leaders and knowledge disseminators, while students are learners and knowledge recipients, which will take more time to change. Stress on equality, student-centered learning, and the interactive approach of the Western mode of education may not be easily integrated with the traditional GCC cultural practice.

1.5 Education for Development, Diversification, and Sustainability

In the bigger landscape of development, the transformation of the public university in the GCC region, let alone Qatar, takes place while there is a dialectic interplay between these institutions and the continuously changing development strategies. The process thereby involves localizing the role of public universities within the development trajectories, objectives, and challenges that are specific to the region. This finds the development aspiration of the GCC countries to share with them in addressing economic diversification to overcome the uncertainties attached with the hydrocarbon-based economy and fluctuating oil prices.

Considering the given perceptions and explicit references within the national visions of GCC countries, the following are therefore necessary with regard to the role of HEIs. These visions target, in a consistent manner, economic diversification through non-oil development, i.e. the service economy pertaining to tourism and financial institutions. Now, in almost all the visions—with the differing levels of urgency given among the member states—the realization of diversification anchored in research, innovation, and entrepreneurship as KBE and the foundational role of HEIs are visible. For instance, the national visions of Oman and Bahrain particularly associate HEIs to KBE, innovation, and entrepreneurship that draw a parallel with their exigencies due to the depleting oil resources. By way of contrast, Kuwait, Qatar, and the United Arab Emirates display less urgency in the pursuit of this goal. The UAE and Qatar, however, were among the first countries to adopt KBE as a guiding development policy, though their national vision texts detail measures for achieving KBE and the HEI role in less specific terms.

The development plans therefore acknowledge that any effort to improve the living standards of the people in the face of such varying revenues of carbon resources and an underdeveloped motivation and capability of the national workforce is within a very complex set of challenges. This realization of the vital role of the development of local human capital is more evident from the recent development policies, especially after the significant rise in oil prices post-2008, which placed new emphasis on the fact that economic diversification was needed. To this extent, the analyzed development visions provide frameworks that HEIs, mostly represented by public universities, have a prominent role in the economic transformation of Gulf economies. This mapping of development through KBE and the role of HEIs has given birth to new expectations from public universities. This is further elaborated within the paper; universities have also initiated reforms to transform into more engaged actors within the local development environments.

The future of education in Qatar is heavily related to the expectations of this country to transform its economy from an oil and gas-based one to a diversified, sustainable, and knowledge-based (KBE) economy. This strategic shift, within the wider development trajectory followed by GCC countries, indicates a need for significant changes in the educational environment, largely toward fostering innovation and entrepreneurship and aligning learning outcomes with the requirements of the market.

In both transformations, public universities are central. Critical in the double sense of: localizing knowledge production to assist in addressing the particular challenges of the GCC region, on the one hand, and being centrally driving the advancement of national-level expectations, particularly of sustainable development. However, their positive effects on local per capita GDP growth and economic growth show a great variation because of the heterogeneity of higher education systems (HES). This kind of size, with respect to human and financial resources and quality of research, and with a focus on disciplines like STEM, becomes significant in facilitating the enhancement of economic growth.

Universities are the core of new expectations that do not concentrate any more only on sustainability but also on diversification, able to stimulate the economy for its growth through innovations and entrepreneurship. This new re-orientation should better employability with the needs of the labor market, especially those pertaining to sustainability and sustainable development. The last decades have seen the so-called academic revolution: increased global research output and fast development of universities in the emerging Global South. Indeed, there has been a gradual shift in focus over the years to increase the roles of universities from simply learning institutions to actively engaging with industry, society, being gatekeepers of knowledge, and the preservation thereof.

However, with such broad national developmental policymaking of the third mission, Higher Education Institutions in the Gulf—Qatar specifically—have allowed them space to engage further in social and economic issues. However, they keep facing rigid structural obstacles in terms of innovation, entrepreneurship, and sustainability. Public universities in Qatar are strongly associated with the mandates of the state. They increasingly align the roles with the visions of the nation and development strategies based on sustainable development and diversification of the economy. The latter has engendered a further repositioning of public universities, whereby they emerge as more engaged actors within the glocal formation of the Gulf States' development landscape.

Neoliberal reforms to education systems around the world have been driven by this trend under economic factors. Some of the other policies and discourses found in Asia, Africa, the Arabian Gulf, including Qatar, and non-English-speaking countries in the Western world are the likes of governments, intergovernmental organizations, and corporate entities advocating for the promotion of neoliberal policies. The transformation of education into a money-generating business that involves many actors and stakeholders, for instance, from English-medium education to the adoption of Western education products and services (Gupta, 2018; Rizvi & Lingard, 2009). The supposition is that the dominance of the English language within the nation-states can result in global market benefits and can further strengthen education systems therein (Barnawi, 2018).

Tsui and Tollefson (2007) observe that the technological English dependency of globalization has made nations invest to ensure their citizens are perfect in these sections. Neoliberal political, social, and educational norms are, therefore, reshaped with Western-inspired educational means and curricula to be pivotal in the dissemination (Roberts & Peters, 2008). As such, this process does not stand as a one-line activity for the development of the public universities of the Gulf; rather, it stands as a multi-aspect required to be developed on the balance between governmental policies, societal needs, and dynamic requirements for a knowledge-based economy. Longer sustainable perspectives of higher education and policies shall be critical in driving progressive and stabilized development within the university institutions beyond periodic development agendas.

1.6 The Book

Aligned with the strategic objectives in Qatar National Vision 2030 of transforming Qatar into a knowledge-based economy, "Future of Education in Qatar" is a seminal volume which offers a comprehensive update on the country's transformative educational landscape. The book contains 25 incisive chapters, contributed by 45 authors. The compilation was inspired by the realization of a bevy of experts engaged in revolutionizing educational reforms and practices that made Qatar their home, some known to us prior to this venture, and others whose acquaintance we made during this enlightening journey. Of the 45 authors, most were based in Qatar at the time of writing, with several others having worked in the country in the past. This wide diversity reflects the whole lot of work toward education and research in Qatar by some eminent organizations such as Qatar University, Hamad Bin Khalifa University (HBKU), the Qatar Foundation, and several institutions and organizations pertaining to education and research.

This project envisions "Future of Education in Qatar" as a fully Open Access book, which is made possible by the NPRP grant #12C-0804-190009 from the Qatar National Research Fund (a member of the Qatar Foundation) titled SDG Education and Global Citizenship in Qatar: Enhancing Qatar's Nested Power in the Global

Arena. In this open-access book, such rich content that includes the gamut of educational reforms and strategies remains available to students, educators, policy formulators, and all stakeholders at any point in time. The chapters touch on current practices, forward-looking policy options, and recommendations. This aspect was important to us, recognizing that access to such comprehensive and updated insights in education is often limited for decision-makers, given the prohibitive costs of academic publications. This is highly important for this kind of content to facilitate informed decision-making.

We are the first to admit that, despite handling the subject of the "Future of Education in Qatar" with the kind of breadth and depth that it deserves, there are areas which may not have been given the right amount of consideration or indeed have been left out. This is quite a visible gap for a book which aims to deliver policy implications and emphasizes the interlinked nature of reforms with the purpose of providing more coherence. From our point of view, essential areas of further exploration emerge: the changing roles of educators in a digital age, the impact of globalization on curriculum development and for special education needs, vocational and technical education training, gender disparities in education, and the challenges of fast modernizing education in maintaining cultural identity. We hope that the book will stimulate more research and discussion of these and other changing topics in an effort to fill out the rich picture of educational development that is taking place in Qatar and extends to many other places around the world.

In the end, this book seeks to be of great help to practitioners and policymakers, students, and researchers who may wish to gain insight into the educational dynamics of Qatar. "The Future of Education in Qatar" is an eye-opener expedition of the daring journey of Qatar in the quest of transforming the education arena. It offers a blueprint and inspiration to other nations that may wish to change their systems of education by overhauling them with modernity but not disposing of the cultural values. The book highlights the relevance of adaptability, innovation, and global integration in education, giving the picture of a vivid ivory tower of education that does everything but define the future of a nation and its place in the global picture.

1.7 Chapters

1.7.1 Section I-Teaching

In detail, Chaps. 2–10 bring out the subtlety of teaching methodologies and practices within the Qatari educational framework. They mention the efforts that are being put in place by the country toward evolving teaching strategies that will not be dynamic but culturally tuned so that educators are in a good place to mold the students toward modern-day challenges. All these chapters are attentive to enrich the experience of teaching in its use both in class and across the digital platforms with the worldwide cultural identity of Qatar. This also extends to the social role of teaching: it goes on to model an education system that is responsive to the local context of Qatar and, at the same time, adapts to global educational trends.

Chapter 2, Teaching Sustainability in Qatar: Challenges for Teacher Readiness. In this chapter, the critical success factor in implementing sustainability education is examined, and several issues are underlined in this chapter. Among the key challenges identified in this respect include the lack of preparation in teacher education regarding sustainability and hence a gap in teachers' preparation of skills and knowledge. There are cultural barriers because the teacher must negotiate between the global value of sustainability and the local, deep tradition. Another aspect that supports or hinders this incorporation is the school culture. The environment, culture, and resources within schools are necessary for teacher support. The authors had pointed out the need to customize the training that will ensure Qatari educators get trained to have the skills and knowledge relevant to teaching sustainability within the cultural and educational context of Qatar.

Chapter 3, International Standardised Tests and Qatar's Educational Policy Development. This chapter looks at the effect of international assessment tools, like TIMSS and PISA, on the educational policies of Qatar. Such standardized tests become a reference point for policymakers in assessing the education performance in Qatar and thus providing meaningful inputs for policy intervention. These tests, in essence, allow the making of "data-driven" decisions that help to identify strengths and weaknesses of the system. However, it is a great challenge that the global standards of education, which are measured by these tests, be balanced with the context of education in Qatar. This may mean the calibration of international benchmarks and local needs in education and cultural relevance that will afterward ensure the policies formulated serve their purpose within the Qatari setting.

Chapter 4, Problematising PBL in Higher Education in GGC Countries: The Case of Qatar. This chapter looks at both the challenges of adaptation and implementation that Problem-Based Learning (PBL) offers at institutions of higher learning in Qatar. Among the dominating issues that pose a major challenge to the adaptation and implementation of PBL in the higher education curriculum include educational borrowing, in that the very process of importing and incorporating largely Western educational practices, such as that of PBL, into the Qatari education system, brings with it a series of cultural and contextual challenges. Another important component is the engagement of educator readiness, given the fact that the success of PBL would be derived from enabling practitioners of education to utilize such a teaching module effectively within Qatar. From this point of view, the chapter proposes culturally responsive, contextually relevant measures that may alleviate those issues and allow the effective implementation of PBL.

Chapter 5, Qatar's Transformation to Digital Learning: Challenges, Opportunities, and Policy Recommendations. This chapter analyzes the transition to digital learning in Qatar and its impact on developing twenty-first-century skills. The transformation faces challenges, including addressing language diversity to ensure that digital learning resources are accessible and effective for a multilingual student population. Cultural nuances also need to be considered, with digital learning platforms and content needing to be sensitive to and reflective of Qatar's cultural values and norms. However, this transformation also presents opportunities, such as leveraging technology to enhance learning experiences and equipping students with vital digital competencies. The chapter emphasizes the importance of creating sustainable, inclusive, and culturally sensitive digital learning environments that support the development of twenty-first-century skills through effective policy development.

Chapter 6, Human-Centred Artificial Intelligence in Progressive Education: Unravelling the Benefits and Challenges in Qatar's HEIs. The authors discuss the innovative use of algorithmically generated personas (AGPs) in higher education institutes (HEIs) in Qatar. They argue that AGPs, as humanized representations of key stakeholders like applicants, students, alumni, faculty, and staff, can significantly improve decision-making in education. These personas provide valuable insights into the needs, preferences, and behaviors of various user groups, fostering empathy, personalization, and inclusivity within the education system. The chapter highlights the transformative impact of using personas in education, particularly in addressing the diverse and evolving needs of learners, educators, and communities in Qatar's HEIs. The authors recommend several key applications of AGPs in education. They emphasize that integrating personas in the education workflows of HEIs offers Qatar a roadmap to a more inclusive, adaptable, and student-centric education system. This approach is seen as crucial for nurturing a generation equipped with the skills, knowledge, and empathy to thrive in a global society, aligning with Qatar's aspirations for educational excellence.

Chapter 7, Reinventing Education Strategic Planning and Policymaking Through the Lenses of Futures Thinking. The chapter offers a critical conceptualization of the way futures thinking might enhance strategic planning for education. The chapter highlights the use of both imaginative and creative thinking, coupled with empirical data, in trying to foresee the dimension of emerging disruptions and promote the development of adaptive education systems. The present review also identifies trends like horizon scanning and scenario planning, which are important for developing agility and resilience at educational institutions. According to the authors, these are the "Futures Thinking" principles to be adopted by the educational institute and Ministry of Education in Qatar to enhance the strategic planning and policy development within institutes.

Chapter 8, Financial Literacy in the Modern Economy: Bridging the Knowledge Gap for Entrepreneurs and Beyond. As highlighted in this chapter, research conducted by the OECD and S&P has pointed to a glaring lack of financial knowledge on an international level. The chapter discusses subjects like entrepreneurial financial planning, the socio-economic determinants affecting the financial choices, and the immediacy of gender-neutrality for financial literacy. The conclusion places this complexity of the financial literacy environment underlines, but with the conclusion, major problems remain for developing countries. The chapter has also underlined the gender differences in financial literacy and highlighted that there are even specific issues faced by women, such as wage differences and little formalized access to banking. The chapter represents an important contribution to the discussion on financial literacy, especially in its relationship with entrepreneurship and gender issues. It sheds valuable light on how challenging the landscape to deliver financial education is and how there must be tailored strategies that can account for a variety of needs.

Chapter 9, The Impact of AI tools on Universities' Educational Strategies and Student Success: Sultan Qaboos University and Qatar University as a Case Study. This chapter is an analysis of just how exactly AI tools are going to have a bearing on educational strategies and the outcomes in the settings of a university with examples drawn from Qatar University and Sultan Qaboos University. It reviews how AI could be utilized in supporting the long-term strategic missions of research, teaching, and community services upon integration with the systems of educational institutions. It will also discuss the strategies of universities vis-à-vis the integration of AI with paramount importance attached to the need for using AI to hasten the implementation of strategies. The chapter observes that AI has both positive and negative impacts on the university strategies, affecting various areas of strategies in operations and management.

Chapter 10, Nurturing the Home Learning Environment: Evidence from Qatar Reads Programme. This chapter evaluates the role of the Qatar Reads program in fostering literacy, language, and children's affective development in the home learning environment. They have observed that it has a positive effect on the development of children in general and that the role of home libraries and parent–child activities is of great importance. This, in fact, suggests that although the program will have a bearing on children of both genders, the gendered difference in parental involvement reduced in the case of older children, hence needing tailored differentiated approaches to handle the two cohorts. It discusses implications this has on policymakers, educators, and families, calling for family-oriented programs, suggesting that educational policies support such initiatives, and generalizing these results very carefully, considering the relatively small sample size used and the biases that might be introduced by recall data.

1.7.2 Section II—Inclusive Education

The following chapters of inclusive education provide a complete view of how Qatar stands up to the challenge in education. From multilingualism to gender equity and from the expansion of private education to making provision for the inclusion of children with special needs, these chapters present a very interesting roadmap that Qatar follows in its pursuit of developing conditions for an inclusive educational setup. This is very much in line with the big national vision for Qatar in developing a knowledge-based economy and society where everyone can "make it."

Chapter 11, The Expanding Private School Market in Qatar: Exploring the Consequences. This chapter provides an account of remarkable growth in the private school sector in Qatar, which now serves relatively large numbers of students from Qatari and non-Qatari backgrounds. Some of these issues include how access, quality, and maintaining cultural heritage in the face of such growth could be done and the kind of regulatory arrangements that would be necessary for such a sector with a great diversity. It also discusses how the voucher system and initiatives like "Outstanding Schools" bring more negative consequences for the increase in private education. This also throws up the challenges before private schools like strong regulations and financial survival pressures bringing forth quality issues in education and teacher pay.

Chapter 12, Multilingual Students in Qatar's Education City International branch campuses: Advancing a social justice-centered policy agenda. The chapter highlights the challenges that multilingual students face with reference to international branch campuses in Qatar's Education City. In the process, students, as it turns out, struggle with the linguistic environment, more so in the shifting of the curriculum medium to English. This struggle affects them at a very emotional and academic level. First, early exposure to higher education and the perceived higher standard offered in the West, and then how it affects students in the university environment, are the issues focused on this chapter. This demonstrates the call to have strategic pathways forward and policy recommendations for improvement in relation to the question of diversity, inclusion, and equity of the experiences of its students.

Chapter 13, Learning Ecosystems for Equitable Education and Lifelong Learning. This chapter presents an overview of the 5-year research and development work conducted by WISE/Qatar Foundation into learning ecosystems. This work has included one of the first international studies into the concept, setting out a definition which has since been adopted by other researchers and practitioners, and the identification of key characteristics and impacts, which has supported global growth in interest at both practice and policy levels. Recently, WISE has focused on understanding the potential of this concept in Qatar, in particular looking at how learning ecosystems could drive efforts toward more equitable and inclusive education and the promotion of lifelong learning. Our work makes the case for recognizing all learning and connecting the learning experiences that young people have inside and outside school, enabling young people in Qatar today to develop the core knowledge, skills and capabilities that they will need to navigate an ever-changing world.

Chapter 14, Undergraduate Qatari Students' University Choice in Qatar. This chapter will focus on the various factors that are considered by Qatari students while choosing their universities. In particular, it will try to find out why they choose a given university. The factors include family influence, cost, location, institution reputation, programs of their choice, and financial aid. Moreover, this chapter takes into account the strange reasons for opting for local universities (LU) and international branch campuses (IBC), like the application processes, university admission, and institution reputation, among other reasons. It gives implications of the perspectives of students to policymakers and recommends to them.

Chapter 15, Gender Inequality in Higher Education Institutions in the GCC States: Current Status and Ways Forward. The chapter provides the reader with the gender inequality situation of higher education among countries of the Gulf Cooperation Council with special attention to the situation of female students and female academics. It further talks about the kind of struggles mothers going for higher education have to go through and where they are placed as female academics at the global level. The consideration for the development of these partnerships is to see the promotion of gender equality and the empowerment of women and to see gender-transformative pedagogy in teaching.

Chapter 16, The Journey Towards Inclusive Education in Qatar. It comprehensively explains the journey of inclusive education in Qatar, focusing on the education of children with disabilities stipulating Qatar's commitment to the implementation of strategies based on inclusion regarding the UN Convention on the Rights of Persons with Disabilities. It cites the government's efforts in the enhancement of the opportunities in education and diagnostic services for children living with disabilities. It also discusses the historical context of the prevailing current programs within Qatari schools, discussing limits, and the outlook of future progress in inclusive education.

Chapter 17, Public Universities and Educational Sustainability in Qatar: A Case Study of Qatar University. With focus on Qatar University from an institutional perspective and how it positions itself toward leading higher education development in Qatar. The chapter assesses the future-orientation of the institution by referring to how Qatar University (QU) positions itself as committed to gender inclusivity, outreach overtures to the community, and how it forges further ahead in a knowledge-based economy relating with the Qatar National Vision 2030. In Qu, it also shows the strategic role it poses toward collaboration, partnership, research, and innovation contribution for an emerging hub in higher education. Sustainability looks at the future for public universities and their stand relating to shaping the educational landscape and underpinnings for national and worldwide initiatives.

1.7.3 Section III—Sustainability

The following two chapters present the book with broad perspectives: one on relevance in education systems at the national level within Qatar and another of a broader regional context of the GCC. They underline that good transformation toward sustainable development through education calls for strategic policy development, collaboration, and innovative teaching approaches.

Chapter 18, The Evolution and Future of Education for Sustainable Development in Qatar. The chapter examines the place of education in the realization of sustainable development goals within the framework of Qatar's national development agenda. It outlines on ESD in society and the national system with reference to placing oneself within the socio-economic progression of this country. The article is then concluded with some future perspectives regarding ESD in Qatar by providing recommendations so that it can further be developed and incorporated into the educational policy at the national level. It also pointed out that Qatar has committed to the principles of sustainable development through such plans as national development, involving preparation of citizens for a diversified and sustainable economy.

Chapter 19, Higher Education and Sustainability in the GCC Countries, the Case of Qatar. The chapter critically examines the role of higher education institutions (HEIs) in advancing sustainable development (SD) in Qatar and the broader Gulf Cooperation Council (GCC) region. It highlights that while HEIs engage in various sustainability-related activities, there is a lack of specific, focused policies and insufficient evaluation of their impact on sustainable development. Authors argue for a more integrated approach, emphasizing the need for HEIs to not only transform internally but also contribute to societal change. Key issues include the need for better collaboration among HEIs, the development of a unified strategic framework for sustainability initiatives, and the promotion of student-centered learning approaches that foster critical thinking and responsible citizenship. The chapter also underscores the challenges in measuring the real-world impact of HEIs on sustainability goals.

1.7.4 Section IV—Career Pathways and Life Skills

Together, the following chapters provide valuable insights into the development of career pathways and life skills in Qatar and the Gulf region. They emphasize the need for educational reforms that cater to the demands of the modern economy, foster entrepreneurial mindsets, address gender disparities in workforce participation, and enhance financial literacy. This collective reflection underscores the importance of aligning educational outcomes with market needs and societal expectations, ensuring that students are equipped with the necessary skills and competencies to thrive in a dynamic and evolving economic landscape.

Chapter 20, Entrepreneurial Education in the GCC: Review and Prospective Directions. This chapter on entrepreneurial education in the Gulf Cooperation Council (GCC) countries focuses on integrating entrepreneurial education into higher education institutions. It emphasizes the need for a strategic paradigm shift in university governance to foster an entrepreneurial ethos, highlighting challenges such as cultural attitudes toward failure and the need for agility and adaptation. The chapter advocates for innovative teaching, rewarding pioneering curricular endeavors, and establishing budgetary frameworks that support such innovations. It also stresses the importance of training and developing entrepreneurial educators and suggests the institutionalization of such training within a holistic framework. Additionally, the chapter proposes accreditation for universities excelling in entrepreneurial education and recommends integrating technological advancements into curricula to enhance digital entrepreneurship and interdisciplinary programs.

Chapter 21, Navigating Entrepreneurship Ecosystems in the UAE, Qatar, and the UK. This chapter explores the state of entrepreneurship education in the GCC, focusing on the UAE and Qatar, and compares it with the UK. It highlights the significance of entrepreneurship in these regions' economic strategies and examines the entrepreneurial ecosystems, considering cultural influences, government interventions, and educational approaches. The chapter underscores the role of networking, cultural norms, and stakeholder support in shaping entrepreneurial mindsets. It suggests that the GCC nations, especially the UAE and Qatar, should enhance their entrepreneurial ecosystems, particularly in higher education, by incorporating experiential learning and drawing inspiration from initiatives like the UK's

'Help to Grow'. The chapter concludes by emphasizing the importance of adapting to the evolving dynamics of artificial intelligence in entrepreneurship education and sets the stage for further discussion on policy implications in the following chapter.

Chapter 22, Higher Education and Workforce Participation in Qatar: Unravelling the Disparity. This chapter examines the discrepancy between women's education and workforce participation in Qatar. Despite high female educational attainment, their engagement in careers is limited. The study explores factors contributing to this, including societal expectations and the dichotomy of education as a private endeavor versus work as a public one. It highlights the challenges women face in patriarchal societies like Qatar in balancing family and career. The chapter also discusses the impact of cultural and religious interpretations on women's choices and underscores the need to reconcile government empowerment efforts with societal norms and expectations.

Chapter 23, Charting the Future of GCC Entrepreneurship: Insights for Education and Incubation Centers from the UK, Turkey, and the Region. The chapter looks at the concept of entrepreneurship education and incubation centers in the GCC and compares them with those from the UK and Turkey. Therefore, it realized the tailor-made policies fortifying GCC entrepreneurship ecosystems toward sustainable economic growth. This chapter looks at some of the key roles that the government plays, from creating an enabling environment for entrepreneurship to the role of the government in ensuring smooth policy implementation; finally, it looks at the complexities that accompany policy implementation. It suggests systemic policy design, experimentation in policy implementation, and careful monitoring and evaluation. Additionally, in this chapter, it is also laid out to encourage academic research in entrepreneurship.

Chapter 24, Arabic Language Proficiency in Higher Education: A Study of Employers' Expectations and Programme Improvements in Qatar. This chapter presents the outcomes of a mixed-method research project that sought to explore employer perceptions of Arabic language competences of graduates from an International Branch Campus (IBC) in Qatar. It will be an important study within the Gulf Cooperation Council (GCC) countries in a way that it will aim at bridging the outcomes of higher education in Qatar with the needs of the job market in matters related to the Arabic language. This will mean that the Arabic study curriculum will have to change and be re-oriented toward that which targets professional careers in the sense of effectively preparing students for the modern job market. Such areas include adaptation of curricula to meet professional requirements, stress on interdisciplinary integration, provision of high-level language courses, and reforming teaching methods in the different aspects of learning. These are items identified with ensuring preparation for appropriate entry into the job market because of the great focus given to the development of communication, research, and policy analysis techniques.

Bibliography

Alayan, S., Rohde, A., & Dhouib, S. (2012). *The Politics of Education Reform in the Middle East: Self and Other in Textbooks and Curricula*. Berghahn Books.

Alpen Capital, *GCC Education Industry* (Alpen Capital, 2018), https://argaamplus.s3.amazonaws.com/9e55ad53-477f-48f0-b202-2c2f03d6a03d.pdf.

Amin, H., & Cochrane, L. (2023). The development of the education system in Qatar: assessing the intended and unintended impacts of privatization policy shifts. *British Journal of Middle Eastern Studies*, 1–21.

Barnawi, O. Z. (2017). *Neoliberalism and English Language Education Policies in the Arabian Gulf*. Routledge.

Gelvin, J. L. (Ed.). (2021). *The Contemporary Middle East in an Age of Upheaval*. Stanford University Press.

Gupta, A. (2018). How neoliberal globalization is shaping early childhood education policies in India, China, Singapore, Sri Lanka and the Maldives. *Policy Futures in Education, 16*(1), 11–28.

Kirkpatrick, R. (2017). *Policy in the Middle East and North Africa.*

Knight, J. (2015). Financing Education Hubs: Who Invests? International Higher Education, 79, Article 79. https://doi.org/10.6017/ihe.2015.79.5836

Mommani, R., Zaidan, E., and Al-Saidi, M. (2024, forthcoming). Knowledge-based economy and the education system in the Gulf: Reflections from reform legacies in Qatar

Rizvi, F., & Lingard, B. (2009). *Globalizing education policy*. Routledge.

Romanowski, M. H., & Du, X. (2022). Education transferring and decentralized reforms: The case of Qatar. *Prospects, 52*(3), 285–298. https://doi.org/10.1007/s11125-020-09478-x

UNDP. (2016). Chapter 3: Education and the transition to work. In *Arab Human Development Report 2016: Youth and the Prospects for Human Development in a Changing Reality*. United Nations Publications. Retrieved from https://arab-hdr.org/wp-content/uploads/2020/11/adhr-report-2018-full-report.pdf

Zaidan, E., Belkhiria, E., & Wazen, C. (2023). Universities of the Future as Catalysts for Change: Using the Sustainable Development Goals to Reframe Sustainability-Qatar University as a Case Study. *The Sustainable University of the Future: Reimagining Higher Education and Research* (pp. 1–23). Springer International Publishing.

Zaidan, E., Momani, R., & Al-Saidi, M. (2024). Entrepreneurial universities and integrated sustainability for the knowledge-based economy: Self-perception and some structural challenges in the Gulf region. *Humanities and Social Sciences Communications, 11*(1), 1–14.

Part I
Teaching

Chapter 2
Teaching Sustainability in Qatar: Challenges for Teacher Readiness

Jennifer Bruder, Noor Hamwy, Abdellatif Sellami, and Michael H. Romanowski

Abstract The core obstacle in imparting knowledge about sustainability lies in teachers' readiness, which encompasses an internal motivation and commitment towards sustainability, adequate knowledge, and teaching techniques, as well as an educational setting that fosters and maintains the incorporation of sustainability within the local cultural context. This chapter explains teacher readiness and explores salient challenges to teacher readiness within Qatar. Notable challenges include inadequate teacher education, professional readiness, educational borrowing, cultural influences, teachers' attitudes, and school culture. The chapter offers a range of localised recommendations based on research findings. Providing teachers with sufficient and specific ESD and GCED training can result in favourable outcomes regarding their attitudes, experiences, and confidence towards teaching sustainability.

Keywords Teacher readiness · Education for sustainable development · Global citizenship education · Challenges · Educational borrowing · Qatar · Teacher education

2.1 Introduction

Many economic, environmental, and social challenges afflict the contemporary global landscape. Consequently, there is an immediate call from leaders and policymakers for proactive measures that contribute to the conservation of natural resources, the safeguarding of ecosystems, and the improvement of the overall quality of life and welfare for forthcoming generations, both at a global and local level (Seddon et al., 2021). Emerging from this call, Education for Sustainable Development (ESD) and Global Citizenship Education (GCED) have become the mainstay aimed at promoting

J. Bruder (✉) · N. Hamwy
Carnegie Mellon University in Qatar, Doha, Qatar
e-mail: jbruder@andrew.cmu.edu

A. Sellami · M. H. Romanowski
Education Research Center, College of Education, Qatar University, Doha, Qatar

E. Zaidan et al. (eds.), *The Future of Education Policy in the State of Qatar*, Gulf Studies 21, https://doi.org/10.1007/978-981-97-9667-0_2

sustainable development (Garcia et al. 2017), with particular emphasis on SDG 4.7, which stresses that:

> …by 2030, all learners will acquire the knowledge and skills needed to promote sustainable development, including, among others, through education for sustainable development and sustainable lifestyles, human rights, gender equality, promotion of a culture of peace and non-violence, global citizenship, and appreciation of cultural diversity and culture's contribution to sustainable development (United Nations 2015, p. 17).

To promote sustainable development (SD), educational institutions and teachers must play a crucial role in fostering ESD and GCED (Garcia et al. 2017). Teachers are agents of change that shape attitudes, beliefs, and behaviours, and they must be prepared to deliver instruction on sustainability (Lovren, 2017). Given that teachers are responsible for equipping future generations to confront these challenges, it is imperative that teacher readiness be considered the foundation of ESD and GCED (García-González et al., 2020).

Since ESD and GCED have become crucial domains in addressing monumental challenges from a grassroots level, the development and readiness of educators have emerged as a priority (Anra & Yuliandra, 2022; Zguir et al., 2022). Teachers' readiness refers to the preparedness level demonstrated by educators equipped with the necessary skills, knowledge, commitment, and confidence to initiate teaching activities, to promote change and improvement within the school setting (Sulaeman et al., 2022). Readiness is the ideal state of proficiency that encompasses both comprehension of a specific subject matter and the capability and willingness of the instructor to engage in instructional and educational activities (Pribudhiana et al., 2021). In the context of ESD and GCED, readiness manifests itself as the teacher's ability to provide effective sustainability instruction, and an eagerness to strive towards offering avenues for student involvement, thereby affording them opportunities to acquire knowledge and practise the principles of SDGs (Sustainable Development Goals). More importantly, teacher readiness is strongly associated with student improvement (Lynch et al., 2017) due to its ability to facilitate students' educational progress (Baharuldin et al., 2019). Therefore, the readiness of teachers in terms of sustainability education is critical.

The importance of teacher readiness and its influence on effective instruction and change cannot be overestimated (Anra & Yuliandra, 2022). One recurring theme in the literature on teachers' readiness to teach SDGs indicates that teachers who are not adequately prepared and lack the necessary knowledge and professional education in SDG pedagogies, can impede effective instruction and learning (Conway et al., 2021). According to Norizan (2010), inadequate teacher preparation contributes to the ineffectiveness of sustainability education. Researchers argue that future teachers must be adequately prepared to design and implement effective environmental education curricula (e.g., Cutter-Mackenzie & Smith, 2003; Norizan, 2010). There are challenges to the development of teacher readiness, which reduces the effective teaching of ESD and GCED (Hamwy et al., 2023).

With that in mind, this chapter is structured as follows. First, we define ESD and GCED and briefly provide details of some of the initiatives occurring in Qatar.

Next, we address teacher readiness as a challenge to teaching sustainability and how several factors influence readiness. Finally, we present research-based and localised recommendations on how to address teachers' preparation for ESD and GCED in Qatar.

2.2 ESD and GCED in Qatar

ESD imparts to students the necessary knowledge, comprehension, abilities, competencies, and qualities related to protecting environmental, social, and economic welfare, for current and forthcoming generations (HEA 2014). GCED seeks to provide individuals across all age groups with the necessary knowledge and abilities to actively cultivate peaceful, tolerant, inclusive, and secure societies locally and globally. GCED is founded upon three fundamental domains of learning: cognitive, socio-emotional, and behavioural (UNESCO, 2020).

Several developmental agendas in Qatar demonstrate a commitment to ESD and GCED teaching and integration. For example, the document Qatar National Vision 2030 (QNV 2030) aimed to facilitate the transformation of Qatar into a knowledge-based society, thereby guaranteeing its citizens a high standard of living (GSDP 2008). According to Al-Thani et al. (2021), QNV 2030 includes goals related to SD, ethical principles for preserving and protecting the environment, and the promotion of environmental consciousness among the populace. Qatar has displayed a specific inclination towards addressing sustainability issues and propagating SD principles, within the realm of education. As per international efforts, Qatar incorporates SD and SDGs into its kindergarten through twelfth-grade educational curricula.

Many initiatives focus on ESD and GCED within Qatar's public and private education sectors. These initiatives comprise various approaches, including official top-down efforts, responses to different global frameworks, and structured and unstructured educational activities within formal and informal settings (Sever & Tok 2022). The UNESCO Associated Schools Network is essential in promoting ESD in Qatar. The primary objective of the UNESCO Associated Schools Network is to establish connections among educational institutions worldwide, aiming to achieve SDG target 4.7, which pertains to GCED and ESD. In addition, the Eco-Schools Programme is an important initiative in Qatar that promotes ESD in K-12 education. The primary goals of the Eco-Schools Programme encompass, but are not restricted to, cultivating students' expertise, abilities, and enthusiasm for addressing current issues, while also enhancing comprehension and dedication to the SDGs. Ultimately, localised endeavours, establishments, and unofficial or non-establishment agents in Qatar actively and indirectly bolster the goals and objectives of ESD. Numerous entities are involved in the educational ecosystem ESD in Qatar, such as the Arab Climate Youth Movement, Qatar Youth Power, Greener Future, SDGeneration Network, and Qatar National Library, among others.[1]

[1] See Sever and Tok (2022) for a detailed account of these activities.

When analysing the Qatar National Curriculum Framework (MEHE, 2016), it can be deduced that integrated sustainability topics are organised under six prominent themes: environment and sustainability, life skills, safety and security, health education, values education, and family education (Zguir et al., 2022). This document demonstrates that these pervasive concepts are intended to be incorporated across various subjects at different educational levels, from kindergarten to twelfth grade (Zguir et al., 2022). Finally, as stated in a 2021 United Nations report, Qatar's curriculum and teacher training "somehow" reflects the implementation of the 1974 recommendations on ESD (United Nations 2021). As the implementation of the SDGs increases in Qatar, we argue that teacher readiness is a significant challenge in Qatar, and is influenced by the several factors that follow.

2.3 Content and Professional Knowledge and Readiness

Wang et al. (2014) define professional knowledge as a body of knowledge used in educational practice. Sexton (2007) highlights the concept of professional practice, explicitly pertaining to practitioner teachers, who possess the requisite knowledge, skills, and practices essential for delivering services of a certain quality within their profession. Educators must possess profound professional expertise encompassing sustainability knowledge and pedagogical and assessment strategies for effectively imparting sustainability concepts to their students.

Teachers must also have specialised content knowledge, which is required to teach sustainability. The lack of sustainability knowledge is among the elements of ill-prepared teachers (Ewiss et al., 2018; Hamwy et al., 2023). For instance, in the UK (Cebrian et al. 2015) and Malaysia (Bakar et al., 2021), a lack of comprehension and knowledge are among the factors that impede the advancement of ESD. This lack of understanding hinders teachers from effectively teaching themes related to ESD and GCED (Alkaher & Carmi, 2019). UNESCO examined teacher readiness across 144 countries using its Global Survey for Teachers on ESD and GCED (SDG 4.7). Of 58,000 teachers surveyed, 90% considered these themes essential, and over 80% wanted to continue learning about them. One in four felt they needed more support and training (Education International and UNESCO Report 2021). This lack of knowledge prevents teachers from effectively incorporating ESD and GCED themes into instruction (Alkaher & Carmi, 2019).

This readiness demands that teachers possess a thorough and extensive professional practice with well-developed pedagogical knowledge, encompassing pedagogical and assessment skills (Darling-Hammond, 2018). In Qatar, Bruder et al. (2023) reveal that teachers in public and private schools reported having comparable or elevated levels of knowledge and teaching self-assurance, compared to the participants of UNESCO's recent worldwide survey (2021). In line with these findings, several research studies conducted in Qatar have revealed that teachers should receive more extensive training and actively participate in professional development programmes to improve their effectiveness in teaching ESD (Al-Thani et al., 2021;

Bruder et al., 2023; Du & Chaaban, 2020; Zguir et al., 2022). Teachers' preparedness may further be hampered by insufficient access to educational materials and resources about the SDGs (Kioupi & Voulvoulis, 2019; Waltner et al., 2020). Educators need help identifying appropriate illustrations for integrating SDGs into their instructional approaches (Borg et al., 2012). Furthermore, they have grappled with the challenge of an overloaded curriculum (Kioupi & Voulvoulis, 2019; UNESCO, 2021; Zguir et al., 2022) and an absence of coverage of themes related to ESD and GCED in the curriculum (UNESCO, 2021). Research investigations have consistently demonstrated that subject expertise and pedagogical knowledge positively affect teachers' job performance (Darling-Hammond, 2018; OECD, 2016).

Regarding pedagogical skills, student assessment holds significant importance in education, yet there is a valid concern regarding the insufficient preparedness of teachers to assess ESD and GCED. Assessment methods and feedback evaluate students' abilities and level of achievement in meeting desired objectives, while aiding in the advancement of their learning process through continuous feedback. As per the findings of UNESCO (2021), 40% of teachers who participated in the survey faced challenges in assessing ESD and GCED, with more than one-third of educators expressing a requirement for more assessment tools or guidelines about interdisciplinary subjects. Over a third of respondents said they needed additional resources or procedures to assess multidisciplinary topics. Further research revealed that educators were inadequately equipped to understand and implement assessment methodologies aligned with the SDGs (Ssossé et al., 2021).

Concerns about assessing the SDGs have also been reported in Qatar (Al-Kuwari et al., 2022; Hamwy et al., 2023). Al-Kuwari et al. (2022) report that teachers' ability to adopt novel pedagogical methods or assessment techniques, such as formative assessment, within classrooms in Qatar government schools is regarded as an impediment. Hamwy et al. (2023) discovered that private school educators perceive more significant obstacles in evaluating ESD and GCED themes than their counterparts in public schools. Specifically, teachers identified the challenges in assessing cultural diversity and fostering tolerance as the most formidable.

2.4 Teacher Education and Readiness

It is widely acknowledged that achieving SDG 4 objectives relies heavily on teacher education, and a lack of adequate teacher preparation is a significant contributing factor to deficiencies in environmental education (Bourn et al., 2017; Wolff et al., 2017). Inadequate teacher preparation contributes to the shortcomings observed in environmental education, despite an overall positive attitude towards environmental issues (Norizan, 2010). In Finland, the readiness of teachers to promote ESD is hindered by a deficiency in their training and education, which fails to prioritise sustainability as a fundamental concept (Wolff et al., 2017).

Similarly, in Greece, pre-service primary teachers with limited environmental knowledge exhibited a low implementation rate of environmental programmes in school, despite their interest in environmental protection (Spiropoulou et al., 2007). According to Tuncer et al. (2009), teachers with a strong understanding of the environment, and positive attitudes towards it, are likelier to foster environmentally literate students. Furthermore, UNESCO (2021) reports that, according to educators, teacher training programmes inadequately address sustainable consumption and production and climate change.

Teacher preparedness extends beyond knowledge and access to resources. A research study examining the approaches of primary school teachers in Norway towards environmental education revealed that, although instructional units for students aged 10–13 covered various environmental topics; the teachers missed opportunities to foster students' critical thinking and argumentation skills (Munkebye et al., 2020). This implies that simply delivering information to students is insufficient to effectively promote the necessary competencies to achieve SDG 4.7 (Seatter & Ceulemans, 2017). A transformative pedagogy that facilitates learning is therefore required. Scholars contend that prospective teachers must receive adequate training, to develop and implement effective environmental education curricula (e.g. Norizan, 2010). Several studies have indicated that teachers have expressed a need for professional development opportunities that assist them in integrating sustainable values and concepts into their teaching, rather than additional policy layers on SDGs and ESD. They also require greater support and capacity building in sustainability education (Waltner et al., 2020). Therefore, it is imperative to integrate this aspect into teacher education. A more transformative pedagogical approach is required for effective sustainability learning.

2.5 Educational Borrowing and Teacher Readiness

Educational borrowing can be characterised as transferring knowledge pertaining to policies, frameworks, and practices from one context to another to achieve similar outcomes (Dolowitz & Marsh, 2000). According to Phillips and Ochs (2003), borrowing involves policymakers utilising foreign examples in initiating and implementing educational change. These borrowed educational policies create situations where the cultural values and norms of the countries implementing the policies and the governments they borrowed from are challenged through cultural exchanges and conflicts (Suarez-Orozco & Qin-Hilliard, 2004).

Implementing borrowed practices is widely regarded as credible due to its practicality and ability to bridge the gap between theoretical concepts and practical application. The legitimacy is derived from acknowledging that these practices have already demonstrated success in a distinct context, providing a model for others to replicate (Bernstein & Van der Ven, 2017). This thinking is based on and shaped by the fallacy of universal best practices, a key factor to consider when transferring educational best practices. This concept involves the misconception of universalism, which is inherent in believing that a practice yielding positive results in one context will inevitably produce similar outcomes in another (Rose, 1991). This notion is particularly prevalent in contemporary educational policies and practices, particularly in the Global South. The assumption is that a practice that proves effective in one setting automatically gains legitimacy, and can be readily replicated in another location (Bernstein & Van der Ven, 2017).

One essential criticism of educational borrowing is that it fails to consider the cultural aspects of an education system. Educational policymakers often adopt policy blueprints of leadership practices, management structures, and professional development programmes (PDPs) derived from various cultural contexts, but fail to adequately consider their compatibility with the prevailing cultural norms and values (Dimmock & Walker, 2000). McDonald (2012) states that "educational policy borrowing is endemic, but without careful attention to the key ideas of contextualisation and ownership, its value is less likely to be achieved" (p. 1871). Perry and Tor (2008) suggest that when educational policy borrowing is imposed, borrows encounter significant challenges, and borrowed programmes tend to be ineffective due to their failure to consider the specific circumstances and factors present in the local context. For instance, Hayes (2017) examined Bahrain's policymaking process in the GCC region to evaluate the reforms' effectiveness and contends that implementing borrowed practices could have been improved by addressing the contradiction between the local policy process and the imported strategies, consequently hindering the achievement of desired outcomes.

The degree of educational borrowing, educational products, policies, systems, and practices prevalent in Qatar is significant (Alkhater, 2016; Nasser, 2017). Many have been implemented through a top-down approach to policies and practices (Nasser, 2017; Romanowski & Du, 2020). Despite incorporating several borrowed educational reforms and practices, their efficacy has not been as anticipated. According to Alkhater (2016), the implementation of borrowed reforms in Qatar was hindered due to the influence of the local context, despite originating from a genuine need for reform. For example, conflicts have occurred in implementing a decentralised education system which is contradictory to a centralised government, such as that of Qatar (Abdel-Moneim, 2020). Teachers in Qatar have faced pedagogical conflicts regarding Problem-Based Learning (Du & Chaaban, 2020), child-centred pedagogy (Portnoi, 2016), and professional licensures (Ellili-Cherif et al., 2012), among others. These conflicts can be attributed to both educational borrowing and cultural factors.

Since teachers are usually the implementers of educational reform, their belief systems reflect the established educational traditions in a particular context regarding teaching methods, which, in turn, influence teachers' pedagogical approaches (Handal & Herrington, 2003). When faced with borrowed products, borrowers will likely be impacted by ideas and cultural influences from other regions (Phillips, 2005). This can result in tensions, conflicts, and dilemmas for teachers. Studies contend that educators frequently need more preparation to effectively implement a specific borrowed instructional approach. According to Romanowski et al. (2022), it is commonly observed that when teachers implement borrowed practices from external sources, their capacity or readiness to cultivate and maintain the practice effectively becomes constrained. In Qatar, for example, Abdel-Moneim (2020) reports that teachers were ill-equipped to fulfil their anticipated duties during the borrowed education reform *Education for a New Era,* which is often cited as a reason for the failure of the reform (Abou-El-Kheir & MacLeod, 2019; Alkhater, 2016; Alnaimi, 2019). Du and Chaaban (2020) found that teachers in Qatar required a greater level of preparedness to implement a problem-based learning programme adapted from another context effectively. It can be inferred that education borrowing can also negatively impact the readiness of teachers to teach ESD and GCED.

2.6 Cultural Influences on Readiness

There is little doubt that culture profoundly affects educational values and behaviours, as they are intricately intertwined with culture and cultural traditions. How does culture impact teachers' readiness and the understanding and implementation of SDGs in education? Because culture is a fundamental driver of human choices and behaviours and becomes a framework when defining and understanding sustainable development, culture, and sustainability become intricately interconnected (Streimikiene & Kiausiene, 2019). The way the SDGs are understood, as well as the way they should be taught, are significantly influenced by culture.

It is essential to note that sustainability and the SDGs are written words and a form of discourse that requires active engagement and teacher interpretation. According to McGregor (2004), discourse is how people produce knowledge. Discourse limits acceptable knowledge and is fundamental to which perspectives can be employed to interpret reality. In this way, discourse plays a pivotal role in generating norms and regulations by creating boundaries and restrictions on what is considered legitimate knowledge, and the range of perspectives that can be used to interpret reality.

The SDGs and the language of ESD and GCED can be regarded as types of discourse, because they represent a collection of agreed-upon statements that define sustainability and SDGS without dispute. However, discourse formation is always influenced by external factors, such as culture, that shape meanings (Laclau 1996). This flexibility allows for the development of floating signifiers, or descriptors, that carry excessive meaning because of various interpretations within different contexts (Torfing, 1999). The discourse of sustainability and floating signifiers allow for structural changes that vary among individuals, as they can represent multiple or unlimited meanings depending on the views and understandings of their interpreters (Chandler, 2007).

Regarding sustainability and SDGs, meanings seem fixed, yet are quite fluid as they move within diverse cultures or contexts, developing different meanings. Defining the meanings of concepts such as human rights, gender equality, cultural diversity, and tolerance is complex, because of the different cultural views, beliefs, values, and understandings that may be connected to the terms. These concepts are sites of struggle with different views, beliefs, values, and understandings used to establish meaning, often floating or shifting from context to context, resulting in different outcomes.

As an illustration, SDG 5 seeks to achieve gender equality and empower all women and girls (United Nations, 2022). One element of SDG 5 includes ending discrimination and violence against women and girls worldwide. These terms are floating signifiers, lacking a clear-cut or common understanding, since culture shapes the definition of these terms and will be interpreted in many ways. UNESCO's definition of specific SDG components, i.e. gender equality, is understood differently in the Global South than in the Global North. This will influence the readiness among teachers and principals, and how they interpret and implement SDGs. (UNESCO, 2021). For example, Yli-Panula et al. (2022) discovered that Finnish educators encountered difficulties imparting knowledge and fostering an understanding of sustainability due to conflicting cultural values, attitudes, and beliefs. Moreover, these conflicting factors considerably impacted the values held by students and the instructional approaches employed within the classroom setting. Therefore, it can be inferred that culture affects how teachers view and understand sustainability and related themes and concepts.

2.7 Readiness and Teachers' Attitudes Towards Sustainability

Teacher readiness requires that teachers not only accept favourable perspectives on aspects of sustainability, including human rights and gender equality, but also consider them valuable components of their teaching methods. Singh (2016) reports that educators who demonstrate a strong understanding of environmental issues exhibit a more favourable disposition towards their active role in disseminating environmental content. Tuncer et al. (2009) suggest that teachers who comprehensively understand the environment and exhibit positive attitudes in this respect are more likely to produce environmentally literate students. Consequently, educators' positive environmental attitudes can potentially cultivate positive environmental attitudes in young children (Esteve-Guirao et al., 2019).

Concerning the preparation of teachers, Malhotra and Upadhyay (2021) assert that an effective means of cultivating pre-service teachers' favourable disposition towards sustainability is through education on ESD. In ESD, it is presupposed that aspiring educators must not only cultivate a comprehensive set of competencies and acquire a profound understanding of SD, but are expected to exhibit favourable dispositions towards sustainability concerns, as these attributes are essential prerequisites for effectively implementing and promoting this form of pedagogy (UNESCO, 2017). Hence, a teacher's disposition towards environmental consciousness becomes significant in facilitating the advancement of environmental education across various educational stages, including primary, secondary, and senior secondary levels (Singh, 2016).

Teachers' perspectives on sustainability and attitudes towards educational borrowing significantly impact the programme's effectiveness (Liu & Feng, 2015).

In his work, Hu (2002) emphasises the significance of an ideological attitude, in influencing and regulating the processes and results of educational borrowing. This ideological attitude acknowledges the presence of culturally embedded diversity, and dismisses the idea of universally applicable methods of instruction and learning (Hinkel, 1999). According to Hu (2002), the widespread global resistance towards foreign pedagogical methods is evidence of an "ideological" standpoint (p. 102).

Again, Hu (2002) argues that the ideological attitude of those involved in education reflects the dominant framework, lifestyle, assumptions, and reasoning within a particular cultural setting. When educational policies are transferred to a different context, the borrowed ideas and practices encounter the influence of pre-existing theories, and, at times, conflicts arise, shaping how sustainability is taught. Consequently, teachers' attitudes play a crucial role in educational borrowing, and positive attitudes towards sustainability and borrowing cannot simply be instilled into teachers.

2.8 School Culture

Just as teachers must be ready to teach sustainability, school culture is essential for effective SDG instruction, since it may hinder or support teachers' readiness and opportunities to teach sustainability. Deal and Peterson (1999) define school culture as an "underground flow of feelings and folkways [wending] its way within schools, dragging people, programmes, and ideas toward unstated purposes" (p. 3). School culture refers to the underlying beliefs and assumptions, rituals and ceremonies, history and stories, and physical symbols that are often overlooked but greatly influence individuals' actions, and shape their attitudes and behaviours over an extended period. It significantly impacts their interpretation of various interactions and occurrences that they encounter daily (Deal & Peterson, 1999). The school's culture shapes its values and guides its daily operations.

There is a notable relationship between teachers' attitudes towards implementing ESD and GCED in their schools and the prevailing school culture. The nature of the school culture plays a crucial role in achieving long-term sustainability and sustainable development (Yli-Panula et al., 2022). For instance, if a school's organisational and leadership culture has traditionally adopted a top-down approach to implementing change, teacher autonomy can be hampered. It can hinder the effectiveness of the implementation process, ultimately impeding the success and potentially leading to the programme's failure. Among many factors, top-down approaches often result in a culture of resistance from educational leaders and teachers, making it challenging to promote sustainability (Forssten Seiser et al. 2022). If the school's culture does not embrace and support ESD and GCED, it will be difficult for teachers to involve students in sustainability.

2.9 Recommendations

The development of ESD and GCED for each nation must move beyond the reliance on standardised solutions proposed by the United Nations (Alam et al., 2021). Every country must consider its cultural background, and actively pursue innovative cultural practices tailored to its specific context rather than adopting uniform educational curricula predominantly derived from more developed contexts (Alam et al., 2021). Therefore, to enhance the sustainability pedagogy, it is insufficient for countries to adopt theories detached from contextual social norms and practices. Instead, societal and contextual factors should shape ESD and GCED (Guryanova et al., 2020). Sever and Tok (2022) rightly argue that it is imperative that ESD is tailored to the specific circumstances, limitations, requirements, and capabilities of societies, and is implemented in a localised manner. They suggest that ESD is integrated with Qatar's cultural and societal norms and principles and that the implementation of localised teaching materials and resources in the native language, and incorporation of sustainability topics relevant to the student's surroundings are essential components. With that in mind, it is highly recommended that the educational leadership in Qatar refrain from borrowing foreign educational practices, and instead focus on utilising domestic resources to foster the development of ESD and GCED. In what follows, we offer several research-based and localised recommendations.

2.9.1 Innovate K-12 Teacher Education Programmes

It is important to acknowledge that the success of the UNESCO initiatives heavily hinges on the willingness and preparedness of educators to impart knowledge and promote principles and ideas associated with SD and the SDGs (Zguir et al., 2022). Regarding teachers, there is a widely acknowledged consensus that teacher education plays a pivotal role in the achievement of SDG 4 goals (Sagdic & Sah, 2019), while inadequate teacher training is seen as a contributing factor to deficiencies in environmental education (Knapp 2000; Wolff et al., 2017). Zachariou and Valanides (2006) argue that there is a need to restructure teacher education programmes at the university level, to foster a cultural shift that enables future teachers to recognise and appreciate the advantages and potential of ESD.

Various investigations have pinpointed crucial elements of SD that should be integrated into teacher education curricula, to adequately prepare instructors for attaining SDG 4. The pertinent features are comprehension of sustainability, tactics for managing behaviour, regulations and policies, comprehensive principles, and proficiency in pedagogical skills (Jetly & Singh, 2019; Sunthonkanokpong & Murphy, 2019). Educators with greater educational attainment and teaching experience express greater aptitude and self-assurance in instructing on ESD and GCED subjects. Therefore, those who develop teacher education programmes must develop ones that integrate ESD within the preparation capacities of teachers.

Chaaban and her colleagues (2023) conducted a study on teacher education programmes in Qatar, focusing on incorporating top-down frameworks for implementing ESD. The research aimed to develop and assess a framework that outlines the essential skills and knowledge that future teachers in Qatar need in order to implement ESD effectively. The study analysed the perspectives and findings of various stakeholders, revealing the challenges in reaching a consensus on the competencies required for pre-service teachers to implement ESD. These differing viewpoints stem from divergent beliefs regarding the responsibility for designing the curriculum and defining competencies. Chaaban and others emphasise the importance of effectively integrating ESD principles into the curriculum, and stress the need for all education stakeholders to agree on the integration process. This includes determining the role of teachers in designing lessons and units, which will translate the curriculum documents into tangible practices for students. This also points out the need not to use borrowed programmes, since teacher participation in the development is vital.

Teacher education can raise K-12 pre-service teachers' understanding of various themes embedded in SDGs. According to Bourn and other researchers (2017), teacher education should include themes such as gender equality, peace education, human rights, and cultural diversity. According to Spahiu & Lindemann-Matthies, 2015, comprehensive pre-service teacher education can enhance teachers' confidence in integrating and implementing ESD. Furthermore, such education and training endeavours can improve teachers' ability to develop contextually appropriate ESD programmes and projects. Finally, pre-service teacher education is especially effective in implementing new approaches such as ESD (Spahiu & Lindemann-Matthies, 2015). Therefore, it is paramount to enhance comprehension of SD and its incorporation into teacher education programmes, and to provide future teachers with essential pedagogical abilities.

Finally, Sever and Tok (2022) point out that it is imperative for policymakers and educators to actively engage in the identification, comprehension, and resolution of prevailing obstacles that impede the effective dissemination of sustainability teachings. As a result, it is necessary to enhance the alignment between the K-12 curricula and the SDGs and sustainability, incorporate them into teaching and learning practices, and develop educational programmes that effectively raise awareness about sustainability challenges, in particular locally significant ones such as water, food, and energy issues. Additionally, policymakers must establish educational curricula and training initiatives that align with Qatar's current and future needs.

2.9.2 Useful and Localised Professional Development Programmes

There is little doubt that the mainstay of teacher development is professional development. External professional development is widely recognised as an effective method for enhancing teachers' expertise and knowledge, and for promoting information sharing among all teaching staff members (Al-Thani et al., 2021). Several studies have revealed that teachers express a need for professional development that can assist them in integrating SDGs and ESD values and concepts into their teaching practices, rather than additional layers of policy (e.g. Waltner et al., 2020). In Qatar, Hamwy and other researchers (2023) reported that teachers thought professional development or workshops were needed to aid them in teaching SD and/or global citizenship.

Regarding the teaching of sustainability in Qatar's schools, it is vital for those who develop and implement professional development programmes for teachers at all levels, to be knowledgeable about Qatar's schools and culture, and start developing a comprehensive understanding of the teachers' perspectives towards ESD and GCED, about the knowledge and pedagogical skills they possess and/or lack in terms of teaching sustainability in Qatar, and the specific barriers that they face when teaching sustainability in Qatar's schools.

Gaining insights into the perspectives and needs of educators' pedagogical approaches and integrating ESD and GCED may increase the amount of teaching of these themes. Specific guidelines and attached resources that are comprehensive to meet the needs of each grade can help achieve easier access, and simultaneously respond to the limited time teachers have on a given day. Having to add on any teaching or training of students in ESD and GCED through after-school programmes and extracurricular activities may be helpful. Still, it would require teachers to sacrifice additional time and may not be the incentive that ignites sustainable change. Teachers cite a need to make sustainability an integrative part of the curriculum.

Zguir et al. (2022) argue that numerous frameworks provide, integrate, and embed SDGs and ESD values, and that one could design one primarily for a Qatari context. The issue of borrowing arises in this context. Qatar's educational policymakers must avoid simply adopting a framework and subsequently imposing it on teachers in a top-down manner. More specifically, Zguir et al. (2022) offer a framework that aids teachers in integrating SDGs and SD within curricula which is worth viewing (see Zguir et al., 2022 Fig. 7 p. 16). The authors argue that the framework is built on design thinking, and is aligned with sustainability competencies and SDG values, which is crucial for assisting teachers in developing and implementing lessons that promote SDG values and competencies among students.

2.9.3 Conclusion

This chapter has focused on the readiness of teachers to teach sustainability. The chapter discusses several areas that, if not developed, will hamper teachers' readiness to teach ESD and GCED, rendering programmes ineffective. Programmes designed to educate on ESD/GCED necessitate implementation on a local scale, and this chapter offers some guidance on the preliminary considerations that must be considered before initiating any ESD and GCED initiative. Given Qatar's past experiences with educational borrowing, which have been challenging, it could be argued that attaining

SDGs will entail some degree of borrowing, and that Qatar's teachers may find this challenging. The potential reliance on borrowed practices to train ESD and Global Citizenship Education GCED teachers is particularly concerning. Providing teachers with sufficient and specific ESD and GCED training can result in favourable outcomes regarding their attitudes, experiences, and confidence in teaching sustainability.

Bibliography

Abdel-Moneim, M. A. (2020). Between global and national prescriptions for education administration: The rocky road of neoliberal education reform in Qatar. *International Journal of Educational Development, 74*, 102160. https://doi.org/10.1016/j.ijedudev.2020.102160

Abou-El-Kheir, A. & MacLeod, P. (2019). The struggle for balance: policy borrowing and continuous reform in the practice of English language teaching in the Gulf Cooperation Council (GCC) Countries. In: Gao, X. (eds) *Second Handbook of English Language Teaching* (pp. 31–49). Springer International Handbooks of Education. Springer. https://doi.org/10.1007/978-3-319-58542-0_3-1

Alam, G. M., Roslan, S., Al-Amin, A. Q., & Leal Filho, W. (2021). Does GATS' influence on private university sector's growth ensure ESD or develop city 'Sustainability crisis' — policy framework to respond COP21. *Sustainability, 13*(8), Article 8. https://doi.org/10.3390/su1308 4520

Alkaher, I., & Carmi, N. (2019). Is population growth an environmental problem? Teachers' perceptions and attitudes towards including it in their teaching. *Sustainability, 11*(7), 1994. https://doi.org/10.3390/su11071994

Alkhater, L. R. M. (2016). Qatar's borrowed K-12 education reform in context. In L. R. M. Alkhater, M. E. Tok, & L. A. Pal (Eds.), *Policymaking in a transformative state: The case of Qatar* (pp. 97–130). Palgrave Macmillan.

Al-Kuwari, M. M., Du, X., & Koç, M. (2022). Performance assessment in education for sustainable development: A case study of the Qatar education system. *Prospects, 52*(3–4), 513–527. https://doi.org/10.1007/s11125-021-09570-w

Alnaimi, T. N. (2019). Reflections on education reform in Qatar: A look at independent school experience. *Multi-Knowledge Electronic Comprehensive Journal for Education and Science Publications (MECSJ), 20*, 1–19.

Al-Thani, W. A., Ari, I., & Koç, M. (2021). Education as a critical factor of sustainability: case study in Qatar from the teachers' development perspective. *Sustainability, 13*(20), Article 20. https://doi.org/10.3390/su132011525

Anra, E., & Y. & Yuliandra, L. (2022). Analysis of the influence of organizational culture, transformational leadership behavior, and psychological capital on readiness for change in elementary school teachers. *American Journal of Humanities and Social Sciences Research, 6*(10), 60–74.

Baharuldin, Z., Jamaluddin, S., Shahril, M., Shaharom, N., Mohammed, S., & Zaid, R. (2019). The role of teacher readiness as a mediator in the development of ICT competency in Pahang primary school. *Journal of Educational Research and Indigenous Studies, 2*(2), 15.

Bakar, R. N. A. R. A., Zakaria, S. Z. S., Isa, N. M., Majid, N. A., & Razman, M. R. (2021). Teachers' perception and roles regarding global citizenship for sustainable development. *Ecology, Environment and Conservation 27*, 209–215. ISSN 0971-765X

Bernstein, S. & Van der Ven, H. (2017) Best practices in global governance. *Review of International Studies 43*(3): 534–556. https://https://doi.org/10.1017/S0260210516000425.

Borg, C., Gericke, N., Höglund, H. O., & Bergman, E. (2012). The challenges encountered by teachers implementing education for sustainable development: Discipline bound differences and teaching traditions. *Research in Science & Technological Education, 30*(2), 185–207. https://doi.org/10.1080/02635143.2012.699891

Bourn, D. Hunt, F. & Bamber, P. (2017). A review of education for sustainable development and global citizenship education in teacher education. *Background paper for the 2017/8 Global Education Monitoring Report.*

Bruder, J., Hamwy, N., Sellami, A. & Romanowski, M. H. (2023). Exploring Qatar's private and school teachers' perspectives toward readiness for teaching education for sustainable development and global citizenship. *Globalisation, Societies and Education.*

Chandler, D. (2007). *Semiotics: The basics* (2nd ed.). Oxford.

Conway, B., Leahy, K., & McMahon, M. (2021). Design education for sustainability: Identifying opportunities in Ireland's second level education system. *Sustainability, 13*(16), 8711. https://doi.org/10.3390/su13168711

Cutter-Mackenzie, A., & Smith, R. (2003). Ecological literacy: The 'missing paradigm' in environmental education (Part One)". *Environmental Education Research, 9*(4), 497–524. https://doi.org/10.1080/1350462032000126131

Darling-Hammond, L. (2018). *Teaching is the profession on which all other professions depend: Linda Darling-Hammond on transforming education.* Stanford University.

Deal, T. E., & Peterson, K. D. (1999). *Shaping school culture: The heart of leadership.* Jossey-Bass. (pp. 1–3).

Dimmock, C., & Walker, A. (2000). Developing comparative and international educational leadership and management: A cross-cultural model. *School and Leadership Management, 20*(2), 143–160.

Dolowitz, D. & Marsh, D. (2000). Learning from abroad: The role of policy transfer in contemporary policymaking, *Governance, 13* (1), 5–24. https://https://doi.org/10.1111/0952-1895.00121.

Du, X., & Chaaban, Y. (2020). Teachers' readiness for a statewide change to PjBL in primary education in Qatar. *Interdisciplinary Journal of Problem-Based Learning, 14*(1). www.ijpbl.org (ISSN 1541-5015)

Education International & UNESCO Report. (2021). *Teachers motivated to teach sustainable development and global citizenship but need more support. Education International.* https://www.ei-ie.org/en/item/25601:education-international-and-unesco-report-teachers-motivated-to-teach-sustainable-development-and-global-citizenship-but-need-more-support.

Ellili-Cherif, M., Romanowski, M. H., & Nasser, R. (2012). All that glitters is not gold: Challenges of teacher and school leader licensure licensing system in Qatar. *International Journal of Educational Development, 32*(3), 471–481. https://doi.org/10.1016/j.ijedudev.2011.11.010

Esteve-Guirao, P., García, M. J., & Banos-González, I. (2019). The interdependences between sustainability and their lifestyle that preservice teachers establish when addressing socio-ecological problems. *Sustainability, 11*(20), 5748. https://doi.org/10.3390/su1120574

Ewiss, M. Z., & Afifi, S. (2018). Education enhancement for sustainable development in developing countries. In *ICERI2018 Proceedings* (pp. 8808–8816). IATED.

García-González, E., Jiménez-Fontana, R., & Azcárate, P. (2020). Education for sustainability and the sustainable development goals: Pre-service teachers' perceptions and knowledge. *Sustainability, 12*(18), 7741. https://doi.org/10.3390/su12187741

Guryanova, A. V., Krasnov, S. V., & Frolov, V. A. (2020). Human transformation under an influence of the digital economy development. In S. Ashmarina, A. Mesquita, & M. Vochozka (Eds.), *Digital Transformation of the Economy: Challenges, Trends and New Opportunities* (pp. 140–149). Springer International Publishing. https://doi.org/10.1007/978-3-030-11367-4_14

Hamwy, N., Bruder, J., Sellami, A., & Romanowski, M. H. (2023). Teacher challenges to education for sustainable development and global citizenship education in Qatar. *Sustainability, 15*(15), 11479. https://doi.org/10.3390/su151511479

Handal, B., & Herrington, A. (2003). Mathematics teachers' beliefs and curriculum reform. *Mathematics Education Research Journal, 15*, 59–69. https://doi.org/10.1007/BF03217369

Hayes, A. (2017). Deconstructing the 'magnetic' properties of neoliberal politics of education in Bahrain. *Globalisation, Societies and Education, 15*(2), 175–187.

Higher Education Academy (HEA) 2014. Education for Sustainable Development: Guidance for UK Higher Education Providers. https://s3.eu-west-2.amazonaws.com/assets.creode.advanc

ehe-document-manager/documents/hea/private/resources/education-sustainable-development-guidance-june-14_1568036951.pdf
Hinkel, E. (1999). Introduction. In E. Hinkel (Ed.), *Culture in second language teaching and learning* (pp. 1–7). Cambridge University Press.
Hu, G. (2002). Potential cultural resistance to pedagogical imports: The case of communicative language teaching in China. *Language, Culture and Curriculum, 15*(2), 93–105.
Jetly, M., & Singh, N. (2019). Analytical study based on perspectives of teacher educators in India with respect to education for sustainable development. *Journal of Teacher Education for Sustainability, 21*(2), 38–55. https://doi.org/10.2478/jtes-2019-0016
Kioupi, V., & Voulvoulis, N. (2019). Education for Sustainable Development: A Systemic Framework for Connecting the SDGs to Educational Outcomes. *Sustainability, 11*(21), Article 21. https://doi.org/10.3390/su11216104
Laclau, E. (1996). *Why do empty signifiers matter to politics?* Emancipation(s), Verso.
Liu, S. N., & Feng, D. M. (2015). How culture matters in educational borrowing? Chinese teachers' dilemmas in a global era. *Cogent Education, 2*(1). 1046410.https://doi.org/10.1080/2331186X.2015.1046410
Lovren, V. O. (2017). Promoting sustainability in institutions of higher education: the perspective of university teachers. In: Leal Filho, W., Azeiteiro, U., Alves, F., & Molthan-Hill, P. (eds) *Handbook of theory and practice of sustainable development in higher education. World sustainability series* (pp. 475–490). Springer, Cham. https://doi.org/10.1007/978-3-319-47877-7_32
Lynch, D., Smith, R., Provost, S., Yeigh, T., & Turner, D. (2017). The correlation between "teacher readiness" and student learning improvement. *International Journal of Innovation, Creativity and Change, 3*(1), 1–12.
Malhotra, R. K., & Upadhyay, R. (2021). Teaching sustainability in technology education: A comparison of perception and practice. *Ilkogretim Online, 20*(1), 7328–7342.
McDonald, L., (2012). Educational transfer to developing countries: Policy and skill facilitation. *Procedia Social and Behavioral Sciences, 69 (*Iceepsy), 1817–1826.
McGregor, S. (2004). *Critical discourse analysis – a primer.* www.kon.org/archives/forum/15-1/mcgregorcda.html (accessed July 6, 2015).
Ministry of Education and Higher Education. MEHE. (2016). Qatar National Curriculum Framework.
Munkebye, E., E. Scheie, A. Gabrielsen, A. Jordet, S. Misund, T. Nergård & A. Bergljot Øyehaug. (2020). Interdisciplinary primary school curriculum units for sustainable development. *Environmental Education Research 26*(6): 795–811. https://https://doi.org/10.1080/13504622.2020.1750568
Nasser, R. (2017). Qatar's educational reform past and future: Challenges in teacher development. *Open Review of Educational Research, 4*(1), 1–19. https://doi.org/10.1080/23265507.2016.1266693
Norizan, E. (2010). Environmental knowledge, attitude and practices of student teachers. *International Research in Geographical and Environmental Education, 19*(1), 39–50.
OECD. (2016). *Supporting teacher professionalism: Insights from TALIS 2013* (p. 2016). OECD.
Perry, L. B., & Tor, G. (2008). Understanding educational transfer: Theoretical perspectives and conceptual frameworks. *Prospects, 38*(4), 509–526. https://doi.org/10.1007/s11125-009-9092-3
Phillips, D. (2005). Policy borrowing in education: Frameworks for analysis. In J. Zajda (Ed.), *Second international handbook on globalisation, education and policy research* (pp. 23–24). Springer.
Phillips, D., & Ochs, K. (2003). Processes of policy borrowing in education: Some explanatory and analytical devices. *Comparative Education, 39*(4), 451–461.
Portnoi L. M. (2016). Patterns of educational reform and policy borrowing. In *Policy Borrowing and Reform in Education.* Palgrave Macmillan: New York. URL: https://link.springer.com/content/pdf/10.1057%2F978-1-137-53024-0.pdf

Pribudhiana, R., Bin Don, Y., & Bin Yusof, M. R. (2021). Determining the influence of teacher quality toward teacher readiness in implementing Indonesian education policy. *Eurasian Journal of Educational Research*. https://eric.ed.gov/?id=EJ1300016

Romanowski, M. H., Alkhateeb, H., & Sellami, A. (2022). Problematizing K-12 best practices. *Equity in Education & Society, 1*(3), 388–402.

Romanowski, M. H., & Du, X. (2020). Education transferring and decentralized reforms: Qatar's decentralized and recentralized education system. *Prospects*. https://doi.org/10.1007/s11125-020-09478-x

Rose, R. (1991). Comparing forms of comparative analysis. *Political Studies, 39*, 46–462.

Sagdic, A., & Sah, E. (2019). An assessment of Turkish elementary teachers in the context of education for sustainable development. *Journal of Teacher Education for Sustainability, 21*(2), 91–104. https://doi.org/10.2478/jtes-2019-0019

Seatter, C. S., & Ceulemans, K. (2017). Teaching sustainability in higher education: Pedagogical styles that make a difference. *Canadian Journal of Higher Education, 47*(2), 47–70.

General Secretariat for Development Planning. (GSDP). (2008). *Qatar national vision 2030*. Doha, Qatar. https://www.psa.gov.qa/en/qnv1/Documents/QNV2030_English_v2.pdf.

Seddon, N., Smith, A., Smith, P., Key, I., Chausson, A., Girardin, C., House, J., Srivastava, S., & Turner, B. (2021). Getting the message right on nature-based solutions to climate change. *Global Change Biology, 27*(8), 1518–1546. https://doi.org/10.1111/gcb.15513

Sever, S. D. & Tok. M. E. (2022). Education for sustainable development in Qatar. In Cochrane, L. & Al-Hababi, R. (Eds.) *Sustainable Qatar: social, political and environmental perspectives*, pp. 329–347. Singapore: Springer Nature.

Sexton, M. (2007). Evaluating teaching as a profession — Implications of a research study for the work of the teaching council. *Ir. Educ. Stud., 26*, 79–105.

Singh, A. (2016). The impact of environmental awareness on teachers' attitude towards environmental education. *ESSENCE Int. J. Env. Rehab. Conserv, 7*(2), 69–72.

Spahiu, M. H., & Lindemann-Matthies, P. (2015). Effect of a toolkit and a one-day teacher education workshop on ESD teaching content and methods — A study from Kosovo. *Sustainability, 7*(7), Article 7. https://doi.org/10.3390/su7078051

Spiropoulou, D., Triantafyllia, A., Kontaxakaki, S., & Bouras, S. (2007). Primary teachers' literacy and attitudes on education for sustainable development. *Journal of Science Education and Technology, 16*, 443–450.

Ssossé, Q., Wagner, J., & Hopper, C. (2021). Assessing the impact of ESD: methods, challenges, results. *Sustainability, 13*(5), Article 5. https://doi.org/10.3390/su13052854

Streimikiene, M., & Kiausiene. (2019). The Impact of Value Created by Culture on Approaching the Sustainable Development Goals: Case of the Baltic States. *Sustainability, 11*(22), 6437. https://doi.org/10.3390/su11226437

Suarez-Orozco, M., & Qin-Hilliard, D. B. (2004). *Globalization: Culture and education in the new millennium*. University of California.

Sulaeman, N., Efwinda, S., & Putra, P. D. A. (2022). Teacher readiness in STEM education: Voices of Indonesian Physics Teachers. *Journal of Technology and Science Education, 12*(1), 68. https://doi.org/10.3926/jotse.1191

Sunthonkanokpong, W., & Murphy, E. (2019). Quality, equity, inclusion and lifelong learning in preservice teacher education. *Journal of Teacher Education for Sustainability, 21*(2), 91–104. https://doi.org/10.2478/jtes-2019-0019

Torfing, J. (1999). *New theories of discourse: Laclau, Mouffe and Zizek*. Wiley Blackwell: Hoboken, NJ.

Tuncer, G., Tekkaya, C., Sungur, S., Cakiroglu, J., Ertepinar, H., & Kaplowitz, M. (2009). Assessing pre-service teachers' environmental literacy in Turkey as a mean to develop teacher education programs. *International Journal of Educational Development, 29*(4), 426–436. https://doi.org/10.1016/j.ijedudev.2008.10.003

UNESCO. (2017). *Education for Sustainable Development: Learning Objectives*. Paris: UNESCO, http://unesdoc.unesco.org/images/0024/002474/247444e.pdf

UNESCO. (2020). *What is global citizenship education?* https://en.unesco.org/themes/gced/definition#:~:text=Global%20Citizenship%20Education%20(GCED)%20aims,%2C%20socio%2Demotional%20and%20behavioural.

UNESCO (2021). *Teachers have their say: Motivation, skills and opportunities to teach education for sustainable development and global citizenship.* UNESCO and Education International, Brussels, Belgium.

United Nations (2022). *Sustainable development goals. Goal 5: Achieve gender equality and empower all women and girls.* https://www.un.org/sustainabledevelopment/gender-equality/

Waltner, E. M., Scharenberg, K., Hörsch, C., & Rieß, W. (2020). What teachers think and know about education for sustainable development and how they implement it in class. *Sustainability, 12*(4), 1690. https://doi.org/10.3390/su12041690

Wang, L., Lai, M., & Lo, L. N. (2014). Teacher professionalism under the recent reform of performance pay in Mainland China. *Prospects, 2014*(44), 429–443.

Wolff, L. A., Sjöblom, P., Hofman-Bergholm, M., & Palmberg, I. (2017). High performance education fails in sustainability? — A reflection on Finnish primary teacher education. *Education Sciences, 7*(1), 32. https://doi.org/10.3390/educsci7010032

Yli-Panula, E., Jeronen, E., & Mäki, S. (2022). School culture promoting sustainability in student teachers' views. *Sustainability, 14*(12), Article 12. https://doi.org/10.3390/su14127440

Zguir, M. F., Dubis, S., & Koç, M. (2022). Integrating sustainability into curricula: Teachers' perceptions, preparation and practice in Qatar. *Journal of Cleaner Production, 371*, 133167. https://doi.org/10.1016/j.jclepro.2022.133167

Chapter 3
International Standardised Tests and Qatar's Educational Policy Development

Muhammad Mubashir Ehsan

Abstract Qatar's education system has evolved over the last two decades, and it envisions itself as a knowledge-based society aligned with Qatar National Vision 2030. Since the early 2000s, Qatar has transitioned towards integrating international standards into its education system. This chapter investigates the significance of international standardised tests for Qatar's educational policy development. The chapter presents a descriptive analysis based on TIMSS and PISA, and comments regarding Qatar's global position and its neighbours, namely the United Arab Emirates and Saudi Arabia. Also presented is a decline in the Harmonised Learning Outcomes (HLOs) for Qatar during the pandemic. The chapter discusses the limitations and significance of understanding the interpretation of this type of test. Overall, it provides a comprehensive overview of how international standardised tests can be used to inform educational policy development in Qatar. It emphasises the importance of balancing global standards with local context, to achieve sustainable educational achievement.

3.1 Introduction

In pursuing excellence and innovation, the State of Qatar has worked towards a transformative journey to reshape its educational landscape. Aligned with its Qatar National Vision 2030, the country endeavours to foster a knowledge-based society with the skills and knowledge to increase interconnectivity in this globalised day and age. To reach this goal, along with investment in the education sector, Qatar has incorporated international standardised testing to inform educational policy development. This chapter investigates the pivotal role of standardised international tests in Qatar's educational development, shedding light on why these assessments are valuable, and giving evidence of how they have been used to improve educational planning within Qatar.

M. M. Ehsan (✉)
Education Above All (EAA), Qatar Foundation, Doha, Qatar
e-mail: mubashir.ehsan993@gmail.com

E. Zaidan et al. (eds.), *The Future of Education Policy in the State of Qatar*,
Gulf Studies 21, https://doi.org/10.1007/978-981-97-9667-0_3

Qatar's educational journey from the early 2000s illustrates its commitment to providing its citizens with an international standard of educational opportunities, and preparing its youth to tackle the complex challenges of the twenty-first century. International standardised tests, such as the Program for International Student Assessment (PISA) and the Trends in International Mathematics and Science Study (TIMSS), have assisted Qatar in measuring its student achievement compared with different countries.

This chapter explores the multifaceted significance of standardised international testing for Qatar's educational development. It explores these assessments' benefits, including benchmarking Qatar's educational performance against global standards, facilitating data-driven policy decisions, ensuring educational quality, and enhancing global recognition. The chapter also addresses the limitations of international standardised testing in measuring success in a cross-country context. Additionally, it highlights the challenges related to these tests being integrated into Qatar's educational system, emphasising the importance of striking a balance between global standards and local context. The chapter also discusses how Harmonised Learning Outcomes (HLOs) were used in COVID-19 to measure the learning losses.

As we investigate, it becomes evident that standardised international tests serve as more than just evaluative tools; they could be considered catalysts for change and progress, driving Qatar towards its educational vision, and shaping the future of its citizens. This chapter provides a holistic overview of how interpreting these two international standardised tests (TIMSS and PISA) can provide insights regarding educational policy development. Moreover, the chapter provides the use of Harmonised Learning Outcomes as the most recent comparable data for education. The chapter focuses on the importance of the factors of language, diversity, and culture that must be incorporated into policy development for sustainable educational achievement.

3.2 The Emergence of International Standardised Testing

Large-scale international studies of educational achievement have a long history, dating back to the 1960s. These studies typically compare students' performance in different countries on standardised tests in maths, science, and reading. They provide valuable insights into the strengths and weaknesses of different education systems, and can help policymakers identify areas where improvement is needed (Wagemaker, 2020).Over the last six decades, there have been twelve different prominent international maths, science, or reading tests, including various countries that voluntarily participated in these assessments (Table 3.1). The assessments contain 36 possible test combination scores (e.g. science for students of grade 8 in 1972 as part of the First International Science Study or maths of 15–16-year-olds in 2000 as a part of the Program on International Student Assessment).

The assessments are designed to identify a standard set of expected skills, which are then tested in the native language. Mathematics seems a sound comparable option as it is numeracy-based. On the other hand, science scores are also manageable as

Table 3.1 The list of tests across different timelines (OECD, IEA at different time intervals)

	Abbreviation	Study	Year	Subject	Age,	Countries	Organisation	Scale
1	FIMS	First International Mathematics Study	1964	Maths	13,FS	11	IEA	PC
2	FISS	First International Science Study	1970–71	Science	10,14,FS	14,16,16	IEA	PC
3	FIRS	First International Reading Study	1970–72	Reading	13	12	IEA	PC
4	SIMS	Second International Mathematics Study	1980–82	Maths	13,FS	17,12	IEA	PC
5	SISS	Second International Science Study	1983–84	Science	10,13,FS	15,17,13	IEA	PC
6	SIRS	Second International Reading Study	1990–91	Reading	9,13	26,30	IEA	IRT
7	TIMSS	Third International Mathematics and Science Study	1994–95	Maths/ Science	9(3 + 4), 13(7 + 8),FS	25,39,21	IEA	IRT
8	TIMSS-Repeat	TIMSS-Repeat	1999	Maths/ Science	13(8)	38	IEA	IRT
9	PISA 2000/02	Program for International Student Assessment	2000–02	Reading/ Maths/ Science	15	31 + 10	OECD	IRT
10	PIRLS	Progress in International Reading Literacy Study	2001	Reading	9(4)	34	IEA	IRT
11	TIMSS 2003	Trends in International Mathematics and Science Study	2003	Math/ Science	9(4),13(8)	24,45	IEA	IRT
12	PISA 2003	Program for International Student Assessment	2003	Reading/ Maths/ Science	15	40	OECD	IRT

they could be translated into the native language. However, the reading proficiency scores are based on learning English as the main language; hence, implementation and score estimation is a complex process for the non-native English population. One of the main standardised reading tests is the PIRLS (IEA, n.d.).

One of the most comprehensive efforts to understand student performance is the PISA, which provides data on family and institutional factors. The test has been developed to consider a variety of relevant skills and competencies. A standard procedure is that each subject is tested across a wide sample, enabling a detailed understanding of student performance.

PISA is not the first international comparative survey of student achievement. Others have been conducted over the past four decades, mainly developed by the International Association for the Evaluation of Educational Achievement (IEA) and the Education Testing Service's International Assessment of Educational Progress (IAEP). These surveys have focused on outcomes linked directly to those parts of the curriculum that are essentially common across the countries involved in these assessments. Characteristics of the curriculum which are unique to one country, or a smaller number of countries, have usually not been considered in the assessments, regardless of how important those parts of the curriculum are for the countries participating. Each country has different socioeconomic backgrounds, such as culture, religion, political stability in the area, language, and customs.

On the other hand, TIMSS (Trends in International Mathematics and Science Study) was created to engage more countries worldwide, so as to impact mathematics and science learning positively. The data collected by TIMSS is at two age group levels: 4th grade and 8th grade. TIMSS collects other data related to quality, quantity, and instructional contents (Mullis et al. 2020). The fundamental difference between PISA and TIMSS is that TIMSS does not incorporate reading tests, but PISA does.

3.3 Methodology

A descriptive analysis is presented in the chapter based on three different data sources. Firstly, the analysis conducted using the TIMSS data for Qatar from 2007 to 2019 (EdStats n.d.). The second data source is from the PISA, the data taken from 2006 to 2018; the PISA data is also used to develop a comparative analysis with average OECD and the GCC countries (The UAE and Saudi Arabia) (EdStats n.d.) (OECD, 2018). In order to understand socioeconomic and cultural diversity factors, the analysis uses grey literature presented in reports produced by OECD, IEA and the World Bank.

Lastly, to understand the changes during the COVID-19 Pandemic, Harmonised Learning Outcome[1] data is used from 2010 to 2020 (2010, 2017, 2018, 2020). One

[1] Variable description: Harmonised Test Scores from major international student achievement testing programmes. They are measured in TIMMS-equivalent units, where 300 is minimal attainment and 625 is advanced attainment. The most recent estimates are used. The year of the most

key limitation of the analysis that is due to the data limitation aspects, there is no recent comparable data on education after 2020.

The analysis presents the overall scores across different subjects: in the case of Qatar it does not aim to present an in-depth analysis on the contents of assessment.

3.4 Qatar's Progress in International Standardised Testing

In this section, we will investigate how the scores have changed over the last decade for Qatar based on two main standardised tests, PISA and TIMSS. This section provides a descriptive analysis of both test measures.

3.4.1 TIMSS

Qatar was part of TIMSS in four cycles of their testing: 2007, 2011, 2015, and 2019. The tests allowed the education system to gain insights into skills and knowledge in two subjects, mathematics and science (Kelly et al. 2020). Moreover, TIMSS was vital to inform the government regarding the strengths and weaknesses faced by learners at different grade levels. The following Tables 3.2 and 3.3, illustrate TIMSS scores over the four cycles i.e., Cycle 1 = 2007, Cycle 2 = 2011, Cycle = 2015 and Cycle 4 = 2019.

The tables above represent changes across four cycles, illustrating the trend for Maths and Science subjects. Investigating each cycle, it is clear that the most significant improvement occurred from the first cycle to the second cycle. For changes in the first two cycles, the mean score for mathematics grade 4 was higher than that of science at the same grade level. However, in the case of grade 8, the science score was higher than that for mathematics.

The score did not increase exponentially after the first two cycles. Even the changes within genders only show a slight difference. Based on the tables, it can be concluded that there has been improvement in the scores, but at a slower rate for the last two

recent estimate is shown in the data notes. Test scores from the following testing programmes are included: • TIMSS/PIRLS: This refers to the average of test scores from TIMSS (Trends in International Mathematics and Science Study) and PIRLS (Progress in International Reading Literacy Study), both carried out by the International Association for the Evaluation of Educational Achievement. Data from each PIRLS round, moved to the year of the nearest TIMSS round and averaged with the TIMSS data. • PISA: Refers to test scores from the Program for International Student Assessment • PISA + TIMSS/PIRLS: Refers to the average of these programmes for countries and years where both are available • SACMEQ: Refers to test scores from the Southern and Eastern Africa Consortium for Monitoring Educational Quality • PASEC: Refers to test scores from the Programme on the Analysis of Education Systems • LLECE: Refers to test scores from the Latin American Laboratory for Assessment of the Quality of Education • EGRA: Refers to test scores from Early Grade Reading Assessments.

Table 3.2 Changes in TIMSS score for the 4th Grade (2007–2019). (Authors work)

Grade 4 tests	Change cycle 1–2	Change cycle 2–3	Change cycle 3–4
TIMSS: Mean performance on the mathematics scale for fourth grade students, female	113.053	20	10
TIMSS: Mean performance on the mathematics scale for fourth grade students, male	121.96	31	11
TIMSS: Mean performance on the mathematics scale for fourth grade students, total	116.73	26	10
TIMSS: Mean performance on the science scale for fourth grade students, female	100.87	40	8
TIMSS: Mean performance on the science scale for fourth grade students, male	101.00	42	19
TIMSS: Mean performance on the science scale for fourth grade students, total	99.60	42	13

Table 3.3 Changes in TIMSS score for the 8th Grade (2007–2019). (Authors work)

Grade 8 Tests	Change cycle 1–2	Change cycle 2–3	Change cycle 3–4
TIMSS: Mean performance on the mathematics scale for eighth grade students, female	89.52	25	7
TIMSS: Mean performance on the mathematics scale for eighth grade students, male	116.03	30	6
TIMSS: Mean performance on the mathematics scale for eighth grade students, total	103.21	27	6
TIMSS: Mean performance on the science scale for eighth grade students, female	78.17	39	17
TIMSS: Mean performance on the science scale for eighth grade students, male	122.38	35	20
TIMSS: Mean performance on the science scale for eighth grade students, total	100.15	38	18

cycles. The lowest rate of change in the latest available data is for 8th-grade mathematical mean scores. Interestingly, the 2015 TIMSS figures provided some scope for recommendations for the Ministry of Education and Higher Education to develop a plan to ensure that all international standards are met within Qatar's schools. These included three key elements that could improve the curriculum and standards so as to achieve international standards, i.e. academic criteria, awareness, and motivation (Mullis et al., 2016) Understanding how to evaluate the incorporation of such recommendations in the long run is crucial.

In the short run, a marginal improvement suggests a slower growth rate performance based on the most recent TIMSS score. The long-term impact of exploring the changes would be seen ten years from the introduction of the changes. It is crucial to understand the factors which have reduced the growth rate of the scores. Another vital aspect would be to examine the impact of the lockdowns and closures on these test scores, for which the data will be published in 2024.

3.4.2 PISA

In order to evaluate its performance through a cross-country comparable lens, it should be noted that Qatar participated in PISA assessment in 2006 for the first time. Since 2006, Qatar has participated in five testing cycles i.e., Cycle1 = 2006, Cycle2 = 2009, Cycle3 = 2012, Cycle 4 = 2015 and Cycle5 = 2018. For the most recent test in 2018, 92% of the total population of children of 15 years from 188 schools participated in the test (OECD, 2018). Table 3.4 illustrates the change in score in these five cycles over the time period.

The table above illustrates that the strongest progress occurred in the first two cycles for Mathematics and Reading, however the highest positive change for science was observed between the third and fourth cycle. However, the estimates from the fourth to five cycles show a lower positive change in scores for all subjects, especially science, where there was only an increase of in single figures. Figures 3.1, 3.2 and 3.3 shows the trends for all three subjects in comparison to the OECD countries.

Figures 3.1, 3.2 and 3.3 show that Qatar's PISA scores have increased over the five cycles across all subjects. However, they have remained lower than the OECD average. In 2006, Qatar's PISA science score was 349, below the OECD average of 495. Over the last five cycles, the gap between the OECD average and the Qatar PISA score for science has narrowed to a difference of 70 points. Similar trends are

Table 3.4 Change in PISA Scores across three subjects from 2006–2018. (Authors work)

Test subjects	Change cycle-2	Change cycle 2–3	Change cycle 3–4	Change cycle 4–5
PISA Maths	50	8	26	12
PISA Science	30	5	34	1
PISA Reading	60	16	14	5

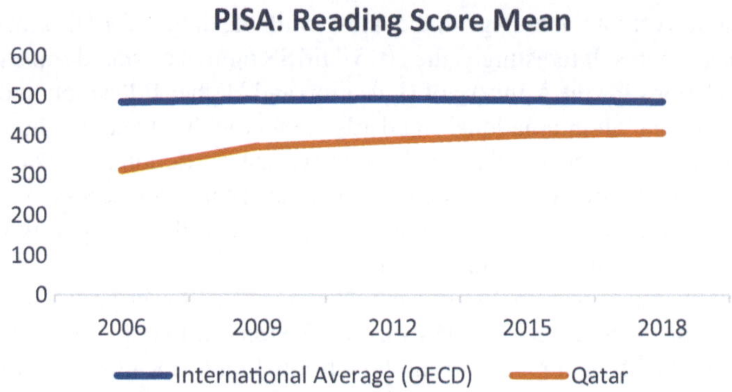

Fig. 3.1 Comparison of PISA Reading mean scores between OECD countries and Qatar (OECD, 2018)

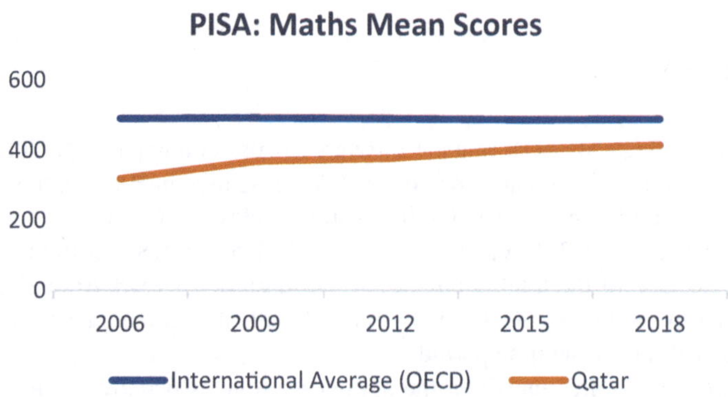

Fig. 3.2 Comparison of PISA Mathematics mean scores between OECD countries and Qatar (OECD, 2018)

evident across the other two subjects. This increase in Qatar's PISA science score is likely due to several factors, including increased investment in education, and teacher professional development. Qatar has also significantly invested in early childhood education, which is vital for long-term academic success. A crucial factor to understand over the last decade is the change in demographics with regard to an increase in immigrant students in the Qatar's education system. In the most recent estimates, the PISA scores highlight the fact that the number of students with an immigrant background who participated in the test increased by 11% in 2018. Another nuanced factor is that one in five of the students belonged to socioeconomically disadvantaged families. If we compare the reading scores of the immigrants in the OECD with the immigrants in Qatar, 36% of them outperformed the non-immigrants, that is 19% higher than the case in the OECD (OECD, 2018). So based on this, it could be highlighted that schools in Qatar were effectively assisting the immigrant students with disadvantaged background in achieving these scores.

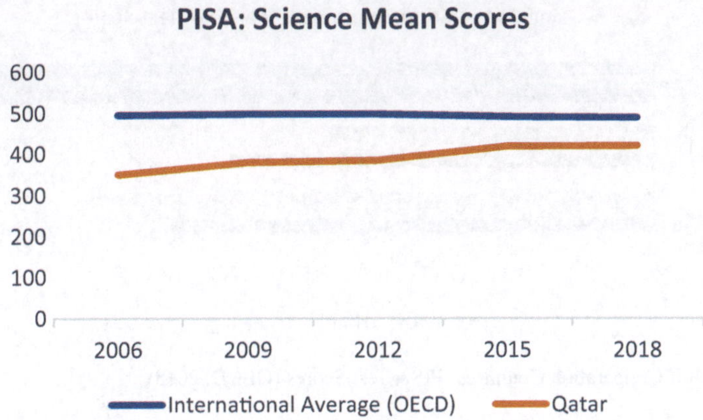

Fig. 3.3 Comparison of PISA Science mean scores between OECD countries and Qatar (OECD, 2018)

The increase in Qatar's PISA science score is a positive sign that the country is improving its students' education quality. However, in Qatar's Education Ministry's Strategic Plan 2018–2022, the desired targets for the international scores for both PISA and TIMSS were to achieve an increase of 30 points on average; no results for any test or subject actually met the target. At the global level, a similar trend was observed in most countries.

In comparison, exploring the overall progress of the OECD, the rate of increase in scores for them is also not rapid compared to the initial years. The most recent document regarding the PISA comparison between 2015 and 2018 illustrates that student performance in reading, mathematics, and science across OECD countries remained relatively unchanged between 2015 and 2018. However, some individual countries had a more significant variation (OECD, 2019). For instance, for the reading tests, Georgia and Indonesia saw the most significant declines in performance; on the other hand, North Macedonia and Turkey saw higher improvements. For mathematics, there were comparatively few declines in performance, but Turkey, Albania, and North Macedonia all saw substantial improvements. In the case of the science subjects, there was no significant change in performance across OECD countries, but Turkey, Albania, and North Macedonia again saw some significant improvements (OECD, 2019). Overall, the results suggest no clear trend in student performance across the OECD countries between 2015 and 2018. However, there is some evidence that Turkey, Albania, and North Macedonia are making progress in improving student performance in all three subjects.

In the case of comparison in the region, the following Fig. 3.4 bar graph illustrates the PISA standing across all three subjects.

Figure 3.4 shows where Qatar stands compared to other countries in the region. Based on the recent test scores in 2018, the graph illustrates that Qatar's performance in all subjects is better than that of Saudi Arabia. However, the UAE has a far higher performance level based on PISA scores. However, all three countries performed

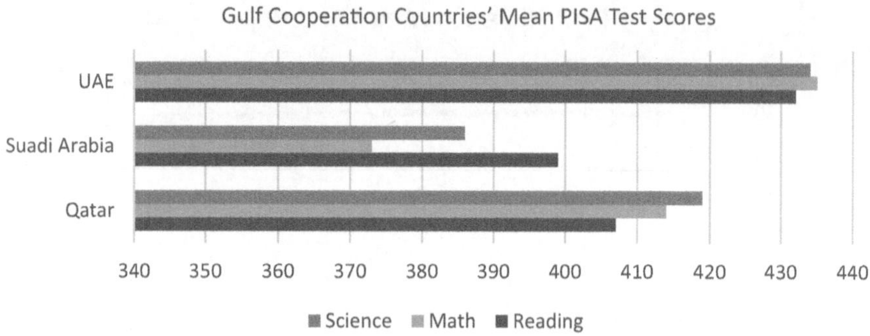

Fig. 3.4 Gulf Cooperation Countries' PISA Test Scores (OECD, 2018)

Table 3.5 Level of girls' out-performance of boys based on PISA scores (EdStats, n.d.)

Sr#	Country	Maths	Science	Reading
1	Qatar	24	39	24
2	UAE	9	26	57
3	Saudi Arabia	13	29	54

below the OECD average. Interestingly, in the case of the GCC context, gender differences play a significant role; the OECD gender data on PISA illustrates that in all three selected countries, as illustrated in Table 3.5, girls outperformed boys in all subjects.

This table shows how girls outperformed boys in three GCC countries across all three subjects. The highest difference is for mathematics subjects in the case of Qatar, while girls in the UAE strongly outperformed the boys in reading. These results align with the literature on gender studies in the region, underlining factors related to the MENA paradox, whereby women tend to be higher achievers in education. However, those numbers do not translate into higher female participation (Participation at school? Or in jobs maybe? It's unclear. Editor).

3.5 Harmonised Learning Outcomes (HLO)

The main intention behind adding Harmonised Learning Outcomes to this chapter is to understand the most recent trends i.e. during and after the COVID-19 pandemic. Both PISA and TIMSS have not published any test results since COVID-19, but HLO has become a useful instrument to understand the quality of education based on student learning. Before getting into the trends for Qatar, it is crucial to understand how HLOs are developed. According to Harry, (2021) the HLO "database includes results from seven different tests, which cover between 10 and 72 countries each. The results of these tests have been combined and made comparable so that researchers can easily compare them". Harmonised Learning Outcomes are utilised in a few recent studies that have attempted to understand the learning losses due to the COVID-19 restrictions and lockdown policies in various countries. In Qatar's case, there have been four years for which that the data has been collated (2010, 2017, 2018, 2020) (Fig. 3.5).

The diagram is a line graph showing Harmonised Learning Outcomes before and after the COVID-19 pandemic. The graph shows that these outcomes were higher before the COVID-19 pandemic. The HLOs in 2010 were gauged at a total of 385, which increased to 432 in 2017 and remained in the 430 s in 2018. In 2018, the Harmonised Learning Outcomes began to decline, and they reached 427 in 2020.

Several possible explanations exist for the decline in these outcomes after the COVID-19 pandemic. Factors could be addressed through the nature of structural issues at a macro level, and the individual issues based on existing characteristics of the demographics (Engzell et al., 2021). There is emerging literature based on studies that have pinpointed mental health as one of the main issues contributing to changes in behaviour towards education (Viner et al., 2022); other issues include the lack of access to internet services and the lower concentration span of students learning through online mediums. However, the purpose of this chapter is not to delve into the factors contributing to learning losses in Qatar, but to identify measures to assess the level of educational progress against other countries.

This section has provided descriptive insights with a brief cross-country analysis across the countries in the region. Qatar has made some improvements with regard to the international standardised test score; however, the recent trends estimate slower growth in the score improvement. However, it is crucial to understand that the analysis does not claim that the overall student performance is attributed to learning losses or test scores, as there are other factors that determine the performance such as individual factors related to mental health, emotional intelligence, and socioeconomic

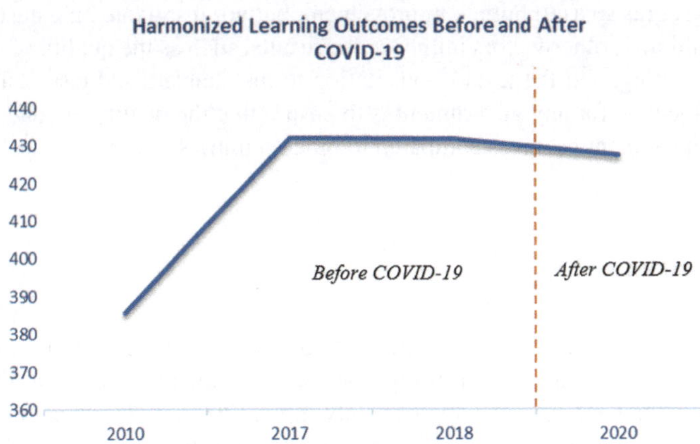

Fig. 3.5 Before and after HLOs for Qatar. (World Bank Education Statistics, 2021)

background. The following section discusses what these scores mean, and explains how to interpret the results of standardised tests.

3.6 Comparison

If we examine these scores in comparison to the 8th-grade TIMSS scores, the trend is not similar. The PISA score for mathematics is higher than that for TIMSS, and the science score is lower in PISA compared to TIMSS. Therefore, this brings us to understand that these tests are not comparable in the sense that the aim of TIMSS is to reflect skills and knowledge based on the curriculum taught in schools. On the other hand, PISA evaluates students' ability to apply their knowledge in order to unravel "real-world" problems.

Going back to Table 3.3, based on an understanding of the distinction between the tests, it can be argued that the students' ability in problem-solving skills in a real-world context is low; however, they perform better on curriculum-based learning, as is evident from the PISA scores. Another factor could be the compatibility between the curriculum and the PISA testing. For instance, countries such as Finland are known to have a more compatible curriculum with the PISA (Andrews et al., 2014). This concept of mirroring the curriculum could enhance the country's reputation for achieving high international education scores, specifically in PISA. However, it may have long-term systematic implications regarding traditional conceptual learning.

A lesson learned from a Finnish study illustrates how changes to the curriculum for mathematics, and over-dependency on using calculators in class resulted in a lower mathematical score for a similar test that was conducted in 1981 and repeated in 2003 (Martio, 2009). This lesson is crucial for Qatar and other countries facing the dilemma of lower scores on international standardised tests, that they should reflect on the test scores so as to initiate improvement, but not just fixate on adjustments to the curriculum, as other factors influence the results, such as the quality of teaching, mode of learning, and the level of adaptation to the standardised tests. Ultimately, the core objective for any government is to ensure that the quality of education and student achievement progress compared to other countries.

3.7 Discussion

A main concern in the case of countries with multilingual composition of its population is generally related to ensuring that the medium of instruction allows the learner to understand concepts appropriately. Ejieh (2004, p. 73) stated that "when … communication takes place in a language known to the student, the chances of achieving understanding are high compared to when it takes place in a language with which the student is not familiar". Qatar has faced this challenge in the development of its curriculum, with regard to the medium of instruction. In the last two decades, two major developments that occurred in Qatar that highlight the issue of language in rolling out the curriculum. Having a multilingual society makes it difficult to resort to a single language as the main mode of learning, especially when the native language holds a significant value based on identity and cultural values. Qatar's mathematical curriculum development has faced similar constraints, as before 2001, the schools taught mathematics in the Arabic language. In 2001, Education of the New Era was introduced, replacing Arabic with English as the medium of instruction for mathematics, science and technology (Mustafawi & Shaaban, 2019).The consequences of this transition were not fruitful, as student achievement did not improve, rather it seemed to lead to reduced scores across all English medium subjects.

It is crucial to understand the limitations of the international standardised tests, especially in the case of PISA, which aims to help students to solve real-world problems; however, the real-world programmes are not uniform across all countries, as context plays a major role in understanding the problems and providing solutions. Moreover, it is crucial to acknowledge the path dependencies in education practice, especially when related to societal resistance to change, as was the case in 2001 after the introduction of English as a medium of instruction in Qatar. Decontextualisation of educational policy may challenge the idea of a culturally bounded system (Steiner-Khamsi, 2004, p. 5). Some studies, such as Romanowski et al. (2018), emphasise that borrowing education policy from other countries can be ineffective if cultural, historical, and other contexts are neglected.

Another major limitation is the interpretation of the PISA results, especially when making claims regarding major successes. For example, in the case of Vietnam, a high level of success was reported based on the results; however what was not brought out was that only 56% of 15-year-olds attended school (Sellar et al., 2017, p. 44). Moreover, it is crucial to emphasise that there are critiques regarding PISA testing that explain that the questions are developed and framed especially for the reading test with a bias towards Western countries, which as a consequence creates a greater gap between the Indo-European and Non-Indo-European languages (Asil & Brown, 2016). This issue raises questions with regard to the cross-country comparison of the PISA tests.

3.8 Recommendations

The recommendations in this section serve two purposes: (1) to ensure effective educational strategy, so as to address the multilingual concerns related to subjects such as science and mathematics; (2) to understand the use of different types of international standardised tests in education policy development.

3.8.1 A Multilingual Approach to Pedagogical Practices

In order to target higher mathematical proficiency in multilingual societies, it is crucial to understand the integrating techniques that could ensure that all students from different backgrounds have access to equal opportunities to understand the concepts in mathematics and science lessons. Content and Language Integrated Learning (CL/IL) is an approach that provides an innovative solution for cultural integration and curriculum development. This approach ultimately has the aim of enhancing conceptual learning in mathematics.

However, the case of Qatar is complex, due to its diverse population structure. An example from Finland can provide a simplistic understanding of the application of (CL/IL), where an experimental study was conducted to see the difference between contents when learning a foreign language, with an experimental group and native Finnish speakers (the Control group) across two subjects: mathematics and science (Jäppinen 2005). (This is not in the list of references. Editor) The study aimed to investigate the consequences of analogical CLIL reason systems, as a measure of cognitive development. One relevant finding from the study was:

"In some cases, in the second age group, the cognitional development of the experimental group seemed to be even faster than that in the control group. One thus can conclude that analogical CLIL reasoning systems work in the Finnish CLIL environments where the mother tongue of all learners is Finnish and they are taught through English, French, or Swedish." (Jäppinen 2005). In Qatar's case, a longitudinal study could be conducted, in order to understand the impact of CLIL. This could be developing treatment and a control group at the education level. (This sentence does not make sense. Editor) This experimental design could help explain student competency in mathematics and science in Qatar. These experiments can provide vital insights into the educational policy that can be developed to improve competency in mathematics and science. This policy might not be the complete roadmap for educational reform, but its insights could provide policymakers with an opportunity to understand the effectiveness of such policies in the context of Qatar.

3.8.2 Educational Policy Development

The policy narrative around the international standardised tests should focus on the nuances of the results, not just the differences from the initial benchmark, but deeper introspection on factors contributing to a higher or lower score. However, in the above section, there is a deeper discussion of what the different results mean. This chapter only focuses on the two tests: PISA and TIMSS.

At the global level, Education Agenda 2030 emphasises the importance of learning outcomes as a major part of the international education monitoring framework (UNESCO 2015). Moreover, these international standardised tests contribute to informing Sustainable Development Goal 4.1 which focuses on "the percentage of students who meet a minimum proficiency level in mathematics (a) in early grades, (b) at the end of primary education and (c) at the end of lower secondary education" (IEA & UNESCO 2020). The achievement of the SDGs is one of the commitments for Qatar (Aguilar 2017). However, there is a need for the policy narrative to be developed, based on identifying specific factors contributing to the student's competency level. Supply-side factors such as the number of teachers and the quality of teaching, methodology, curriculum, and the language of instruction all play a vital role in student competency development.

The interpretation of TIMSS is more aligned with the curriculum and teaching material covered in the syllabus. Regarding the results, TIMSS reflects knowledge based on the school curriculum. Therefore, what can be identified from the results is how well the teachers are able to instruct students based on the curriculum, and how well the students can absorb it. At the policy level, lower scores can identify deficiencies in the education systems, and other structural issues in the education sector, however based on the current evidence from analysis in this chapter, we cannot clearly attribute the lower scores to the deficiencies of the education system or structural issues; there can be several other factors that may influence the scores, such as individual factors related to the students. In the case of PISA, the interpretation depends on how well the students can apply the knowledge in real-world scenarios.

There should be a holistic approach in which educational policy is developed, ensuring that insights from both these tests are considered in policy development. Here the holistic approach means developing a system that ensures that international standardised tests and school levels tests are monitored, and that annual analysis is done based on those scores. The Ministry of Education and Higher Education, Qatar, can develop a team to take the lead on this initiative. This initiative could focus on the analysis of the international standardised tests and the school level results, based on how education policies can be shaped. Moreover, the policymakers in Qatar must set clear, tangible short-term and medium-run goals that could feed into their commitment towards Qatar National Vision 2030, and at the global level to Education Agenda 2030, which incorporates achievement of the SDGs.

3.9 Conclusion

In summary, Qatar has made overall progress since the benchmarking based on international standards; however, the recent improvement in the international standardised test scores does not meet the target set by the ministry. Moreover, the scores have been lower than the OECD average throughout the various cycles of testing. Having said that, it is crucial to understand that compared to the neighbouring countries, Qatar has performed at an average level, as it has done better than Saudi Arabia but scored lower than the United Arab Emirates.

The chapter has acknowledged the limitations of international standardised tests and emphasised the importance of culture and the prestige of the language used in certain countries. Moreover, it is vital to note and understand that these standardised tests have been developed in the Western context, borrowing policies from the West, while neglecting history, context, and culture, which may not be sustainable, especially for the local population in Qatar. Another key limitation of the analysis is that the most recent data that is published for standardised test is from 2020, but that more updated data will be available in 2024 for these selected standardised tests.

The chapter has highlighted how there has been increase in the number of immigrant students, and with the impact of their inclusion on the standardised tests, specifically PISA, the author acknowledges that there is a need for an in-depth analysis that could be carried out based on a primary level research study, based on using surveys and interviews with the key informants within the education sector.

One of the significant challenges that Qatar and most Arab countries face is that the students come from a variety of linguistic backgrounds. The chapter has discussed the implications of such a population structure on learning competencies. To address this challenge, Qatar needs to develop effective educational strategies that can cater to the needs of all students, regardless of their linguistic background; a potential recommendation highlighted in the chapter is Content and Language Integrated Learning (CL/IL).

Another area where Qatar can improve is the way it uses international standardised tests, and the development of policy narratives that nudge the country towards a better quality of education. Qatar and other countries in the region use these tests to compare their performance over the years and between the countries, based on specific recommendations by the international organisations conducting these assessments. However, the policymakers should not focus on targeting higher scores but rather explore why certain countries do better on PISA and not that well on TIMSS. As illustrated in the chapter, an example of this is Finland, which has previously aligned its curriculum with the PISA, an alignment which has assisted the country in eliciting high scores on this test. In this regard, the countries should understand their priorities, either to focus on initiatives that fixate on increasing test scores by mirroring the curriculum with the PISA or to have a more holistic approach that ensures long-term sustainability in the quality of education. Moreover, the chapter has presented the most recent student achievement results through the Harmonised Learning Outcomes that illustrate learning losses during the COVID-19 lockdown.

Qatar has made significant progress in improving its educational system in recent years. However, there are still some areas where the country can improve, such as multilingualism and the use of international standardised tests in policy narratives.

Bibliography

Asil, M., & Brown, G. T. L. (2016). Comparing OECD PISA reading in english to other languages: identifying potential sources of non-invariance. *International Journal of Testing, 16*(1), 71–93. https://doi.org/10.1080/15305058.2015.1064431

EdStats. (n.d.). *Education Statistics | Learning Outcomes*. Retrieved October 29, 2023, from https://datatopics.worldbank.org/education/wDashboard/dqlearningcnty

Ejieh, M. U. C. (2004). Attitudes of student teachers towards teaching in mother tongue in Nigerian primary schools: implications for planning. *Language, Culture and Curriculum, 17*(1), 73–81. https://doi.org/10.1080/07908310408666683

Engzell, P., Frey, A., & Verhagen, M. D. (2021). Learning loss due to school closures during the COVID-19 pandemic. *Proceedings of the National Academy of Sciences, 118*(17), e2022376118. https://doi.org/10.1073/pnas.2022376118

Harry, A. P. (2021, March 15). HLO—Harmonized Learning Outcomes. *Education, Economics and Public Policy*. https://hpatrinos.com/2021/03/15/hlo-harmonized-learning-outcomes/

IEA. (n.d.). *PIRLS*. Retrieved October 13, 2023, from https://www.iea.nl/studies/iea/pirls

Mustafawi, E., & Shaaban, K. (2019). Language policies in education in Qatar between 2003 and 2012: From local to global then back to local. *Language Policy, 18*. https://doi.org/10.1007/s10993-018-9483-5

OECD (2018). Programme for international student assessment (PISA): Results from PISA 2018 Qatar. OECD Publishing. https://www.oecd.org/pisa/publications/PISA2018_CN_QAT.pdf

OECD. (2019). *PISA 2018 Results (Volume I): What Students Know and Can Do*. Organisation for Economic Co-operation and Development. https://www.oecd-ilibrary.org/education/pisa-2018-results-volume-i_5f07c754-en

Romanowski, M. H., Alkhateeb, H., & Nasser, R. (2018). Policy borrowing in the gulf cooperation council countries: cultural scripts and epistemological conflicts. *International Journal of Educational Development, 60*, 19–24. https://doi.org/10.1016/j.ijedudev.2017.10.021

Sellar, S., Thompson, G., & Rutkowski, D. (2017). *The Global Education Race: Taking the Measure of PISA and International Testing*. Brush Education.

Steiner-Khamsi, G. (2004). *The Global Politics of Educational Borrowing and Lending*. Teachers College Press.

Viner, R., Russell, S., Saulle, R., Croker, H., Stansfield, C., Packer, J., Nicholls, D., Goddings, A.-L., Bonell, C., Hudson, L., Hope, S., Ward, J., Schwalbe, N., Morgan, A., & Minozzi, S. (2022). School Closures During Social Lockdown and Mental Health, Health Behaviors, and Well-being Among Children and Adolescents During the First COVID-19 Wave: A Systematic Review. *JAMA Pediatrics, 176*(4), 400–409. https://doi.org/10.1001/jamapediatrics.2021.5840

Wagemaker, H. (Ed.). (2020). *Reliability and Validity of International Large-Scale Assessment: Understanding IEA's Comparative Studies of Student Achievement*. Springer Nature. https://doi.org/10.1007/978-3-030-53081-5

World Bank Education Statistics (2021). *Average learning outcomes*. Our World in Data. https://ourworldindata.org/grapher/average-harmonized-learning-outcome-scores

Chapter 4
Problematising PBL in Higher Education in GGC Countries: The Case of Qatar

Youmen Chaaban, Hessa Al-Thani, and Xiangyun Du

Abstract In this chapter, we use Foucault's problematisation approach to analyse the ways in which the concept of Problem- and Project-based Learning (PBL) is constructed and contested within the GCC higher education context, with a particular focus on the Qatari case. The discussion centres on how the discourse around PBL has been shaped in the particular historical, social, and political context, and the ways in which it has been framed as a solution to certain educational problems. Along these lines, three main themes are illustrated, namely (1) false universalism and pedagogic hegemony, (2) from fidelity of implementation to systems-thinking perspectives, and (3) going beyond the individual and into the institutional space. By analysing the ways in which PBL may become problematic within the Qatari HE context, this chapter concludes with suggestions for understanding the limitations and challenges in implementing a PBL approach in a non-Western context, and finding local solutions to local problems that build on indigenous knowledge and ways of knowing.

4.1 Introduction

Educational research and practice are in the midst of an exciting shift from a focus on best practice to contextualised practice, from fidelity of implementation to local adaptations, and from policy borrowing to policy learning and enactment (Du et al., 2019a, 2019b; Romanowski & Du, 2020). These shifts have taken into account the diverse social, cultural, and political characteristics of particular contexts, and have

Y. Chaaban (✉)
Educational Research Center, College of Education, Qatar University, Doha, Qatar
e-mail: ychaaban@qu.edu.qa

H. Al-Thani
College of Education, Qatar University, Doha, Qatar

X. Du
Department of Planning, UNESCO Center for PBL, Aalborg University, AalborgAalborg, Denmark

E. Zaidan et al. (eds.), *The Future of Education Policy in the State of Qatar*, Gulf Studies 21, https://doi.org/10.1007/978-981-97-9667-0_4

steered away from prescriptions that offer the same solutions to different educational problems. The persistent force of the global education industry; a diverse and multifaceted sector that drives educational solutions from the Global North to the rest of the world, is nothing short of overpowering (Mohamed & Morris 2021). However, recent years have witnessed growing awareness among educators and policymakers of the influence of adopting Western notions of education (Romanowski et al., 2018; Wu & Tarc, 2016). Described as a form of neo-colonialism and neo-imperialism, the imposition of Western educational models and values on non-Western sectors has been under the spotlight lately (Romanowski 2020). Several researchers allude to a number of unintended consequences, including the marginalisation of local cultures and traditions, the perpetuation of inequalities, and the neglect of non-Western epistemologies and ways of knowing (Romanowski et al., 2018; Steiner-Khamsi, 2016; Wu & Tarc, 2016).

This has not always been the case for Gulf Cooperation Council (GCC) countries, which include Bahrain, Kuwait, Oman, Qatar and Saudi Arabia. Over the past few decades, GCC countries have avidly implemented Western models of education reform and so-called *best practices,* to overhaul their educational systems (Romanowski et al., 2018). Particularly in the Higher Education (HE) context, this has involved hiring Western educators, importing textbooks and curricula, and developing partnerships with Western HE institutions (Ibnouf et al., 2014). With large investments, universities in these countries have competed to obtain Western certification and accreditation of their programmes, to transform their instructional practices, and to increase access to technological innovations (Romanowski, 2021). While these efforts have proven effective in improving educational outcomes, there is also criticism that the focus on Western models and evidence-based practices may have overlooked local cultural and social contexts (Mohamed & Morris 2021). In the case of Qatar, several empirical studies and conceptual papers have revealed the challenges and problems of adopting such models in HE (Du et al., 2020a, 2020b; Mohamed & Morris 2021; Romanowski, 2021). Similar to other GCC countries, Western models have permeated almost every aspect of HE in Qatar, including quality assurance measures, accreditation procedures, and instructional design.

One particular point of interest in recent years has been the adoption of Project- and Problem-based Learning (PBL) into the Qatari HE context (Du et al., 2021). In this chapter, we use the two terms interchangeably with the acronym of PBL, although we acknowledge that they may differ in certain aspects. In short, PBL consists of active engagement and practical applications of knowledge, involving students working together in teams over extended periods of time in solving complex problems related to particular topics or disciplines. They are encouraged to take ownership of their learning, engage in the deep understanding of concepts, and develop skills in research, analysis, and synthesis (Du et al., 2021; Guo et al., 2020). Several researchers argue that PBL has been evolving since its inception as a pedagogical approach in the late 1960s, and that unique issues may arise when PBL is implemented in non-Western contexts, where instructors and students may have no prior experience of PBL, and where institutional support may be lacking (Naji et al., 2020; Sabah & Du, 2018).

In this chapter, we use Foucault's problematisation approach to analyse the ways in which the concept of PBL is constructed and contested within the GCC HE context, with a particular focus on the case of Qatar. Specifically, we use the context of several initiatives implemented at Qatar University, which have aimed to facilitate faculty professional learning and PBL implementation in course redesigns and classroom practices. This discussion involves identifying the power relations that have shaped the discourse around PBL in this particular historical, social, and political context, and the ways in which PBL, like other student-centred approaches, has been framed as a solution to certain educational problems. While PBL has often been promoted as an approach that enhances students' collaborative, critical thinking, and problem-solving skills, yet it is equally important to recognise that the implementation of PBL may be shaped by a range of factors that may result in unintended outcomes based on the characteristics of the particular context in which it is implemented (Du et al., 2020a, 2020b). By analysing the ways in which PBL may become problematic within the Qatari HE context, we can better understand the potential limitations and challenges in implementing this approach, and offer local solutions to local problems.

4.2 Problematisation

To develop a critical understanding of PBL and its implementation in the Qatari HE context as a Western-based pedagogical approach, we applied Foucault's notion of problematisation. Problematisation is a process of critically examining a particular phenomenon to realise its underlying assumptions and power dynamics. It involves revisiting and analysing an established concept or approach within its historical, socio-cultural, and political contexts, and revealing the ways in which it is constructed and sustained by different actors and factors (Foucault, 1990). Foucault's notion of problematisation constitutes understanding the emergence of a field of knowledge, and gaining new insights into how ideas and practices come into existence. When applied to education, it can uncover the complexity of enacting change in instructional practices, the intricacy of policy-making, and the dynamics of power structures within educational institutions, and it encourages more nuanced and more inclusive discussions and solutions among actors.

According to Foucault and Lawrence (1988), problematisation can function both as an object of inquiry and a distinctive form of critical analysis. The former allows tracking the way that certain objects develop into problems and how they are examined, understood, and represented. In the case of PBL, problematisation offers an opportunity to understand how its processes and principles were first developed in a Western context, and later imported into a non-Western context. As a method of analysis, problematisation offers the opportunity to understand why and how certain phenomena emerged as problems in a particular context. Regarding PBL, problematisation explores how it developed into an accepted student-centred approach in HE, and why its implementation has generated certain problems.

Befitting this chapter is the notion of problematisation with its emphasis on context; meaning the way it targets certain practices, policies, or concepts, and how they have become problematic, given the features and characteristics of certain contexts. Therefore, one of the main foci of a problematisation analysis is to consider "that which has already become problematic" (Rainbow & Rose, 2003, p.13) in a particular social, cultural, and political context, while further grounding "the problem" by returning to its historical roots. This critical process anchors present conditions within their historical journey, documenting what occurred in the past and how this has influenced the present. According to Webb (2014), problematisation allows the identification of the conditions that result in problems and their potential solutions.

This chapter adopts Foucault's problematisation guidelines, including three central dimensions (see Romanowski, 2021). One dimension entails identifying the phenomenon in a certain historical context, and investigating how it has become problematic. This involves engaging in a historical analysis by reproducing a genealogy of the object of inquiry, to uncover how and why it became problematic. The second dimension considers questions about power and knowledge, which can be addressed using moral reasoning, scientific knowledge, and political analysis. Accordingly, this dimension examines the ways in which the phenomenon has been constructed and regulated through dominant discourses, and the language used to describe and regulate it. It also entails analysing the power relations involved in regulating the phenomenon, including the roles and interests of different actors, such as educational institutions, a certain social group, or individuals. The third dimension includes proposing solutions that may be effective within the practical context where the phenomenon became problematic, and documenting the solutions that have been proposed to address it.

4.3 The Evolution of PBL in Higher Education

PBL, known as a student-centred learning approach emphasising students working on real-life problems in teams, has a history of more than fifty years in higher education (Servant-Miklos et al., 2019a, 2019b). The history of PBL is well-known for its origins in McMaster University's pioneering undergraduate Medical Doctor programme in 1968 in Canada. At McMaster University, PBL was developed as a response to traditional lecture-based teaching, with the aim of promoting higher order thinking inherent to active learning and problem-solving among students (Servant-Miklos, 2022). Over the decades, the concept of PBL has spread to a wide range of academic disciplines, including engineering, business, law, psychology, the humanities, and teacher education, among others (Hung et al., 2019). While becoming a well-recognised approach in universities worldwide, a high level of diversity has surrounded the development of PBL in both theory and practice, as a wide range of PBL models has been synergised (Savin-Baden, 2014).

The initial historical development of PBL was also based in three universities in North America and Europe, which observed a process of exploration by experience and experimentation in their historical contexts. While the initial McMaster PBL model, graduating its first cohort in 1968, was not embedded in any educational theories, it was mainly derived from a group of open-minded practitioner doctors' experimentation of a new medical programme blending basic and clinical sciences, using problems as the starting point of the learning process from day one, and assessing student learning using primarily formative methods (Servant-Miklos 2022). Following rising concerns about student engagement in learning, John Dewey's experiential learning (1993) and Bruner's (1956) notion of intrinsic motivation were later theorised to support the further practice of this PBL model (Schmidt, 1993). PBL development in Maastricht University, graduating its first cohort in 1974 in medicine and other disciplines, was furthered by its well-known Seven-Jump Method as a structured guideline for PBL sessions, thus framing clear roles for tutors and introducing a progressive means of assessment, both formative and summative (Schmidt, 1993). The Maastricht PBL model development also laid down the solid theoretical association of PBL with constructivist works, setting the foundations for PBL research (Schmidt et al., 2009).

Meanwhile, the concept of a problem-oriented project-work approach to learning began to evolve in Denmark, derived from the student revolts of 1968, through the growth of a Marxist critique of bourgeois education (Servant-Miklos & Spliid, 2017). The Danish political debates led to the establishment of two new types of university in the early 1970s, highlighting students' sense of democracy through their active participation in defining curriculum contents and modes. Rooted in the socialist political perspective, project work was adopted as the major characteristic of the learning approach in the Danish model (Illiris, 1992). Both universities were based on the form of new curricula in which students spent a large amount of their study time on solving problems through team projects. Their learning outcomes were largely assessed through conversation-focused and group-based oral exams. Respectively, Roskilde University, graduating its first cohort in 1972, focused on the use of PBL in different humanity and social science disciplines, while Aalborg University, graduating its first cohort in 1974, had a PBL model unique to technical and engineering disciplines initially, and later expanding into a comprehensive university. While the initial development of the Danish practice was also built on practitioners' experiences, the scholarly debates were often linked to the critical pedagogy of Paulo Freire, and the cultural history perspective of Vygotsky (Illiris, 1992). Whether the Danish model of highlighting projects should be considered as "PBL" or not has been debated for decades. Kolmos (1996) argued for a model merging problem- and project-based learning for its further enhancement in engineering education. Such a hybrid model was also associated with the theoretical foundations integrating learning theories from cognitive, constructivist, and socio-cultural perspectives (Kolmos, 1996).

Commonalities among the initial four PBL universities were that PBL began as a form of experimentation by open-minded educational practitioners, and later work progressed towards the need for solid theoretical foundations and educational research (Hung et al., 2019). Over the decades, this so-called "PBL" approach has celebrated its 50-year history in medicine and health, and demonstrated a wide-ranging diversity in practices across disciplines globally (Servant-Miklos et al., 2019a, 2019b). Tremendous efforts have been made on documenting the effective influence of PBL on diverse aspects of student learning, including knowledge, skills, and competence development. While certain consensus on PBL's benefits has been reached by several review studies, previous conclusions after debates on "Does PBL work?" have recently been challenged, due to empirical sources from medicine disciplines in North American and European contexts. The discussion has also moved further on to question: "How does PBL work in different specific contexts?" (Hung et al., 2019). This is echoed in a recent systematic review various forms of PBL implementation in engineering education across the decades by Chen et al. (2020). While multifaceted practices were found across the nine PBL categories suggested by Sabin-Baden (2014), the majority of current practices have been documented at the course level without achieving a systemic and systematic level of change in PBL, similar to that of the initial four PBL universities (Chen et al., 2020). This review study also pointed out that contextual factors, such as institutional conditions, educational systems, and societal history were observed as either supports or constraints for the success of PBL practice. This finding is in line with Kwan's (2019) summary of PBL practices in health disciplines, and Du et al. (2020a, 2020b) summary of PBL practices in teacher education, in that the ideology of the teacher being the major authorised source of knowledge may constrain the changing roles of teaching in a PBL environment.

Despite the great diversity of PBL practices that can be associated with the term "PBL," what has been commonly highlighted, theoretically speaking, are the characteristics of self-directed learning, constructive learning, collaborative learning, and contextual learning (Dolmans, 2019). In their proposal for the next wave of PBL research, Hung et al. (2019) suggested that while PBL practice is largely diversified according to disciplines, institutions, and countries, as well as through hybrid models, the question should become: "Why does PBL with particular implementation characteristics and for specific outcomes work or not work in the contexts where it is implemented?" In their commentary on reimagining education, Servant-Miklos et al. (2023) offered the provocative thought that any PBL models should be redesigned to address the current global challenges that are related to sustainability and to problematise the reality for all disciplines. If only being used to teach technical knowledge and solving single discipline problems, PBL risks becoming irrelevant, and fails to create change in the face of emergent crises, which are complex and demanding of a systemic and systematic level of change. In sum, as it was once born in history, PBL in the current global context demands further radical review as to how it can be appropriately practised, with both students and teachers playing a proactive role.

4.4 The Introduction of PBL into Qatar University

Under the pressing need to modernise, several initiatives at Qatar University mark its advocacy for transforming the teaching and learning processes towards a PBL-driven curriculum. Most notably, PBL has been implemented as a student-centred approach on several academic programmes, particularly in the health-related and engineering-related fields, in addition to some implementation on teacher education programmes. Through continued efforts, higher administration has taken notice of these initiatives, and approved a university-wide pedagogical development (PD) programme for disseminating PBL principles and practices across a wider spectrum of academic programmes. The implementation of PBL at Qatar University has been largely successful, with several studies reporting students' higher levels of engagement and learning emanating from the approach (Du et al., 2019a, 2019b). PBL has been particularly effective in fostering teamwork, communication, and collaboration among students, as well as improving their critical thinking and problem-solving skills (Du et al., 2020a, 2020b). Based on empirical findings (Du et al., 2019a, 2019b, 2020a, 2020b; Naji et al., 2020), the implementation of PBL did not always lead to the expected outcomes, as several factors were also found to hinder student learning through PBL. To further explore how PBL can be more effective in the Qatari context, Naji et al. (2020) proposed a framework exemplifying institutionalised ways of implementing PBL at the course level. This framework provided a blueprint for overcoming some of the pitfalls in PBL implementation at Qatar University, while supporting the progressive transformation from a lecture-based learning environment to PBL, starting with course-level change.

Before we problematise PBL in this context, we will provide further context in the form of a succinct summary of three major published studies that have documented faculty perceptions on implementing PBL from the colleges of medicine and engineering, as well as the outcomes of the PD programme that involved faculty from all colleges at Qatar University. The three studies summarised below offer an overview of these initiatives, and provide context along these common threads: (1) they provided emic—or insider—perspectives that supported a nuanced understanding of the contextual factors influencing the adoption of Western models; (2) they integrated diverse perspectives that allowed for a deeper understanding of the processes, people, cultures, resources, and external systems which interact to influence the desired educational outcomes; and (3) they were undertaken at Qatar University, which provided a common backdrop for the contextual factors influencing the faculty implementation of PBL in their classrooms.

A PBL approach was adopted in the newly-established medical school in 2015, and a few studies were conducted to explore how student learning and professional competences could be supported in such an initiative (Du et al., 2016, 2019a, 2019b; Kassab et al., 2019). A similar PBL model was adopted in the dental medicine programs which was established in 2020. A few studies have been conducted to reveal the PBL benefits for students' perceptions of their learner agency (Ali et al., 2022a, 2022b; Du et al., 2023), the development of self-directed learning (Ali et al., 2022a, 2022b), and career anticipation (Daud et al., 2022). However, these studies revealed how students from an Arab culture struggled to get used to the PBL approach, and to figure out the new roles of teachers and learners.

Further, Du et al. (2022), the authors of this chapter, examined the faculty's perceptions of their professional agency in negotiating their PBL facilitator roles in the college of medicine. Using Q methodology, the study found five significantly different viewpoints on the perceived sources of professional agency, including a mixture of positive and negative viewpoints. There were several decisive factors that influenced participants' viewpoints, namely institutional resources, intrapersonal relationships with students and colleagues experienced in PBL, and intrapersonal characteristics on aligning with PBL. The authors highlighted the latter, indicating that misalignments of participants' pedagogical beliefs with PBL may lead to faculty's resistance to change. They further revealed the complexity of faculty facilitation in a PBL setting, which is shaped by the interplay of personal traits and capacities, interactions with others, and socio-cultural conditions.

In another study conducted at the College of Engineering, Du et al. (2021) explored faculty's professional agency in implementing PBL, using a longitudinal qualitative study. Three engineering members of faculty, who were new to adopting PBL in their engineering courses, took part in this study that explored how they developed agency to manage the challenges involved in the process of change. Several variations in the nature of the course, schedule, faculty experiences, and institutional factors meant that PBL was practised differently on each course. However, common themes emerged in their motivation for change, beliefs about PBL, and readiness for change. During the implementation of PBL, participants enacted multiple forms of professional agency to align PBL with their students and environments, as well as attempting to overcome contextual challenges; that is through their engagement in critical reflection and existing scholarship on teaching and learning.

Following on from these college-level initiatives, Chaaban et al. (2023) explored a university-wide PD programme that involved thirty members of faculty across all colleges, and aimed to encourage the implementation of PBL in course design. Through a systems-based thinking perspective, this evaluation study revealed participants' perceptions of their professional learning, conceptualised as the change to student-centred beliefs and/or practices, as well as the factors that supported or hindered their learning and change within individual, socio-cultural, and structural systems. Differences in participants' perceptions of belief and attitudes to practice change were documented, with some of them indicating the incompatibility of PBL with their disciplines. Another important note on change was evident in the way participants used their agency to make decisions about engagement, to what extent they wanted to change their practices and in which direction. In some cases, participants opted out of change altogether. The findings supported the need to examine the interrelatedness and intricacies of factors that may influence faculty's professional study of PBL principles and approaches, with a particular emphasis on the interplay of top-down and bottom-up change mechanisms.

While these initiatives did not particularly call for critical adaptations of PBL within the host culture, the three studies implicitly revealed the cultural conflicts that can occur when Western education models are adopted in a GCC context. This chapter is an attempt to bring to the fore these cultural conflicts arising from the implementation of PBL, as evident from these studies and others conducted in a GCC HE context. When Western models are transferred, borrowed, or even willingly adopted, this will necessarily require a process of sifting through a cultural lens which fosters realisation of their relevance to a particular context, and builds on its indigenous traditions. These notions are discussed further in the context of problematising PBL in a GCC HE context, with a particular focus on the Qatari case.

4.5 Problematising PBL in HE in the GCC

4.5.1 False Universalism and Pedagogic Hegemony

Several researchers attribute the adoption of Western models of education, such as PBL, in non-Western contexts, to false universalism (i.e., the belief that a practice originating in one context can be applied in another with similar outcomes), and pedagogic hegemony (i.e., subtle forms of dominance which perpetuate value systems, ideologies and mindset through education) (Nguyen-Phuong et al., 2012; Romanowski, 2022). Western educational models are based on a particular set of cultural assumptions and values, that may not be applicable or relevant to other cultures (Nguyen-Phuong et al., 2012). Their implementation in a non-Western context is not a neutral or straightforward matter, and involves a process of assimilation with the local indigenous cultural system that shapes individual's actions and perceptions (Romanowski & Karkouti, 2021). Proponents of the "best practices" movement argue that there are certain methods, strategies, and approaches that have been proven to work effectively in one context and can be successfully transferred to other contexts intact (see Romanowski, 2021). While these practices are grounded in empirical research and evidence, yet their transferability is still contingent on the characteristics of the socio-cultural context in which they are implemented. Researchers agree that the socio-cultural context plays a defining role in the effective implementation of any practice, whether that be a local development or one borrowed from other educational environments (Du et al., 2020a, 2020b; Mohamed & Morris 2021; Wu & Tarc, 2016).

In analysing educational cultural scripts in the GCC, Romanowski et al. (2018) point out the dominant cultural factors that influence university students' behaviours and understandings. The authors describe three main types of cultural scripts in the GCC, including epistemic scripts related to students' beliefs about the nature of knowledge, teaching scripts related to students' beliefs about teaching and teaching approaches, and cognitive flexibility scripts related to students' ability to adjust learning practices to address new and unexpected challenges. These cultural scripts challenge the adoption of PBL in HE institutions in the GCC (Romanowski & Karkouti, 2021). For example, PBL prioritises critical thinking and problem-solving skills, which may not be particularly valued in a non-Western context, where students may prefer passive forms of learning, such as listening to lectures and rote memorisation (Du et al., 2021; Nguyen-Phuong et al., 2012; Wu & Tarc, 2016). Therefore, the adoption of PBL may cause students to struggle with the cognitive load or face disorienting dilemmas. As PBL requires students to learn through solving complex, real-world problems, manage their own learning, and work collaboratively in teams, those students who prefer passive modes of learning may find this level of interdependence challenging, time-consuming, and even frustrating (Chaaban et al., 2021).

The adoption of PBL in non-Western contexts may be further shaped by particular relational dynamics and power relations. Relational dynamics among students shape how they work together in teams to accomplish common goals. In PBL teams, students must devise a plan, take the initiative to seek out and acquire new skills and competencies, and monitor and reflect on their progress towards their goals (Chaaban et al., 2021). All the while, students must build positive relationships with other team members, or risk failing to accomplish tasks. Therefore, the quality of these relationships and the ability of students to overcome challenges as a team can affect the success of PBL as an instructional approach. While some studies document success stories for students working collaboratively in teams (Du et al., 2019a, 2019b), other studies reveal several challenges, with students lacking positive attitudes towards teamwork, lacking the necessary experience of working together, and lacking effective communication and conflict resolution skills (Chaaban et al., 2021; Nguyen-Phuong et al., 2012).

Furthermore, relational dynamics and power relations are manifested through faculty-student relationships. In GCC contexts, where traditional teacher-centred classrooms dominate the educational scene, faculty are often seen as authority figures who own mastery over the contents, and control the learning process (Sabah & Du, 2018). In contrast, PBL requires a shift in power relations, as students take on a more active role in their learning, and may reach conclusions and outcomes unknown to faculty (Du et al., 2019a, 2019b). This last point adds further complexity to power relations, as faculty must be willing to give up such control, and allow students to take the lead in their learning. This requires them to become facilitators of learning and to trust in students' ability to take responsibility for their own learning (Du et al., 2021). When faculty are unable to let go of their traditional roles, and provide too much guidance or structure, they risk limiting students' autonomy and reinforcing traditional power relations (Wu & Tarc, 2016). In PBL courses, power must be shared among faculty and students, and all voices should be valued.

On pedagogic hegemony, several researchers describe the domination of Western educational models over non-Western educational systems and practices. These models are based on Western epistemologies, or ways of knowing, which may not be compatible with local epistemologies, and often disrupt indigenous knowledge systems (Nguyen-Phuong et al., 2012; Romanowski et al., 2018). Adopting PBL in HE can be seen as a form of pedagogic hegemony, or the imposition of a particular pedagogy on all faculty in HE, regardless of the nature of the discipline or student population (Nguyen-Phuong et al., 2012). As an invisible power, it often works as a subtle form of educational governance (Romanowski, 2022). As a term used to describe the way dominant cultural and social values are promoted through a particular educational system, pedagogic hegemony can lead to reinforcing the status quo and reproducing existing power relations. Driven by false universalism, pedagogic hegemony imposes university pedagogy as a technology that "misleadingly indicates that there can be a common pedagogy for the entirety of the university's educational activities," and "that staff can be trained to follow with a handful of courses" (Magnússon & Rytzler, 2022, p. 3). This point is taken up in the following section.

4.5.2 *From Fidelity of Implementation to Systems-Thinking Perspectives*

In GCC HE contexts, PBL has been brought in as an effective pedagogy for enhancing student learning and engagement. Introducing PBL in this context requires faculty to acquire and demonstrate a new set of intrapersonal and relational skills, as well as to receive institutional support throughout its implementation (Nguyen-Phuong et al., 2012). Accordingly, professional development (PD) initiatives become imperative and should be designed to offer faculty the opportunity to learn about PBL in a collaborative manner, as well as to design, experiment, and reflect on the implementation of their newly designed courses (Bond & Blevins, 2020; Strom & Viesca, 2020). Similar initiatives have been implemented at Qatar University, with varying degrees of adoption and attitudes towards PBL (Chaaban et al., 2023). Such variation brings to the fore the concept of fidelity of implementation, which can also be problematised in this context.

With any intervention, whether exported from a foreign context or brewed locally, there is always the expectation that faculty adhere to the principles and practices of the intervention with a high degree of fidelity (Magnússon & Rytzler, 2022). In simple terms, fidelity of implementation refers to the extent to which an intervention is implemented as intended or designed, and in a manner that is consistent with its original goals, design, and underlying theory (Strom & Viesca, 2020). The aim is to reduce local interpretations and limit different sense making by faculty, while having little tolerance for modifications or adaptations that compromise its expected effectiveness. If such conditions are not maintained, then the intervention takes a new form and ceases to look like its prototype (Chaaban et al., 2023). For example, contradictions may arise when faculty structure their courses based on PBL principles, but also decide to present weekly chapters from an assigned textbook, in the name of contents coverage. Such contradictions raise questions as to whether PBL is truly guiding teaching and learning processes in such courses, or is merely a fancy word used eloquently in course syllabi.

Before we question the legitimacy of fidelity of implementation, let us continue with this line of thought. Supposing that PBL can be learned and implemented with little interpretation and sense making by faculty, there will remain the influence of indigenised and local cultural scripts. Inherent to GCC cultural scripts is the way faculty hold views of instructors as the purveyors of knowledge, adhere to direct instruction methods, and possess behaviourist beliefs, while students seem to welcome authority, encourage knowledge-transmission methods, and approve traditional assessment strategies that rely mostly on memorisation and surface understanding (Du et al., 2020a, 2020b; Romanowski et al., 2018). Among these cultural scripts are university policies that restrict the fidelity of implementation, specifically when faculty professional agency is constrained and they are prevented from making decisions that influence their teaching and assessment practices (Chaaban et al., 2021). For example, faculty may have the readiness to implement PBL in their courses, yet face institutional policies that may restrict their complete adherence to its principles, including assessment policies that dictate the need to implement paper-based assessments and promotion of policies that celebrate traditional research over innovative teaching among faculty (Chaaban et al., 2023). Faculty may also be restricted by the physical layout of the classroom, large class sizes, heavy workloads, and a debilitating assessment culture (Chaaban et al., 2023). The commonly debated agency-structure dilemma, such that faculty are subjected to structural control and supervision, and may be discouraged from exploring new ideas or taking risks in their teaching, continues to challenge PD initiatives (Chaaban et al., 2023).

However, recent theorisations of professional development have refuted the expectation for the fidelity of implementation, and have emphasised the complexity of taking newly-acquired teaching skills and applying them intact in different classroom contexts (Bond & Blevins, 2020; Strom & Viesca, 2020). Such complexity is further exacerbated when the newly acquired skills are imported from Western contexts which differ noticeably from local contexts (Nguyen-Phuong et al., 2012; Sabah & Du, 2018). Several researchers have alluded to the inevitable adaptations to local contexts, and encouraged PD providers to adopt a systems-thinking perspective in regard to faculty professional learning. Such perspectives allow for deeper understanding of the systems of influence on faculty professional learning; that is by examining the multiplicity of factors that interact to influence desired faculty adoption of new pedagogic approaches (Bond & Blevins, 2020; Nguyen-Phuong et al., 2012).

4.5.3 Going Beyond the Individual and into the Institutional Space

Despite the importance of faculty buy-in and student readiness, organisational cultures and structures play a decisive role in the extent to which PBL can progress from the individual to the institutional space, specifically by acknowledging the need for a balance between bottom-up and top-down approaches. In Western contexts, PBL has been mostly adopted using a systemic, top-down approach. This means PBL is the guiding principle of programme and curriculum design, scheduling and timetabling, and faculty development, with teaching and learning processes following suit. Most popular among these systemic-level adoptions there include the engineering programme at Aalborg University in Denmark, and the medicine programme at McMaster University in Canada. Other Western universities have implemented PBL more selectively, incorporating the approach into certain courses, yet with a high degree of institutional support. The point to be made here is that the implementation of PBL in Western HE is widely variable, yet it is generally advocated at a system level, as many universities continue to explore ways to integrate PBL into their overall educational philosophies (Guo et al., 2020).

The adoption of PBL in GCC HE has been more limited compared to its use in Western HE, and follows a prevalent bottom-up approach. There are some notable examples of PBL implementation in this region. One example is Qatar University, which has adopted PBL in its medical programme. The university has integrated PBL into the curriculum as a core instructional approach, emphasising the importance of collaboration, self-directed learning, and critical thinking (Du et al., 2021). Another example is United Arab Emirates University, which has implemented PBL in its engineering and technology programmes, and used PBL as a collaborative, student-centred and deep learning approach (Chowdhury, 2015). A further example can be found at the college of medicine at King Saud University, which follows a PBL curriculum based on a Western model that is contextualised to the Saudi setting (Almulhem & Almulhem, 2022).

These examples show that in the majority of cases in the GCC HE context, the colleges of medicine and engineering tend to be in the forefront. Despite differences in their implementation models, these examples follow similar patterns as in Western HE contexts, such that systemic advocacy and leadership support have been available for the implementation of PBL. For example, at the college of medicine at Qatar University, PBL has been institutionalised as the pedagogical approach guiding programme and curriculum design, and enacted in a top-down fashion (Du et al., 2022). At the college of engineering, the adoption of PBL has taken a reverse mode. Beginning with a few members of faculty, PBL has grown organically in a bottom-up manner as more of faculty have become interested in the approach (Du et al., 2021). In both cases, senior leadership support for the approach has proven to be indispensable for the provision of resources and infrastructure, faculty development opportunities, guidance in curriculum design, and supportive structures and cultures that encourage experimentation and innovation in teaching and learning.

However, such support for PBL may not always be available at HE institutions in the GCC. The required institutional support for PBL involves the process of making significant changes to the policies, practices, and structures that influence change with relation to PBL at a department- or college-level (Chaaban et al., 2023). In many instances, curriculum change is a lengthy process that involves extended deliberations and reflections among members of faculty, curriculum coordinators, and heads of department, which may slow down the process or even prevent change from occurring altogether. Furthermore, quality professional development opportunities based on social constructivist theories of adult learning are only rarely available to faculty, and require a concerted effort by senior leadership to ensure their sustainability (Chaaban et al., 2023).

On a final note, systemic support for PBL is not always consistent nor persistent over extended periods of time and across different HE spaces (e.g., colleges and departments). Especially in the early stages of the adoption of PBL, unforeseeable problems and conflicts may emerge, and faculty may face a number of challenges that hinder their PBL implementation (Chaaban et al., 2023). Far from being a quick fix, the implementation of PBL may encounter faculty resistance to change, lack of student readiness, micromanagement procedures, and a lack of leadership support (Chaaban et al., 2023). Upon careful deliberation, these challenges can be transformed into opportunities, with the right mix of top-down and bottom-up mechanisms of change, as faculty enact professional agency in implementing change supported by leadership vision, resource allocation, and policy-making. In the attempt to institutionalise Western pedagogies, and ensure their seamless contextualisation within the social, cultural, and political arena of HE in the GCC, several recommendations may be taken into consideration, as discussed next.

4.6 Looking Inward, Outward and Forward

Despite the problematisation practice documented in this chapter, there is already promising evidence that PBL has potential in a GCC HE context. Evidence from Du and colleagues (2019a, 2019b, 2021) suggests that the implementation of PBL can improve teaching and learning processes, and have a desirable influence on student learning, engagement, and the acquisition of complex skills. However, inattention to the three arguments presented above on how PBL has become problematic within the GCC HE context may hamper the sustainability of its implementation in the longer term and on a larger scale. The following recommendations are proposed to direct further policy, research, and practice within the contentious field of educational borrowing.

According to socio-cultural theories of learning, such as those developed by Lev Vygotsky, the significance of social and cultural factors in shaping human development and learning have been emphasised (Chaaban et al., 2023). These theories provide a useful framework for understanding how cultural scripts play a role in determining how borrowed educational approaches may not lead to their intended outcomes when implemented in a different cultural context. These cultural scripts reflect deeply engrained ways of doing, and it may become necessary to promote localised forms of professional agency among faculty, and learner agency among students. As argued above, any change to the dynamics of faculty-student relationships, teaching methods, and learning strategies must emanate from identifying local problems and proposing local solutions. Promoting localised agency means allowing the right conditions for professional agency and learner agency to flourish, and prompting fossilised cultural scripts to begin shifting in new directions. We acknowledge the many challenges such endeavours may generate, such as a natural resistance to change, the complexity of socio-cultural classroom dynamics, time and effort, as well as the ability to strike a balance between cultural preservation and innovation. However, change must begin from somewhere, and we would argue that it should be a system-wide change. Therefore, educational policies should be devised to prompt the initiation of a context-sensitive PBL implementation at the school level for students; that is, long before they enter HE, as it provides a foundation of necessary competencies and proficiencies from whence to begin constructing further skills and abilities. Notwithstanding the importance of faculty and student readiness, this should not be seen as a pre-requisite for the adoption of PBL, but can be used as a way to support its maturity for both faculty and students. A localised PBL may be designed to support such maturity, then building on acquired skills, faculty and students can become more agentic and proactive in the teaching and learning processes. Our problematisation of PBL does not aim to eliminate this approach from HE contexts in the GCC, but to acknowledge that maturity levels are needed.

A second recommendation pertains to the necessity of attending to relational and contextual dimensions of institutional change, without neglecting the *individual*. Adopting Western models of education, with careful assimilation into the host context, requires systems-thinking perspectives that support an integrated approach of implementing PBL in HE in the GCC. Policy-making should adopt systems-thinking perspectives that emphasise the interconnectedness and interdependence of various components within a system. Rather than focusing solely on isolated problems relating to the adoption of PBL in GCC HE contexts, systems-thinking encourages a holistic view that considers the relationships, interactions, and unexpected consequences that can arise during implementation. Such understanding enables university leaders and policymakers to make better informed decisions, and anticipate potential impacts across the educational system as a whole. In the HE context, it is not merely PBL or other student-centred approaches that are adopted uncritically from foreign educational contexts. Policy-making should also be cognisant of multiple other competing discourses and power relations arising from neoliberal influences on HE, which create educational systems based on market-oriented principles, policies, and structures. Therefore, systems-thinking perspectives remain a necessary approach to providing a comprehensive and holistic understanding of the complex interactions and dynamics within the HE system.

Lastly, we propose the implementation of iterative processes of change that develop gradually, from adopting PBL on individual courses to systemic and systematic change. One example has been recently launched at the College of Education at Qatar University, where lessons learned from previous implementations of PBL in the engineering and medicine colleges, as documented above, have been embedded into its Curriculum Transformation Programme. One such lesson encouraged the creation of ripple effects with a smaller number of faculty who have the buy-in and willingness to take part in course-level change towards PBL, while being supported through top-down leadership advocacy. This is an example of policy-making that has taken into consideration some of the lessons learned in this chapter. The pilot of this programme is in its infancy, and it remains to be seen just what other lessons will continue to arise, even when so many previous lessons have been considered. This is not to say that the adoption of PBL will be problem-free, otherwise what will be the use of PBL without the *P for problem*?

4.7 Conclusion

This chapter uses Foucault's problematisation approach to explore the way PBL has become problematic in its implementation in a non-western context, such as in the GCC. This exploration has revealed three principal themes emerging from the problematisation activity, namely challenging the universal applicability of foreign models of education and the potential for pedagogic hegemony, rethinking the fidelity of implementation, and going beyond individual factors in the search for the reasons for failed PBL implementation. The analysis has revealed the influence of cultural scripts at play when Western-based pedagogies are borrowed uncritically in the GCC. By specifically drawing on empirical research from Qatar, this chapter concludes by suggesting strategies for recognising the constraints and difficulties in adopting a PBL approach, as well as the policy-making needed to mitigate some of these challenges.

In recent years, there has been a growing recognition of the need to decolonise instructional approaches and promote more inclusive and culturally responsive approaches to teaching and learning (Nguyen-Phuong et al., 2012). This recognition has been driven by a growing body of research that demonstrates the value of incorporating diverse perspectives and knowledge systems into instructional practices (Romanowski, 2021, 2022). This chapter has called for a growing awareness among policymakers and educators of the influence of Western notions of education, and suggested a number of recommendations that can begin to promote localised agency, attend to individual, relational, and contextual dimensions of systemic change, and implement incremental and iterative processes of change.

Decolonising Western epistemology involves recognising and challenging the ways in which Western knowledge systems and ways of thinking have historically dominated other forms of knowledge and ways of knowing. Future research may use certain theoretical foundations, such as critical theory and decolonial theory, to further explore dominant epistemologies and the preservation of indigenous knowledge systems. Through such research, it may become possible to engage in a critical examination of the assumptions and biases inherent in Western knowledge systems, and to recognise the diverse ways in which knowledge can be produced and transmitted across different cultures and communities. This also includes recognising the importance of indigenous knowledge systems, and the problems that may be generated when Western pedagogies are adopted without critical assimilation within the host culture.

Bibliography

Ali, K., Daud, A., Hattab, R., Philip, N., Matoug-Elwerfelli, M., Anweigi, L., Khabuli, J., & Du, X. (2022a). Dental students' development of self-regulated learning in PBL. *European Journal of Dental Education*. https://doi.org/10.1111/eje.12820

Ali, K., Du, X., et al. (2022b). Exploring dental students' perceptions of learner agency enactment in problem-based learning using Q methodology. *European Journal of Dental Education*. https://doi.org/10.1111/eje.12872

Almulhem, M. A., & Almulhem, J. A. (2022). Evaluation of problem-based learning implementation in a College of Medicine, Kingdom of Saudi Arabia: A cross sectional comparative study. *BMC Medical Education, 23*, 311. https://doi.org/10.1186/s12909-022-03347-1

Bond, M. A., & Blevins, S. J. (2020). Using faculty professional development to foster organizational change: A social learning framework. *TechTrends, 64*(2), 229–237. https://doi.org/10.1007/s11528-019-00459-2

Chaaban, Y., Qadhi, S., & Du, X. (2021). Student teachers' enactment of agency in teamwork settings. *Eurasia Journal of Mathematics, Science and Technology Education, 17*(1), 1–15. https://doi.org/10.29333/ejmste/10978

Chaaban, Y., Al-Thani, H., & Du, X. (2023). A systems-thinking approach to evaluating a university professional development programme. *Professional Development in Education*. https://doi.org/10.1080/19415257.2023.2193199

Chen, J. B., Kolmos, A., & Du, X. (2020). PBL in engineering education—Forms of implementation and challenges: A review of literature. *European Journal of Engineering Education*. https://doi.org/10.1080/03043797.2020.1718615

Chowdhury, R. K. (2015). Learning and teaching style assessment for improving project-based learning of engineering students: A case of United Arab Emirates University. *Australasian Journal of Engineering Education, 20*(1), 81–94. https://doi.org/10.7158/D13-014.2015.20.1

Daud, A., Elwerfelli, M., Du, X. Y., Ali, K., et al. (2022). Why dentistry? A qualitative enquiry into dental students' perceptions of dentistry as a career choice in the State of Qatar. *BMC Medical Education*. https://doi.org/10.1186/s12909-022-03522-4

Dewey, J. (1998) (1933). *How We Think: A Restatement of the Relation of Reflective Thinking to the Educative Process*. Houghton Mifflin.

Dolmans, D. H. (2019). How theory and design-based research can mature PBL practice and research. *Advances in Health Sciences Education*. https://doi.org/10.1007/s10459-019-09940-2

Du, X., Ebead, U., Sabah, S., Ma, J., & Naji, K. K. (2019). Engineering students' approaches to learning and views on collaboration: How do both evolve in a PBL environment and what are their contributing and constraining factors? *EURASIA Journal of Mathematics, Science and Technology Education, 15*(11), 1–15. https://doi.org/10.29333/ejmste/106197

Du, X., Ali, K., et al. (2023). Dental students' development of professional identity in PBL. *Journal of Dental Education., 87*(1), 93–100. https://doi.org/10.1002/jdd.13092

Du, X., Chaaban, Y., Sabah, S., Al-Thani, A., & Wang, L. (2020a). Active learning engagement in teacher preparation programs: A comparative study from Qatar, Lebanon and China. *Asia-Pacific Journal of Education, 40*, 1–16. https://doi.org/10.1080/02188791.2020.1717436

Du, X., Kassab, S. E., Al-Moslih, A. M., Abu-Hijleh, M. F., Hamdy, H., & Cyprian, F. S. (2019a). Identifying essential competencies for medical students. *Journal of Applied Research in Higher Education, 11*(3), 352–366. https://doi.org/10.1108/JARHE-07-2018-0114

Du, X., Massoud, W., Al-Banna, N. A., Al-Moslih, A. M., Abu-Hijleh, M. F., Hamdy, H., & Cyprian, F. S. (2016). Preparing foundation-year students for medical studies in a problem-based learning environment: Students' perceptions. *Health Professions Education, 2*(2), 130–137. https://doi.org/10.1016/j.hpe.2016.06.001

Du, X., Naji, K., Ebead, U., & Ma, J. (2021). Engineering instructors' professional agency development and identity renegotiation through engaging in pedagogical change towards PBL. *European Journal of Engineering Education, 46*(1), 116–138. https://doi.org/10.1080/03043797.2020.1832444

Du, X., Naji, K. K., Sabah, S., & Ebead, U. (2020b). Engineering students' group-based strategy use, forms of collaboration and perceptions of assessment in team projects—a case study in Qatar. *International Journal of Engineering Education, 36*(1), 296–308.

Du, X., Nomikos, M., Ali, K., Lundberg, A., & Abu-Hijleh, M. (2022). Health educators' professional agency in negotiating their problem-based learning (PBL) facilitator roles: Q study. *Medical Education, 56*(8), 847–859. https://doi.org/10.1111/medu.14792

Foucault, M. (1990). *Politics, Philosophy, Culture: Interviews and Other Writings*, 1977–1984. (1st ed.). Routledge.

Foucault, M., & Lawrence, K. (1988). Politics, philosophy, culture: Interviews and other writings, 1977–1984. *MLN, 104*(4), 255–567. https://doi.org/10.2307/2905276.ISSN0026-7910.JSTOR2905276

Guo, P., Saab, N., Post, L. S., & Admiraal, W. A review of project-based learning in higher education: student outcomes and measures. *International Journal of Educational Research, 102,* 101586. https://doi.org/10.1016/j.ijer.2020.101586

Gyamera, G. O., & Burke, P. J. (2018). Neoliberalism and curriculum in higher education: A post-colonial analysis. *Teaching in Higher Education, 23*(4), 450–467. https://doi.org/10.1080/13562517.2017.1414782

Hung, W., Dolmans, D. H. J. M., & van Merriënboer, J. J. (2019). A review to identify key perspectives in PBL meta-analyses and reviews: Trends, gaps and future research directions. *Advances in Health Sciences Education.* https://doi.org/10.1007/s10459-019-09945-x

Ibnouf, A., Dou, L., Knight, J., (2014). The evolution of qatar as an education hub: moving to a knowledge-based economy. In: Knight, J. (Ed.), *International Education Hubs: Student, Talent, Knowledge-Innovation Models.* Springer: Netherlands, pp. 43–61. https://doi.org/10.1007/978-94-007-7025-6_4

Illeris, K. (1992). The Organization of studies at Roskilde University, the concept, practice and problems of project organization. *CRE-Seminar* 19–24 June. Roskilde University.

Kassab, S. E., Du, X., Toft, E., Cyprian, F., Al-Moslih, A., Schmidt, H., & Abu-Hijleh, M. (2019). Measuring medical students' professional competencies in a problem-based curriculum: A reliability study. *BMC Medical Education, 19*(1), 155. https://doi.org/10.1186/s12909-019-1594-y

Kolmos, A. (1996). Reflections on project work and problem-based learning. *European Journal of Engineering Education, 21*(2), 141–148. https://doi.org/10.1080/03043799608923397.

Kwan, C. Y. (2019). A thorny path: The developmental course of problem-based learning for health sciences education in Asia. *Advances in Health Sciences Education.* https://doi.org/10.1007/s10459-019-09920-6

Magnússon, G., & Rytzler, J. (2022). *Towards a Pedagogy of Higher Education: The Bologna Process, Didaktik and Teaching.* Routledge.

Mohamed, M., & Morris, P. (2021). Buying, selling and outsourcing educational reform: the Global Education Industry and 'policy borrowing' in the Gulf. *Compare: A Journal of Comparative and International Education, 51*(2), 181–201. https://doi.org/10.1080/03057925.2019.1607255

Naji, K. K., U,E Ebead, A. Al-Ali, and X. Y. Du. (2020). Comparing Models of Problem and Project-Based Learning (PBL) Courses and Student Engagement in Civil Engineering in Qatar. *EURASIA Journal of Mathematics, Science and Technology Education, 16*(8). em1867. https://doi.org/10.29333/ejmste/8291

Nguyen-Phuong, M., Terlouw, C., & Pilot, A. (2012). Cooperative learning in Vietnam and the West-East educational transfer. *Asia Pacific Journal of Education, 32*(2), 137–152. https://doi.org/10.1080/02188791.2012.685233

Rabinow, P., & Rose, N. (2003). *The essential Foucault: Selections from essential works of Foucault, 1954–1984.* The New Press.

Romanowski, M. H. & Karkouti, I. (2021). Transporting problem-based learning to the gulf cooperation council countries (GCC): using cultural scripts to analyze cultural complexities. *The Interdisciplinary Journal of Problem-based Learning, 15*(1). https://doi.org/10.14434/ijpbl.v15i1.28793

Romanowski, M. H. (2021). Internationalization and CAEP accreditation: Replicating US teacher education programs abroad. *Teaching Education.* https://doi.org/10.1080/10476210.2021.1948991

Romanowski, M. H. (2022). CAEP accreditation: Educational neocolonialism and non-US teacher education programs. *Higher Education Policy, 35*, 199–217.

Romanowski, M. H., Alkhateeb, H., & Nasser, R. (2018). Policy borrowing in the Gulf Cooperation Council Countries: Cultural scripts and epistemological conflicts. *International Journal of Educational Development, 60*, 19–24.

Sabah, S., & Du, X. (2018). University faculty's perceptions and practices of student centered learning in Qatar: Alignment or gap? *Journal of Applied Research in Higher Education, 10*(4), 514–533. https://doi.org/10.1108/JARHE-11-2017-0144

Savin-Baden, M. (2014). Using problem-based learning: New constellations for the 21st century. *The Journal on Excellence in College Teaching, 25*(3&4), 197–219.

Schmidt, G. H. (1993). Foundations of problem-based learning: Some explanatory notes. *Medical Education, 27*(5), 422–432.

Schmidt, H. G., Van der Molen, H. T., Te Winkel, W. W., & Wijnen, W. H. (2009). Constructivist, problem-based learning does work: A meta-analysis of curricular comparisons involving a single medical school. *Educational Psychologist, 44*(4), 227–249.

Servant-Miklos, V. F., Norman, G. R., & Schmidt, H. G. (2019). A short intellectual history of problem-based learning. *The Wiley Handbook of Problem-Based Learning*, 3–24

Servant-Miklos, V. (2020). Problem-oriented project work and problem-based learning: "Mind the gap!". *Interdisciplinary Journal of Problem-Based Learning, 14*(1)

Servant-Miklos, V. F., & Spliid, C. M. (2017). The construction of teaching roles at Aalborg University Centre, 1970–1980. *History of Education, 46*(6), 788–809.

Servant-Miklos, V. F., Woods, N. N., & Dolmans, D. H. (2019b). Celebrating 50 years of problem-based learning: Progress, pitfalls and possibilities. *Advances in Health Sciences Education, 24*, 849–851.

Steiner-Khamsi, G. (2016). New directions in policy borrowing research. *Asia Pacific Education Review, 17*(3): 381–390. https://doi.org/10.1007/s12564-016-9442-9

Strom, K. J., & Viesca, K. M. (2020). Towards a complex framework of teacher learning-practice. *Professional Development in Education*, 1–16. https://doi.org/10.1080/19415257.2020.1827449

Vygotsky, L. S. (1986). *Thought and Language*. The MIT Press.

Webb, P. T. (2014). Policy problematization. *International Journal of Qualitative Studies in Education, 27*(3), 364–376. https://doi.org/10.1080/09518398.2012.762480

Wu, X., & P. Tarc. (2016). Translations and Paradoxes of 'Western' Pedagogy: perspectives of English Language Teachers in a Chinese College. *L2 Journal, 8*(4), 55–75. https://doi.org/10.5070/L28430214.

Chapter 5
Qatar's Transformation to Digital Learning: Challenges, Opportunities, and Policy Recommendations

Mary L. Newsome and Khalid A. Al-Ali

Abstract The State of Qatar, known for its significant contributions to the global energy sector, has made substantial strides in the field of education over the past two decades. This chapter delves into Qatar's transformation to digital learning by highlighting seminal educational reforms, strategic investments in education, the impact of COVID-19, and current digital and e-learning initiatives. The chapter sheds light on Qatar's aspiration to build a robust knowledge economy, underpinned by the urgent need to develop highly skilled lifelong learners capable of thriving in an increasingly competitive digital world. Discussion revolves around the concept of Twenty-First-Century Skills and Qatar's progress towards developing these skills among learners. Qatar's digital learning landscape, however, faces challenges that require careful consideration. By highlighting these challenges and opportunities, the authors provide recommendations for policy and practice offering valuable insights for stakeholders in the education sector, particularly those interested in Qatar's ongoing digital learning transformation.

Keywords Digital learning · Twenty-first-century skills · Knowledge economy · Education policy · Qatar

5.1 Introduction

The State of Qatar, located along the western coast of the Arabian Gulf, is most notable as one of the world's largest producers of oil and natural gas. However, over the past two decades, the small peninsula has also gained recognition for its investment and expansion in education. For example, according to reports from Qatar

M. L. Newsome (✉)
Assistant Dean, Office of Graduate Studies, Qatar University, Doha, Qatar
e-mail: mary.newsome@qu.edu.qa

K. A. Al-Ali
Director of Higher Education Institutional Affairs, Ministry of Education and Higher Education, Doha, Qatar

E. Zaidan et al. (eds.), *The Future of Education Policy in the State of Qatar*, Gulf Studies 21, https://doi.org/10.1007/978-981-97-9667-0_5

Development Bank, QAR, 22.1 billion riyals of the government's 2020 budget was allocated to the education sector, which represents nearly 11% of its total expenditure (Qatar Development Bank, 2021). Consequently, the country has become an educational hub within the region, with more than 500 public and private schools, as well as 34 higher education institutions, despite its relatively small population of 2.6 million inhabitants. Complementing its expansion in the education sector, Qatar has been able to successfully attract global companies such as Google, Microsoft, and Thales, to its free trade zones, due to its unique geographic location, which provides easy access to local and regional markets. Such investments in education, paired with Qatar's unique business ecosystem signals its desire for building a strong knowledge economy. Consequently, there is a need to develop highly skilled lifelong learners to meet the demands of a rapidly evolving digital world. These Twenty-First-Century Skills are composed of three broad categories: (1) Learning Skills (critical thinking, creativity, communication, collaboration), (2) Literacy Skills (information literacy, digital literacy, media literacy), and (3) Life Skills (flexibility, leadership, initiative, productivity, social skills). Consequently, learning environments should integrate these skills into their curricula, to ensure learners are acquiring a holistic set of skills essential for life in the digital age. Qatar has made significant progress in the right direction with investments in e-Learning platforms, smart classrooms, virtual learning environments, and online higher education, as well as curricular reforms with a focus on STEM education, innovation and entrepreneurship, project-based learning, and communication skills. However, there are still challenges affecting the digital learning landscape of the country that bear careful consideration, including teacher preparation, curriculum integration, data transparency and data sharing, cultural concerns, and issues around equity and access. This chapter focuses on Qatar's digital learning transformation, and its current initiatives to equip learners with Twenty-First-Century Skills. The authors also address the challenges and opportunities as well as policy recommendations. This work is of particular interest to students, faculty, researchers, administrators, and policymakers in education and higher education in the region, specifically in the State of Qatar.

5.2 Roots of Digital Learning in Qatar

Digital learning has been defined as any learning that is facilitated by technology and gives learners some control over time, place, path, and/or pace (Manzoor, 2016). Although the groundwork for digital learning can be traced back to the 1960s, with the use of "computer-assisted instruction" (CAI) and "computer-based training" (CBT), the term "digital learning" began gaining popularity in much of the world during the late 1990s, as technology such as the Internet and personal computers became a more integral part of society (Keegan, 2020). The advent of CD-ROMs, interactive educational software, and the availability of online resources, opened new pathways allowing educators and practitioners to explore the potential of technology and digital tools to enhance learning (Haleem et al., 2022). Since then, digital learning has not only become accepted as an effective way of enhancing the learning experience, but also essential to it. Digital learning can ease access to instruction and content, provide multimedia learning experiences, afford students avenues for receiving rapid feedback, enhance engagement, and tailor instruction to students' specific strengths and weaknesses (Social Research, 2015).

In Qatar, the roots of digital learning can be found in the educational reforms of the early 2000s, which followed governmental concerns that the country's educational system was "not producing high-quality outcomes, was rigid, outdated, and resistant to reform" (Brewer et al., 2007, p. iii). Consequently, Qatar consulted the RAND Corporation, an American institution that assists organisations' policy-making decisions, which resulted in a massive reform of K-12 and higher education called *Education for a New Era* (EFNE). EFNE aimed to develop human capital and ensure that citizens would be able to contribute to the socio-economic development of the country (Karkouti, 2016). In a further effort to develop a knowledge-based society and address the country's shortage of professional workers, the Supreme Council of Information and Communication Technology (ictQATAR) was established under Emiri Decree Law no. 36 of 2004. This entity later evolved into the Ministry of Communication and Information Technology. The ministry launched various national initiatives for digital inclusion, digital literacy, and ICT skills, and is the primary body responsible for supporting and enhancing the country's technology literacy (Karkouti, 2016, p. 186). The government-initiated educational reforms of the past two decades clearly acknowledge Qatar's understanding of the global shift towards a digital society (Nasser et al., 2017). Furthermore, Qatar National Vision, 2030 (QNV 2030) reiterates a commitment to the development of human potential and sustainable growth, through investment in education, specifically the development of Twenty-First-Century Skills to prepare Qatari citizens and residents for the challenges of the modern world.

5.3 Twenty-First-Century Skills

The Digital Age, also known as the Information Age, is characterised by rapid technological advancements, and the proliferation of digital information. Such advancements have ushered in transformative changes in how we live, work, and communicate (Said et al., 2021). For example, technology and the Internet have necessitated an ability to use digital tools and devices to access, evaluate, and utilise a vast amount of information that is freely and readily available online. The Digital Age has also redefined education, with the availability of online learning and digital learning platforms. Likewise, social media networking sites have changed how people interact and share their lives, in addition to shaping public discourse and prompting social movements. Similarly, the Digital Age has expanded the boundaries of healthcare, by enabling the development of telemedicine, and allowing patients to consult with healthcare providers remotely. To navigate this dynamic landscape, new skills are required. These skills are commonly referred to as Twenty-First-Century Skills.

Several prominent international organisations have released frameworks outlining Twenty-First-Century Skills within basic and higher education. Noteworthy among these organisations are the Organization for Economic Co-Development (Ananiadou & Claro, 2009), the US National Research Council (National Research Council, 2012), Next Generation Science Standards for States, by States (2013), and the Partnership for Twenty-First-Century Skills (Trilling & Fadel, 2009). Additionally, the Association of American Colleges and Universities (2007), as well as the Accreditation Board for Engineering and Technology (2021) have both issued Twenty-First Century Skills frameworks for education. Table 5.1 below broadly summarises the skills highlighted in these frameworks.

An escalating demand for these survival skills has been acknowledged globally by scholars, educational institutions, and economic entities alike (Bentur et al., 2019; National Research Council, 2012; World Economic Forum, 2016). Qatar Second National Development Strategy 2018–2022 also acknowledged the development of these skills as necessary for life and employability in the modern world. However, cultivating Twenty-First-Century Skills among learners has proved a formidable task. To illustrate, global estimates suggest that by 2025, 50% of all employees will need reskilling (Li, 2022). Furthermore, research has shown that fresh graduates are not equipped for the demands of their field (Badcock et al. 2022) and that a gap exists between students' skills/competencies and the skills/competencies needed in the workplace (Abdulwahed & Hasna, 2017; Alshare & Sewailem, 2018), which suggests a lack of effective communication between academic institutions and employers. Qatar Second National Development Strategy 2018–2022 has also identified concerns regarding some important pre-requisites for the acquisition of Twenty-First-Century Skills, such as students' poor performance in Maths, Science, and English language in international exams, as well as a high rate of absenteeism, lack of professional qualifications among teachers, and lack of data for monitoring and policymaking. Perhaps more alarming is evidence of a lack of operational planning to achieve education and training outcomes in the country (Qatar Second National Development Strategy, 2018–2022). Despite many educational reforms, investment in education, and a general consensus concerning the skills necessary for participation in the digital era, evidence suggests a long journey ahead (Fadlelmula et al., 2022).

5.4 COVID-19 and Qatar's Accelerated Digital Transformation

Like elsewhere in the world, Qatar's transformation to digital learning was greatly accelerated because of the global COVID-19 Pandemic and the emergency learning conditions that resulted from it. The pandemic shone a light on the gaping spaces between educational planning and practice, and revealed many vulnerabilities within online and distance learning, such as those in student and faculty development,

Table 5.1 Summary of twenty-first-century skills

Category	Skill	Description
Learning skills	Critical thinking	The ability to analyse information, identify patterns, and make informed decisions
	Creativity and innovation	The capacity to generate new ideas, approaches, and solutions to problems
	Communication skills	The ability to effectively convey ideas, thoughts, and information through various mediums
	Collaboration/teamwork	Working effectively with others, often in diverse teams, to achieve common goals
	Problem solving	The skill of identifying, defining, and solving complex problems, using critical thinking and creativity
Literacy skills	Digital literacy	Proficiency in using digital technologies and tools to access, evaluate, and create information
	Information literacy	The ability to find, evaluate, and use information from various sources critically
	Media literacy	Understanding and critically analysing media messages and their impact on society
Life skills	Flexibility/adaptability	The capacity to adapt to changing circumstances and learn from new experiences
	Global awareness	Knowledge and awareness of global issues, cultures, and perspectives
	Civic/social responsibility	A sense of duty to contribute positively to society and address social issues
	Ethical decision-making	Making moral and ethical choices in personal and professional life
	Leadership	The ability to guide, motivate, and inspire others to achieve common goals
	Entrepreneurship	The skill of identifying and pursuing opportunities to create value, often in a business context
	Resilience	The ability to bounce back from setbacks and adapt to adversity
	Adaptability	Readiness and willingness to embrace change and learn new skills
	Cultural competency	Understanding and respecting diverse cultures and perspectives
	Time management	Effective organisation and prioritisation of tasks to maximise productivity
	Emotional Intelligence	Recognising, understanding, and managing one's own emotions and those of others
	Self-directed learning	Taking initiative and responsibility for one's own learning and personal development

(continued)

Table 5.1 (continued)

Category	Skill	Description
	Data analysis	The ability to collect, analyse, and interpret data to inform decision-making
	Sustainability awareness	Recognising the importance of environmental and social sustainability

Source Compiled by the authors

accessibility, and availability of support. Local studies conducted during the early stages of the pandemic suggested that teachers at all levels were not adequately equipped with the skills necessary for the implementation of online learning environments (Alshaboul et al., 2022; Moyo & Maifala, 2022; Thomas et al., 2022). For example, a study of more than 1,500 high school students and 600 teachers across 39 secondary schools found that students and teachers alike lacked essential digital literacy skills, and faced significant barriers managing online coursework (Fadlelmula et al., 2022). Likewise, large-scale national studies indicated a perceived lack of administrative and ministerial support for online learning among educators (Chaaban et al., 2022). Furthermore, studies of students' readiness for and satisfaction with online learning also revealed that gender (Elkhouly, 2021; Newsome 2022), educational level (Alshaboul, 2021), discipline (Newsome et al. 2022), and pedagogical mode (Naji et al. 2022) were major mitigating factors. The pandemic has also greatly impacted employability. As virtually all work was forced online, a lack of essential skills among employees became more evident, which has forced institutions to work towards better aligning curricula to market needs (Qatar Development Bank, 2021). More than three years of this global crisis have left its mark on education, with academic institutions drastically rethinking curriculum development, assessment and evaluation, student engagement, and faculty development. While the negative impact of the pandemic on education is clear, this period also marked a heightened focus on educational research, global experiments in distance learning, the sharing of ideas and experiences, an increase in educational data and data transparency, as well as a willingness to rethink the outdated model of education.

5.5 Current Digital Learning Initiatives in Qatar

The Ministry of Education and Higher Education (MOEHE) has been actively implementing an e-learning strategy, which aims to enhance Qatar's education system by providing students, teachers, and parents with digital technology for learning. Prior to the COVID-19 pandemic, the MOEHE utilised an LMS system, Qatar Education System (QES), to monitor education at a national level. However, the forced shift to online learning during the pandemic revealed several weaknesses in the LMS in areas such as generating reports, managing educational e-content, facilitating parent follow-up, the limited customisation capabilities, lack of a smartphone app, incompatibility with different browsers, and technical problems when integrating with other ministry systems. These challenges highlighted the urgent need for a comprehensive electronic system that could efficiently manage and support the educational process. As a result, the MOEHE developed its current e-Learning strategy, which was launched in 2022 and will run through 2027. The e-learning strategy was developed from an extensive review of the relevant literature and global best practices, with considerable attention focused on adapting findings to the local context. The strategy supports public education at all levels and has five main objectives: (1) to support the achievement of education outcomes in the country, (2) to coordinate the contribution of different entities to the education system, (3) to improve students' experience in the education process, (4) to upskill stakeholders academically and digitally, and (5) to ensure inclusive education for all parties. The e-learning strategy has an embedded electronic system, which includes an executive dashboard capable of following the progress of the strategy, including its performance indicators and initiative completion rates. This strategy signals a qualitative leap in adopting advanced e-Learning methodologies and modern educational technologies to improve education outcomes within Qatar's educational system. Figure 5.1 below highlights the stakeholders that have been upskilled to date under the newly launched e-learning strategy.

While nearly 13,500 individual stakeholders (i.e., parents, teachers, principals, e-learning coordinators, vice principals, pioneer teachers, and supervisors) have received training in e-learning tools under the strategy, there is considerable work remaining to achieve the MOEHE's strategic e-learning objectives. To illustrate, for the 2022–23 academic year alone, nearly 210,000 user accounts were synced to the ministry's electronic e-learning system, which suggests that only 6% of registered users have received training. Furthermore, the data represented in Fig. 5.1 (above) does not include higher education, which will be initiated in a later phase. The first stage of the e-learning strategy and the system's progress will be assessed in 2024.

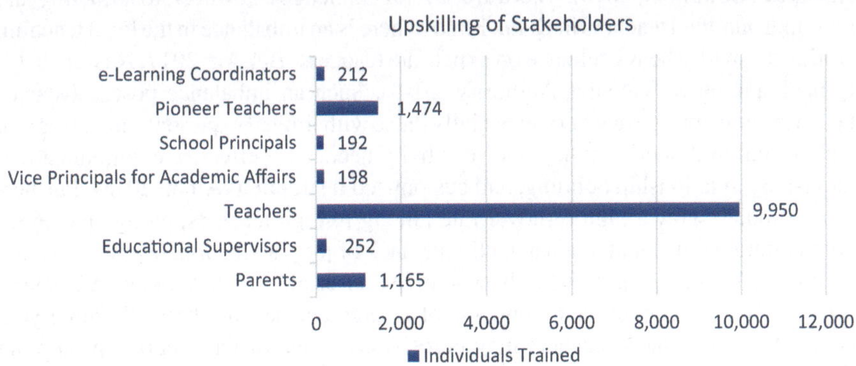

Fig. 5.1 Individuals trained under Qatar MOEHE e-learning strategy by category to date (Compiled by the authors from Qatar Ministry of Education and Higher Education 2022/27 e-Learning Strategy)

Table 5.2 below summarises some of the other key initiatives that are driving digital learning in Qatar.

These initiatives reflect Qatar's commitment to digital learning and to preparing students for the challenges and opportunities of the digital age. However, leveraging advanced tools and technologies to achieve quality education for all in the country will require a concerted effort, significant investment, continuous assessment, and multi-stakeholder collaboration. Perhaps more importantly, consistent, resilient, and sustainable education policy and frameworks are needed at a national level. The sections that follow address some of the major challenges and opportunities regarding digital learning in Qatar, followed by policy recommendations.

5.6 Challenges and Opportunities for Qatar's Digital Learning Landscape

5.6.1 Attracting and Retaining Qualified Teachers

Qualified teachers may be the most important aspect of developing learners' Twenty-First-Century Skills (Musa & Alhanan, 2013). However, a global shortage of teachers equipped for twenty-first-century instruction has caused significant competition in attracting and retaining teachers (Vanderlinde et al., 2016). Like elsewhere, Qatar is experiencing a teacher shortage, and, therefore, a crisis in the development of Twenty-First-Century Skills among its students (Alzahrani, 2019; Nasser, 2017). This is mirrored in the comparatively low academic performance and poor skills of graduates in the Arab world on international assessments (Alhothali, 2021; Qatar Second National Development Strategy 2018–2022), as well as in the dissatisfaction of graduates' competencies among employers (Abdulwahed & Hasna, 2017; Alshare & Sewailem, 2018). There are several challenges, however, to attracting and retaining qualified teachers in Qatar. Firstly, there is an imbalance in the local teaching workforce, with a heavy reliance on expatriate teachers (Bel-Air, 2017; Nasser, 2017; Qatar Planning & Statistics Authority, 2019). Such an imbalance poses obstacles because many foreign teachers, especially those with limited experience in the region, lack the cultural understanding and sensitivity needed for effective communication, inclusivity, relationship building, and customised instruction (Altan, 2018). Furthermore, studies show a high turnover rate among foreign teachers, citing challenges with cultural adjustment, the harsh climate, lack of job satisfaction, heavy workload, or finding better opportunities elsewhere (Alloh et al., 2019; Wilkins & Annabi, 2023). Also, the employment contracts of foreign teachers are typically three-year contracts, which many teachers may perceive as a lack of job security, prompting them to look elsewhere for more permanent conditions. Recruiting educators from abroad can also be difficult because of bureaucratic challenges like the issuance of work visas and lengthy immigration processes for certain nationalities.

Table 5.2 Summary of digital learning initiatives and enablers in Qatar

Digital learning initiative/enabler	Description
National ICT Plan 2015	Aims to create a vibrant ICT sector, in Qatar; to become an ICT regional hub; develop world-class ICT infrastructure, provide high-speed broadband fibre to home networks
Memorandum of Understanding (MoU) with MOEHE, Microsoft, and "Starlink" Company	Three-party agreement to accelerate the introduction of technology in Qatar schools; to avail all students and teachers of the opportunity to obtain Microsoft educational devices and technologies
World Innovation Summit for Education (WISE)	Tech hub, focused on transforming education through innovation; addresses ed-tech trends; encourages investment in educational technology
Qatar Science & Technology Park (QSTP)	Innovation hub for education, research, and development; includes incubation centre and Arab Innovation Academy
e-Learning Platforms	Blackboard, Microsoft Teams, Moodle, Canvas, etc. Effective digital enablers widely used in Qatar schools and universities
e-Learning Tools	Google classroom, Coursera, eduTechnoz, SchoolPedia, Century, etc. Represent the fragmented use among Qatar schools and universities
Qatar University Digitally Enriched Education Booklet (digital pedagogy)	Handbook created for QU educators to facilitate digitally enriched educational practices
Digital Libraries	Qatar National Library, Qatar University Library, Museum of Islamic Art Library
Virtual Labs and Simulations	Weill Cornell Medicine—Qatar Clinical Skills and Simulation Lab (CSSL); Qatar University Virtual Reality Lab (builds students' VR and 3D skills); Itqan Clinical Simulation and Innovation Center-Hamad Corporation (education and training facility for healthcare professionals); MADA—Qatar Assistive Technology Center's flight simulation lab for children with autism and their parents

(continued)

Table 5.2 (continued)

Digital learning initiative/enabler	Description
Online Resources for Special Needs Students	MADA Inclusive Digital Education Portal, Qatar University Special Needs Support Center (trains students on the use of assistive technology)
Cybersecurity and Data Privacy	Qatar 2022 Cybersecurity Framework; Qatar National Cyber Security Strategy; "Educational Cybersecurity Curricula" Project (raises awareness among children and adolescents in Qatar of concepts related to digital safety and cyber security); Qatar National Cyber Security Agency (workshop series on Personal Data Privacy Protection)

Source Compiled by the authors

The complex demands regarding teacher qualifications (i.e., requirements for various certifications and licences), top-down policy mandates, and increased professional development requirements among in-service teachers, have also added to the challenge of retaining qualified teachers in Qatar (Al-Naimi et al., 2020; Al-Thani, Ari, & Koç, 2021; De Wever et al., 2016). Perhaps worse is evidence suggesting a lack of professional development opportunities for local teachers, and that current professional development programmes have no direction, purpose, or progress (Al-Thani, Ari, & Koç, 2021). Likewise, Abu-Tineh et al. (2023) concluded the poor retention rates among Qatari teachers is largely attributed to a lack of career advancement opportunities. These factors, paired with a perception of the teaching profession as overworked and underpaid (Al-Thani, Ari, & Koç, 2021; Lee & Zahir, 2021), translate to a continued shortage of local teachers, and a crisis for twenty-first-century education, if left unattended.

Two important local initiatives to combat the challenges mentioned above include the Empowering Leaders of Learning (ELL) Programme (a partnership between MOEHE and WISE), and Teach for Qatar (TFQ). The ELL initiative is an education leadership programme that nurtures the development of the knowledge and skills needed to foster systematic change and improve education outcomes. In addition, it provides a platform for education leaders in Qatar to share best practices, and collaborate on innovative solutions to education challenges. Although an official ELL report entitled *Post-pandemic National Educational Investments: School Leadership Development through Innovative Learning Designs* alludes to the positive impact the programme has had on student performance, (Cosner & Alfadala, 2021, p. 25), no concrete evidence is presented in the report. Furthermore, an inspection of the annual education statistics published on the MOEHE website shows no significant difference in the exam success rates of secondary students in public and private schools since 2017 (Education Annual Statistics 2017–2021, Ministry of Education and Higher Education). Nonetheless, it is likely to be too early to gauge the full extent to which the ELL Programme has impacted student performance and education outcomes in the country. With continuous assessment of the initiative, however, a measurable impact is likely. Furthermore, the qualitative feedback from programme participants will surely prove valuable for policy and planning going forward.

Teach for Qatar (TFQ), part of the Teach for All Network, was established in 2013, and is also a strategic partnership with MOEHE, Qatar Petroleum, Qatar Foundation, Exxon Mobil Qatar, and Qatar National Bank. The aim of TFQ is to address educational inequalities and improve educational outcomes by recruiting and developing individuals, or "Fellows" from various non-teaching backgrounds to teach in Qatari government schools. TFQ was established to meet the need for qualified teachers in Qatar and "provide solutions for the challenges facing students in the government school system" (Romanowski & Qadhi, 2022, p. 1). TFQ Fellows receive comprehensive pre-service training and support, to become effective teachers and classroom leaders. The programme focuses especially on the development of innovative teaching and the use of modern tools and technologies for learning. Despite the unavailability of systematic data and reports regarding the programme's impact, qualitative feedback suggests that students of TFQ Fellows become more responsible for their own learning, show an increased willingness to learn, develop more respectful relationships with teachers and peers, and improve academically (Teach for Qatar, 2022). Both ELL and TFQ are important initiatives that are designed to systematically address the national shortage of highly skilled local teachers, an obvious prerequisite to learners' acquisition of Twenty-First-Century Skills.

Policy Recommendations:

- Enhance national capacity-building initiatives to attract and retain local teachers by offering more scholarship assistance opportunities, signing on bonuses, tailored professional development programmes, and career advancement pathways. Develop recruitment strategies that target local schools, universities, and organisations.
- Develop a governance framework for the development/implementation of professional development programmes to ensure alignment with best practices and the country's desired education outcomes.
- Establish a variety of national recognition and appreciation initiatives that celebrate the efforts and accomplishments of local educators, particularly those who are excelling in twenty-first-century teaching.
- Invest heavily in technology and tools that can reduce teachers' administrative load, including digital tools for assessment and evaluation.
- To attract and retain highly qualified foreign teachers, government authorities and MOEHE may work to establish smoother visa processes.
- Dedicated initiatives to integrate foreign teachers into the society and into professional communities of practice are also needed to retain foreign teachers.

5.6.2 Curriculum Integration

Curriculum integration of Twenty-First-Century Skills is essential for students' success in the digital world. However, curricular change poses challenges for both educators and educational institutions (Granville, 2014; Ibrahim, 2023). First and foremost, institutions need to address any ambiguity about what Twenty-First-Century Skills entail, and how those skills might be integrated into the curriculum. For example, Van den Akker (2010) describes three separate curricula: *the intended curriculum* (the documents designed to guide what should be taught), *the implemented curriculum* (what actually happens in the classroom), and the *attained curriculum* (the learning process as perceived by students including the skills and content they retain). The integration of Twenty-First-Century Skills, according to this view, requires extensive awareness-raising, professional development, and continuous assessment, to bring these three aspects into harmony. Similarly, a study conducted by Al-Kuwari et al. (2022) of Qatar's education system revealed a lack of alignment between current educational goals, instructional and assessment practices, and Sustainable Development Goals (for which the acquisition of Twenty-First-Century Skills is vital).

Another challenge for curriculum integration, which has received little attention among reform stakeholders is determining the age-appropriateness of Twenty-First-Century Skills and how they should be introduced at different grade levels (Güney, 2023; Plevyak, 2022; Woodward & Kimmons, 2018). Furthermore, assessing twenty-first-century skills can be quite complex, which may present obstacles to accurately measuring students' progress (Webb et al., 2018). In fact, there is a growing body of research dedicated to the development of digital tools to aid in the assessment of twenty-first-century skills (Herde et al., 2016; Siddiq et al., 2017). Also, integrating additional skills into curricula that are challenging to assess requires significantly more time and resources, which already overworked teachers may struggle with (Rusdin & Ali, 2018). Ultimately, teachers, administrators, and policymakers may view such reform as disruptive and be resistant to change. Therefore, the provision of opportunities designed for educators and other stakeholders to share their experiences, voice their concerns, and learn from data-driven best practices is paramount.

Policy Recommendations:

- Establish expert committees/sub-committees under the supervision of the MOEHE to review, revise, and make recommendations regarding integration of twenty-first-century skills across the curriculum (at all levels).
- Develop models for mapping progression in the acquisition of twenty-first-century skills. By understanding the developmental stages and milestones associated with such skills, teaching methods can be better tailored to support student learning.
- At the ministerial level, mechanisms for collecting, compiling, and sharing data regarding curricula should be enhanced to differentiate between various aspects of curriculum integration, and to clearly delineate the different educational levels

(i.e., primary, secondary, undergraduate, graduate, etc.). The availability of clear and reliable data is essential for curriculum enhancement.

- A centralised entity for research and continuous improvement may be established to (1) conduct and report research on the effectiveness of twenty-first-century skill integration, and suggest national research priorities for such integration as part of a long-term strategy.

5.6.3 *The Need for a Community of Learners for Twenty-First-Century Instruction*

While there is no universal definition of a learning community, also known as a Community of Learners (CoL), it essentially refers to a collaborative and interactive learning environment where individuals come together to engage in shared learning experiences, exchange knowledge, and support each other's growth and development (Stoll et al., 2006a). CoLs can exist in many educational environments such as online, in classrooms, within institutions, or in professional development settings. Learning communities have been referred to as key features of twenty-first-century education (Watkins, 2005; Zhao & Kuh, 2004). Regarding in-service teacher development, CoLs, or professional learning communities (National Education Standards, 2020), have also been shown to be more effective than externally prescribed professional development programmes, which are often perceived as leading to an increased workload rather than as a support system, further adding to teacher isolation and burnout (Carrol et al., 2010; Stoll et al., 2006a). According to Bielaczyc et al. (2013), there are four characteristics that a CoL must have: (1) diversity of expertise among members, (2) a shared objective to continually advance the collective knowledge and skills, (3) an emphasis on learning how to learn, and (4) mechanisms for sharing what is learned (p. 2). The presence of a strong professional learning community can be highly effective in helping teachers prepare for twenty-first-century instruction by fostering professional development, innovation and creativity, collaboration, and student-centeredness (Seashore, Anderson, & Riedel, 2003; Stoll et al., 2006b). In addition, professional learning communities can support technology integration, cultural competence, and data-driven instruction, as well as teacher well-being (Bryk et al., 1999; Cordingley et al., 2003; National Education Standards, 2020). Ultimately, through these communities, teachers can practise the kind of collaborative learning and knowledge construction they expect of their students, and that is at the core of twenty-first-century competencies (Carroll 2010, p. 2). Because Twenty-First-Century Skills are not discipline-specific and instead apply to the whole person, CoLs for twenty-first-century teaching and learning can integrate participants from various roles and sectors, such as teachers, students, instructional designers, school leaders/administrators, teacher educators, ministries, parents/caregivers, educational researchers, policy makers, community organisers, and business/industry representatives. Such professional learning environments can exponentially advance teaching and learning in the country. There are several well-established professional learning communities currently active in Qatar, including the International School of London Qatar (ISL Qatar), the Education Development Institute (EDI) under Qatar Foundation, Empowering Leaders of Learning (ELL) programme, and Agile Leaders of Learning Network (ALL-IN). However, studies show that professional learning communities in Qatar need reform, because of a lack of established norms, trust and communication, and support from administrators (Khalid, 2021). Consequently, clear frameworks for designing effective learning communities in the current context are needed.

Policy Recommendations:

- MOEHE support is needed to promote and normalise professional learning communities in education and higher education, and to streamline their efforts. This could be done by taking the lead to develop clear frameworks for designing effective learning communities in the current context, as well as standards for professional learning, and methods for measuring the outcomes of professional learning communities.
- Schools and universities may be expected or required to have a CoL (or facilitate teachers' access to CoLs) as part of the institution's community building and professional development obligations.
- Job descriptions for education positions should include clear expectations regarding applicants' participation in CoLs.
- Teacher performance and evaluation policies should be revised to include and incentivise participation in CoLs, and be reflected in teacher appraisals.

5.6.4 Hesitation Towards Big Data, Data Analytics, and Machine Learning

The digitalisation of education and advancements in AI tools have resulted in various methods of generating, collecting, and analysing educational data that can be harnessed to guide decision-making and improve education outcomes (Vieira, 2018). Big data in education (structured and unstructured data that is large in volume, contains great variety, and generated at a high velocity), can be analysed to gain insight into student and faculty performance, curriculum design, instructional strategies, resource allocation, and policy development, for example (Yağcı, 2022). Additionally, AI applications like machine learning, which utilise advanced algorithms to process big data and continue students learning independently, provide support for innovations that can significantly enhance student performance and the learning experience (Vaidya & Saini 2020). Some examples of these innovations include predictive data analytics used in e-Learning software solutions, assistive technologies, chatbots and AI tutors, educational robots for children, real-time translation, task-automation, and fraud detection, among others. Studies have shown that big data, data analytics, and machine learning are useful for improving education outcomes and will be widely used at universities and institutions in the future (Manohar et al., 2016). Unfortunately, Qatar has been slow to harness the power of AI in the education sector. Article 14 of Emiri Decision No.16 of 2014, gave authority to the Ministry of Information and Communications Technology in Qatar to supervise, regulate, and develop the ICT sector, in alignment with the requirements of the national development goals; however, there has been little evidence of collaboration with the Ministry of Education and Higher Education, to achieve digital learning outcomes in the country. In fact, in a national study of higher education institutions in Qatar, Abdelrahman (2023) concluded that there are no predictive capabilities built into the current education system, and that Qatar has not reported any use of data analytics in the field of education to date (p. 14).

Because AI has the potential to transform the way students learn, and to support education decision-making, teachers and education administrators need the knowledge and skills to harness this technology effectively. Unfortunately, Qatar University's College of Education (the only teacher training programme in the country) does not currently have undergraduate or graduate degree offerings in the fields of educational or instructional technology, AI in Education, or Educational Data Science. Consequently, the gaps created by this lack of preparedness translate into increased professional development mandates for in-service teachers. More teacher preparation programmes that focus on AI in education may be useful in better preparing educators to meet the demands of twenty-first-century teaching and learning. Furthermore, more collaboration is needed between government entities and the Qatar education system, to establish an effective data infrastructure, data-sharing protocols, and a culture of data-driven decision-making that is supported by AI technology. It is important to note that the integration of AI into education is not a substitute for quality human teaching. Rather, the suggestions here are merely so that educators can work in symbiosis with these powerful technologies to enhance the quality of education and students' learning experience.

Policy Recommendations:

- An AI Certification system should be established as part of pre-service teacher preparation programmes, to ensure educators have the necessary knowledge and skills to teach AI concepts and utilise AI-powered tools for teaching and learning.
- Dedicate funding for the development of PD programmes and initiatives in schools and universities on AI technology and its integration in the classroom.
- MOEHE should include an "AI in Education" plan, as part of the licencing, accreditation, and renewal processes for universities.

5.6.5 Culture and Society

There are many aspects of Qatar's traditional, conservative culture that likely have a positive impact on digital learning and twenty-first-century instruction. For example, within the local culture, there is a strong sense of respect for education and strong governmental support. Furthermore, Qatar has adopted an approach towards modernisation that aims to preserve tradition, as highlighted in Qatar National Vision (2030). Nonetheless, other aspects of the culture pose significant challenges. For example, rigid protocols and organisational structures pose obstacles to cultivating the kinds of learning ecosystems necessary for twenty-first-century education (Clayton et al., 2021). In addition, the spread of globalisation and the large expatriate population in Qatar has led many locals to perceive a threat to their national identity, language, and culture (Abu-Shawish et al., 2021), which may add to resistance against the current educational transformation among some segments of the population. Because English is widely considered the language of technology and the Internet, there is an urgency for learners to acquire strong English language skills for survival in the digital world (Lütge & Merse 2022). Consequently, and because language and identity are intrinsically connected, there is concern among many locals that prioritising Twenty-First-Century Skills may signal a de-emphasis on Arabic (Mustafawi et al., 2022). It can be difficult to balance such cultural and linguistic factors in digital learning, particularly when relying on digital tools and online resources that are primarily available in English.

Policy Recommendations:

- Increase Arabic language initiatives in schools and universities.
- Invest in the development of more online tools and resources in Arabic for teaching and learning environments.

5.6.6 *Inclusion, Diversity, Equity, and Access*

Inclusion, diversity, equity, and access are crucial for twenty-first-century teaching and learning because equity-focused strategies help close achievement gaps among students from marginalised groups, an intrinsic good given that education is a basic human right. IDEA, an acronym used to refer to issues regarding inclusion, diversity, equity, and access, has gained increasing prominence in education and social justice in recent decades. The acronym encapsulates the principle of combating structural inequities among underserved communities. Put simply, *Inclusion* refers to environments where all are welcome and valued, *Diversity* embraces the ways in which we differ, *Equity* ensures that everyone has the opportunity to fully participate, and *Access* ensures that anyone of any ability be included. While Qatar has made significant progress in improving equity in education, there are still groups that face considerable barriers to experiencing quality education. These include individuals with disabilities, as well as children and adolescents from low-income and migrant families. Scholars such as Al-Hendawi et al. (2023) have conducted a thorough analysis of the ideals and realities of special education in Qatar, describing the extent to which such ideals are being achieved. One notable achievement is in the "Doha Declaration" established by the Qatar Foundation for Social Work in 2019, which advocated that the rights of individuals with disabilities be included in policies involving education, employment, and health (Qatar Foundation for Social Work, 2019). Advocating for rights is an essential first step; however, meeting the needs of underserved learners is fraught with challenges. Keller et al. (2016), for example, highlighted how Arabic as the language of instruction in special education programmes, may actually be contributing to a shortage of highly skilled teachers in the country. Similarly, there is a shortage of specialised schools to support individuals with disabilities (Rocha, 2014), and private schools that allow enrolment for learners with special needs often lack the assistive technology to accommodate their needs (Tlili et al., 2020). Although the MOEHE has reassigned 70 government schools to be centred for integrating students with disabilities, known as "integration schools", as part of their efforts to support persons with disabilities, the lack of publicly available data on learners with special needs at private schools, institutes, and centres in Qatar significantly hinders advancements in equity and inclusion (Al-Hendawi et al., 2023).

Likewise, students from migrant families often face language and cultural barriers that make it difficult to adjust to a new education system that may not be equipped to fully support them (Edrosolam et al., 2020). Children and adolescents from low-income families also face barriers to accessing education, such as the cost of tuition, uniforms, transportation, school supplies, and learning devices. In Qatar, low-income schools receive assistance from the government, to ensure that all students have access to quality education. Such support includes providing professional development training to teachers, as well as support for technology integration. However, as hybrid and distance learning become more integrated in schools, students from low-income families may struggle to keep pace with the high volume of at-home work that relies on access to high-speed internet and personal devices (i.e., tablets, PCs). In short, there is a need for more opportunities to explore and understand how implicit biases and historical perceptions about various communities can infiltrate teaching, learning, and decision-making at all levels.

Policy Recommendations:

- Establishing an IDEA advisory committee or IDEA checklist to guide education decision-making, programming, policy, and procedures.
- Increase collaboration within and across organisations and institutions to raise awareness, share best practices, and generate resources and support for IDEA initiatives.
- Prioritise and incentivise research and data collection relevant to underserved communities, as well as accountability measures to better support their needs.

5.7 Conclusion

An inclusive education system that provides quality education for all is vital to Qatar's ambitions as a knowledge-based society. While the past two decades have reflected the country's remarkable transformation towards digital learning as a means to support twenty-first-century instruction, the roadmap for the way forward is full of challenges heightened by dwindling resources, emerging needs, and a racing clock. Some of the most evident challenges include attracting and retaining qualified teachers, preparing pre-service teachers, curricula reform, establishing professional learning communities, integrating AI in education, and cultural and societal challenges, as well as issues of inclusion and access. The Ministry of Education and Higher Education plays a pivotal role in shaping and overseeing all aspects and levels of education in the country. Consequently, achieving the nation's desired education outcomes will entail the ministry's strong leadership and guidance. However, a successful approach will surely involve a variety of stakeholders in the process of developing, implementing, and monitoring education policy, planning, regulation, and development. This chapter has sought to address the question of the pertinent education reforms and investments, current digital learning initiatives, and the relevant literature, and to attempt an examination of the challenges and opportunities as well as policy recommendations that may be relevant to education stakeholders in Qatar and elsewhere.

Bibliography

Abdelrahman, M. (2023). Data analytics in education: factors affecting the utilization of data analytics for education in Qatar. Unpublished master's thesis, Qatar University.

Abdulwahed, M., & Hasna, M. (2017). Engineering and technology talent for innovation and knowledge-based economies. https://doi.org/10.1007/978-3-319-46439-8

Abu-Shawish, R., Cherif, M. & Romanowski, M. (2021). Teachers' experiences and challenges with student diversity in Qatar's government schools. *The International Journal of Diverse Identities, 21*, pp. 13–27. https://doi.org/10.18848/2327-7866/CGP/v21i01/13-27

Abu-Tineh, A., et al. (2023). Career advancement, job satisfaction, career retention, and other related dimensions for sustainability: A perception study of Qatari public school teachers. *Sustainability, 15*, 4370. https://doi.org/10.3390/su15054370

Accreditation Board for Engineering and Technology, (2021). 2022–2023 criteria for accrediting engineering programs. Engineering Accreditation Committee. https://www.abet.org/wp-content/uploads/2022/01/2022-23-EAC-Criteria.pdf

Al-Hendawi, M., Keller, C., & Khair, M. (2023). Special education in the Arab Gulf countries: an analysis of ideals and realities. *International Journal of Educational Research Open, 4*. https://doi.org/10.1016/j.ijedro.2022.100217

Alhothali, H. (2021). Inclusion of 21st century skills in teacher preparation programs in the light of global expertise. *International Journal of Education and Practice, 9*(1), pp. 105–127. https://doi.org/10.18488/journal.61.2021.91.105.127

Al-Kuwari, M., Du, X., & Koç, M. (2022). Performance assessment in education for sustainable development: A case study of the Qatar education system. *Prospects, 52*, 513–527. https://doi.org/10.1007/s11125-021-09570-w

Alloh, M., Hasan, M., Du, X., & Romanowski, M. (2019). Burnout of primary teachers in Qatar government schools during an era of educational reform. *International Journal of Learning, Teaching and Educational Research, 18*(10), 1–19.

Al-Naimi, S., Romanowski, M., & Du, X. (2020). Novice teachers' challenges and coping strategies in Qatari government schools. *International Journal of Learning, Teaching and Educational Research, 19*(9), 118–142. https://doi.org/10.26803/ijlter.19.9.7

Alshaboul, Y., et al. (2021). COVID-19 enforced shift to distance education: readiness and challenges. *Journal of Education and e-Learning Research, 8*(3), pp. 349–359. https://doi.org/10.20448/journal.509.2021.83.349.359

Alshaboul, Y., et al. (2022). University instructors and students' attitudes toward distance education: The case of Qatar. *Journal of Positive School Psychology, 6*(8), 7940–7959.

Alshare, K., & Sewailem, M. (2018). A Gap analysis of business students' skills in the 21st Century: A case study of Qatar. *Academy of Educational Leadership Journal, 22*, 1–22.

Altan, M. (2018). Intercultural Sensitivity: a study of pre-service English language teachers. *Journal of Intercultural Communication, 46*.

Al-Thani, A., Ari, I., & Koç, M. (2021). Education as a critical factor of sustainability: case study in Qatar from the teachers' development perspective. *Sustainability, 13*(20). https://doi.org/10.3390/su132011525

Alzahrani, A. (2019). A proposal for developing teaching practices for mathematics teachers' perception of 21st century skills. *Umm Al-Qura University Journal of Education and Psychology Sciences, 11*(1), 1–47.

Ananiadou, K. & Claro, M. (2009). 21st century skills and competences for new millennium learners in OECD countries. http://www.oecd.org/officialdocuments/publicdisplaydocumentpdf/?cote=EDU/WKP(2009)20&doclanguage=en

Association of American Colleges and Universities. (2007). *College learning for the new global century*. Report of the Association of American Colleges and Universities.

Badcock, P., Pattison, P., & Harris, K. (2010). Developing generic skills through university study: A study of arts, science and engineering in Australia. *Higher Education, 60*, 441–458.

Bel-Air, De F. (2017) "Demography, Migration, and Labour Market in Qatar," Explanatory Note No. 3/2017, GLMM, http://gulfmigration.eu.

Bentur, A., Zonnenshain, A., Nave, R., & Dayan, T. (2019). Education of engineers in the 21st century: paradigms, insights and implications to Israel. Samuel Neaman Institute for National Policy Research.

Bielaczyc, K., Kapur, M., & Collins, A. (2013). Cultivating a community of learners in K-12 classrooms. In C. E. Hmelo-Silver, A. M. O'Donnell, C., Chan, & C. A. Chinn (Eds.), *International Handbook of Collaborative Learning*, pp. 233–249. Routledge.

Brewer, D., et al. (2007). *Education for a new era: Design and implementation of k-12 education reform in Qatar*. Rand Corporation.

Bryk, A., Camburn, E., & Louis, K. S. (1999). Professional community in Chicago elementary schools: Facilitating factors and organizational consequences. *Educational Administration Quarterly, 35*, 751–781.

Carrol, T., Fulton, D., & Doerr, H. (2010). Team up for 21st century teaching and learning: what research and practice reveal about professional learning. National Commission on Teaching and America's Future, pp. 1–26.

Chaaban, Y., Sawalhi, R., & Du, X. (2022). Exploring teacher leadership for professional learning in response to educational disruption in Qatar. *Professional Development in Education, 48*(3), 426–443. https://doi.org/10.1080/19415257.2021.1973070

Clayton, R., Amaral, A., & Shafique, A. (2021). Developing local learning ecosystems in Qatar to advance equity, inclusion, and social cohesion. World Innovation Summit for Education. Qatar Foundation.

Cordingley, P., Bell, M., Rundell, B. & Evans, D. (2003, September). How does CPD affect teaching and learning? Issues in systematic reviewing from a practitioner perspective. In annual conference of the *British Educational Research Association*, Edinburgh. Retrieved from http://www.leeds.ac.uk/educol/documents/00003231.html.

Cosner, S., & Alfadala, A. (2021). Post-pandemic national educational investments: school leadership development through innovative learning designs. World Innovation Summit for Education. https://www.wise-qatar.org//app/uploads/2023/05/wise-ell-dissemination-web-version-.pdf.

Ministry of Education and Higher Education. (2019). Curriculum Framework for the State of Qatar. (December 1, 2019).https://www.edu.gov.qa/en/MediaCenter/Pages/MediaCenter/Publications.aspx

Edrosolam, J., et al. (2020). Unity in diversity: The assimilation of culture to Filipino students in Qatar. *International Journal of Research Publications, 62*(1), 1–16.

Elkhouly, M. (2021). Perceptions of undergraduate students, lecturers, and lab assistants towards the use of online laboratory in science education during COVID-19 pandemic. [Unpublished master's thesis, Qatar University]. http://hdl.handle.net/10576/21201

Fadlelmula, F., Sellami, A., Abdelkader, N. et al. (2022), A systematic review of STEM education research in the GCC countries: trends, gaps and barriers. *International Journal of STEM Education, 9*(2). https://doi.org/10.1186/s40594-021-00319-7

Granville, G. (2014). The impossibility of curriculum change in the mind of someone educated: Shark, Sabre-Tooth and Junior Certificate. *Pathways to Innovation and Development in Education*, 20.

Güney, K. (2023). Considering the advantages and disadvantages of utilizing social media to enhance learning and engagement in K-12 Education. *Research in Social Sciences and Technology, 8*(2), pp. 83–100. https://doi.org/10.46303/ressat.2023.13

Haleem, A., Javaid, M., Qadri, M., & Suman, R. (2022). Understanding the role of digital technologies: A review. *Sustainable Operations and Computers, 3*(4), 275–285. https://doi.org/10.1016/j.susoc.2022.05.004

Herde, C., Wüstenberg, S., & Greiff, S. (2016). Assessment of complex problem solving: What we know and what we don't know. *Applied Measurement in Education, 29*(4), 265–277. https://doi.org/10.1080/08957347.2016.1209208

Ibrahim, Y. (2023). Moving from the subject-based curriculum to the skills-based curriculum in Abu Dhabi schools: does it lead to reform? A theoretical analysis & case study paper in one of Abu Dhabi private schools. In: Al Marri, K., Mir, F., David, S., Aljuboori, A. (eds) *BUiD Doctoral Research Conference 2022. Lecture Notes in Civil Engineering*, vol. 320. Springer, Cham. https://doi.org/10.1007/978-3-031-27462-6_9

Karkouti, I. (2016). Qatar's educational system in the technology-driven era: Long story short. *International Journal of Higher Education, 5*, 182–189. https://doi.org/10.5430/ijhe.v5n3p182

Keegan, L. (2020, September 10). *Complete history of eLearning from 1923—Present Day*. Skill Scouter. https://skillscouter.com/history-of-elearning

Keller, C., Al-Hendawi, M., & Abuelhassan, H. (2016). Special education teacher preparation in the Gulf Cooperation Council countries. *Teacher Education and Special Education, 39*, 194–208. https://doi.org/10.1177/0888406416631125

Khalid, T. (2021). The influential role of professional learning communities on teachers' performance. *International Journal of Scientific & Technology Research, 10*(2), 74–78.

Lee, S., & Zahir, L. (2021). Applying global models of teacher development to improve student outcomes: insights from a teacher development & leadership program in Qatar. *Gulf Education and Social Policy Review.* pp. 46–68.

Li, L. (2022). Reskilling and upskilling the future-ready workforce for industry 4.0 and beyond. *Information Systems Frontiers*. July 2022. https://doi.org/10.1007/s10796-022-10308-y

Lütge, C., & Merse, T. (2022). *Digital Citizenship in Foreign Language Learning and Teaching.* https://doi.org/10.4324/9781003183839-14

Manohar, A., Gupta, P., & Priyanka, V. (2016). Utilizing big data analytics to improve education. *Journal of Engineering Science and Technology*. Special Issue on ICCSIT, July (2018), pp. 229–241.

Manzoor, A. (2016). Media literacy in the digital age: literacy projects and organizations. In: Yildiz, M., & Keenwe, J. (eds) *The Handbook of Research on Media Literacy in the Digital Age.* pp. 249–274. https://doi.org/10.4018/978-1-4666-9667-9.ch012

Moyo, C., & Maifala, S. (2022). Navigating the new COVID-19 normal: the challenges and attitudes of teachers during online learning in Qatar. *International Journal of Learning, Teaching and Educational Research, 21* (8), pp. 368–383. https://doi.org/10.26803/ijlter.21.8.21

Musa, O., & Alhanan, T. (2013). A proposal for developing the competence of Arabic language and history teachers in the light of the requirements of knowledge society. *Assiut Educational Journal, 2*(1), 1–2.

Mustafawi, E., et al. (2022). Perceptions and attitudes of Qatar University students regarding the utility of Arabic and English in communication and education in Qatar. *Language Policy, 21*, 75–119. https://doi.org/10.1007/s10993-021-09590-4

Naji, K., et al., (2020). Engineering students' readiness to transition to emergency online learning in response to COVID-19: a case in Qatar. *EURASIA Journal of Mathematics, Science and Technology Education, 16* (10), https://doi.org/10.29333/ejmste/8474

Nasser, R. (2017). Qatar's educational reform past and future: Challenges in teacher development. *Open Review of Educational Research, 4*(1), 1–19.

National Research Council. (2012). Education for life and work: developing transferable knowledge and skills in the 21st Century. http://mhalabs.org/wp-content/uploads/2014/04/Pellegrino_Report_Education-for-Work-and-Life.pdf

National Education Standards (2020). The standards for professional learning. leaning forward: the professional learning association. https://standards.learningforward.org/.

Newsome, M., et al. (2022). The effect of learners' sex and STEM/non-STEM majors on remote learning: a national study of undergraduates in Qatar. *The Electronic Journal of e-Learning, 20* (4), pp. 360–373. https://doi.org/10.34190/ejel.20.4.2262

Plevyak, L. (2022). *Teaching towards green schools: Transforming K–12 education through sustainable practices*. Routledge.

Qatar Second National Development Strategy 2018–2022. (2018). Ministry of Development Planning and Statistics. https://www.psa.gov.qa/en/knowledge/Documents/NDS2Final.pdf

Qatar Planning and Statistics Authority. (2019). Education in Qatar: Statistical Profile 2019Doha, Qatar. https://www.psa.gov.qa/en/statistics/Statistical%20Releases/General/StatisticalAbstract/2019/Education_Statistical_Profile_2019_En.pdf

Qatar Foundation for Social Work. (2019). Qatar launches declaration to promote rights of over 1.5 billion people with disabilities. Disabled World (December 8). https://www.disabled-world.com/news/middle-east/qatar.php

Qatar Development Bank (2021). Education Sector in Qatar. https://www.qdb.qa/en/Documents/Education_Sector_in_Qatar_EN.pdf

Qatar National Vision 2030. (July 2008). General Secretariat for Development Planning. https://www.gco.gov.qa/wp-content/uploads/2016/09/GCO-QNV-English.pdf

Rocha, P. (2014). *The perception and experiences of expatriate mothers regarding professional support services for children with special needs in Qatar* [Doctoral dissertation, Walden University].

Romanowski, M., & Qadhi, S. (2022). Teach for Qatar: Teachers' challenges and resources in Qatar's government schools. *SAGE Open, 12* (2). https://doi.org/10.1177/21582440221099523

Rusdin, N., & Ali, S. (2018). Implementation of 21st century learning and the challenges. *Proceedings of the Qualitative Research Conference (QRC)*. July 2018. Melaka, Malaysia.

Said, Z., Pavlova, M., & Wheeler, L. (2021). 21st century skills' challenges to postsecondary TVET institutions in Qatar. In: *Paper presented at the 14th International Technology, Education and Development Conference, Valencia, Spain, 2–4 March*. https://doi.org/10.21125/inted.2020.1534

Seashore, K., Anderson, A. & Riedel, E. (2003). Implementing arts for academic achievement: the impact of mental models, professional community and interdisciplinary teaming. In: *Paper presented at the Seventeenth Conference of the International Congress for School Effectiveness and Improvement, Rotterdam, January*.

Siddiq, F., Gochyyev, P., & Wilson, M. (2017). Learning in digital networks – ICT literacy: A novel assessment of students' 21st century skills. *Computers & Education, 109*, 11–37. https://doi.org/10.1016/j.compedu.2017.01.014

Social Research. (2015). Literature review on the impact of digital technology on learning and teaching: children, education, and skills. The Scottish Government. https://dera.ioe.ac.uk/id/eprint/24843/1/00489224.pdf

Next Generation Science Standards for States, by States (2013). Available at: https://www.nextgenscience.org/

Stoll, L., Bolam, R., McMahon, A., Thomas, S., Wallace, M., Greenwood, A., & Hawkey, K. (2006b). *Professional learning communities: Source materials for school leaders and other leaders of professional learning*. Innovation Unit, DfES, NCSL and GTC.

Stoll, L., Bolam, R., McMahon, A., Wallace, M., & Thomas, S. (2006a). Professional learning communities: A review of the literature. *Journal of Educational Change, 7*(4), 221–258.

Teach for Qatar. (2022). Who are we? https://www.teachforqatar.org/en/who-we-are.php

Thomas, M., et al. (2022). A cross-sectional survey on the effect of COVID-19 related restrictions on undergraduate and postgraduate medical education in Qatar. *BMC Medical Education, 22*, 212. https://doi.org/10.1186/s12909-022-03268-z

Tlili, A. et al. (2021). Remote special education during crisis: COVID-19 as a case study. In: Burgos, D., Tlili, A., Tabacco, A. (eds) *Radical Solutions for Education in a Crisis Context. Lecture Notes in Educational Technology*. Springer, Singapore. https://doi.org/10.1007/978-981-15-7869-4_5

Trilling, B., & Fadel, C. (2009). *21st century skills: Learning for life in our times*. Jossey-Bass.

Vaidya, A., & Saini, J. (2021). A Framework for implementation of learning analytics and educational data mining in traditional learning environment. In: Fong, S., Dey, N., Joshi, A. (eds) *ICT Analysis and Applications. Lecture Notes in Networks and Systems*, vol 154. Springer: Singapore. https://doi.org/10.1007/978-981-15-8354-4_11

Van den Akker, J. (2010). Building bridges: how research may improve curriculum policies and classroom practices. In S. M. Stoney (Ed.), *Beyond Lisbon 2010: Perspectives from research and development for educational policy in Europe* pp. 175–195. CIDREE.

Vanderlinde, R., Tuytens, M., De Wever, B., & Aelterman, A. (2016). Professional learning of teacher educators, teachers, and student teachers: an introduction. In: *Professional learning in education—Challenges for teacher educators, teachers, and student teachers*. Gent: Academia Press.

Vieira, C. (2018). Visual learning analytics of educational data: A systematic literature review and research agenda. *Computers & Education, 122*, 119–135. https://doi.org/10.1016/j.compedu.2018.03.018

Watkins, C. (2005). Classrooms as learning communities: A review of research. *London Review of Education, 3*(1), 47–64. https://doi.org/10.1080/14748460500036276

Webb, M.E., Prasse, D., Phillips, M. et al. (2018). Challenges for IT-enabled formative assessment of complex 21st century skills. *Technology Knowledge & Learning, 23*, pp. 441–456.https://doi.org/10.1007/s10758-018-9379-7

De Wever, B., Tuytens, M., Vanderlinde, R., & Aelterman, A. (2016). *Professional Learning in Education*. Academia Press.

Wilkins, S., & Annabi, C. A. (2023). Academic careers in transnational higher education: The rewards and challenges of teaching at international branch campuses. *Journal of Studies in International Education, 27*(2), 219–239. https://doi.org/10.1177/10283153211052782

Woodward, S., & Kimmons, R. (2018). Religious implications of social media in education. *Religion & Education*, 1–24. https://doi.org/10.1080/15507394.2018.1510249

World Economic Forum. (2016). *The future of jobs: Employment, skills, and workforce strategy for the fourth industrial revolution*. Global Challenge Insight Report.

Yağcı, M. (2022). Educational data mining: Prediction of students' academic performance using machine learning algorithms. *Smart Learning Environments, 9*, 11. https://doi.org/10.1186/s40561-022-00192-z

Zhao, C.-M., & Kuh, G. D. (2004). Adding value: Learning communities and student engagement. *Research in Higher Education, 45*(2), 115–138. https://doi.org/10.1080/14748460500036276

Chapter 6
Human-Centred Artificial Intelligence in Progressive Education: Unravelling the Benefits and Challenges in Qatar's HEIs

Bernard J. Jansen, Soon-gyo Jung, Ali Farooq, Joni Salminen, Kholoud Aldous, Pilira Stella Msefula, Amani Alabed, Salar M. Khan, and Richard O'Kennedy

Abstract Algorithmically generated personas (AGPs) can provide actionable insights into the needs, preferences, and behaviours of stakeholders, including Applicants, Students, Alumni, Faculty, and Staff (ASAFS) for progressive education at higher education institutes (HEIs) in Qatar. AGPs empower data-based decision-making, and contribute to progressive education by representing stakeholders in an empathic format. This chapter sheds light on the pivotal role of alumni, showing how

B. J. Jansen (✉) · S. Jung · A. Farooq · K. Aldous
Qatar Computing Research Institute, Hamad Bin Khalifa University, Doha, Qatar
e-mail: jjansen@acm.org

S. Jung
e-mail: sjung@hbku.edu.qa

J. Salminen
School of Marketing and Communication, University of Vaasa, Vaasa, Finland
e-mail: jonisalm@uwasa.fi

P. S. Msefula
Higher Education, Qatar Foundation, Doha, Qatar
e-mail: pmsefula@qf.org.qa

A. Alabed
College of Business, University of Doha for Science and Technology, Doha, Qatar
e-mail: amani.alabed@udst.edu.qa

S. M. Khan
Academyati, Qatar Foundation, Doha, Qatar
e-mail: sakhan5@qf.org.qa

R. O'Kennedy
Research, Hamad Bin Khalifa University Research, Doha, Qatar

Dublin City University, Dublin, Ireland

R. O'Kennedy
e-mail: richardokennedy@gmail.com

E. Zaidan et al. (eds.), *The Future of Education Policy in the State of Qatar*, Gulf Studies 21, https://doi.org/10.1007/978-981-97-9667-0_6

the innovative approach of using personas enables the "beyond education" concept, while providing strategic foresight for HEIs in Qatar. Machine learning models can generate AGPs as humanised representations of these key stakeholders. Building on our multi-year experience in studying AGPs, we argue that these personas can significantly aid HEI decision-makers in designing more learner-centred and inclusive curricula, pedagogy, and assessment via empathic understanding. While AGPs have been prevalent in marketing or design, their potential remains largely unexplored in HEIs.

6.1 Introduction

This chapter's key themes and focus are the advocacy for applying human-centred artificial intelligence (HCAI) and data-driven persona generation in the context of progressive education within Qatari **higher education institutions** (HEIs). We focus on the potential of humanised representation of analytics data, using algorithmically generated personas (Nielsen, 2013b; Nielsen et al., 2015; Park & Kang, 2022; Pruitt & Adlin 2010) in the context of Qatar's progressive education landscape. Rooted in student-centred learning, progressive education is a pedagogical approach to interdisciplinary instruction and developing critical thinking skills (Howlett, 2013; Levi & Marchionini, 2019; Reese, 2001; Ylimaki, 2011). The chapter examines the benefits and challenges of using **algorithmically generated personas** (AGPs) to represent key stakeholders, such as **Applicants, Students, Alumni, Faculty, and Staff** (ASAFS), and its implications for practice, policy, and future research. Despite the growing interest in progressive education, the benefits and challenges of the student-personas approach, particularly in the context of HEIs that form a strategic imperative for Qatar's societal development, remain largely unaddressed.

This chapter attempts to fill this gap by highlighting the benefits and challenges of comprehensively studying one aspect of progressive education at HEIs in Qatar. Specifically, we focus on employing AGPs to express the experiences of key stakeholder groups (i.e., the ASAFS) at HEIs in Qatar of our proposed conceptual foundation of the progressive education journey discussed below (Alves et al., 2010; Falqueto et al., 2020; Ferrero-Ferrero et al., 2018; Logermann & Leišytė, 2015). By examining the perspectives and needs of these ASAFS groups, we can gain a deeper understanding of the effective implementation and continuous improvement of progressive education and "beyond education" that refers to areas, concepts, or ideas that extend past the scope of formal learning, encompassing broader aspects of life, society, and personal development (Cunningham, 1992; Grace & Grace, 1995; Meyerhoff, 2019) practices in Qatari HEIs. Figure 6.1 shows an AGP listing of the personas representing alumni of a Qatari HEI who responded to a survey.

Much of the foundation of this chapter is based on our extensive and world-leading research on data-driven AGPs, with our work published in esteemed outlets like the MIT Technology Review and multiple first-class ACM conferences and journals. Our expertise, initially invested in these contexts, has been catering to the unique

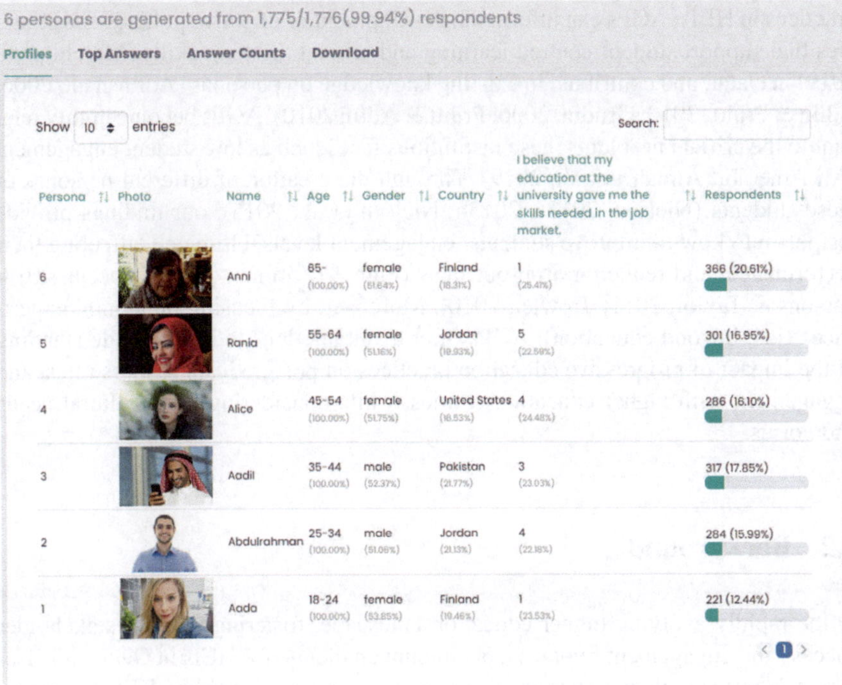

6 personas are generated from 1,775/1,776(99.94%) respondents

Profiles | Top Answers | Answer Counts | Download

Show 10 entries | Search:

Persona	Photo	Name	Age	Gender	Country	I believe that my education at the university gave me the skills needed in the job market.	Respondents
6		Anni	65- (100.00%)	female (61.64%)	Finland (18.31%)	1 (25.41%)	366 (20.61%)
5		Rania	55-64 (100.00%)	female (51.16%)	Jordan (19.93%)	5 (22.59%)	301 (16.95%)
4		Alice	45-54 (100.00%)	female (51.75%)	United States (16.53%)	4 (24.48%)	286 (16.10%)
3		Aadil	35-44 (100.00%)	male (52.37%)	Pakistan (21.77%)	3 (23.03%)	317 (17.85%)
2		Abdulrahman	25-34 (100.00%)	male (51.06%)	Jordan (21.13%)	4 (22.18%)	284 (15.99%)
1		Aada	18-24 (100.00%)	female (53.85%)	Finland (19.46%)	1 (23.53%)	221 (12.44%)

Fig. 6.1 An AGP listing (i.e., cast) of personas representing alumni of a Qatari HEI who responded to a survey. The personas were generated using Survey2Persona, an AGP system developed at Qatar Computing Research Institute

needs of HEI stakeholders. We have received previous recognition for our work, including a Facebook-funded research grant to build data-driven MENA personas, and a grant in the education domain via a project funded by Hamad Bin Khalifa University (HBKU). In 2020, our work on AGPs was honoured with an "Innovation of the Year" nomination, and an award from the Qatar Foundation (QF).

In this chapter, we illustrate the use of a mixed-methods approach, combining both qualitative and quantitative data with the application of artificial intelligence (AI) and machine learning (ML) tools, to construct AGPs as humanised data representations of the key stakeholders. The mixed method combined both numerical and textual data. The AI and ML approach leverages large language models (LLMs), and sentiment and topic classification methods. These personas are invaluable tools for decision-makers in Qatari HEIs to develop and refine key performance indicators (KPIs) based on the needs, behaviours, and shortcomings expressed in the personas (Waheed et al., 2011; Wang et al., 2023).

The use of AGPs is not merely theoretical but also holds profound implications for practice, policy, and future research in the realm of progressive education in Qatar. By identifying the key factors contributing to the success of progressive education practices in HEIs, AGPs can inform the development of effective policies and strategies that support student-centred learning and critical thinking skills (Marchionini, 2019) in Qatar, and contribute to existing knowledge on personas (Adlin et al., 2006; Adlin & Pruitt, 2010; Grudin, 2006; Pruitt & Adlin 2010). AGPs become highly relevant to the critical problems these institutions face, such as low student engagement (Ali Amer Jid Almahri et al., 2019). Through the creation of different personas of these students (Nielsen, 2013a, 2013b; Nielsen et al., 2015), our findings provide insights into how to improve students' engagement levels, ultimately affecting their performance, and reducing drop-out rates (Finn & Zimmer, 2012; Groccia, 2018; Parsons & Taylor, 2011; Trowler, 2010). Moreover, by focusing on alumni experiences (i.e., beyond education), AGPs offer a suitable lens to enable understanding of the impact of progressive education practices on postgraduate success rates, and engagement with higher education entities, while considering local cultural needs and norms.

6.2 Background

In the rapidly evolving higher education landscape, fostering ASAFS stakeholder success and engagement remains a paramount challenge for HEIs in Qatar and elsewhere. New technologies such as genomics and AI are now enabling HEIs to provide much enhanced diagnostics and, thus, personalised interventions that are more appropriate and effective. The emergence of student-centred approaches has shifted the focus from a one-size-fits-all educational model towards a dynamic **personalised learning experience** (Bhutoria, 2022; Patrick et al., 2013; Yang et al., 2019). This approach tailors educational experiences to students' needs, interests, and learning styles, to optimise their engagement and understanding. Within this transformative landscape, one innovative strategy is proposed: the use of AGPs. While personas are a well-established design thinking tool in marketing and user experience (UX) research, the application of personas in higher education is limited. Understanding the implications of employing such tools becomes imperative, as they have the potential to enhance student-centric approaches, ultimately leading to improved learning outcomes and overall ASAFS stakeholder satisfaction. By understanding the diverse needs, preferences, and motivations of their ASAFS stakeholders, especially students, institutions can tailor curricula, support services, and campus experiences, to cater to individual learning styles, thus nurturing a more inclusive learning environment, conducive to success.

Existing theories and models have been developed to describe the stages of student development (Patton et al., 2016) during their college years (Evans et al., 2010; Kegan, 1998). We draw upon two prominent theories in this regard, mainly Perry's intellectual and ethical development theory and Kegan's theory of self-evolution.

These theories offer invaluable insights into how college students perceive knowledge, values, and authority, and how they develop cognitive complexity and self-regulation. Perry's intellectual and ethical development theory is widely recognised and influential in understanding student development. According to Perry, college students progress through four stages (i.e., dualism, multiplicity, relativism, and commitment) that reflect their evolving perspectives on knowledge, values, and authority (Evans et al., 2010). Initially, students may hold dualistic beliefs, perceiving knowledge as absolute and authority as external. As they advance, they move towards more nuanced views, recognising the complexity and subjectivity of knowledge and developing their ethical frameworks. In turn, Kegan's theory of self-evolution complements Perry's theory by focusing on the cognitive complexity and self-regulation of college students (Kegan, 1998). Kegan proposes that students transition from a socialised mind, where they rely heavily on external validation and conform to societal expectations, to a self-authoring mind. In the self-authoring stage, students develop a more autonomous sense of self, taking responsibility for their beliefs, values, and actions.

HEIs worldwide have witnessed a transformation in their pedagogical practices, driven by the realisation that learners are not homogenous entities, but individuals with unique backgrounds, goals, and learning preferences (Garces-Bacsal et al., 2021; Gu & Lee, 2019; Hamdan Alghamdi, 2014). During the recent COVID-19 pandemic, new bewildering challenges related to learning and engagement emerged (Laato et al., 2022). As colleges and universities strive to create enriching educational experiences for their diverse stakeholder populations, the integration of personas as a strategic planning tool is gaining considerable attention. By leveraging personas, educators, and administrators can transcend traditional demographics-based segmentation, and unearth a more profound understanding of their students' needs, habits, and pain points (and those of other stakeholders), paving the way to implementing targeted interventions and personalised support systems. This chapter aims to push forward further the emerging use of personas in higher education, notably within Qatari HEIs, and provide insights into successful implementation strategies across various institutional settings. HEIs can build a strong foundation for elevating student success and enriching the overall academic experience, by harnessing personas to embrace stakeholder diversity. For example, personas help HEIs understand students' different cultural perspectives, languages, and learning preferences, allowing them to create a more inclusive and culturally sensitive learning environment.

By developing AGPs representing different ASAFS groups, Qatar's education systems can tailor educational programmes to meet each group's specific needs and goals. This personalisation can lead to more effective learning outcomes and increased stakeholder engagement. Personas can inform many aspects of the progressive education process, such as integrating technology into classrooms, and ensuring that digital tools and platforms align with students' and teachers' technological skills and preferences. By understanding the multifaceted needs and aspirations of students, teachers, and communities, Qatar can strengthen its education sector, and progress towards its long-term development objectives.

6.3 Defining Personas

Personas are fictional characters created often through a data-driven synthesis of qualitative and quantitative student information, representing specific student archetypes that share common characteristics, behaviours, and aspirations. These personas are ideally based on real data and research, such as surveys, interviews, and observations, to accurately reflect the targeted stakeholder group's characteristics, behaviours, and needs. As a humanised representation of the data, personas are an empathetic approach to representing segments of the target populations by presenting this data in the form that most people can relate to—another person (Jansen et al., 2020). Personas are widely developed in many domains and organisations, using various methods, including qualitative and algorithmic approaches (Volden & Wattne, 2023). Personas have been evaluated in multiple manners, including statistic evaluation for segmentation identification tasks (Salminen et al., 2020a), with various positive benefits reported in the literature (Hemzo, 2023). A persona is typically presented in a one- to two-page document called a persona profile (Pruitt & Grudin, 2003); however, personas can also be presented algorithmically in data-driven ML or AI systems (Jiang et al., 2020). Personas are used for various tasks within teams and organisations, including stakeholder communication, planning, and objective setting (Pruitt & Adlin 2010).

6.4 Defining Progressive Education

The term progressive education encompasses a broad range of ideas and reforms (Davies, 2002), and encompasses a range of teaching and educational approaches. Due to its vast and sometimes vague generalities (McNear, 1978), it becomes difficult to construe the precise meaning. For this reason, our focus is on progressive pedagogy and practices.

Progressive education pedagogy includes several related concepts that promote effective learning and teaching: *Action Learning* (a professional development approach that involves learning through real-world projects and reflective group discussions), *Authentic Learning* (a teaching and learning approach that integrates real-world experiences and tasks, to develop practical skills and meaningful understanding), *Interdisciplinarity* (the integration and collaboration of knowledge and methodologies from different academic disciplines, to address complex issues and challenges), *Project-based Learning* (an instructional approach where students learn by actively engaging in projects that foster critical thinking, problem-solving, and practical application of knowledge), *Reflective Learning* (a process of self-examination and thoughtful consideration of experiences, enabling the development of insights and personal growth), *Self-regulated Learning* (the ability of individuals to independently manage their own learning process, including setting goals, monitoring progress, and adapting strategies, to achieve optimal learning outcomes),

Flipped Learning (an educational approach where traditional in-class lectures are replaced with independent learning activities at home, while interactive discussions and problem-solving take place during class time), and many more (Tippett & Lee, 2019). Most of these progressive pedagogies are visible in HEI practices nowadays. For example, service learning is a progressive pedagogy visible in the HEI's co- and extra-curriculum. For this chapter, we will consider practices of progressive pedagogy as progressive education. These practices are fully or partially part of the Qatari HEIs.

6.5 The Use of Personas in Education

Using personas in education involves creating fictional representations of people who belong to specific groups within the educational context. These personas help educators, administrators, and other stakeholders to better understand and address the diverse needs and preferences of students, teachers, parents, and other educational community members. There are several benefits to creating ASAFS personas in education. First, creating personas of ASAFS groups helps provide a richer account to stakeholders. This, in turn, allows designers and developers to maintain a student-centric approach during the technology design process (Greer et al., 2015). Second, creating personas helps facilitate an understanding of the HEI context of ASAFS groups, and supports the design and development of learning tools (Lilley et al., 2012). Third, creating student personas makes personal information about ASAFS groups more accessible (Sun & Rosson, 2017). Fourth, creating personas helps better represent the ASAFS groups' profiles, and expresses students' goals, needs, learning objectives, behaviours, and challenges (Dittmar & Forbrig, 2019). Fifth, the ASAFS groups' personas could reveal similarities and differences between students (Sim et al., 2019). Sixth, creating ASAFS personas fosters a shared understanding of the stakeholder group around which the design process is built. This can help prioritise design considerations, for example, in the context of student learning (Peters et al., 2021). Seventh, creating ASAFS personas cultivates empathy, facilitating an understanding of user expectations, goals, needs, experiences, and behaviours (Ekşioğlu et al., 2018). Eighth, ASAFS personas could enable the addressing of issues for individuals with learning difficulties or other disparities, drawing attention to special learning needs.

In summary, ASAFS personas are indispensable assets within education. They enable educators and administrators to understand better their target stakeholders, design more effective and personalised learning experiences, foster inclusive practices, and make informed decisions to enhance the overall educational journey for all stakeholders involved. ASAFS personas are powerful educational tools that promote personalised learning experiences, improve curriculum design, target marketing efforts, and foster empathy and inclusion. By using AGPs personas, academic institutions can better cater to their students' and staff's diverse needs and preferences, ultimately leading to more effective and successful learning outcomes.

6.6 Educational Ecosystem—Qatar. The Country's Background

Qatar, officially known as the State of Qatar, is a small, prosperous country on the northeastern coast of the Arabian Peninsula in Western Asia, specifically the Middle East and North Africa (MENA) region (Zahlan, 2016). Qatar is also a member of the Gulf Cooperation Council (GCCs). It shares its only land border with Saudi Arabia to the south, while the Persian Gulf surrounds it on the remaining sides. The capital city is Doha, the country's largest city and economic hub. As of 2023, Qatar's population was around 2.8 million. Most of the population comprises expatriates, with a relatively small percentage of Qatari citizens. Arabic is the official language of Qatar, but English is widely spoken, and used for business and communication. Qatar has a hot desert climate, with scorching summers, and temperatures often exceeding 40 °C (104°F), and mild winters, with temperatures ranging from 15 °C to 25 °C (59°F to 77°F) (Asia & Lanka, 2023).

Qatar has garnered international attention (Peterson, 2006), notably through its ambitious sports initiatives (Matzarakis & Fröhlich, 2015). It hosted the FIFA World Cup in 2022, making it the first MENA country to do so. The FIFA World Cup 2022 event has driven significant infrastructure development and investment in the country. Qatar boasts one of the world's highest Gross Domestic Product (GDP) figures per capita, primarily attributed to its vast natural gas and oil reserves. The country has established itself as one of the leading exporters of liquefied natural gas, and plays a significant role in the global energy market. Additionally, the government has been actively diversifying its economy into finance, tourism, real estate, and the sports sector. Qatar has also been promoting itself as a tourist destination, attracting visitors with its modern architecture, cultural heritage, desert landscapes, and luxury shopping experiences (Asia & Lanka, 2023).

Qatar is an absolute monarchy, with the ruling Al Thani family having held power since the mid-1800s. The country's political system is centred around the Emir, who wields executive authority, appoints ministers, and governs the nation. Qatar's culture is deeply rooted in its Bedouin heritage and Islamic traditions. The country places high importance on preserving its cultural identity, while embracing modernity. Qatar has seen significant development and transformation in recent decades, and plays a prominent role in the Gulf region and the global economy (Asia & Lanka, 2023).

6.7 Emphasis on Education

Qatar's education system has undergone significant development and improvement (Ibnouf et al., 2013). The country has prioritised education, ambitiously aiming to transform it to align with its National Vision 2030. Qatar has consistently invested in its education sector, dedicating significant financial resources to enhance infrastructure, educational facilities, and access to quality education for all citizens (Ben

Hassen, 2021). Qatar has achieved near-universal access to primary and secondary education, ensuring that most of its population can receive a basic education. The country has seen an increasing number of internationally recognised universities and educational institutions setting up campuses in Qatar. These partnerships enhance the quality of the country's higher education and research opportunities. It has been implementing comprehensive education reforms to improve the quality of teaching and learning, including updating curricula, incorporating technology into classrooms, and promoting innovative teaching methods (Berrebi et al., 2009).

Qatar has strategically invested in research and innovation, to position itself as a regional knowledge hub (Knight, 2017). This focus on research contributes to advancements in various fields, and fosters a culture of continuous learning. Given its multicultural population, Qatar places importance on multilingual education. English and Arabic are the primary languages of instruction, with efforts to promote bilingualism among students. The country has been actively working towards empowering women in education and promoting gender equality through measures such as increasing access to education and opportunities for women and girls (Coleman, 2004; Inglehart & Norris, 2003; Mitchell et al., 2015). Qatar attracts a growing number of international students due to the reputation of its educational institutions, and the diverse learning environment it offers.

In alignment with its 2030 Qatar National Vision (QNV), Qatar has placed a strong emphasis on education, recognising its role in nurturing human capital, innovation, and economic diversification. It has invested heavily in building world-class educational institutions and facilities and has made significant strides in developing its higher education sector, transforming it into a hub of academic excellence in the MENA (Knight, 2017). Notably, Qatar boasts many prominent higher education entities that play a crucial role in shaping the Gulf's educational landscape. Its commitment to investing in higher education entities has significantly improved the country's educational landscape. The nation continues to attract top-tier faculty and students, while fostering an environment conducive to research, innovation, and academic excellence. As Qatar looks to diversify its economy and develop a knowledge-based society, higher education remains a central pillar in achieving these ambitious goals.

As of 2020, there were a total of 32 HEIs in Qatar that are recognised by the Ministry of Education and Higher Education in Qatar. Of these, ten are public, six are military, eight are QF-affiliated Universities, and nine are privates, offering over 366 educational programmes catering to a broad spectrum of academic interests. Over 41,000 students from more than 100 nationalities were registered in these HEIs in the 2020–2021 academic year.[1]

One of the most notable institutions in Qatar is Qatar University, which was established in 1973, and remains the country's oldest and largest public university. It offers undergraduate and graduate programmes spanning various disciplines,

[1] Qatar Planning and Statistics Authority. (2021). Education Statistics Chapter 4 - 2020. https://www.psa.gov.qa/en/statistics/Statistical%20Releases/Social/Education/2021/Education_Chapter_4_2021_AE.xlsx

including engineering, business, social sciences, and the humanities. Qatar University has been instrumental in producing skilled graduates and nurturing research initiatives, contributing to the nation's development.

Moreover, the Qatar Foundation (QF) is at the forefront of shaping the country's Education, Science, Community Development, and stimulating research (Rostron, 2009). It has established partnerships with leading international universities, creating branch campuses of reputable institutions within QF's Education City (Walsh, 2019). These branch campuses provide world-class educational opportunities to Qatari students, and attract learners from across the globe, further enhancing the country's position as an education destination. Other notable institutions of QF include Hamad Bin Khalifa University (HBKU), now over ten years old, which focuses on responding to the country's research and workforce requirements (Eggeling, 2020). HBKU primarily provides postgraduate programmes with a major emphasis on research and innovation. In addition, its research institutes and colleges have significant research programmes, aligned with Qatar's 2030 Strategic Vision. In addition, various research centres within QF's Education City conduct pioneering research and innovation. Initiatives like these enhance the education ecosystem, and enrich students' learning experiences, providing them with cutting-edge opportunities, and exposure to groundbreaking advancements.

6.8 Using AGPs in Qatari HEIs

The joint sponsors of QF and HBKU funded a pilot project to act as a prototype for the initial premises, and the methodologies were refined in a Facebook Research grant for creating personas for privacy. According to the QF Strategy Plan, Progressive Education is defined as "*Innovatively re-imagining and transforming education to unlock the potential of learners and prepare them for the world of tomorrow by encouraging creativity and student agency*." This is further highlighted in QF's Higher Education vision: *"Our aspiration is to build the future of Qatar by nurturing capable, well-rounded and value-driven human beings, who are committed to leading and shaping their communities. We provide them with world-class, personalized, relevant learning in an inclusive, innovative ecosystem."* AGPs support this goal, and progress in progressive education will enable Qatari HEIs to achieve their strategic objectives[2] and progressive education aims (Howlett, 2013; Reese, 2001; Ylimaki, 2011).

QF's Education City (EC) (Asquith, 2006; Ibnouf et al., 2013; Khodr, 2011; Weber & City, 2012) offers a wealth of information that can generate rich data sets which can be leveraged in creating AGPs with a significant impact. The student population is remarkably diverse, comprising more than 100 nationalities, reflecting the substantial expat community and international non-residential students. Furthermore, the convergence of student communities from eight universities hosted in

[2] Cultivating Multiversity Qatar Foundation Strategy Refresh 2022-2032.

EC, pursuing different programmes yet using shared facilities such as the National Library, housing, or attending similar extra-curricular activities, can provide valuable insights into areas of interest, such as student engagement and motivation. In addition, cross-registration (i.e., a system allowing students at one university to take individual courses for credit at another institution), as well as the presence of non-academic centres, formulate unique learner experiences and create individualised student pathways, enriching the data pool that can be utilised in creating AGPs to further support decision-making. The process of leveraging in EC can then be extended to other Qatari HEIs.

Understanding student development theories provides invaluable insights for educators and policymakers into students' learning and growth processes (Marchionini, 2010, 2019). These theories provide frameworks for comprehending the students' diverse perspectives, needs, and challenges through their educational journey. Educators can leverage this knowledge to tailor their teaching approaches, curriculum design, and support systems, so as to meet student development needs. For example, Perry's theory informs curricula design and instructional strategies that promote intellectual growth and ethical reasoning. By recognising students' positions, educators can engage them in meaningful discussions, challenge their assumptions, and foster critical thinking skills. Embedding AGPs (Spiliotopoulos et al., 2020) can enhance our understanding of the complexities we face in our education systems (Ozkan et al., 2019). However, modern education approaches have yet to address the needs of the current or future workforce with diverse demographic, behavioural, experiential, and social profiles (Toldson, 2021) in different journey phases. Our method takes a novel approach by proposing a progressive education journey (see Fig. 6.1) modelled on extensive business and marketing literature research (Jansen & Schuster, 2011; Lemon & Verhoef, 2016; Tueanrat et al., 2021) applied to HEIs. The rationale behind using business and marketing literature to model the cycle of lifelong learning arises from the rationale that terminologies such as the personalisation of education are based on marketing theory and are now an emerging "movement" within the field of education. Our education cycle modifies the customer's journey phases to resemble that shown in Fig. 6.2, with stakeholders at and within each stage having unique goals and needs to be represented by personas.

- **Exploration**. Prospective students seek information about colleges and programmes to achieve their educational goals. HEIs can raise students' awareness and interest in their offerings and values.
- **Application**. Prospective students decide to apply and submit the required documents and fees, and HEIs can streamline the application process and provide clear communication and timely feedback.
- **Enrolment**. Prospective students accept admission offers and complete the necessary steps like enrolment, payment of tuition, and course registration. HEIs guide for a smooth transition.
- **Engagement**. Enrolled students participate in various academic and co-curricular activities and experiences, gaining knowledge, skills, and competencies. HEIs

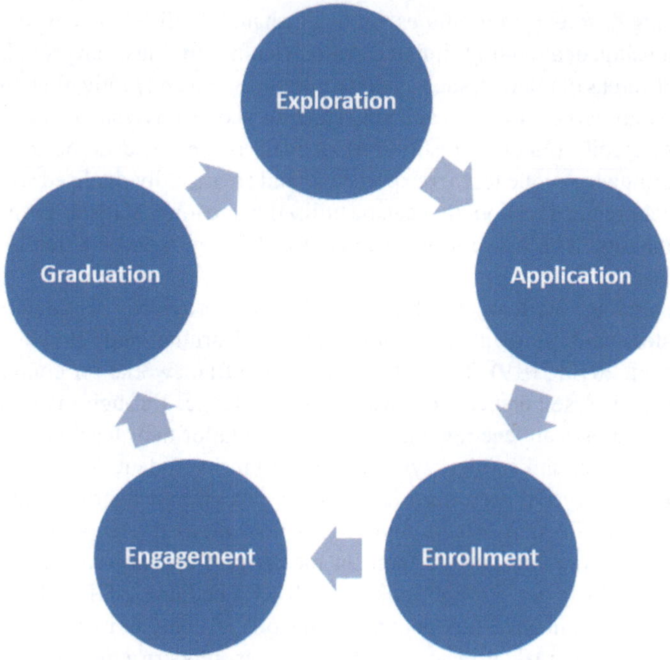

Fig. 6.2 Theoretical model of education cycle presented as a five-stage continuous cycle of lifelong learning

help to foster a sense of belonging and community among students, and provide learning and development opportunities.

- **Graduation**. Students complete their degree requirements, celebrate their academic achievement, and become alumni, joining a network of graduates. HEIs help recognise their accomplishments, and maintain a lifelong relationship with them.

Using AGPs in progressive education can provide valuable insights into the needs and expectations of ASAFS stakeholders. Personas (Nielsen, 2011, 2013a, 2013b; Nielsen & Storgaard Hansen, 2014) are *humanised representations of data in an easily digestible form.* By developing these personas (Grudin, 2006), Qatar's progressive educational HEIs can better understand their stakeholders' diverse needs and preferences, and tailor programmes and services to meet them (Jansen et al., 2021a, 2021b; Salminen et al., 2020b; Salminen, Guan, et al., 2021a, 2021b; Salminen, Jung, et al., 2021a, 2021b; Salminen, Nielsen, et al., 2022a, 2022b). Personas can also help HEIs in Qatar to develop targeted recruitment and retention strategies. For example, personas can shed light on the factors influencing prospective students' decisions to apply and enrol in a programme, and the barriers that may prevent them from doing so. This information can enhance marketing campaigns and support services for different student populations.

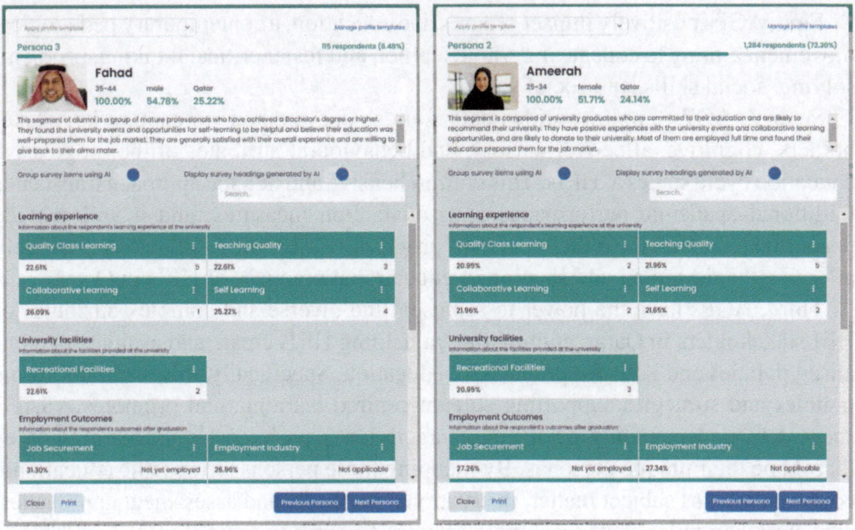

Fig. 6.3 Two AGPs from Survey2Persona of Qatari HEIs Alumni. The names and images are auto-loaded from meta-tagged data. The descriptions are created from Generative AI using survey responses. The survey questions are also grouped and labelled using AI

Similarly, personas can yield insights into the factors that influence alumni engagement, such as the types of activities and events that alumni find most meaningful and the barriers that may prevent them from staying connected with the institution. At the graduation stage, these AGPs extend to the alumni of the progressive education cycle and the related stakeholders. This information can be used to design KPIs for more effective alumni engagement strategies that foster long-term relationships and support the Qatari HEIs institutional goals. Figure 6.3 below illustrates two personas that can be created to assist Qatari HEI decision-makers.

Overall, using AGPs can aid HEIs within Qatar, in promoting more inclusive and equitable educational experiences in progressive education. By understanding different stakeholders' goals, motivations, and behaviours, Qatar HEIs can design or refine more targeted KPIs, programmes, and services that satisfy their specific requirements and preferences, including addressing local cultural needs (Akinci, 2020; Hopkyns, 2020; Ruipérez-Valiente et al., 2020).

6.9 Discussion

The use of AGPs by Qatari HEIs represents a novel, original, creative, and potentially transformative approach in several profound ways.

First, AGPs positively impact progressive education, a contemporary pedagogical movement aiming to educate the whole learner, and foster critical thinking, problem-solving, social skills, and social values.

Second, AGPs employ various data sources and methods to capture key stakeholders' cognitive, affective, sensory, and behavioural attributes of the progressive education cycle stages at HEIs. This comprehensive and holistic approach transcends traditional academic performance and satisfaction measures, and it will provide deeper insights into the effectiveness of progressive education, ultimately positively affecting the designing and shaping of educational practices of HEIs in Qatar.

Third, AGPs have the power to represent the diverse and complex attributes of HEI stakeholders in Qatar, culminating in helping HEIs create and evaluate personalised policies and KPIs for progressive education. Specifically, this method informs policies and strategies supporting student-centred learning, and promotes ASAFS success, by understanding the diverse characteristics, preferences, and learning styles that shape their unique personas. By mapping these personas to specific educational contexts, such as subject matter, instructional methods, and assessment approaches, educators can tailor their teaching strategies to cater to their students' needs and learning preferences.

Fourth, the use of AGPs builds on the existing literature and theories of progressive education, persona generation, and human–AI interaction (An et al., 2018; Jansen et al., 2021a, 2021b; Salminen, Guan, et al., 2021a, 2021b; Salminen, Jung, et al., 2022a, 2022b) but also contributes fresh insights and frameworks that can advance the fields of education, computer science, and psychology. This research can potentially transform how HEIs in Qatar design and implement educational programmes to enhance stakeholder learning outcomes and overall well-being.

6.9.1 Implications

This chapter proposes and endorses an exciting opportunity to advance the field of persona usage in progressive education, by applying AI and ML techniques to generate personas of ASAFS stakeholders at HEIs in Qatar. AGPs can address stakeholders' dynamic and diverse needs, enriching policymakers with a holistic understanding of the key players' preferences in various stages of the progressive education cycle within HEIs in Qatar. This understanding empowers policymakers to set and evaluate KPIs that align with the stakeholders' needs (Chandran et al., 2023). AGPs also set several expected scientific outcomes, contributing to theory and practice by improving educational quality and designing effective policies using AGPs. These implications of using AGPs are listed below in the Expected Outcomes (EO):

- **EO1**: A comprehensive and holistic understanding of the diverse and complex stakeholders' attributes in the progressive education cycle stages at HEIs in Qatar.
- **EO2**: AGPs representing realistic HEI stakeholders in Qatar, using state-of-the-art AI and ML techniques in an HCI, human-centred representation.

- **EO3**: Insights into how AGPs of critical stakeholders can inform and design policies, pedagogy, assessment practices, and user-centric curricula for progressive education at HEIs in Qatar.
- **EO4**: A significant contribution to the theory and practice of progressive education, providing evidence-based and user-centred tools and methods for improving the quality and impact of education for learners, educators, administrators, and policymakers.

This call to action for using AGPs aligns with Qatar's vision of empowering the next generation of scientists, by providing them with progressive education for their physical, emotional, and intellectual growth, through its Research, Development and Innovation (RDI) ecosystem[3]. The ASAFS stakeholders are crucial to Qatari HEIs regarding financial support, recruiting, and classroom engagement. The RDI ecosystem aims to enhance Qatar's ability to address challenges and seize opportunities with innovative and progressive solutions based on innovative research and development. Developing AGPs of the stakeholders of HEIs in Qatar will significantly contribute to Qatar's RDI ecosystem by:

- **RDIC1**: Identifying the gaps in the education system and the ASAFS stakeholders' perspectives, enabling the development of personalised strategies to address these challenges.
- **RDIC2**: Providing a user-centred perspective that guides the design and development of RDI solutions, to address the needs and challenges of different sectors and domains with HEIs.
- **RDIC3**: Enhancing the communication and collaboration among RDI stakeholders across government, industry, and academia, by creating a common language and understanding of ASAFS user groups by aligning with HEI strategy planning.
- **RDIC4**: Facilitating the evaluation and improvement of RDI outcomes and impacts by measuring feedback and engagement through shared empathic and humanised representation of HEI stakeholders.
- **RDIC5**: Enabling the identification and exploration of new opportunities and markets for HEI products and services that align with the national vision and goals.

AGPs in progressive education at HEIs in Qatar present an unprecedented opportunity to understand key ASAFS stakeholders' needs, enabling evidence-based strategies for designing user-centric policies and practices. By aligning with the national priority to develop, adapt, and deploy new digital technologies to increase the growth and competitiveness of Qatar's vital economic sectors, this research presents practical solutions to national challenges.

[3] https://qrdi.org.qa/

6.9.2 Policy Implications

Some critical policy implications for the use of AGPs by HEIs are:

- **Embrace Data-Driven Approaches**: HEIs in Qatar should formally integrate AGPs into their institutional policy framework. These personas can provide insights into various stakeholders' needs and behaviours, and they can become an essential tool in the strategic planning and decision-making processes, ensuring that policies are aligned with the actual needs of the ASAFS community.
- **Strengthening Alumni Engagement**: Given the emphasised role of alumni in the abstract, HEIs should create policies that bolster alumni engagement. This could involve creating alumni-centric platforms or programmes that enable continuous feedback, or setting up dedicated alumni relations departments, to maintain close ties and facilitate a "beyond education" experience.
- **Investment in Technology and Training**: Recognising the importance of AGPs suggests a need for policies prioritising investments in technology (e.g., ML and LLMs) and the training required for staff to utilise and interpret AGP insights effectively. Ensuring that personnel can harness the full potential of these tools will be crucial for their effective implementation.
- **Inclusive and Learner-Centred Curriculum Design**: As AGPs can aid in understanding the needs and challenges faced by various stakeholders, HEIs should adopt policies that encourage a curriculum design process that is more inclusive and learner-centred. This could involve regularly reviewing and updating the curriculum based on AGP insights and ensuring diverse stakeholder representation in curriculum development committees.
- **Expanding Research and Development in AGPs**: Recognising the unexplored potential of AGPs in higher education, HEIs should consider implementing policies that encourage research and development in this area. This could involve setting up dedicated research centres or funding initiatives, fostering collaborations with technology and data science departments, or encouraging interdisciplinary research that bridges the gap between pedagogy and technology.

Incorporating these policy implications would assist HEIs in Qatar in staying ahead of the curve, ensuring that they remain responsive to stakeholder needs, and provide a progressive, inclusive, forward-thinking education experience.

6.9.3 Conclusion

Using personas, specifically AGPs, in education, offers a powerful and transformative approach to addressing the diverse and evolving needs of learners, educators, and communities in Qatar's HEIs. By creating these fictional representations of various user groups, educational stakeholders can gain invaluable insights, fostering empathy, personalisation, and inclusivity within the education system.

The implications of utilising personas in Qatar's education sector are profound and far-reaching. By embracing this approach, Qatari educational institutions can proactively respond to their diverse student body's unique cultural, linguistic, and learning preferences, ensuring an inclusive and supportive learning environment for all. Customised educational programmes, tailored to the needs and goals of specific personas, will lead to more engaged and successful students.

Moreover, the integration of personas can guide the strategic use of educational technology, aligning it with students' and educators' technological literacy and preferences. This harmonious blending of traditional pedagogies and innovative digital tools will foster a future-ready generation prepared to navigate Qatar's technology-driven landscape.

As Qatar aspires to evolve into a knowledge-based economy and achieve its ambitious National Vision 2030, the role of personas in education becomes even more critical. By investing in continuing education and professional development, customised to meet the aspirations of educators and their personas, the nation can cultivate a proficient and empowered teaching workforce.

The implications of using personas in education extend beyond the classroom, fostering stronger connections with parents and communities. By understanding and addressing their expectations, the education system can build robust partnerships, fostering a collective commitment to the success of Qatar's students.

In conclusion, integrating personas in the education workflows of HEIs offers Qatar a roadmap to a more inclusive, adaptable, and student-centric education system. By leveraging this powerful tool, Qatar can nurture a diverse and empowered generation equipped with the skills, knowledge, and empathy to thrive in an interconnected global society. As the nation embarks on its journey towards educational excellence, embracing personas will undoubtedly contribute to realising its ambitious aspirations and transforming its education landscape.

Bibliography

Adlin, T., Pruitt, J., Goodwin, K., Hynes, C., McGrane, K., Rosenstein, A., & Muller, M. J. (2006). Putting personas to work. *CHI'06 Extended Abstracts on Human Factors in Computing Systems*, 13–16.

Adlin, T., & Pruitt, J. (2010). *The essential persona lifecycle: Your guide to building and using personas*. Morgan Kaufmann.

Akinci, I. (2020). Culture in the 'politics of identity': Conceptions of national identity and citizenship among second-generation non-Gulf Arab migrants in Dubai. *Journal of Ethnic and Migration Studies, 46*(11), 2309–2325.

Ali Amer Jid Almahri, F., Bell, D., & Arzoky, M. (2019). Personas design for conversational systems in education. *Informatics, 6* (4), 46.

Alves, H., Mainardes, E. W., & Raposo, M. (2010). A relationship approach to higher education institution stakeholder management. *Tertiary Education and Management, 16*, 159–181.

An, J., Kwak, H., Jung, S., Salminen, J., Admad, M., & Jansen, B. (2018). Imaginary people representing real numbers: Generating personas from online social media data. *ACM Transactions on the Web (TWEB), 12*(4), 1–26.

Asia, S., & Lanka, S. (2023). *The World Fact Book*. Central Intelligence Agency.
Asquith, C. (2006). Accepted into education city. *Diverse Issues in Higher Education, 23*(8), 22.
Ben Hassen, T. (2021). The state of the knowledge-based economy in the Arab world: Cases of Qatar and Lebanon. *EuroMed Journal of Business, 16*(2), 129–153.
Berrebi, C., Martorell, F., & Tanner, J. C. (2009). Qatar's labor markets at a crucial crossroad. *The Middle East Journal, 63*(3), 421–442.
Bhutoria, A. (2022). Personalized education and artificial intelligence in the United States, China, and India: A systematic review using a human-in-the-loop model. *Computers and Education: Artificial Intelligence, 3*, 100068.
Chandran, V., Fadia, A., Isherwood, T., Shah, N., & Soni, K. (2023). *The state of AI in GCC countries—and how to overcome adoption challenges | McKinsey*. https://www.mckinsey.com/capabilities/mckinsey-digital/our-insights/the-state-of-ai-in-gcc-countries-and-how-to-overcome-adoption-challenges
Coleman, I. (2004). The payoff from women's rights. *Foreign Affairs, 83*, 80.
Cunningham, D. J. (1992). Beyond educational psychology: Steps toward an educational semiotic. *Educational Psychology Review, 4*, 165–194.
Davies, S. (2002). The paradox of progressive education: a frame analysis. *Sociology of Education*, 269–286.
Dittmar, A., & Forbrig, P. (2019). Integrating Personas and Use Case Models. In D. Lamas, F. Loizides, L. Nacke, H. Petrie, M. Winckler, & P. Zaphiris (Eds.), *Human–Computer Interaction—INTERACT 2019* (Vol. 11746, pp. 666–686). Springer International Publishing. https://doi.org/10.1007/978-3-030-29381-9_40
Eggeling, K. A. (2020). Cultural diplomacy in Qatar: between 'virtual enlargement', national identity construction and elite legitimation. In *Cultural Diplomacy and International Cultural Relations: Volume I* (pp. 59–73). Routledge.
Ekşioğlu, M., Güler, H., Terzi, F., Yıldırım, H. S., & Yücel, B. (2018). UXD for a prototype campus information Kiosk. In: *proceedings of the 2nd International Conference on Computer-Human Interaction Research and Applications*, 176–182. https://doi.org/10.5220/0007229801760182
Evans, N., Forney, D., Guido, F., Patton, L., & Renn, K. (2010). Perry's theory of intellectual and ethical development. In: *Student Development in College: Theory, Research, and Practice*, 82–98.
Falqueto, J. M. Z., Hoffmann, V. E., Gomes, R. C., & Onoyama Mori, S. S. (2020). Strategic planning in higher education institutions: What are the stakeholders' roles in the process? *Higher Education, 79*, 1039–1056.
Ferrero-Ferrero, I., Fernández-Izquierdo, M. Á., Muñoz-Torres, M. J., & Bellés-Colomer, L. (2018). Stakeholder engagement in sustainability reporting in higher education: An analysis of key internal stakeholders' expectations. *International Journal of Sustainability in Higher Education, 19*(2), 313–336.
Finn, J. D., & Zimmer, K. S. (2012). Student engagement: What is it? Why does it matter? *Handbook of Research on Student Engagement*, 97–131.
Garces-Bacsal, R. M., Tupas, R., Alhosani, N. M., & Elhoweris, H. (2021). Teachers' perceptions of diversity and 'others' in United Arab Emirates (UAE) Schools. *Pedagogy, Culture & Society*, 1–19. https://doi.org/10.1080/14681366.2021.2011774
Grace, G. R., & Grace, G. (1995). *School leadership: Beyond education management: An essay in policy scholarship*. Psychology Press.
Greer, J. E., Frost, S., Banow, R., Thompson, C., Kuleza, S., Wilson, K., & Koehn, G. (2015). The student advice recommender agent: SARA. *International Workshop on Personalization Approaches in Learning Environments*.
Groccia, J. E. (2018). What is student engagement? *New Directions for Teaching and Learning, 2018*(154), 11–20.
Grudin, J. (2006). *Why Personas Work: the Psychological Evidence. The persona lifecycle: keeping people in mind throughout the product design*. Morgan Kaufmann.

Gu, M. M., & Lee, J.C.-K. (2019). "They lost internationalization in pursuit of internationalization": Students' language practices and identity construction in a cross-disciplinary EMI program in a university in China. *Higher Education, 78*, 389–405.

Hamdan Alghamdi, A. K. (2014). The road to culturally relevant pedagogy: Expatriate teachers' pedagogical practices in the cultural context of Saudi Arabian higher education. *McGill Journal of Education, 49*(1), 201–226.

Hemzo, M. A. (2023). Defining the Persona: Segmentation and Targeting. In *Marketing Luxury Services: Concepts, Strategy, and Practice* (pp. 75–94). Springer.

Hopkyns, S. (2020). *The impact of global English on cultural identities in the United Arab Emirates: Wanted not welcome*. Routledge.

Howlett, J. (2013). *Progressive education: a critical introduction*. A & C Black.

Ibnouf, A., Dou, L., & Knight, J. (2013). The evolution of Qatar as an education hub: moving to a knowledge-based economy. In *International education hubs: Student, talent, knowledge-innovation models* (pp. 43–61). Springer.

Inglehart, R., & Norris, P. (2003). *Rising tide: gender equality and cultural change around the world*. Cambridge University Press. https://books.google.com/books?hl=en&lr=&id=Gzvt11ztO7IC&oi=fnd&pg=PR7&dq=Qatar+has+been+actively+working+towards+empowering+women+in+education+and+promoting+gender+equality+through+measures+such+as+increasing+access+to+education+and+opportunities+for+women+and+girls&ots=oe2KKdSB8M&sig=Jfu8ahe48rgyY7O7oAkUwR9-LF8

Jansen, B. J., Jung, S., & Salminen, J. (2021). Making better decisions with big data personas | MIT Technology Review. *MIT Technology Review*. https://www.technologyreview.com/2021/03/11/1020207/making-better-decisions-with-big-data-personas/

Jansen, B. J., Salminen, J. O., & Jung, S. (2020). Data-driven personas for enhanced user understanding: Combining empathy with rationality for better insights to analytics. *Data and Information Management, 4*(1), 1–17. https://doi.org/10.2478/dim-2020-0005

Jansen, B., Salminen, J., Jung, S., & Guan, K. (2021b). Data-driven personas. *Synthesis Lectures on Human-Centered Informatics, 14*(1), i–317.

Jansen, B. J., & Schuster, S. (2011). Bidding on the buying funnel for sponsored search and keyword advertising. *Journal of Electronic Commerce Research, 12*(1), 1.

Jiang, B., Zhou, W., Yang, J., Yang, C., Wang, S., Pang, L., Scott, D., Bel, N., & Zong, C. (2020). Pednet: a persona enhanced dual alternating learning network for conversational response generation. *Proceedings of the 28th International Conference on Computational Linguistics*, 4089–4099. https://doi.org/10.18653/v1/2020.coling-main.361

Kegan, R. (1998). *In over our heads: The mental demands of modern life*. Harvard University Press.

Khodr, H. (2011). The dynamics of international education in Qatar: Exploring the policy drivers behind the development of Education City. *Journal of Emerging Trends in Educational Research and Policy Studies, 2*(6), 514–525.

Knight, J. (2017). Education hubs: international, regional and local dimensions of scale and scope. In *Space, Place and Scale in the Study of Education* (pp. 118–131). Routledge.

Laato, S., Farooq, A., Vilppu, H., Airola, A., & Murtonen, M. (2022). Higher education during lockdown: Literature review and implications on technology design. *Education Research International, 2022*, e7201043. https://doi.org/10.1155/2022/7201043

Lemon, K. N., & Verhoef, P. C. (2016). Understanding customer experience throughout the customer journey. *Journal of Marketing, 80*(6), 69–96.

Levi, M. D., & Marchionini, G. (2019). Fruitful collaborations: integrating research and practice. In *Human-Centered Computing* (pp. 799–803). CRC Press.

Lilley, M., Pyper, A., & Attwood, S. (2012). Understanding the Student Experience through the Use of Personas. *Innovation in Teaching and Learning in Information and Computer Sciences, 11* (1), 4–13. https://doi.org/10.11120/ital.2012.11010004

Logermann, F., & Leišytė, L. (2015). Students as Stakeholders in the policy context of the European Standards and Guidelines for Quality Assurance in Higher Education Institutions. *The European Higher Education Area: Between Critical Reflections and Future Policies*, 685–701.

Marchionini, G. (2010). The many meanings of information. In *Information Concepts: From Books to Cyberspace Identities* (pp. 1–9). Springer.
Marchionini, G. (2019). Search, sense making and learning: Closing gaps. *Information and Learning Sciences, 120*(1/2), 74–86.
Matzarakis, A., & Fröhlich, D. (2015). Sport events and climate for visitors—The case of FIFA World Cup in Qatar 2022. *International Journal of Biometeorology, 59*, 481–486.
McNear, D. B. (1978). A critical examination of definitions for progressive education. *The Clearing House, 52*(1), 37–40.
Meyerhoff, E. (2019). *Beyond education: Radical studying for another world*. U of Minnesota Press.
Mitchell, J. S., Paschyn, C., Mir, S., Pike, K., & Kane, T. (2015). In majaalis al-hareem: The complex professional and personal choices of Qatari women. *DIFI Family Research and Proceedings, 4*, 1–12.
Nielsen, L. (2011). Personas in co-creation and co-design. *Proceedings of the 11th Danish Human-Computer Interaction Research Symposium (DHRS2011)*, 38–40.
Nielsen, L. (2013b). *Personas—user focused design* (Vol. 1373). Springer.
Nielsen, L. (2013a). Personas. *The Encyclopedia of Human-Computer Interaction, 2.*
Nielsen, L., & Storgaard Hansen, K. (2014). Personas is applicable: A study on the use of personas in Denmark. In *Proceedings of the SIGCHI Conference on Human Factors in Computing Systems*, 1665–1674.
Nielsen, L., Hansen, K. S., Stage, J., & Billestrup, J. (2015). A template for design personas: Analysis of 47 persona descriptions from Danish industries and organizations. *International Journal of Sociotechnology and Knowledge Development (IJSKD), 7*(1), 45–61.
Ozkan, D. S., Reeping, D., McNair, L. D., Martin, T. L., Harrison, S., Lester, L., Knapp, B., Wisnioski, M., Patrick, A., & Baum, L. (2019). Using personas as curricular design tools: Engaging the boundaries of engineering culture. *IEEE Frontiers in Education Conference (FIE), 2019*, 1–7.
Park, D., & Kang, J. (2022). Constructing data-driven personas through an analysis of mobile application store data. *Applied Sciences, 12*(6), 2869.
Parsons, J., & Taylor, L. (2011). Improving student engagement. *Current Issues in Education, 14* (1).
Patrick, S., Kennedy, K., & Powell, A. (2013). Mean what you say: defining and integrating personalized, blended and competency education. *International Association for K-12 Online Learning*. https://eric.ed.gov/?id=ED561301
Patton, L. D., Renn, K. A., Guido, F. M., & Quaye, S. J. (2016). *Student development in college: Theory, research, and practice*. John Wiley & Sons.
Peters, M., Simo, L. P., Amatller, A. M., Linan, L. C., & Kreitz, P. B. (2021). Toward a user-centred design approach for AR technologies in online higher education. In *2021 7th International Conference of the Immersive Learning Research Network (iLRN)*, 1–8. https://doi.org/10.23919/iLRN52045.2021.9459409
Peterson, J. E. (2006). Qatar and the world: branding for a micro-state. *The Middle East Journal*, 732–748.
Pruitt, J., & Grudin, J. (2003). Personas: practice and theory. *Proceedings of the 2003 Conference on Designing for User Experiences*, 1–15. https://doi.org/10.1145/997078.997089
Pruitt, J., & Adlin, T. (2010). *The persona lifecycle: Keeping people in mind throughout product design*. Elsevier.
Reese, W. J. (2001). The origins of progressive education. *History of Education Quarterly, 41*(1), 1–24.
Rostron, M. (2009). Liberal arts education in Qatar: Intercultural perspectives. *Intercultural Education, 20*(3), 219–229.
Ruipérez-Valiente, J. A., Halawa, S., Slama, R., & Reich, J. (2020). Using multi-platform learning analytics to compare regional and global MOOC learning in the Arab world. *Computers & Education, 146*, 103776.

Salminen, J., Jung, S., Chowdhury, S., Sengün, S., & Jansen, B. J. (2020b). Personas and analytics: a comparative user study of efficiency and effectiveness for a user identification task. In *Proceedings of the 2020 CHI Conference on Human Factors in Computing Systems*, 1–13.

Salminen, J., Jung, S., Chowdhury, S., Sengün, S., & Jansen, B. J. (2020a). Personas and analytics: a comparative user study of efficiency and effectiveness for a user identification task. In *Proceedings of the 2020 CHI Conference on Human Factors in Computing Systems*, 1–13.

Salminen, J., Guan, K., Jung, S.-G., & Jansen, B. J. (2021). A survey of 15 years of data-driven persona development. *International Journal of Human–Computer Interaction, 37*(18), 1685–1708.

Salminen, J., Jung, S.-G., & Jansen, B. (2022). Developing persona analytics towards persona science. In *27th International Conference on Intelligent User Interfaces*, 323–344.

Salminen, J., Nielsen, L., Bahloul, M., Jørgensen, R. G., Santos, J. M., Jung, S.-G., & Jansen, B. J. (2022). Persona preparedness: a survey instrument for measuring the organizational readiness for deploying personas. *Information Technology and Management*, 1–26.

Salminen, J., Jung, S.-G., Chowdhury, S., Robillos, D. R., & Jansen, B. (2021b). The ability of personas: An empirical evaluation of altering incorrect preconceptions about users. *International Journal of Human–Computer Studies, 153*, 102645.

Sim, G., Shrivastava, A., Horton, M., Agarwal, S., Haasini, P. S., Kondeti, C. S., & McKnight, L. (2019). Child-Generated Personas to Aid Design Across Cultures. In D. Lamas, F. Loizides, L. Nacke, H. Petrie, M. Winckler, & P. Zaphiris (Eds.), *Human–Computer Interaction—INTERACT 2019* (Vol. 11748, pp. 112–131). Springer International Publishing. https://doi.org/10.1007/978-3-030-29387-1_7

Spiliotopoulos, D., Margaris, D., & Vassilakis, C. (2020). Data-assisted persona construction using social media data. *Big Data and Cognitive Computing, 4*(3), 21.

Sun, N., & Rosson, M. B. (2017). Connection Enablers in Online Learning Community: from Informative Online Personae to Meaningful Social Space. In *Companion of the 2017 ACM Conference on Computer Supported Cooperative Work and Social Computing*, 315–318. https://doi.org/10.1145/3022198.3026348

Tippett, T. P., & Lee, J. J. (2019). Looking back to move forward: understanding progressive education in the 21st century. *Journal of Applied Learning in Higher Education, 8*, 79–98.

Toldson, I. (2021). *The Post-Reform Education Era: Mitigating Institutional Racism in Education at the Human-Technology Frontier*. The QEM Network, Washington, D.C. https://qem.org/events/mitigating-institutional-racism-in-education-at-the-human-technology-frontier/

Trowler, V. (2010). Student engagement literature review. *The Higher Education Academy, 11*(1), 1–15.

Tueanrat, Y., Papagiannidis, S., & Alamanos, E. (2021). Going on a journey: A review of the customer journey literature. *Journal of Business Research, 125*, 336–353.

Volden, F., & Wattne, O. E. (2023). Wayfinding and navigation in the outdoors: quantitative and data driven development of personas and requirements for wayfinding in nature. In *International Conference on Human–Computer Interaction*, 199–210.

Waheed, B., Khan, F. I., & Veitch, B. (2011). Developing a quantitative tool for sustainability assessment of HEIs. In *International Journal of Sustainability in Higher Education.*

Walsh, P. (2019). Establishment of an American Branch-campus model of higher education: Qatar's early goals, rationales, and challenges. *Athens Journal of Education, 64*(4), 271–289.

Wang, K., Li, B., Tian, T., Zakuan, N., & Rani, P. (2023). Evaluate the drivers for digital transformation in higher education institutions in the era of industry 4.0 based on decision-making method. *Journal of Innovation & Knowledge, 8* (3), 100364.

Weber, A. S., & City, E. (2012). Inclusive education in the gulf cooperation council. *Journal of Educational and Instructional Studies in the World, 2*(2), 85–97.

Yang, S., Tian, H., Sun, L., & Yu, X. (2019). From one-size-fits-all teaching to adaptive learning: the crisis and solution of education in the era of AI. *Journal of Physics: Conference Series, 1237* (4), 042039. https://iopscience.iop.org/article/https://doi.org/10.1088/1742-6596/1237/4/042039/meta

Ylimaki, R. M. (2011). *Critical curriculum leadership: A framework for progressive education*. Routledge.
Zahlan, R. S. (2016). *The creation of Qatar*. Routledge.

Chapter 7
Reinventing Education Strategic Planning and Policymaking Through the Lenses of Futures Thinking

Fawaz Hazza Abu Sitta

Abstract Educational Institutes plan their strategies and policies on the assumption that tomorrow mirrors today and yesterday, relying heavily on historical data to forecast the future. This approach ignores weak signals and wild cards, limiting their responsiveness to disruptive innovations. Furthermore, these entities operate under the notion of a certain future, with a single strategic plan and no consideration of alternative scenarios. Conversely, Strategic Futures Thinking employs diverse techniques and methods to anticipate plausible discontinuities and surprises. This chapter discusses the core differences between Futures Thinking and Strategic Planning, emphasising that integrating them enhances agility and resilience.

7.1 Introduction

Navigating the futures of educational institutes in an era of unprecedented technological advancements and innovative disruption requires a creative Futures Thinking approach. Organisations develop their strategic plan through the lenses of forecasting and trend analysis (Conway 2016). Although these methods offer valuable insights into the present, they fall short of projecting the major dynamic changes and radical shifts of what lies ahead. To better anticipate disruptions and radical changes, educational institutions should integrate and embrace Futures Thinking methods into their strategic planning and policy development efforts. Methods such as Horizon Scanning, Systems Thinking, and Scenario Planning are essential for promoting agility and resilience.

Educational institutions in the gulf region develop and/or revise their strategic directions in response to mega-trends and trends rather than exploring the impact of weak signals and wild cards. It is essential for those institutes to conduct and explore the plausible impact of newly emerging weak signals and wild cards, to better

---

F. H. A. Sitta (✉)
UNESCO Chair in Anticipatory, Foresight Research Director at the Center for Futures Studies, University of Dubai, Dubai, United Arab Emirates
e-mail: fawazabusitta@ud.ac.ae

E. Zaidan et al. (eds.), *The Future of Education Policy in the State of Qatar*, Gulf Studies 21, https://doi.org/10.1007/978-981-97-9667-0_7

anticipate the future, and build an education system that is adaptive and resilient in encountering futures discontinuity and supersizes.

It is recommended that educational institutes and the ministry of education in Qatar adopt Futures Thinking principles and models while developing their strategic plans and articulating the education policy, to promote their agility and resilience.

This chapter aims to explore the core differences between strategic planning and Futures Thinking. The objective is to emphasise the importance of incorporating Futures Thinking as a core agenda in strategic planning and policymaking. The main activities and models of Futures Thinking will be discussed. In addition to that, for illustrative purposes, a number of weak signals and wild cards that could influence the future of learning in the region will be presented. Although the list is not comprehensive, it highlights how plausible emerging changes are frequently neglected in the development of strategic plans and policymaking.

7.2 Futures Thinking Versus Strategic Planning

The aim of Futures Thinking is not to predict the future but rather to influence it. Education institutions in the region update their strategies to enhance the education sector in the short term (3–5 years) (Qatari Ministry of Higher Education, 2018; The UAE Ministry of Education, 2017; The KSA Ministry of Education, 2020; The Sultanate of Oman Education Council, 2018). They use past performance and historical data to forecast the future. They conduct trend analysis to identify key trends and mega-trends that might influence their future. Strategic planning relies on the fact that these institutes base their forecasting on empirical and historical data to understand the patterns of change.

This approach is excellent for environments that are stable and not facing radical changes. However, education systems today encounter radical changes and disruption, such as the one we have witnessed during the COVID-19 Pandemic, and more recently the disruption created by Generative AI. Educational institutes need to adopt new methods to anticipate uncertainties. Futures Thinking as a methodology becomes an essential tool for these institutes to anticipate future discontinuity and surprises. Furthermore, Futures Thinking is not merely about building adaptive and resilient educational institutions, but it also supports educational institutes' quest to shape and design their own futures.

The core differences between strategic planning and Futures Thinking can be summarised in the following points:

1- Strategic planning focuses on the trends while Futures Thinking uses horizon scanning to investigate and explore weak signals and wild cards.
2- Strategic planning focuses on linear thinking: today and tomorrow are reflections of yesterday with small variations. The Futures Thinking approach changes into nonlinear thinking and uses systematic thinking as a value to anticipate future changes.

3- Strategic thinking bases their forecasting on empirical data alone, while Futures Thinking integrates both data, imaginations, and creative thinking in the process, in an effort to transform organisations from data driven to knowledge and immigration-driven organisations.

Consequently, educational institutions in the region and globally should infuse Futures Thinking into their strategic planning efforts, to build education systems to anticipate future discontinuity and disruptions. UNESCO has established chairs for anticipatory systems and Futures Thinking to promote future literacy (UNESCO, 2012).

7.3 Futures Thinking Approach

Integrating Futures Thinking methods and techniques into the strategic planning effort enriches and enhances the agility and adaptability of educational institutes (Hines & Bishop, 2015). Policymakers and strategists should consider incorporating Futures Thinking and horizon scanning, while planning for the promotion of creativity and resilience in the education sector, since organisations that adopt strategic foresight tend to outperform their competitors (Conway, 2017).

It has been observed that a limited number of Futures Thinking methods have been considered or adopted while developing education policies and strategies in the region. The core components of Futures Thinking methods will be discussed in this section. The methods and models presented in this section are not comprehensive; they are provided for illustrative purposes. The aim of this section is also to emphasise the significance of integrating Futures Thinking models into strategic planning efforts.

The framework discussed in this section has been adopted at various educational institutes such as the University of Dubai, Rochester Institute of Technology—Dubai Campus, University of Cambridge Hughes Hall, and Cambridge University Press and Assessment. In addition to that, this approach has been implemented for several government entities in the UAE and was presented in the Empowering Futures—Long-Term Governance, Democracy and Futures Research Conference in Finland 14–16 June 2023 (Alkhoori & Abu Sitta, 2023) and the UNESCO-PMU Symposium 2023: Towards Evaluation Frameworks for Futures Literacy and Foresight in Paris 20–21 July 2023 (Abu Sitta, 2023).

The Futures Thinking framework consists of six phases: Framing, Understanding, Exploring, Realising, Designing, and Shaping, as depicted in Fig. 7.1. This framework is drawn from various foresight and Futures Thinking models. In the following sections, a detailed description of each phase will be discussed:

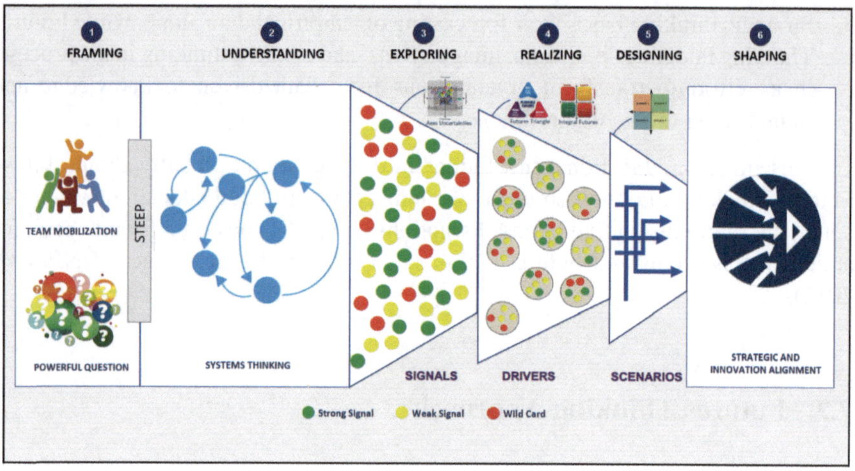

Fig. 7.1 Illustrates the essential phases of Futures Thinking

7.3.1 Framing

The steps undertaken in this phase are partially derived from the Houston Foresight Framework (Hines & Bishop, 2013). Two primary steps executed in this phase are team mobilisation and articulating the domain question. Team mobilisation entails two pivotal steps: first, establishing the team, and second, outlining the scope of the study (or project). A detailed project plan is crafted at this stage, identifying, and assigning the types of tools and models that will be used.

The second step in this phase focuses on articulating the domain question (future question). This step is essential since it focuses on eliminating cognitive biases. The aim of this step is to develop a domain question, while imagining and immersing ourselves in the future. The core objective is to articulate a question that does not carry any assumptions from today. For example, if the study is focusing on the possible futures of education, the domain question should not be written in a format that includes today's perspective; rather it should promote curiosity and it should not be biassed for or against a certain image. Instead of asking "what is the future of schools in 2050?" it would be more stimulating if the question is presented in the context of "How will knowledge be constructed by 2050?"

It is essential that articulating the domain question is achieved by conducting a 2-h workshop. In the workshop, participants are introduced to the principles of Futures Thinking and the core differences between Futures Thinking and strategic planning are highlighted in the workshop. Participants will also be trained in the art of articulating powerful questions.

In strategic planning, usually, the Framing phase is conducted differently from that above. Here, the focus is more towards mobilising the team and scoping the project. Limited time is spent on articulating the domain questions, and in eliminating cognitive biases. The core attention here is focused on what the scope of the strategy is, and how the strategy will be developed from a project management perspective. In addition to that, the approach and the models for strategy development are presented at this stage. Even though strategic planning envisions the future (the next 3–5 years), yet it assumes that tomorrow is a continuation of yesterday, which creates a major limitation in establishing organisations that are adaptive to radical shifts in their environment.

7.3.2 Understanding

This phase focuses on promoting nonlinear thinking. Firstly, key trends and issues impacting this domain are identified and documented. Following that, thorough research is conducted to pinpoint and explore the primary projections affecting this domain. Additionally, the strategies of competitors and other organisations are examined at this stage. The objective of these steps is to form a robust understanding of the domain question.

The second step at this stage is to draw a domain map. This step serves as a valuable tool for multiple purposes, offering numerous benefits. By providing a structured overview of the domain, it details its vital elements, stakeholders, trends, challenges, and opportunities. This visual representation not only aids individuals in comprehending and navigating the domain, but also facilitates a clearer understanding of its complex details. A pivotal benefit of this mapping is its ability to illustrate the relationships and interdependencies between various elements within the domain under study. This insight proves invaluable for foresight practitioners, allowing them to pinpoint potential impact areas, areas of leverage, and concerns, more effectively. Furthermore, when investigating scenario planning, a domain map becomes a foundational tool, ensuring the development of plausible future scenarios through a holistic view of the system. Another notable benefit is its utility as an efficient stakeholder engagement tool. By presenting a clear picture of the domain, it ensures that stakeholders can contribute their perspectives more effectively, and discern where they fit within the broader scheme, fostering collaboration and alignment.

It is recommended that the team involved in the domain mapping step is to undertake systems thinking and complexity management training. It is essential for the team to learn the systems thinking archetypes, and to map systems in a nonlinear manner.

7.3.3 Exploring

The focus here is on conducting horizon scanning to explore weak signals and wild cards that could shape the future of education. The team will not base their research and investigation solely on major trends, but will also emphasise emerging issues.

Moreover, researching the social, technological, economic, environmental, and political (STEEP) domains is crucial for broadening the scope of the research. Scanning hits (a hit refers to a piece of news or information about a specific emerging issue or change) that could evolve into trends in the future are examined.

These hits are categorised under key signals, and then mapped individually in a 2 × 2 axis of uncertainty. Each signal's impact and plausibility are assessed and mapped in the 2 × 2 matrix (Fig. 7.2). The mapping is completed by the team collectively. The aim of this mapping is to identify the signals that must be included in the strategy, and the signals that should be considered in the scenario development. Signals that are predictable and with low impact should be parked while signals that are uncertain with low impact will be mapped in the monitor quadrant. For these two quadrants, no further action is required. While signals that fall in the Act or Focus quadrant should be considered in strategic planning, and action must be in place to address them.

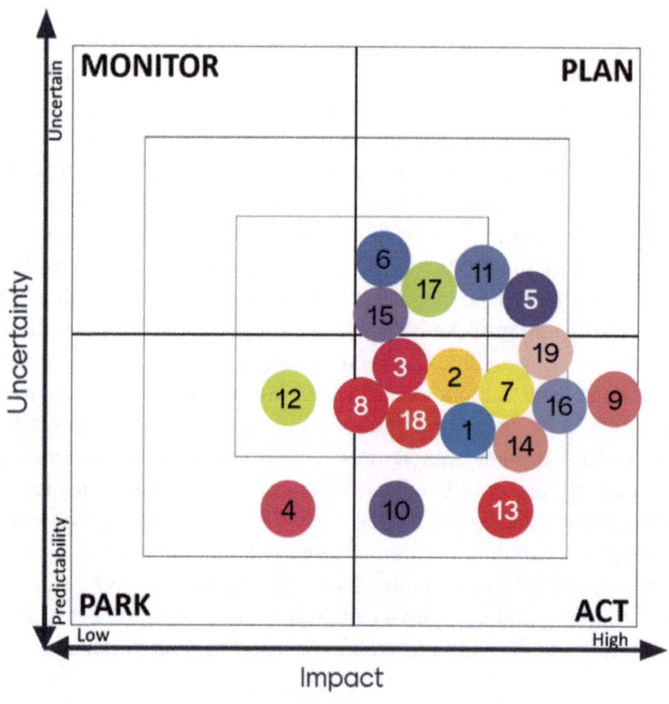

Fig. 7.2 Axis of Uncertainty

7.3.4 Realising

Strategic planning efforts usually focus on the explicit world, and rarely investigate the underlying causes of change. The core objective of this phase is to investigate the underlying causes of change, and to understand the forces of change. Two methods will be discussed in this section. The first is the Futures Triangle, and the second is Causal Layered Analysis (CLA) (Inayatullah, 2008).

The Futures Triangle is a strategic tool employed in foresight and futures studies, offering a comprehensive visualisation of the intricate interplay between the push of the present, the pull of the future, and the weight of the past. At the foundation of this model lies the push of the present, encompassing current trends and driving forces such as technological advancements, societal shifts, and environmental changes, propelling us into the unfolding future. Ascending the triangle, the pull of the future emerges, representing aspirational visions or desired outcomes that beckon society towards particular trajectories. However, embedded within this dynamic is the weight of the past, characterised by legacies, traditions, and historical barriers. These elements, deeply rooted in our collective memory, can sometimes act as impediments, anchoring systems or mindsets in outdated frameworks. Through the lens of the Futures Triangle, foresight practitioners gain the ability to holistically assess these forces, fostering the creation of robust future scenarios and strategies. This understanding empowers organisations and individuals to adeptly navigate potential challenges, all the while harnessing opportunities to craft a more aligned and desired future.

Signals that are allocated in the ACT quadrant in the axis of uncertainty are the input for the pushes of the present, while the signals that are mapped in the PLAN quadrant are input for the pull of the future. The Futures Triangle offers a second iteration for identifying the driving forces.

To realise the underlying causes of change, the Causal Layered Analysis Model of Inayatullah (Inayatullah, 2008) is utilised here. Causal Layered Analysis (CLA) is a technique used in futures studies and strategic foresight, to analyse and deconstruct issues or scenarios across multiple layers of depth. Developed by Sohail Inayatullah, CLA encourages a more profound understanding of the present, and the ways in which alternative futures might emerge. The method challenges taken-for-granted assumptions, and pushes for deeper reflections on underlying causative layers. The CLA comprises four distinct layers:

- **Litany:** This is the surface layer that consists of the observable effects or events. News headlines often reflect this level. It represents the quantitative trends and challenges that are often universally accepted and perceived.
- **Social Causes:** Beneath the litany, this layer delves into the systemic, social, economic, political, and cultural factors causing the surface events. It provides a more in-depth analysis, and uncovers the structures leading to the observed events.
- **Worldview/Discourse:** Deeper still, this layer examines the mindset, narratives, or paradigms that inform the structures identified in the previous layer. It probes

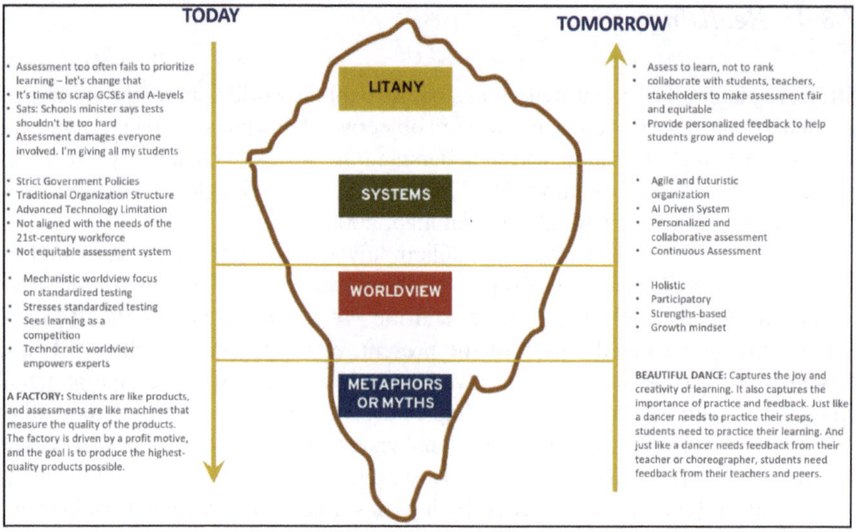

Fig. 7.3 The Realising phase is not limited to these two steps, as there are various other methods that help deepen our knowledge, thinking, and imagination about the domain question. Other Methods such as Integral Futures (Slaughter, 2008) and Theory U (Scharmer, 2007) are good examples of models that could be used in the Realising phase

deeply into the beliefs, values, and ideologies that underpin societal structures and practices.

- **Myth/Metaphor:** This is the deepest layer, focusing on the collective unconscious, the shared stories, myths, and metaphors that provide emotional resonance to the aforementioned layers. These myths and metaphors form the foundational stories a society tells itself.

By progressing through these layers, CLA provides a comprehensive and layered understanding of issues, moving from superficial events to the deeper narratives and myths that shape them. It is a valuable tool for those looking to challenge conventional thinking, unearth deep-rooted beliefs and assumptions, and conceive transformative alternatives for the future. Figure 7.3 presents an illustrative example of CLA, describing students of assessment today by moving downwards, while by moving upwards we are articulating our desired future for students' assessment (Abu Sitta et al., 2023) .

7.3.5 *Designing*

Educational Institutes in the region rarely consider various scenarios while developing their strategic agenda. It is essential that all educational institutions plan for various scenarios that might emerge and cause radical changes. It has been also observed that some organisations claim that they anticipate various scenarios, despite the fact that they use the same variable in each scenario. It is important here that each scenario should add a different perspective from each other. This is essential for establishing organisations that can anticipate various perspectives.

There are different schools for scenario planning. For demonstration purposes, this chapter considers the four archetypal scenarios in future-oriented planning created by Jim Dator in 1979 (Bezold, 2009). This methodology helps in the development of plausible future visions based on distinct perspectives: Baseline, Collapse, Equilibrium, and Transformational. These archetypes serve as narrative frameworks, offering varied interpretations of potential futures rather than exact predictions.

Baseline: This scenario projects a future that extends and intensifies the present trajectory, encompassing both its advancements and its challenges. It closely resembles the "business as usual" concept but diverges slightly, with technological advancements potentially perceived as issues.

Collapsing: Envisioning a future where the prevailing system undergoes a breakdown, this scenario attributes the deterioration either to internal dynamics or external pressures. Typically, this narrative paints a dystopian and less desired picture of the future.

Balance: An adaptation of Dator's original model, this scenario contemplates a future where society achieves stability following a period of decline. Stemming from the previously outlined dystopian narrative, it highlights potential avenues for rejuvenation and novel approaches.

Transformational: This envisions a future that challenges the foundational beliefs of the other scenarios. Such a transformation can stem from profound technological evolutions, material innovations, or a significant shift in societal values and beliefs.

The aim of this phase is to build various mental images of the possible futures, and serves as the core input for developing the preferred future that is being developed in the Shaping Phase.

Peter Schwartz suggests that the development of the four plausible futures can be achieved by articulating the 2 × 2 dimensions of uncertainties (Schwartz, 2013). These can be derived from the forces of change identified while developing the Futures Triangle and CLA. Figure 7.4 demonstrates the two dimensions of uncertainty and the four plausible scenarios for the future of higher education in 2050. This foresight study was developed by the University of Dubai Center for Futures Studies, in collaboration with education experts (Al Dhahri & Abu Sitta, 2021).

7.3.6 Shaping

One of the core purposes of Futures Thinking is not only to build and establish organisations that anticipate the possible futures, but rather to influence and shape those futures. At this stage of the framework, the team focuses on developing the preferred and inspired future. The narrative of the preferred future encompasses the weak signals and wild cards identified during the horizon scanning, as well as the

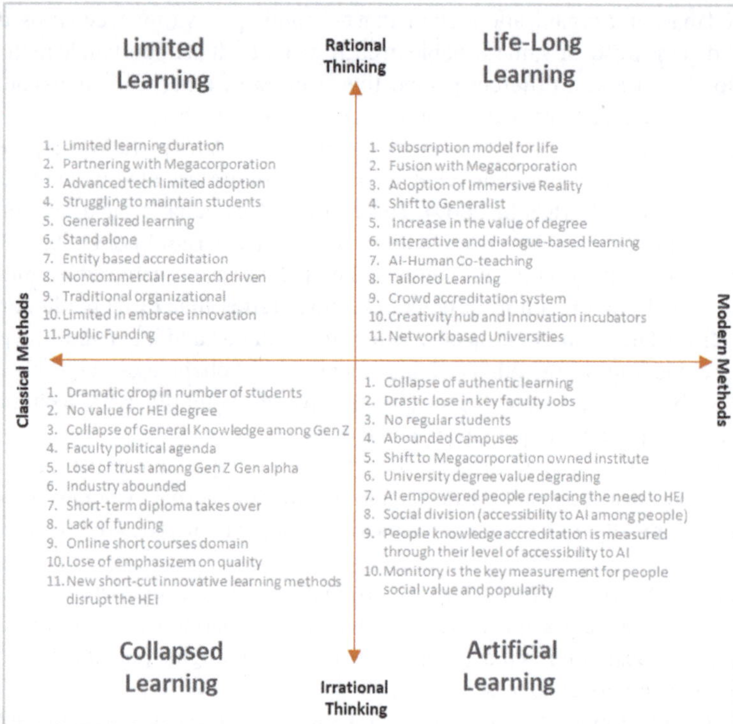

Fig. 7.4 An example of plausible scenarios for Higher Education Institutes

driving forces for change derived from the pushes of the present and the pulls of the futures from the Futures Triangle. In addition to that, the preferred futures narrative is derived from both the CLA Transformational Scenario and from the Plausible scenarios articulated in the Design phase. Developing the preferred future is the first step in shaping and influencing the future.

Backcasting is the second step in this phase. It is a strategic planning method that begins after setting up the preferred future outcome, and then works in reverse, to determine the necessary steps and actions to achieve that outcome. Instead of predicting the future based on current trends, as is done in forecasting, backcasting focuses on the question, "What are the images of the preferred future?" The process starts by clearly defining a desired future, analysing the current situation in relation to it, and then deducing the requisite measures and policies needed to bridge the gap between the present and that envisioned future. Backcasting provides a guide to journey from the current state to the anticipated future, which is then put into action and adjusted as required.

Participants are immersed in the preferred future, and then they backcast the images of the future in a set of intervals of 5 or 10 years. This step is essential as it determines the future initiatives that are needed to be taken today, to anticipate and shape the possible futures.

These initiatives are then aligned with the organisational strategy. It is worth mentioning here that there are always discrepancies between the strategic initiatives identified in the strategic management plan, and the ones that are being identified in the Futures Thinking exercise. Foresight and strategic initiatives must be aligned in order to build future-proof educational institutions and ministries.

7.4 Weak Signals and Wild Cards

Strategists and policymakers often emphasise responding to trends and mega-trends when developing strategic directions and policy agendas. However, a review of a few published strategic agendas of the educational institutes and ministries in the region reveals limited references to weak signals or wild cards. This potentially puts educational institutions in a reactive approach rather than being proactive in anticipating disruptive innovations and major shifts in the education sector.

In the subsequent subsections, examples of weak signals and wild cards that could influence the futures of education are provided. It is worth noting that the list presented is neither exhaustive nor necessarily the most predictable. These signals are chosen primarily for illustrative purposes, underscoring the importance of horizon scanning and scenario planning for educational institutes during strategic planning and policy development.

The omission of weak signals or the lack of horizon scanning in strategy development raises questions about the preparedness of regional educational strategies for anticipating the potential futures. Another noticeable concern is the alignment of these strategies with technological advancements, and plausible shifts in social, environmental, economic, and political domains.

7.4.1 Immersive Learning

Immersive technologies refer to the digital technologies that augment human capabilities, by creating a sense of immersion and presence in a computer-generated environment or experience. These technologies fuse the digital and physical realm, thus creating a new dimension which enables humans see the world in more than three dimensions. For example, Augmented Reality (AR) allows users to combine the physical world with the digital environment, which empowers them to see the world in more than three dimensions. Today, there are a number of educational institutes that are using these technologies for knowledge construction. There are a number of popular AR products, such as Microsoft HoloLens, Magic Leap, and Apple AR, which are being used by corporations to promote productivity and education. These technologies still face major challenges in terms of price and wearability. On the other hand, research centres are working today to make these glasses smaller and more efficient. Virtual reality will also play a major role in transforming the schools of today. Virtual reality (VR) allows users to completely immerse themselves in a simulated environment, isolating them from the physical world, and enabling them to interact with and explore the digital environment. VR today has been able to immerse users by means of their sight, hearing, and feelings.

This technology will have a major impact on the future of knowledge construction and creativity. A very limited number of educational institutes are considering the implementation of these technologies as part of the teaching methods. This is due to limited content availability and the high price of the headsets.

7.4.2 Intelligent Tutor

It is a computer-based educational system that uses artificial intelligence (AI) to provide personalised instruction and feedback to learners. An intelligent tutor can be used to teach a wide range of subjects, including mathematics, science, language skills, social studies, and foreign languages. They can also be used to teach specific skills, such as problem-solving, critical thinking, and test-taking strategies.

Artificial intelligence (AI) is rapidly transforming many industries, and education is no exception. AI has the potential to revolutionise education by providing personalised learning experiences, real-time feedback, and creating more engaging and interactive learning experiences. In a 2022 article for Interesting Engineering, Bill Gates predicted that AI chatbots would be used to teach children how to read and write in the future (Paleja, 2022). He argued that AI chatbots could be more effective than human teachers at providing personalised instruction and feedback. He also believes that AI chatbots can be used to create more engaging and interactive learning experiences. In a 2022 article for EdTech, Andy Viano discusses the potential of AI tutors to transform higher education (Viano, 2022). He argues that AI tutors can be used to provide personalised instruction and feedback to students, freeing up human professors to focus on more creative and strategic tasks. He also believes that AI tutors can be used to create more engaging and interactive activities. In a 2022 article for The Conversation, Patricia A. Young says that the future of college will involve fewer professors (Young, 2022). She cites the rise of online learning and the increasing availability of educational technology as factors that are contributing to this trend. She also claims that the role of professors will change in the future, with them becoming more mentors and coaches than lecturers. In a 2020 article for Forbes, Sameer Maskey (2020) discusses the ways in which AI is transforming education. He argues that AI can be used to personalise learning, provide real-time feedback, and create more engaging and interactive learning experiences. He also believes that AI can be used to improve the efficiency of education systems. In a 2021 article for the Futures Platform, Marianna Mäki-Teeri and Thomas Frey say that AI will play an increasingly important role in education in the future (Maki

Teeri & Frey, 2021). They believe that AI will be used to create personalised learning experiences, provide real-time feedback, and assess student learning.

Intelligent Tutors work by tracking the student's progress, identifying their strengths and weaknesses, and providing tailored instruction and support. They often use games, simulations, and other interactive activities to teach new concepts. ITs can be used in schools and other educational institutions to supplement traditional classroom instruction, or they can be used by students at home to learn at their own pace and in their own time.

Today, organisations such as Duolingo, Carnegie Learning, Knewton, DreamBox Learning, and IBM Watson Education are using some elements of the Intelligent Tutor. This technology will promote personalised learning and will transform the role of teachers. In addition to that, this approach will most likely impact education equality and inclusion, even if not implemented across educational institutes. It has been observed that none of the published education strategies in the region has been addressing the impact of weak signal on education in the region.

7.4.3 Climate Change

Climate change is making it harder for children to go to school and learn. Natural disasters can destroy schools and infrastructure, force families to move, and disrupt students' education. Climate change is also leading to more extreme weather conditions, which can make it difficult or even dangerous for students to travel to and from school. It is important to note that the impacts of climate change on education are not evenly distributed. Students in developing countries and marginalised communities are often the most vulnerable. This is because they are more likely to live in areas that are prone to natural disasters, and have less access to resources to adapt to the effects of climate change (Newsome et al., 2023).

Climate change will affect education in every part of the world. The recent earthquake in Morocco, and the flood caused by Hurricane Daniel that hit Libya at the beginning of the school term have caused hundreds of thousands of children to miss school for months on end. According to a UNICEF (2023a) headline: "Nearly 300,000 children affected in Libya after Storm devastates entire communities", and the earthquake which struck Morocco in September 2023 impacted more than 100,000 children according to (UNICEF, 2023b).

High temperatures can have a significant impact on performance. A study by The Independent found that students in non-air-conditioned environments scored 13% lower in summer exams (O'Brien, 2019). The US National Bureau of Economic Research reported that each 0.55 °C increase in temperature resulted in a 1% drop in learning achievement, with temperatures above 21 °C having significant effects, and those exceeding 38 °C having the most profound impact. This decline was not only seen during school hours. A separate Harvard study found that students in non-air-conditioned rooms performed worse than their counterparts in cooler environments (Park, 2017).

Although these events are not merely a trend and are categorised as a wild card, educational institutes do not consider them or have an agenda to encounter their impact. Incorporating Futures Thinking urges the necessity to navigate the impact of these uncertainties, and their likely major impact on the futures of learning and assessments.

7.4.4 *Teacher Crises*

According to a recent study conducted by Gulp, K-12 teachers have the highest burnout rate in U.S. industry and other fields (Marken & Agrawal, 2022). The National Education Association indicated that 55% of educators are considering quitting the profession sooner than originally planned (Walker, 2022). This decision is due to a number of factors, including low salaries, high workloads, and a lack of respect for the teaching profession. The U.S. is not the only nation grappling with a teacher shortage. In Norway, nearly one-third of teachers are 50 years old or older, indicating that, over the next decade, Norway will need to replace about 30% of its teaching staff (OECD, 2018).

UNESCO's data suggests that to globally attain universal basic education by 2030, there is a requirement for an additional 24.4 million primary education teachers and about 44.4 million secondary education teachers. Sub-Saharan Africa, which has some of the world's most over-crowded classrooms, also contends with the most strained teachers and understaffed educational systems. A staggering 90% of its secondary schools grapple with significant teacher shortages (UNESCO, 2023).

The Arab world is in no better a position in recruiting and retaining teachers. The shortage of teachers is especially pronounced in the Middle East, besides Africa. Meanwhile, in Southern Asia, which has the second-highest shortage, there's an anticipated need for an extra 1.7 million primary school teachers and 5.3 million secondary school teachers by 2030.

There is no strong evidence suggesting that governments and educational institutes are considering this as a situation which requires immediate action. The author argues that integrating futures and systems thinking will enhance the process of policymaking and strategic planning.

7.4.5 *Schools Without Teachers*

The emerging weak signals in vocational learning suggest a shift towards models that could alleviate the shortage of traditional educators. Schools like Paris's Ecole 42, Hive, and Mitra's online platform represent a new wave of educational institutions that rely minimally on traditional teachers. Emphasising internet-based learning and real-world problem-solving, these schools grant students autonomy in crafting their learning paths. Ecole 42, for instance, not only operates without teachers but also confers no formal diplomas, yet its students reportedly find swift employment. This approach, inspired by Ecole 42, has seen global adoption in countries like South Africa, Russia, Indonesia, and Finland.

This emerging trend may be bolstered by the progress of artificial intelligence and the aforementioned AI tutors. These developments could shift the role of educators from direct instruction to mentorship.

7.4.6 *Quantum Learning*

Computers have been programmed according to binary mathematics, with ones and zeros. These systems enable the processing of one task and problem at a time in a particular node of software. With quantum computing, a computer may simultaneously conduct tasks and operations that amount to the total existing calculation power of today. Quantum Neural Networks are the future of computing, and with other advances in AI such as deep learning, the path to profound machine intelligence. The current quantum computers are not able to process enough parameters, lacking the contextual information to do so (Griffin, 2017).

When quantum computing is managed and mass produced, it will not only radically change what computers are, but how they are networked, and how there will be no data and energy loss in transmitting information. This instant fluidity is also a key to building Quantum Neural Networks, which will create a new era of artificial intelligence.

Today, Generative AI has created a major disruption in the education sector and with the advancements in quantum computing new super intelligence might emerge which could reshape the education we know today.

7.4.7 *Hybridisation of Schools*

This term describes the integration of traditional in-person instruction with online learning. The concept has expanded over time to include various models:

- Blended Learning: A mix of face-to-face and online instruction within one course.
- Online Learning: Entire coursework is completed online but may include interactions like video conferencing.
- Flipped Classroom: Students first study materials at home and then practice and collaborate in class.
- Virtual School: Like online learning but offered by a distinct institution, offering more course options and flexibility.

The benefits of hybrid learning include flexibility, increased access, personalisation, and potential cost savings. However, there are challenges like the digital divide, the need for teacher training, and maintaining student engagement. Some current hybrid applications include blended maths courses, online options for struggling students, flipped science classes, and collaborations with virtual schools. This innovative approach to education is rapidly growing in recognition and use.

7.4.8 Internet of Senses

Digital technology currently relies predominantly on sight and sound. Yet advancements in AI, VR, AR, and 5G hyper-connectivity are paving the way for a transition from screen-based to sense-based experiences. By 2030, the "Internet of Senses" (IoS), advocated by Ericsson, is anticipated to immerse us in multisensory digital interactions that mimic real-world experiences. Ericsson's extensive survey underscores the growing anticipation for digital interactions utilising all five senses (Ericsson, 2019).

AR and VR technologies, supported by 5G, are believed to blur the lines between virtual and physical realities. By the end of this decade, sophisticated AR glasses might superimpose lifelike digital elements onto real spaces. Additionally, digital innovations might enable tactile, olfactory, and gustatory interactions. Imagine smelling movie scenes, tasting digital flavours, or experiencing the weight of digital items through wearables. Earphones might soon auto-translate languages or let users focus on specific sounds. Such progress could redefine shopping and tourism, replicating in-store or real-world travel experiences online. Auto-translate applications might lead to a collapse in numbers learning languages, and make students and society less motivated to engage in learning new languages. In addition to that, IoS might also revolutionise communication, with immersive avatars replacing physical interaction, and brain-computer interfaces potentially allowing thought-based communication, bringing futuristic concepts into reality (Futures Platform, 2018).

7.4.9 Laughter and Learning

According to the Curriculum Associations "Laughter can help your students learn and improve your instruction" (Sullivan, 2022). Students' attention, motivation, and learning might improve when teachers use humour as a teaching method, or for interaction or classroom management. Humour can create a more welcoming and interesting learning atmosphere in the classroom, and so support students' academic performance or their behaviour. Humour activates the brain's dopamine reward system, stimulating goal-oriented motivation and long-term memory, which means that humour can improve retention in students of all ages (Henderson, 2015).

The Humor Academy in New York City offers workshops and training programmes on how to use humour in the classroom. The academy's programmes are designed to help teachers create a more engaging and enjoyable learning environment for their students (Association for Applied & Therapeutic Humor, 2023). A

Pew Research poll showed that viewers of humourous satirical news shows such as The Daily Show and The Colbert Report exhibited a higher retention of news facts than those who got their news from newspapers, CNN, Fox News, or other network stations. "When Stephen Colbert ironically declares, 'If we don't cut expensive things like Head Start, child nutrition programs, and teachers, what sort of future are we leaving for our children?', viewers laugh and retain the knowledge of that specific budget issue" (Pew Research Center, 2007).

7.4.10 Intuitive Intelligence

Intuitive intelligence is the ability to understand something immediately, without the need for conscious reasoning. It is a type of intelligence that is not based on logic or analysis, but rather on a feeling or gut instinct. Finland is one country that has made a concerted effort to incorporate intuitive intelligence into its education system. The Finnish Ministry of Education has created a curriculum that emphasises creativity, problem-solving, and self-awareness, and these skills are all essential for developing intuitive intelligence (Lähdemäki, 2019).

It is essential for educational institutes and policymakers in the education sector to consider the adoption of this phenomenon. It has been observed that schools and ministries in the region have a limited view in encouraging the implementation of intuitive intelligence in schools.

7.5 Conclusion

In an effort to build educational institutes that are adaptive and resilient, able to encounter discontinuities and surprises, as well as shape their futures and accelerate innovations, unconventional methods of strategic thinking and policy development should be adopted. Futures Thinking models and frameworks provide an alternative method for anticipating the future proactively. Infusing foresight into strategic planning efforts eliminates cognitive biases and introduces impartiality into the images of plausible futures and in articulating scenarios. Futures Thinking methods, such as Horizon Scanning, Axes of Uncertainty, Scenario Planning, articulating preferred futures, and backcasting, assist organisations in shaping the future. Horizon Scanning helps in identifying emerging signals, such as weak signals and wild cards. Axes of Uncertainty is a collaborative technique which helps in identifying the signals that need to be considered in strategy, or those to be monitored or included in scenario planning.

It is worth highlighting the fact that advanced technologies have their own negative impact on privacy and health. These negative impacts could push back the adoption of technology in schooling. In a report developed by the University of Cambridge, "The Futures of Assessment Navigating Uncertainties through the Lenses of Anticipatory

Thinking" (Abu Sitta et al., 2023) it has been indicated that "Research on the use of pen and paper, and other traditional manual exercises (gesture and movement), have found that these methods activate the brain more than, for instance, taking notes on a digital device. The rise in 'screen time' and perceived negative impacts and the challenges of assessing in the context of generative AI have already led to widespread 'pushback' against technology use in favour of adopting pen and paper exams".

In this article, the principles and framework of Futures Thinking have been discussed. Additionally, a sample of weak signals that might shape the future of education is presented. It is essential for educational institutes and ministries to consider and conduct horizon scanning when developing their strategic directions and policy developments. They should also create scenarios and integrate imagination as a central activity, while planning for the short, medium, and long term.

Futures Thinking is a crucial part of developing strategic plans and developing educational policies. Relying solely on trends while developing educational strategies is insufficient. This process should be complemented by Futures Thinking, systems thinking, horizon scanning, scenario planning, and backcasting efforts.

Bibliography

Abu Sitta, F., Maddox, B., Casebourne, I., Hughes, S., Kuvalja, M., Hannam, J., & Oates, T. (2023). *The Futures of Assessment Navigating Uncertainties through the Lenses of Anticipatory Thinking*. ISBN number: 978-1-7384336-0-5

Abu Sitta, F. H. (2023). *A Foresight Framework for Building an Anticipatory System for Culture and Tourism: The Department of Culture and Tourism: Abu Dhabi Case Study*. Presented at UNESCO-PMU Symposium 2023: Towards Evaluation Frameworks for Futures Literacy and Foresight in Paris, 20–21 July 2023.

Al Dhahri, S., & Abu Sitta, F. (2021). *The Futures of Higher Education by 2050: University of Dubai Foresight Study*. https://www.ecssr.ae/en/ecssr-news/ai-and-future-of-education

AlKhoori, K., & Abu Sitta, F. (2023). *From Fiction to Fact: How Science Fiction Is Shaping the Transformation of Government Public Services – Abu Dhabi Culture and Tourism Case Study*. Presented at Empowering Futures—Long-Term Governance, Democracy and Futures Research conference in Finland, 14–16 June 2023.

Association for Applied and Therapeutic Humor. (2023). *Our Rallying Cry: Humor Helps!* https://www.aath.org/

Bezold, C. (2009). Jim Dator's Alternative Futures and the Path to IAF's Aspirational. *Futures: The Journal of Policy, Planning and Futures Studies*, *41*(3), 197–208.

Conway, M. (2017). *Foresight Infused Strategy Development: A How-to Guide for Using Foresight in Practice*. ISBN 978-1095942482.

Ericsson. (2019). *10 Hot Consumer Trends 2030: Internet of the Senses*. https://www.ericsson.com/4ac661/assets/local/reports-papers/consumerlab/reports/2019/10hctreport2030.pdf

Griffin, A. (2017). IBM to build quantum computers, selling machines millions of times faster than anything made before. *The Independent*. https://www.independent.co.uk/tech/ibm-quantum-computer-how-work-commercial-api-faster-q-systems-computing-a7613271.html

Henderson, S. (2015). Laughter and Learning: Humor Boosts Retention. *Edutopia*. https://www.edutopia.org/blog/laughter-learning-humor-boosts-retention-sarah-henderson

Hines, A., & Bishop, P. (2015). *Thinking about the Future: Guidelines for Strategic Foresight*. ISBN 0996773401.

Hines, A., & Bishop, P. C. (2013). Framework foresight: Exploring futures the Houston way. *Futures, 51*, 31–49.

Inayatullah, S. (2008). Six pillars: Futures thinking for transforming. *Foresight: The Journal of Futures Studies, Strategic Thinking and Policy, 10* (1), 4–21. DOI–https://doi.org/10.1108/14636680810855991.

KSA Ministry of Education. (2020). *Ministry of Education's Strategy 2020–2023.* https://moe.gov.sa/

Lähdemäki, J. (2019). Case Study: The Finnish National Curriculum 2016—A Co-created National Education Policy. In: Cook, J.W. (Ed.), *Sustainability, Human Well-Being, and the Future of Education.* Palgrave Macmillan, Cham. https://doi.org/10.1007/978-3-319-78580-6_13

Maki Teeri, M., & Frey, T. (2021). Future of education after COVID-19: AI becomes the teacher while human mentor and coach. *Futures Platform.* https://www.futuresplatform.com/blog/future-education-after-covid-19-ai-becomes-teacher-while-humans-mentor-and-coach

Marken, S., & Agrawal, S. (2022). *K-12 Workers Have Highest Burnout Rate in U.S. Gallup.* https://news.gallup.com/poll/393500/workers-highest-burnout-rate.aspx

Maskey, S. (2020). Artificial Intelligence in Education Transformation. *Forbes.* https://www.forbes.com/sites/forbestechcouncil/2020/06/08/artificial-intelligence-in-education-transformation/#2cd0f9ab32a4

Newsome, D., Newsome, K. B., & Miller, S. A. (2023). Teaching, learning, and climate change: anticipated impacts and mitigation strategies for educators. *Behaviour & Social Issues.* https://doi.org/10.1007/s42822-023-00129-2

O'Brien, S. (2019). *Do High Temperatures Impact Exam Performance? Education Briefing.* https://education-forum.co.uk/briefing/do-high-temperatures-impact-exam-performance

OECD. (2018). *Norway Teachers and Teaching Conditions.* https://gpseducation.oecd.org/CountryProfile?primaryCountry=NOR&treshold=5&topic=TA

Paleja, A. (2022). Bill Gates says AI Chatbots will teach kids how to read and write. *Interesting Engineering.* https://interestingengineering.com/innovation/ai-chatbots-teach-kids-read-and-write

Park, J. (2017). *Temperature, Test Scores, and Human Capital Production.* Scholar Harvard.

Pew Research Center. (2007). *Public Knowledge of Current Affairs Little Changed by News and Information Revolutions.* https://www.pewresearch.org/politics/2007/04/15/public-knowledge-of-current-affairs-little-changed-by-news-and-information-revolutions/

Futures Platform. (2018). *The Internet of Senses: The Future of Digital Experience is Multi-Sensory.* https://www.futuresplatform.com/blog/internet-senses-future-digital-experience-multi-sensory

Qatari Ministry of Higher Education. (2018). *The Strategic Plan for the Ministry of Education and Higher Education 2018–2022.* www.edu.gov.qa

Rohrbeck, R., & Gemünden, H. G. (2011). Corporate foresight: Its three roles in enhancing the innovation capacity of a firm. *Technol. Forecast. Soc. Change, 78*(1), 231–243.

Scharmer, C. O. (2007). *Theory U: Leading from the Future as It Emerges.* The Society for Organizational Learning Inc.

Schwartz, P. (2013). *The art of the long view: Planning for the future in an uncertain world.* Currency Doubleday.

Slaughter, R. (2008). Integral futures methodologies. *Futures, 40*, 103–108.

Sullivan, D. (2022). Get Your Giggle On! 4 Ways to Use Laughter for Learning. *Context.* https://www.curriculumassociates.com/blog/four-ways-to-use-laughter-instruction

Sultanate of Oman Education Council. (2018). *The National Strategy for Education 2040.* https://www.educouncil.gov.om/

UAE Ministry of Education. (2017). *The UAE Ministry of Education Strategic Plan 2017–2021.* https://www.moe.gov.ae

UNESCO. (2012). *Future Literacy.* https://www.unesco.org/en/futures-literacy/about

UNESCO. (2023). *World Teachers' Day: UNESCO Sounds the Alarm on the Global Teacher Shortage Crisis.* https://www.unesco.org/en/articles/world-teachers-day-unesco-sounds-alarm-global-teacher-shortage-crisis

UNICEF. (2023b). *Earthquake in Morocco.* https://www.unicef.org/emergencies/earthquake-morocco

UNICEF. (2023a). *Nearly 300,000 children affected in Libya after Storm Daniel devastates entire communities.* https://www.unicef.org/mena/press-releases/nearly-300000-children-affected-libya-after-storm-daniel-devastates-entire

Viano, A. (2022). The Future of AI Tutors in Higher Education. *EdTech.* https://edtechmagazine.com/higher/article/2022/09/future-ai-tutors-higher-education

Walker, T. (2022). Survey: Alarming Number of Educators May Soon Leave the Profession. *neaToday.* https://www.nea.org/nea-today/all-news-articles/survey-alarming-number-educators-may-soon-leave-profession

Young, P. A. (2022). Future of college will involve fewer professors. *The Conversation.* https://theconversation.com/future-of-college-will-involve-fewer-professors-166394

Chapter 8
Financial Literacy in the Modern Economy: Bridging the Knowledge Gap for Entrepreneurs and Beyond

Said Elbanna, Mohamed Abdelaziz Eissa, and Loreta Armstrong

Abstract This chapter underscores the critical role of financial literacy, especially for entrepreneurs, in today's rapidly changing economic landscape. The significance of this subject is evident, yet recent studies from the Organization for Economic Co-operation and Development (OECD) and Standard & Poor's (S&P) reveal an alarming deficiency in financial knowledge, with a vast number of adults, including budding entrepreneurs, unable to understand fundamental financial concepts. This was particularly evident during the COVID-19 Pandemic. This knowledge gap affects not just individual entrepreneurs but can also pose systemic risks, especially with emergent challenges like the introduction of digital assets. Our chapter takes a holistic view, exploring topics from entrepreneurial financial planning to the socio-economic factors influencing financial choices. Highlighting gender disparities in financial literacy, we emphasise the urgent need for a gender-inclusive approach to financial education. This approach ensures both individual entrepreneurial success and collective resilience against economic adversities.

Said Elbanna and Mohamed Abdelaziz Eissa are equivalent having contributed 40% each.

S. Elbanna (✉)
Professor of Strategic Management & Strategy Advisor, Qatar University, P.O. Box 2713, Doha, Qatar
e-mail: selbanna@qu.edu.qa

M. A. Eissa (✉)
Associte Professor of Finance, Department of Finance and Economics, College of Business and Economics, Qatar University, P.O. Box 2713, Doha, Qatar
e-mail: m.eissa@qu.edu.qa

L. Armstrong
Research Assistant, Research Assistant, Qatar University, P.O. Box 2713, Doha, Qatar
e-mail: ll1805333@qu.edu.qa

E. Zaidan et al. (eds.), *The Future of Education Policy in the State of Qatar*,
Gulf Studies 21, https://doi.org/10.1007/978-981-97-9667-0_8

8.1 Introduction

Financial literacy, often interchangeably used with financial awareness, stands as an indispensable cornerstone in today's ever-evolving economic landscape, including the realm of entrepreneurship. Deeply woven into the very fabric of our individual and collective decision-making processes, it serves as both a compass for individuals navigating the murky waters of personal finance, and a broader barometer for the socio-economic well-being of entire communities and nations. The data available from a myriad of sources presents a vivid picture of the current state of financial literacy. A report from the Organization for Economic Co-operation and Development (OECD, 2020) uncovers the fact that, in an array of countries, less than 60% of adults possess the capability to answer rudimentary financial questions correctly. Similarly, the Standard and Poor's Ratings Services Global Financial Literacy Survey (2015) reveals that in developing economies, only a paltry 33% of adults are financially literate. These numbers not only underscore individual vulnerabilities but also shine a light on potential systemic risks, since aggregated financial decisions at the household level can significantly sway broader economic directions. The tangible repercussions of these statistics are perhaps most pronounced when juxtaposed against the backdrop of global disruptions. Take, for instance, the unprecedented upheaval brought about by the COVID-19 Pandemic. National Endowment for Financial Education (2020) found that almost two-thirds of Americans grappled with financial challenges in the initial months of the pandemic. Simultaneously, the World Bank (2021) noted that those with limited financial literacy were less adept at adjusting to the abrupt economic downturn, emphasising the direct correlation between financial acumen and resilience, a lesson that resonates profoundly in the entrepreneurial arena.

Yet the challenge is not confined solely to times of crisis. The Global Financial Literacy Excellence Center (GFLEC 2019) rings the alarm bells by revealing that in some so-called advanced economies, nearly 30% of adults routinely struggle to make ends meet. This statistic is not just a reflection of economic hardships, but an indication of a gaping financial literacy deficit, one that entrepreneurs must navigate skilfully to ensure the success of their ventures. Moreover, in times of relative calm in the global economy, numerous households across the globe teeter precariously on the edge, living from paycheck to paycheck. Delving deeper, the World Bank's (2018) categorisation of financial literacy offers a tripartite perspective, encapsulating basic numeracy skills, advanced financial knowledge, and behavioural considerations. The National Bureau of Economic Research (NBER, 2016) further amplifies the significance of the latter two, noting a startling deficiency in the understanding of even fundamental concepts such as inflation, and compounded in developed economies. Shiller's (2019) seminal work on behavioural finance elaborates on the importance of the 'act-how,' postulating that financial decisions are often at the mercy of psychological biases, from herd behaviour to overconfidence. These factors resonate deeply with the entrepreneurial community, as they grapple with investment choices and risk assessment.

The ever-accelerating AI and digital revolution add another layer of complexity to the entrepreneurial landscape (Elbanna & Armstrong, 2023). Cryptocurrencies, blockchain, and a plethora of Fintech solutions are redrawing the financial landscape, presenting both opportunities and challenges for entrepreneurs. Yet, as the Cambridge Centre for Alternative Finance (2020) study highlights, a staggering 70% of respondents admit to a shaky grasp of these digital financial assets. Furthermore, socio-economic disparities in financial literacy have severe, wide-reaching implications for entrepreneurs. Lusardi and Tufano's (2015) research unveils the fact that the financially challenged are often ensnared in detrimental high-cost borrowing cycles, thereby exacerbating their vulnerabilities. This dynamic is particularly relevant for entrepreneurs, in that their access to capital and financial decision-making can significantly impact a business. However, there is a silver lining. Efforts like the OECD/INFE Global Financial Education Campaign and community-centric initiatives, as spotlighted by the African Development Bank Group (2017), are being seen to address these gaps. For entrepreneurs, these initiatives provide resources and educational opportunities to enhance their own financial literacy.

Merging this diverse wealth of information, statistics, personal accounts, and regional specifics, this chapter endeavours to provide an exhaustive exploration of the multifaceted realm of financial literacy and entrepreneurship. By threading through a comprehensive mosaic of research, experiences, and quantitative metrics, our aim is to present an encompassing view of financial literacy's profound influence on individual entrepreneurs and the broader entrepreneurial ecosystem. The world of entrepreneurship carries risks as well as opportunities, and the significance of financial literacy cannot be overstated. The goal is clear: to chart a path towards a future where informed, empowered financial decisions are the universal standard. This endeavour fills a significant void in the existing literature, particularly within the Arab world and GCC region, where research on management, business, and entrepreneurship is notably scarce (El-Bassiouny et al., 2022 Elbanna et al., 2020; Younis et al., 2022).

The next section thoroughly examines 15 essential components of financial awareness, providing a comprehensive guide for readers. Afterwards, we explore how economic resilience can be achieved through financial literacy, particularly in turbulent times, before discussing the importance of comprehensive financial education as a tool to enhance this resilience. A dedicated section addresses the intersection of gender and financial literacy, with a focus on crafting a pathway towards equitable economic resilience. The chapter concludes by outlining key pillars for enhanced financial literacy, and presenting a dynamic financial proficiency matrix for tailored assessment.

8.2 Essential Components of Financial Awareness: A Comprehensive Guide

Colloquially dubbed the cornerstone of economic self-sufficiency, financial awareness encapsulates the nuanced understanding of the modalities surrounding earning, managing, investing, and conserving financial assets—skills that are vital within the entrepreneurial landscape. This multidimensional proficiency is paramount in not only safeguarding one's fiscal well-being but also in the proactive creation and multiplication of wealth, which is a sine qua non to many entrepreneurs. Rooted deeply within the broader ambit of economic sciences, financial awareness bridges the gap between theoretical finance principles and their pragmatic applications in daily life. A connection that is particularly pronounced in the entrepreneurial context, where business decisions often hinge on financial acumen. The integral dimensions delineating financial awareness are manifold. Herein, we shall dissect these dimensions, leveraging insights from academia, empirical studies, and global best practices:

8.2.1 Financial Planning and Cost Management

In the intricate tapestry of financial decision-making, the orchestration of income and expenses stands as an elemental pillar, acting as a barometer for an individual's or institution's financial health. Essentially, the act of balancing one's financial inflows and outflows is not merely transactional but also strategic, aligning short-term actions with long-term objectives (Fisher, 1930). It is particularly relevant for entrepreneurs, as it forms the foundation of their venture's fiscal sustainability and growth. The ability to efficiently administer income and expenses indeed epitomises a fundamental facet of fiscal awareness, a sentiment echoed by Keynes (1936) in his seminal work, where he illuminated the importance of effective financial planning in stabilising economic cycles. Through his lens, the effective management of financial resources transcends mere household or institutional budgeting, influencing broader economic paradigms and underpinning entrepreneurial success.

The detailed practice of tracking income and expenditure, often underpinned by budgetary frameworks, is the cornerstone upon which sustainable financial behaviours are built for entrepreneurs. Meticulous record-keeping not only provides transparency, but also offers invaluable insights into spending patterns, enabling proactive course corrections, and fostering financial discipline (Modigliani & Brumberg, 1954). In the entrepreneurial context, this discipline is indispensable, ensuring that resources are allocated skilfully, and that the business remains agile in response to financial fluctuations. Forecasting future financial obligations, a more advanced dimension of financial planning, necessitates a holistic understanding of microeconomic household factors and macroeconomic trends. It is here that financial literacy becomes indispensable for entrepreneurs. Lusardi and Tufano (2015) underscore the fact that heightened financial literacy leads to better aniticipation and an ability to navigate financial downturns, showcasing the intimate link between fiscal foresight and financial education.

Furthermore, planning based on these forecasts involves creating contingencies, identifying investment opportunities, and managing potential risks, which are all vital aspects of entrepreneurial success. The complexity of these tasks magnifies the importance of financial literacy for entrepreneurs. As highlighted by Campbell (2006), the intricate financial landscape of the modern era, with its myriad of choices, requires entrepreneurs to be financially aware and adept at assimilating and acting upon financial information, making informed decisions that define the trajectory of their ventures.

8.2.2 *Savings and Financial Security*

In the dynamic ecosystem of personal finance, the act of saving is not merely a rudimentary practice, but rather a strategic move that lays the groundwork for sustainable economic resilience—a principle that resonates in the entrepreneurial landscape, particularly within the early stages of entrepreneurship. At its core, the act of saving embodies the essence of deferred gratification, a testament to one's foresight in prioritising future financial health over immediate consumption (Bernheim, 1999). For entrepreneurs, saving is not just about personal finance, it is about securing the financial stability of their ventures, and enabling future growth. Building an emergency fund, and conscientiously setting aside money for critical junctures, such as retirement, not only shields individuals and entrepreneurs from the potential adverse effects of unforeseen financial shocks, but also steers them towards the path of long-term financial independence. Such prudence underscores the vital intersection between savings and broader financial literacy. According to a study by Lusardi and Mitchell (2007), individuals and entrepreneurs with a higher degree of financial literacy, are significantly more likely to plan for the financial sustainability and growth of their businesses, highlighting the profound correlation between financial knowledge and saving behaviours.

This nexus becomes even more pronounced when we delve into more nuanced financial concepts, such as compound interest and the time value of money. Compound interest, often dubbed the "eighth wonder of the world," exemplifies the potent force that time wields in amplifying the growth of savings, emphasising that the earlier one starts saving and investing in their business, the more pronounced the benefits become in the long run (Malkiel, 1996). Meanwhile, the time value of money, a foundational principle in finance, elucidates the evolving worth of money over time, emphasising the economic rationale behind saving today for a potentially higher valued return tomorrow (Ross et al., 2002). To harness these principles effectively, entrepreneurs must possess a certain degree of financial literacy. Financial education arms them with the tools to navigate the intricate labyrinth of fiscal choices, ensuring that they maximise the potential of their savings and any investments in their businesses. According to the Organization for Economic Co-operation and Development (OECD, 2017), enhancing financial literacy can substantially boost savings rates within the entrepreneurial community, and foster a more secure financial future and greater entrepreneurial success.

8.2.3 Investment Strategies

The realm of investments is vast and multifaceted, serving as a testament to the intricate nature of capital allocation and appreciation in contemporary finance—a terrain where entrepreneurs often struggle as they seek to grow and expand their businesses. To successfully navigate this landscape, an individual or institution must not only understand the mechanics of various investment vehicles, but also possess the strategic acumen to leverage them optimally. Central to this proficiency is financial literacy, a keystone in the edifice of investment decision-making (Bodie et al., 2008). Grasping diverse investment options, such as stocks, bonds, mutual funds, and real estate, represents the foundational knowledge of the investment universe for entrepreneurs. Each of these asset classes possesses distinct characteristics, risk profiles, and potential returns, underscoring the importance of being well-versed with their intricacies (Sharpe, 1964). For instance, while stocks offer ownership in a corporation with potential dividends and capital gains, bonds are essentially loans, providing periodic interest payments and principal repayment. Financial literacy equips entrepreneurs with the wherewithal to discern the nuanced differences and synergies among these asset classes, paving the way for informed allocation decisions that directly impact the growth and stability of their businesses.

Equally paramount in this domain is understanding the risk-reward paradigm, since entrepreneurs seek to make strategic investments in their ventures. As elucidated by Markowitz's Modern Portfolio Theory (1952), an inherent trade-off exists between risk and expected returns. Financially literate entrepreneurs can better quantify and qualify the risks associated with particular investments, weighing them against potential returns, and making choices that align with their risk appetites and financial goals. Furthermore, the principle of diversification, often touted as the only "free lunch" in finance (Evans & Archer, 1968), emphasises the dispersion of investments across various assets to reduce risk. Through diversification, the adverse performance of one investment can potentially be offset by the favourable performance of another. This principle underscores the need for a deep understanding of the correlations between asset classes, fostered by heightened financial literacy. Lastly, the 'buy-and-hold' investment approach, championed by investment luminaries such as Warren Buffett, advocates for long-term investment horizons—a perspective that aligns with the strategic thinking of entrepreneurs focused on building enduring businesses. It posits that over extended periods of time, market volatilities tend to average out, potentially offering entrepreneurs the true intrinsic value of assets (Graham & Dodd, 1934). Adopting this perspective requires patience, conviction, and a robust understanding of underlying asset fundamentals, further underscoring the indelible link between investment strategies and financial literacy.

8.2.4 Credit, Liabilities, and Financial Health

Credit and liabilities are opposing sides of the financial coin, each possessing the capability to empower or ensnare individuals and entrepreneurs, based on their management. In our modern economic ecosystem, credit is not merely a tool; it is a pivotal mechanism that can unlock doors to business growth, expansion, and myriad opportunities. Conversely, mishandled liabilities can erect barriers to these avenues, leading to financial distress and limited business opportunities. Central to navigating this dichotomy effectively is financial literacy, a linchpin in credit management and debt mitigation strategies (Modigliani & Brumberg, 1954). For instance, an in-depth understanding of credit cards goes beyond just knowing how to use them for transactions. It involves comprehension of interest rates, grace periods, fees, and the implications of carrying a balance month-to-month (Durkin, 2000). Misuse, resulting from a lack of knowledge, can lead to accumulating debt, skyrocketing interest, and a damaged credit reputation, all of which can impact an entrepreneur's ability to secure financing for business growth. Loans, whether for business expansion, equipment purchase, or working capital, come with their own set of intricacies. The principles of amortisation, variable versus fixed interest rates, and the impact of loan tenure on total interest paid, are areas where financial literacy plays an instructive role for entrepreneurs. A well-informed borrower can better evaluate loan offers, discerning the most favourable terms and conditions, in line with their business's financial standing and goals (Stango & Zinman, 2009).

Debt management, an art in itself, requires strategic planning, discipline, and foresight for entrepreneurs. Beyond merely servicing debt, it involves optimising repayment structures, leveraging tools like debt consolidation, and understanding the implications of different types of debt for financial health (Lusardi & Tufano, 2015). A financially literate entrepreneur is better equipped to prioritise high-interest debts, negotiate with creditors, and chart a clear path out of indebtedness, ensuring that business operations remain financially sustainable. In essence, the nexus between credit, liabilities, and financial health is intricate and interwoven. However, the thread of financial literacy runs through this nexus, offering clarity, direction, and control. As posited by Mian and Sufi (2014), the 2008 financial crisis spotlighted the catastrophic repercussions of mismanaged credit, and lack of financial understanding within the business world. Thus, enhancing financial literacy is not merely an academic endeavour; it is a societal imperative for entrepreneurs, ensuring that credit remains a stepping stone to business prosperity, rather than a stumbling block to financial well-being and entrepreneurial success.

8.2.5 Insurance and Risk Mitigation

In the intricate fabric of financial planning and management, insurance emerges as a quintessential instrument for shielding entrepreneurs from unforeseen adversities and their associated economic repercussions. At its core, insurance is fundamentally about risk transference—a means of shifting the financial burden of potential losses from the individual to an insurance provider, for a fee or premium (Black & Skipper, 2000). This tool, however, extends beyond mere risk transfer. When understood and utilised judiciously, insurance becomes a bulwark against fiscal shocks that could otherwise cripple an individual's or business's financial trajectory. Here, the role of financial literacy becomes invaluable, bridging the gap between recognising risks and astutely mitigating them through insurance mechanisms (Lusardi & Mitchell, 2014).

Diving into the variants, insurance, for instance, serves as a safeguard for entrepreneurs, ensuring financial survival in unforeseen challenges. While the basic premise seems straightforward, the nuances—such as policy types, coverage limits, and deductibles, demand a level of financial literacy to make informed decisions tailored to specific needs, and to safeguard against risks faced by the business (Zietz, 2003). Given the escalating worldwide medical costs, health insurance has metamorphosed to become a vital consideration for many businesses. Beyond merely understanding the extent of coverage and exclusions, financial literacy elucidates the intricacies of deductibles, out-of-pocket maximums, and network constraints. This knowledge equips business owners with what is necessary to optimise their healthcare expenditures, securing the best care for their employees without incurring exorbitant costs (Finkelstein & McGarry, 2006). Property and liability insurance, too, play pivotal roles in asset protection for businesses. While insuring business property safeguards against potential damages from calamities or thefts, liability coverage protects against legal claims. Grasping the differences between various types of coverage, or discerning the implications of claim deductibles, necessitates a basic grounding in financial literacy (Hoyt & Khang, 2000). However, it is not just about understanding the distinct insurance variants. A higher plane of financial literacy equips entrepreneurs with the nous to strategically bundle insurances, avail themselves of discounts, and even comprehend the impact of business credit scores on insurance premiums (Doherty & Schlesinger, 1983).

8.2.6 Tax Literacy

Within the vast domain of financial literacy, tax literacy emerges as an indispensable subset, influencing business growth and broader economic dynamics. To an individual or an entrepreneur, taxation is not merely a statutory obligation; it significantly affects financial posture, dictating choices on business operations, investments, and financial planning (Slemrod, 2003). Understanding the nuances of taxation—the types, rates, deductions, and implications—is fundamental to making astute financial decisions, and safeguarding a business from inadvertent non-compliance, which could lead to penalties and potential legal repercussions.

Tax literacy encompasses several layers. At the most fundamental level, entrepreneurs must grasp the differences between direct and indirect taxes. Direct taxes, such as corporate income tax, directly affect a business's earnings, with implications for profit margins. Knowledge of tax brackets, available deductions, credits, and exemptions can significantly affect a business's tax liability, shaping its investment strategies and financial planning (Chetty et al., 2013). Indirect taxes, on the other hand, such as sales tax or value-added tax (VAT), influence the cost of goods and services, impacting the pricing strategies of a business. A comprehensive understanding of these taxes can inform pricing strategies and expenditures, especially concerning significant expenditures, allowing entrepreneurs to strategise business operations to optimise financial outcomes (Besley & Rosen, 1999). Moreover, for entrepreneurs who venture into investments or exit strategies, being conversant with capital gains and tax literacy become vital. Recognising the differential tax implications of short-term versus long-term capital gains can inform investment decisions and exit timing, potentially optimising post-tax returns and business profitability (Poterba, 1987).

Tax literacy also possesses a broader socio-economic dimension. From a macroeconomic perspective, tax policies can be used to influence broader economic trends, stimulate investments, or even redistribute wealth, directly impacting business operations and strategies (Mankiw et al., 2009). Armed with tax literacy, an informed entrepreneurial community can more effectively participate in economic and policy processes, understanding and advocating for tax policies that align with their business objectives and economic perspectives, emphasising the fact that tax knowledge can also be a strategic tool for business growth and financial sustainability.

8.2.7 Retirement Planning

At the intersection of long-term financial strategy and fiscal prudence lies the intricate domain of retirement planning, a realm important for both ordinary individuals and entrepreneurs. Embodying more than just the act of saving money for the post-work years, this necessitates a comprehensive grasp of various financial instruments, their interplay in the macro-economic landscape, and their implications for individual wealth accumulation and preservation. From the perspective of financial literacy, retirement planning represents a synthesis of several financial tenets, amplifying their importance in ensuring a financially secure future (Lusardi & Mitchell, 2007), a concern that extends to business owners planning for their retirement and succession. To begin with, retirement planning involves understanding different retirement savings vehicles, from employer-sponsored plans like the 401(k) in the United States, to individual retirement accounts (IRAs), or government pension schemes in various countries (Hubbard & Skinner, 1996). Each of these instruments comes with its own set of tax implications, withdrawal rules, and potential employer match contributions, which dictate their suitability and attractiveness for individual investors and business owners. Additionally, understanding the underlying investments within these retirement accounts is equally important for entrepreneurs. Depending on one's appetite for risk and the time horizon until retirement or business exit, the asset allocation between equities, bonds, and other securities can significantly influence the eventual retirement corpus or business evaluation. Herein lies the application of the principles of diversification (Markowitz, 1952), where the aim is to maximise returns for a given level of risk, by strategically allocating assets.

Furthermore, grasping the exponential growth potential of compounded returns over extended periods can guide individuals, including entrepreneurs, in starting their retirement savings early, thus capitalising on the power of time and recurring investments (Bodie et al., 2007). Inflation is another critical factor affecting both individuals and business owners. As Mankiw (1990) notes, inflation can erode the real value of savings over time. Therefore, projecting realistic post-retirement expenses, adjusted for expected inflation, is key to ascertaining the required retirement corpus, or ensuring the business remains profitable and resilient. Additionally, understanding life annuities, healthcare provisions, and the potential need for long-term care in retirement becomes imperative for both individuals and entrepreneurs, especially given the increasing life expectancies worldwide (Brown & Warshawsky, 2001). For entrepreneurs, this also extends to business continuity and succession planning. These considerations play into the broader dynamics of retirement planning, ensuring that individuals or entrepreneurs are financially equipped, not just to retire without fiscal strain, but to smoothly transition their businesses to the next generation while securing their financial futures.

8.2.8 Estate Planning

Deeply enmeshed in the fabric of long-term financial strategy and inter-generational wealth transfer, estate planning emerges as a pivotal domain, requiring acute foresight and an intricate understanding of legal and financial frameworks. Not only does it encompass the disbursement of assets posthumously, but it also extends to ensuring that an individual's wishes, both financial and healthcare-related, are upheld during their lifetime, especially if they are incapacitated (Bernheim et al. 1985). From a financial literacy perspective, estate planning underscores the importance of comprehensive knowledge, as it connects individual wealth management with the broader legal system, economic landscape, and societal norms (Joulfaian, 2004). This also concerns entrepreneurs, as they consider business succession and legacy planning. Central to the realm of estate planning is the concept of a will. A will, essentially a legal document, dictates how an individual's assets will be distributed upon their demise. A grasp of how wills function, their binding nature, and the implications of dying "intestate" (without a will) is crucial (Holtz-Eakin et al., 1993).

Financial literacy equips individuals and business owners to articulate their financial desires clearly, ensuring their assets are divided as they intended, minimising potential disputes among heirs, and facilitating a smooth transition. Trusts, another cornerstone of estate planning, serve as vehicles to manage and protect assets, and extend to safeguarding the future of a business. Understanding the difference between revocable and irrevocable trusts, their tax implications, and the roles of trustees, is paramount for lay individuals as well as entrepreneurs (Gale & Scholz, 1994). Given their ability to bypass probate (the legal process of administering a deceased person's estate), trusts can be established to offer quicker asset transfers to beneficiaries, and be structured to provide income to heirs over an extended period or under specific conditions.

The intricacies of inheritance laws, which vary significantly across jurisdictions, form another layer in this mosaic. These laws define asset distribution in the absence of a will, and can profoundly impact wealth transfers across generations (Modigliani, 1988), including the transfer of business ownership. Financially literate individuals, cognisant of such regulations, can strategically structure their estate or business succession plans to optimise asset transfers, reduce inheritance tax liabilities, and ensure that their descendants' fiscal future aligns with their envisioned legacy. Additionally, financial literacy aids in understanding instruments like the power of attorney (for financial decisions) and advanced healthcare directives (for medical decisions), which determine who will make decisions on behalf of an individual or a business owner should they become incapacitated (Zweig, 1995).

Financial literacy and political-economic awareness converge in a manner that extends beyond mere theoretical constructs; they underscore the practical realities that govern both individual and collective fiscal dynamics (Lusardi & Mitchell, 2014), navigating the complexities of business ownership and succession planning. In an interconnected global environment, understanding the intricate mesh of politics, socio-economic factors, and finance, is quintessential for sustainable wealth management, business continuity, and risk mitigation.

8.2.9 Understanding of Political Systems and Governance

Having a thorough understanding of political systems and governance is essential for anyone looking to comprehend the complex relationship between politics and finance. Fundamentally, significant changes in political directives or governance structures can influence the economy, affecting both individual and institutional investments. As an analogy, portfolio managers consistently evaluate corporate governance during investment considerations. As evidenced by Gompers et al. (2003), there is a clear link between strong corporate governance and increased shareholder value. Additionally, grasping the subtleties of political systems aids in forecasting fiscal measures, such as tax reforms or monetary policies, which in turn influence interest rates, inflation, and currency values. This knowledge is pivotal for individuals during strategic planning, be it for immediate objectives like buying property, or long-term aims like retirement. In our interconnected global landscape, being aware of international political dynamics is vital. Given the interconnectedness of economies, a political incident in one nation can echo across global markets. It is notable that, while understanding politics is universally significant, it holds particular importance in areas like the Arab region, characterised by its prevailing uncertainty, as noted by Elbanna et al. (2020).

8.2.10 Awareness of Political Climate and Stability

The political climate and its stability significantly influence a nation's economic structure. For example, political turmoil can discourage foreign investments, hinder trade partnerships, and even result in economic sanctions in severe situations. Moreover, financial markets, sensitive to such shifts, can exhibit increased volatility at mere whispers of political instability. In a politically stable setting, businesses can better predict regulatory changes, aiding in long-term planning and investments. On the other hand, turbulent political conditions might lead to reluctance in initiating long-term projects, due to uncertain policy futures. Beyond macro-economic considerations, understanding the political atmosphere is also crucial for personal financial choices. For instance, individuals may decide to diversify their investments geographically, shifting assets to more politically stable regions, when their home country faces political uncertainty.

8.2.11 Knowledge of Policy and Legislation

Regulatory changes can mould and even transform economic pathways. Such alterations can present both opportunities and challenges for businesses and individuals, underscoring the need to keep pace with legislative updates. For instance, an update of environmental rules might require manufacturers to adjust their processes, leading to added investments or changes in supply chains. On the other hand, deregulation could foster innovation and competition, opening new growth horizons for businesses. For individuals, policy knowledge can guide decisions, when dealing with everything from tax strategies to real estate purchases. For instance, changes in tax laws might influence investment decisions, while adjustments to property laws can affect choices between buying and renting. Additionally, in our globalised world, international policies are equally significant. Global regulations and treaties, like trade agreements or environmental pacts, can deeply impact local sectors. Moreover, in our digital era, with fast-evolving sectors like Fintech (financial technology), regulations often struggle to keep pace. In such fluid sectors, businesses that foresee and brace for regulatory changes gain a competitive advantage, ensuring smoother adaptations.

8.2.12 Insight into International Relations

International relations profoundly impact economic landscapes, especially in our interconnected global environment. International relations go beyond just official ties between countries; they cover a wide range of interactions, from trade agreements to alliances and conflicts. Such interactions significantly affect global financial markets. For example, a new trade deal between two major economies might boost stock markets, while escalating geopolitical tensions could drive investors towards safe assets like gold. For global businesses, understanding these relations is crucial. Consider supply chains that span multiple countries: a disagreement between two nations could disrupt these chains, leading to delivery delays, higher costs, or even the search for new suppliers. For individual investors, understanding international relations can be a tool for portfolio diversification. Being aware of global dynamics allows for better judgement in allocating assets across regions, sectors, and asset classes.

8.2.13 Understanding of Economic Policies

Macro-economic policies, including fiscal, monetary, and trade policies, are fundamental to a nation's economy, influencing financial markets, interest rates, inflation, employment, and trade balances. As highlighted by Taylor (1995), understanding these policies is vital for informed financial and business decisions. Fiscal policies, related to government revenue and spending, can sway economic growth and individual disposable incomes. Monetary policies, set by central banks, affect the money supply, interest rates, borrowing costs, and investment attractiveness. Trade policies, which determine tariffs and influence trade agreements, have significant implications for international businesses and their competitive strategies. For entrepreneurs and businesses, understanding these policies is essential for strategic planning. Whether considering expansion, investments, or simply day-to-day operations, anticipating the direction of economic policies can provide valuable insights. For example, foreseeing a contractionary monetary policy might lead a business to reconsider a major loan or an individual, to lock in a mortgage rate.

8.2.14 Active Participation

The relationship between politics and economics is dynamic, with both individuals and businesses influencing policies, not just adapting to them. Individuals can actively participate, by attending community meetings, joining advocacy groups, or even venturing into politics. Such platforms allow them to voice concerns and advocate for policies that resonate with their economic interests. Businesses, especially influential ones, can engage in lobbying to advocate for favourable regulations, or to counter unfavourable ones. By actively participating, both individuals and businesses can influence their economic futures, and contribute to a more stable and predictable business environment, reducing the risk of sudden, disruptive policy changes.

8.2.15 Critical Consumption of News

In today's digital age, the abundant access to information is both advantageous and problematic. While immediate news updates can offer crucial insights, they also introduce the risk of misinformation, biases, and speculative narratives. The vast influx of news, combined with the ease of content creation, complicates the distinction between truth and falsehood. For investors and entrepreneurs, falling victim to misinformation can result in substantial financial setbacks. An unfounded rumour might prompt an investor to impulsively trade, possibly leading to losses. Similarly, an entrepreneur may change their business strategy based on an inaccurate economic forecast, wasting resources or missing out on opportunities. Considering the significant risks in the financial sector, and the possibility of bubbles driven by misinformed sentiment, it is vital to critically evaluate information sources. Through careful and informed consumption of news, individuals can make sound financial decisions, safeguarding themselves against the perils of misinformation.

In conclusion, intertwining financial literacy with political and socio-economic cognisance equips individuals, including entrepreneurs, with a holistic perspective, fostering prudent financial decisions. This multifaceted awareness acts as a bulwark against unforeseen financial adversities and amplifies opportunities, to harness economic prospects, irrespective of the geopolitical and macro-economic milieu (Lusardi & Mitchell, 2014).

8.3 Economic Resilience Through Financial Literacy: Navigating Fragile Times

In the intricate tapestry of today's entrepreneurial landscape, episodes of economic turbulence, notably the 2008 financial collapse, and the ongoing ramifications of the COVID-19 Pandemic, have driven home the seminal importance of robust financial literacy as a bastion against such volatile disruptions (Gorton, 2010; McKibbin & Fernando, 2020). For emergent and established entrepreneurs, the assimilation of profound financial sagacity is indispensable, in steering businesses through the quagmire of transient economic constructs and the kaleidoscope of market vicissitudes.

The 2008 US financial crisis, dissected by Lusardi et al. (2011), presented a stark portrait of the dangers posed by a lack of financial insight. Many businesses, especially those founded on shaky financial grounds—and knowledge—found themselves at risk. The consequences were felt globally, with emerging markets like Pakistan and Malaysia displaying vulnerabilities similar to their more developed counterparts (Ali et al., 2020). Historically, economic disturbances have consistently highlighted the value of financial literacy. The 1997 Asian financial crisis stands as a testament to the dangers of inadequate financial knowledge, with countries from Thailand to South Korea suffering the consequences of limited comprehension of foreign debt and currency dynamics. Similarly, the Dot-com bubble of the early 2000s was another painful lesson on the importance of dividing cautious sustainable business models from fleeting trends. The Eurozone crisis, which began in 2009, emphasised the need for robust personal financial management, in the face of systemic economic challenges. The aforementioned historical economic cataclysms underscored the exigency of adopting discerning and meticulous entrepreneurial initiatives and sustainable financial frameworks, and of separating ephemeral market trends from solid, long-term business paradigms.

However, every challenge also presents a learning opportunity. By the time the world had grappled with the COVID-19 Pandemic, the lessons from past crises began to crystallise. Those enterprises and individuals equipped with sound financial knowledge were better positioned to navigate the turmoil, drawing upon principles like comprehending savings, investments, and risk diversification. Their resilience during these uncertain times was noticeable, exemplifying the indispensability of financial literacy in cultivating entrepreneurial resilience during cyclical economic ebbs and flows (Lusardi & Tufano, 2015). For example, by amplifying the importance of financial literacy, troubled nations can arm their populations with the tools necessary to withstand fiscal shocks, and promote sustained economic progress (Hastings et al., 2013).

As global financial systems become more intertwined, and as the choices available to entrepreneurs grow in complexity, financial literacy transitions from a mere asset to an imperative. Beyond simply safeguarding against economic downturns, it paves the way for fostering entrepreneurship and innovation, and ensuring that in an interconnected world, individuals and businesses do not just survive but thrive. Furthermore, financial literacy extends beyond basic money management. It encompasses a vast range of skills, from deciphering breaking financial news to understanding the global economic implications of a singular event (Prasad 2014). In an interconnected global economy, the aftershocks of a decision in one region can reverberate across continents, influencing trade, currency dynamics, and investment climates. Another often overlooked facet of financial literacy is its role as a deterrent against fraud. A savvy individual is less likely to fall prey to financial scams or dubious investment schemes, especially when economic downturns make such deceptive practices rampant (Huston, 2010). Moreover, there is a symbiotic relationship between financial literacy and mental well-being. The security that stems from financial knowledge can significantly alleviate stress and anxiety, especially during economic crises (Shim et al., 2015). As the work paradigm shifts towards gig economies and less traditional forms of employment, financial awareness becomes even more crucial, especially for those who traverse these new, often unpredictable terrains (Manyika et al., 2016).

8.4 Enhancing Financial Resilience Through Comprehensive Financial Education

The synergy between a comprehensive financial education and entrepreneurial fortitude is inextricably linked, creating a resilient foundation for innovative endeavours in a fluctuating economic landscape. Nurturing financial acumen fosters an environment wherein individuals and enterprises are capacitated to steer through economic adversities adeptly, thereby promoting entrepreneurial success. Individuals and businesses with sound financial knowledge have historically navigated economic downturns with greater proficiency. However, Csiszárik-Kocsir et al. (2021) have rightfully identified that fundamental financial knowledge, while indispensable, may not be sufficient. They accentuate the importance of historical context, underscoring the need for integrating lessons from past financial downturns into educational curricula. This enriched understanding marries theoretical knowledge with the invaluable wisdom of experience. By incorporating entrepreneurial education and case studies of resilient entrepreneurs into financial curricula, students can gain practical insights into building and managing businesses during economically challenging times. The intricate relationship between individual behaviour and financial decisions is of paramount importance. Delving into the heart of this, Ottaviani and Vandone (2018) emphasise the profound influence of impulsivity on financial behaviours, which can also impact entrepreneurial decision-making. Such impulsivity, manifested through a penchant for immediate gratification, frequently leads to adverse financial outcomes. Alarmingly, when combined with limited financial literacy, these impulsive traits magnify an individual's vulnerability during financial crises. Recognising the importance of addressing these behavioural nuances becomes imperative, moving beyond a one-size-fits-all educational approach.

Bruhl's (2019) study on the financial literacy of German students underscores a trend that is increasingly noticeable across various global regions. Despite Germany's stature as an advanced economy, the surprising lacunae in basic financial understanding among its students is alarming. This concern is echoed in other parts of the world. Take Japan, for instance. Fujisaki and Nakata's (2018) investigation into Japanese university students revealed similar trends. In a nation renowned for its technological advancements and strong work ethic, many students exhibited limited knowledge of essential financial matters, which can have a tremendous impact on their future entrepreneurial activity and outcomes. What is particularly striking is their discomfort with credit-related concepts, given Japan's deeply embedded cash culture. Meanwhile, in Australia, a study by Wilson and Tinkler (2017) found that many Australian students had a limited understanding of basic financial concepts like inflation and compound interest. This is particularly disconcerting, given the country's strong emphasis on self-managed superannuation funds, and the importance of personal financial planning for retirement. Brazil presents another compelling example. A 2018 study, by Oliveira and Santos, found that despite Brazil's rapid economic growth and increasing middle-class population, financial literacy among Brazilian university students was significantly lower than that of their counterparts in more developed countries. The complexity of Brazil's taxation and finance systems further emphasises the country's need for robust financial education. Sági et al. (2020) indeed highlight a troubling trend—the overreliance of students on external organisations for financial guidance. This trend is not restricted to any one country. For example, in India, a study by Rajput and Agrawal (2019) found that students frequently depend on advice from family elders or friends, often leading to misconceptions about financial matters.

However, not everything is bleak. As seen in Germany, students from certain academic backgrounds such as economics and technology display an enhanced understanding of financial principles. For example, a Canadian study by Lefebvre and Vieira (2020) notes that students enrolled on commerce or engineering programmes outperformed their peers in arts or humanities in terms of financial literacy. This underscores the potential benefits of infusing tailored financial education modules within diverse academic disciplines. Such initiatives could pave the way for a generation better equipped to navigate the intricate landscape of global finance. In crafting a future marked by financial resilience and entrepreneurship, a dynamic and multifaceted approach to financial education is non-negotiable. By intricately weaving lessons from historical financial downturns with behavioural economics insights, and entrepreneurial education into educational frameworks, we stand a chance of creating a financially literate society, capable of entrepreneurial success. As we tread the complex pathways of our global economy, this holistic, in-depth approach to financial education is not merely an advantage, it is an imperative for enduring stability.

8.5 Gender, Financial Literacy, and Crisis Preparedness: Crafting a Path to Equitable Economic Resilience

In the contemporary global economy, understanding the nexus of gender and financial literacy, along with entrepreneurship is essential. The profound influence of this relationship not only shapes a business's resilience against financial tremors, but also contributes to economic development. This real case study illustrates how a lack of relevant financial literacy among women during times of crisis can lead to the failure of their businesses (for the full case study, see Mauji et al., 2022). In Doha, Qatar, Fatma, the founder of Little Birds Kindergarten, faced financial adversity due to the COVID-19 Pandemic. As a passionate advocate of modern educational methods, she introduced a British-style curriculum, and consistently integrated advanced educational technology. Despite gaining an excellent reputation, the pandemic-induced downturn severely impacted enrolments. Struggling to keep her kindergarten afloat, Fatma, like many female entrepreneurs in the GCC, relinquished her salary and utilised her personal savings. Unfortunately, by January 2022, Little Birds Kindergarten had to cease operations. Many parents were facing job losses or salary reductions, making it increasingly difficult for the finance department to collect tuition payments promptly. Fatma acknowledged the uncertainty, wondering if her kindergarten might have had a different fate under different circumstances. Her scenario underscores the importance of such initiatives, illustrating how a combination of unforeseen economic challenges and a potential gap in financial preparedness can endanger even the most dedicated ventures.

A historical lens reveals a recurring theme: women often bear the brunt of financial hardships (Allgood & Walstad, 2016), a theme brought into sharp relief in crises such as the COVID-19 Pandemic (Fonseca et al., 2020). While countries like Canada, Japan, and Sweden are frequently lauded for their advancements in gender equality, beneath the surface, challenges persist. Despite having a higher percentage of educated women, a significant proportion of them in these nations, around 30% according to studies (Lusardi & Mitchell, 2014), still struggle with self-perceived financial competence. This is compounded by the financial sector's biases, evidenced by the fact that women receive 20% fewer offers of credit on average than their male counterparts.

For nations like Zimbabwe, Sudan, Nigeria, and Bangladesh, the issues are more overt and intertwined with socio-cultural norms. Women's financial vulnerabilities there are not just about knowledge but also about access to banking facilities, credit, and even basic financial autonomy. For instance, a significant proportion of women, upwards of 60% in some regions, lack access to basic banking facilities, let alone advanced financial services. Systems in these countries often sideline women's financial perspectives, resulting in literacy programmes that lack relevance to the unique challenges women face (Johnson & Nino-Zarazua, 2011. These challenges can impact the nation's economic development, by limiting its entrepreneurial prospects. However, countries like India, Brazil, Ghana, and Nepal are paving the way with targeted financial literacy programmes. For instance, microfinance initiatives in India have touched the lives of over 50 million women, equipping them with financial tools and training. As a result, there has been a reported 40% increase in female-led startups in regions with these programmes. Brazil's emphasis on women's financial education resulted in a 15% increase in household investments in children's health and education. Similarly, in Ghana and Nepal, commuAgarwalnity-based workshops have not only educated women but have shifted entire community mindsets, leading to a reported 25% increase in women-driven financial decisions (Singer et. al, 2013). Below are some tailored tools for women to use, to build and fortify their financial resilience:

1. **Customised Financial Education**: Beyond generic financial literacy, there is a need to concentrate on areas like entrepreneurship, digital finance, and investment, tailored explicitly for women.
2. **Strengthening Institutional Support**: Banks, credit unions, and microfinance institutions should design products and services, bearing women's financial needs and constraints in mind.
3. **Building Collaborative Networks**: By forging strong partnerships between governments, international organisations, NGOs, and grassroots women's collectives, it is possible to create a cohesive strategy that addresses both macro- and micro-level challenges.
4. **Empirical Research and Policy Integration**: Research should not just highlight problems but guide solutions. Insights into gender-specific financial behaviours, access barriers, and decision-making patterns can provide the foundation for gender-responsive policies.
5. **Cultural and Social Sensitisation**: Socio-cultural beliefs are at the core of many financial barriers for women. It is essential to integrate financial literacy with campaigns that address these deep-seated biases, ensuring that both men and women understand the importance of women's financial empowerment.

8.6 Conclusion and Recommendations

Understanding the landscape of financial fragility and the role of financial literacy necessitates a complex and multifaceted approach. Financial literacy has emerged as a pivotal axis around which modern economies revolve, with its significance accentuated by the inter-dependencies of global markets. A close examination of the OECD 2020 report and the S&P 2015 survey unveils a stark panorama: a global populace that is often ill-equipped to navigate the intricacies of finance. These vulnerabilities were already theoretically apparent, and manifested themselves tangibly during crises such as the COVID-19 Pandemic.

Considering the varied landscapes of developed and developing countries offers illuminating insights. For instance, despite its economic prowess, Germany has revealed startling gaps in financial literacy among its student population. This brings forth an essential nuance—the challenge of financial literacy is not limited to economies with limited resources, but extends to economic powerhouses with sophisticated financial infrastructures. Conversely, the U.S., though it also endured the tremors of the 2008 financial crisis, showcased commendable resilience and adaptability, by significantly reducing its financial vulnerability over the subsequent years. This improvement could be attributed, in part, to conscious interventions and awareness initiatives. In contrast, many developing nations are still grappling with the fundamental challenges of financial literacy, with only a third of their adult population demonstrating adequate comprehension.

However, one cannot discuss financial literacy without considering the gender dynamics at play. Contemporary research underscores the pronounced disparities faced by women in economic spheres. The findings by Yakoboski et al. (2020) and Pepur et al. (2022) emphasise the multiplier effect of financial literacy among women, especially in crisis scenarios. The challenges faced by women, from wage disparities in nations like the U.S., to limited access to formal banking in parts of Africa, highlight the intertwined challenges of gender and finance.

For a truly resilient global economy, a multi-pronged strategy is non-negotiable. This includes a fusion of fundamental financial education to promote the prospects of entrepreneurship growth and success, insights from historical economic downturns, and cutting-edge behavioural economics. It is vital to tailor financial education initiatives, to address diverse demographics and unique challenges. For instance, programmes in Scandinavian countries, known for their gender parity, could focus on refining and fine-tuning financial acumen, whereas in parts of Asia or Africa, basic financial literacy and access to funds could be the initial focus.

In closing, as economies become increasingly intertwined, and financial instruments grow in complexity, the call for comprehensive financial literacy has never been louder or clearer. Addressing this imperative will fortify individuals and entrepreneurs against economic shocks, and pave the way for a more robust, equitable global economy, fostering mutual growth and stability.

Following this conclusion, we will delineate the essential pillars that support the framework for an evolved financial education. Furthermore, we will introduce a dynamic proficiency matrix, a tool tailored for comprehensive financial assessment.

8.6.1 Charting the Path to Enhanced Financial Literacy: Key Pillars

In the evolving world of finance, ensuring robust financial literacy has never been more vital. This empowers individuals to make informed decisions, and strengthens economies, by fostering responsible financial behaviours. As we explore the basic aspects of enhancing financial education, it is pivotal to recognise these crucial elements:

- Comprehensive Financial Education Programmes:
  - o Start Early: Integrate financial education into school curriculums, starting at the elementary level. Early exposure can build the foundation for informed financial decisions in adulthood (Fernandes et al., 2014).
  - o Lifelong Learning: Offer refresher courses for adults, accommodating the evolving nature of the financial market, and newer financial products.

- Tailored Programmes for Vulnerable Groups:
- Gender Focus: Recognising the specific financial challenges faced by women—wage disparities, career breaks for child-rearing, etc., courses should address these nuances (Duflo, 2012).

- For the Marginalised: Programmes should be designed keeping in mind marginalised communities, accounting for challenges like access to banking or credit facilities (Collins et al., 2009).
- Promotion of Socio-economic and Political Cognisance:
- Interdisciplinary Approach: Blend finance education with global affairs, ensuring students understand the broader implications of financial decisions on the world stage (Shiller, 2017).
- Local Context: While global perspectives are crucial, courses should touch upon local socio-political dynamics that can impact personal finance (Hastings et al., 2013).
- Collaborative Initiatives:
- Public–Private Partnerships: Private sectors can bring expertise, technology, and funding to make programmes more effective.
- Engage Community Leaders: In regions where formal education might be limited, community leaders can be pivotal in disseminating financial knowledge (Shiller, 2017).
- Labour Market Reforms:
- Equal Pay Initiatives: Governments should enforce policies ensuring equal pay for equal work, reducing financial vulnerabilities for women (Goldin, 2014).
- Flexible Work Arrangements: Recognising the dual roles many women play (work and home), offering flexible work arrangements can help them balance both effectively (Mas & Pallais, 2017).
- Localised Solutions for Global Problems:
- Cultural Sensitivity: Financial behaviours and decisions are often rooted in local cultures and beliefs. Recognise and incorporate these unique aspects into educational contents (Chen, 2013).
- Regional Case Studies: Use regional financial crises or success stories as case studies, to provide relatable lessons (Reinhart & Rogoff, 2009).
- Regular Assessments and Adaptation:
- Feedback Loops: Create mechanisms to gather feedback from programme participants, to continually refine contents and delivery methods (Drexler et al., 2014).
- Stay Updated: With the rapid evolution of the financial landscape, especially with such innovations as digital assets and the fintech revolution, ensure contents are updated regularly.
- Promote Research:
- Funding: Allocate funds specifically for researching the interplay between financial literacy and economic outcomes.
- Collaboration: Encourage collaboration between universities, thinktanks, and financial institutions for comprehensive research (Campbell, 2006).

8.6.2 The Dynamic Financial Proficiency Matrix: A Tailored Assessment

In practical terms, we advocate for individuals, particularly entrepreneurs, to appraise their financial literacy. To facilitate this, we offer here a versatile matrix that functions as a dynamic template. This tool can be adjusted to align with the distinct requirements of each individual or group. For instance, the average person might find it beneficial to concentrate on the first eight components, whereas investors and entrepreneurs could derive greater advantage from utilising the full array of elements. The weighting attributed to each component can be modulated in accordance with the unique needs and priorities of each group, thereby enhancing the applicability and relevance of the matrix.

It is important to note that the total weighted score can vary between 1 and 10, signifying different degrees of financial acumen. A score as low as 1 indicates very limited financial proficiency, while a score approaching 10 signifies comprehensive financial proficiency. We offer the following classifications for each score range:

1–2: "Inadequate Financial Proficiency"—Those scoring within this range may exhibit limited understanding of financial concepts, and are likely to encounter difficulties making informed financial decisions.
3–4: "Below-Average Financial Proficiency"—Individuals within this score range may possess rudimentary financial knowledge, but still face challenges in effective financial management and decision-making.
5–6: "Average Financial Proficiency"—Falling in this range denotes a reasonable grasp of financial principles, and the ability to make reasonably informed financial decisions.
7–8: "Above-Average Financial Proficiency"—Individuals with scores in this range have a robust understanding of financial concepts, and can confidently make informed and strategic financial decisions.
9–10: "Exceptional Financial Proficiency"—A score in this range represents an advanced mastery of financial matters, with individuals capable of adeptly strategising and navigating complex financial landscapes.

Individualised Financial Proficiency Assessment Matrix

Components of Financial Proficiency (1)	Score (2)	Weight (3)	Weighted score (4)	Comments (5)
1. Financial Planning and Cost Management				
1. Savings and Economic Security				
1. Investment Strategies				
1. Credit, Liabilities, and Financial Health				
1. Insurance and Risk Mitigation				
1. Tax Literacy				
1. Retirement Planning:				

(continued)

(continued)

Components of Financial Proficiency (1)	Score (2)	Weight (3)	Weighted score (4)	Comments (5)
1. Estate Planning				
1. Comprehension of Political Systems and Governance				
1. Awareness of Political Stability				
1. Legislative Acumen				
1. Insight into International Relations				
1. Understanding of Fiscal Policies				
1. Active Participation				
1. Critical Consumption of News				
Total		100%		

Source Authors

Guidelines for Preparing to Use the Matrix

Score (2): Scores range from 1 to 10 for each component, with higher scores indicating greater awareness or proficiency and lower scores signifying less awareness or proficiency.
Weight (3): The weight is distributed among the components, totaling 100%. More weight signifies greater importance to the individual or group, and it should be allocated according to the significance of each component. The sum of the component weights should equal 100%.
Weighted score (4) = It is the product of Score (2) plus Weight (3), and it will range from 1 to 10.
Higher scores (close to 10) indicate advanced financial proficiency, while lower scores (close to 1) denote inadequate financial proficiency. The interpretation of the ranges is as outlined above.
Comments (5): Users should add comments and clarifications to the allocated scores and weights, providing insights into the reasoning behind the assigned values.

By categorising the total weighted scores in this manner, our objective is to provide a clearer, more discernable interpretation of an individual's financial awareness and literacy level, grounded in their self-assessment. This stratification intends to give individuals insightful reflections on their financial proficiency, allowing them to discern their strengths and areas needing enhancement in financial knowledge and application. It not only aids in identifying specific areas of improvement, but also facilitates the formulation of targeted strategies to augment financial literacy. This nuanced approach is especially beneficial for entrepreneurs, as it enables a deeper understanding of financial dynamics, essential for making informed and strategic business decisions, and so ultimately contributing to the sustainability and success of their ventures.

Acknowledgements This chapter was made possible by a Qatar University & Marubeni Corporation Grant # (MQJRC-2020-1). The statements made herein are solely the responsibility of the authors.

Bibliography

African Development Bank Group. (2017). *Community-centric financial education initiatives.* ADB Publications.

Ali, R., Keppler, M., Rahman, H., & Ahmed, I. (2020). The financial literacy conundrum: A comparative study of financial literacy in emerging markets. *Global Finance Journal, 45*, 100517.

Allgood, S., & Walstad, W. B. (2016). The effects of perceived and actual financial literacy on financial behaviors. *Economic Inquiry, 54*(1), 675–697.

Besley, T., & Rosen, H. S. (1999). Sales Taxes and Prices: An Empirical Analysis. *National Tax Journal, 52*(2), 157–178.

Bernheim, B. D., Shleifer, A., & Summers, L. H. (1985). The strategic bequest motive. *Journal of Political Economy, 93*(6), 1045–1076.

Bernheim, B. D. (1999). *Taxation and saving.* National Bureau of Economic Research Working Paper Series, No. 7061.

Black, K., & Skipper, H. (2000). *Life and Health Insurance.* Prentice Hall.

Bodie, Z., Kane, A., & Marcus, A. J. (2008). *Investments.* McGraw-Hill/Irwin.

Bodie, Z., Treussard, J., & Willen, P. (2007). The theory of life-cycle saving and investing. *Public Policy and Aging Report, 17*(3), 28–33.

Brown, J. R., & Warshawsky, M. J. (2001). Longevity-Insured Retirement Distributions from Pension Plans: Market and Regulatory Issues. *Aging Issues in the United States and Japan* (pp. 135–157). University of Chicago Press.

Bruhl, V. (2019). A survey of financial literacy among German students. *Journal of Behavioral and Experimental Finance, 23*, 198–210.

Cambridge Centre for Alternative Finance (2020). *Understanding digital financial assets.* CCAF Research Series.

Campbell, J. Y. (2006). Household finance. *The Journal of Finance, 61*(4), 1553–1604.

Chen, G. (2013). The determinants of financial literacy: Evidence from China. *Applied Economics Letters, 20*(12), 1143–1147.

Chetty, R., Friedman, J. N., & Saez, E. (2013). Using differences in knowledge across neighborhoods to uncover the impacts of the EITC on earnings. *American Economic Review, 103*(7), 2683–2721.

Collins, D., Morduch, J., Rutherford, S., & Ruthven, O. (2009). *Portfolios of the poor: How the world's poor live on $2 a day.* Princeton University Press.

Csiszárik-Kocsir, Á., Medar, L., & Szűcs, R. (2021). Why do they make mistakes?—Financial literacy among Hungarian university students. *Sustainability, 13*(2), 610.

Doherty, N. A., & Schlesinger, H. (1983). Optimal insurance in incomplete markets. *Journal of Political Economy, 91*(6), 1045–1054.

Drexler, A., Fischer, G., & Schoar, A. (2014). Keeping it simple: Financial literacy and rules of thumb. *American Economic Journal: Applied Economics, 6*(2), 1–31.

Duflo, E. (2012). Women's empowerment and economic development. *Journal of Economic Literature, 50*(4), 1051–1079.

Durkin, T. A. (2000). Credit cards: use and consumer attitudes, 1970–2000. *Federal Reserve Bulletin, 86* (9).

Elbanna, S. & Armstrong, L. (2023). Exploring the integration of ChatGPT in education: adapting for the future. *Management & Sustainability: An Arab Review.* (ahead-of-print).

Elbanna, S., Abdelzaher, D. M., & Ramadan, N. (2020). Management research in the Arab World: What is now and what is next? *Journal of International Management, 26*(2), 1–21.

El-Bassiouny, N. M., Abbas-Ali, H., Hammerschmidt, M., Elbanna, S., & Fröhlich, E. (2022). Editorial: Between antiquity and modernity in sustainability management research: Reflections from board members. *Management & Sustainability: An Arab Review, 1*(1), 1–13.

Evans, J. L., & Archer, S. H. (1968). Diversification and the reduction of dispersion: An empirical analysis. *The Journal of Finance, 23*(5), 761–767.

Fernandes, D., Lynch, J. G., & Netemeyer, R. G. (2014). Financial literacy, financial education, and downstream financial behaviors. *Management Science, 60*(8), 1861–1883.

Finkelstein, A., & McGarry, K. (2006). Multiple dimensions of private information: Evidence from the long-term care insurance market. *American Economic Review, 96*(4), 938–958.

Fisher, I. (1930). *The theory of interest*. Macmillan.

Fonseca, R., Mullen, K. J., Zamarro, G., & Zissimopoulos, J. (2020). Financial literacy and the gender gap. *Applied Economics, 52*(1), 76–94.

Fujisaki, Y., & Nakata, H. (2018). Financial literacy and cash holding in Japan. *Journal of the Japanese and International Economies, 50*, 81–89.

Gale, W. G., & Scholz, J. K. (1994). Intergenerational transfers and the accumulation of wealth. *The Journal of Economic Perspectives, 8*(4), 145–160.

Global Financial Literacy Excellence Center. (GFLEC, 2019). *Financial literacy in advanced economies*. GFLEC Reports, Stanford, CA.

Goldin, C. (2014). A grand gender convergence: Its last chapter. *American Economic Review, 104*(4), 1091–1119.

Gompers, P. A., Ishii, J. L., & Metrick, A. (2003). Corporate governance and equity prices. *Quarterly Journal of Economics, 118*(1), 107–155.

Gorton, G. (2010). *Slapped by the invisible hand: The panic of 2007*. Oxford University Press.

Graham, B., & Dodd, D. L. (1934). *Security analysis*. McGraw-Hill.

Hastings, J. S., Madrian, B. C., & Skimmyhorn, W. L. (2013). Financial literacy, financial education, and economic outcomes. *Annual Review of Economics, 5*(1), 347–373.

Holtz-Eakin, D., Joulfaian, D., & Rosen, H. S. (1993). The Carnegie conjecture: Some empirical evidence. *The Quarterly Journal of Economics, 108*(2), 413–435.

Hoyt, R. E., & Khang, H. (2000). On the determinants of enterprise risk management implementation. *Risk Management and Insurance Review, 3*(2), 37–52.

Hubbard, R. G., & Skinner, J. (1996). Assessing the effectiveness of saving incentives. *Journal of Economic Perspectives, 10*(4), 73–90.

Huston, S. J. (2010). Measuring financial literacy. *The Journal of Consumer Affairs, 44*(2), 296–316.

Johnson, D., & Nino-Zarazua, M. (2011). Financial access and exclusion in Kenya and Uganda. *The Journal of Development Studies, 47*(3), 475–496.

Joulfaian, D. (2004). Inheritance and saving. *Economic Inquiry, 42*(2), 297–312.

Keynes, J. M. (1936). *The general theory of employment, interest and money*. Macmillan.

Lefebvre, J., & Vieira, C. (2020). Financial literacy and overconfidence: Evidence from Canada. *Economic Modelling, 89*, 189–201.

Lusardi, A., Schneider, D., & Tufano, P. (2011). Financially fragile households: evidence and implications. *Brookings Papers on Economic Activity*, 83–150.

Lusardi, A., & Mitchell, O. S. (2007). Baby boomer retirement security: The roles of planning, financial literacy, and housing wealth. *Journal of Monetary Economics, 54*(1), 205–224.

Lusardi, A., & Mitchell, O. S. (2014). The economic importance of financial literacy: theory and evidence. *Journal of Economic Literature, 52*(1), 5–44.

Lusardi, A., & Tufano, P. (2015). Debt literacy, financial experiences, and overindebtedness. *Journal of Pension Economics and Finance, 14*(4), 332–368.

Malkiel, B. G. (1996). *A random walk down Wall Street: The time-tested strategy for successful investing*. WW Norton and Company.

Mankiw, N. G. (1990). The savers-spenders theory of fiscal policy. *American Economic Review, 80*(2), 120–125.

Mankiw, N. G., Weinzierl, M., & Yagan, D. (2009). Optimal taxation in theory and practice. *Journal of Economic Perspectives, 23*(4), 147–174.

Markowitz, H. (1952). Portfolio selection. *The Journal of Finance, 7*(1), 77–91.

Mas, A., & Pallais, A. (2017). Valuing alternative work arrangements. *American Economic Review, 107*(12), 3722–3759.

Mauji, I. N., Elbanna, S., & Al Shamari, J. (2022). The COVID-19 pandemic and the case of little birds kindergarten in Qatar: Strategic challenges and the role of leadership. *Emerald Emerging Markets Case Studies, 12*(4), 1–24.

McKibbin, W. J., & Fernando, R. (2020). The global macroeconomic impacts of COVID-19: Seven scenarios. *Asian Economic Papers, 20*(2), 1–30.

Mian, A., & Sufi, A. (2014). What explains the 2007–2009 drop in employment? *Econometrica, 82*(6), 2197–2223.

Modigliani, F., and Brumberg, R. (1954). Utility analysis and the consumption function: an interpretation of cross-section data. *Post-Keynesian Economics*, 388–436.

Modigliani, F. (1988). The role of intergenerational transfers and life cycle saving in the accumulation of wealth. *Journal of Economic Perspectives, 2*(2), 15–40.

National Bureau of Economic Research. (NBER, 2016). *Foundational financial concepts in developed economies*. NBER Working Papers: Cambridge, MA.

National Endowment for Financial Education. (2020). *Financial challenges during the COVID-19 Pandemic*. NEFE Reports: Denver, CO.

OECD. (2017). *Financial literacy in Europe: Assessment methodologies and evidence from European countries*. OECD Publishing.

Organization for Economic Co-operation and Development. (OECD, 2020). *State of global financial literacy*. OECD Studies.

Ottaviani, C., & Vandone, D. (2018). Impulsivity and household indebtedness: Evidence from real life. *Journal of Economic Psychology, 67*, 125–142.

Pepur, S., Rodriguez, L., & Kim, J. (2022). The multiplier effect: Financial literacy and women's empowerment. *Global Finance Review, 38*(1), 12–36.

Poterba, J. M. (1987). Tax policy and corporate saving. *Brookings Papers on Economic Activity, 1987*(2), 455–515.

Rajput, N., & Agrawal, R. (2019). Financial literacy among youth. *Journal of Commerce and Trade, 14*(1), 61–67.

Reinhart, C. M., & Rogoff, K. S. (2009). *This time is different: Eight centuries of financial folly*. Princeton University Press.

Ross, S. A., Westerfield, R., & Jordan, B. D. (2002). *Fundamentals of corporate finance*. McGraw-Hill Education.

Sági, J., Németh, E., & Semenyik, A. (2020). The relationship between financial literacy, financial attitude, and financial well-being among 18–24-year-olds. *Society and Economy, 42*(1), 83–103.

Sharpe, W. F. (1964). Capital asset prices: A theory of market equilibrium under conditions of risk. *The Journal of Finance, 19*(3), 425–442.

Shiller, R. J. (2017). *Narrative economics: How stories go viral and drive major economic events*. Princeton University Press.

Shiller, R. (2019). *Behavioral finance: The importance of the 'act-how.'* Yale University Press.

Shim, S., Serido, J., & Tang, C. (2015). The ant and the grasshopper revisited: The present psychological benefits of saving and future oriented financial behaviors. *Journal of Economic Psychology, 49*, 120–130.

Singer, D., Demirguc-Kunt, A., & Klapper, L. (2013). *Financial inclusion and legal discrimination against women: Evidence from developing countries* (Policy Research Working Paper No. 6416). World Bank, Washington, DC. https://doi.org/10.1596/1813-9450-6416 (License: CC BY 3.0 IGO).

Slemrod, J. (2003). Trust in Public Finance. In *Public Finance and Public Policy in the New Century*. Cambridge, Massachusetts: MIT Press.

Standard and Poor's Ratings Services Global Financial Literacy Survey. (2015). *Financial literacy in developing economies*. S&P Reports.

Stango, V., & Zinman, J. (2009). Exponential growth bias and household finance. *The Journal of Finance, 64*(6), 2807–2849.

Taylor, J. B. (1995). The Monetary Transmission Mechanism: An Empirical Framework. *The Journal of Economic Perspectives, 9*(4), 11–26.

Wilson, F., & Tinkler, J. E. (2017). The gender gap in financial literacy: A game-based approach in a high school setting. *Journal of Consumer Affairs, 51*(2), 280–299.

World Bank. (2018). *A tripartite perspective on financial literacy*. WB Research Series.

World Bank. (2021). *Financial resilience during economic downturns*. WB Analysis Papers.

Yakoboski, P., Smith, A., & Johnson, M. (2020). Gender disparities in financial literacy: A comparative analysis. *Journal of Financial Studies, 45*(3), 250–278.

Younis, H., Dimitratos, P., & Elbanna, S. (2022). International entrepreneurial SMEs in the Muslim world: The role of religion in the GCC countries. *International Business Review, 31*(5), 102023.

Zietz, E. N. (2003). The role and regulation of insurance in personal financial planning. *Journal of Personal Finance, 2*(1), 58–68.

Zweig, P. (1995). *Wriston: Walter Wriston, Citibank, and the rise and fall of American financial supremacy*. Crown Business.

Chapter 9
The Impact of AI Tools on Universities' Educational Strategies and Student Success: Sultan Qaboos University and Qatar University as a Case Study

Talal Al-Awadhi, Amira D. Al-Balushi, Meshal M. Abdullah, and Ammar Abulibdeh

Abstract Sultan Qaboos University (SQU) and Qatar University (QU) are working to adopt short- and long-term plans to ensure leadership in teaching and learning, scientific research, and community service at regional and international levels. Modern technologies, especially artificial intelligence (AI), play an important role in our lives, providing realistic opportunities to accelerate business implementation and improve the final value of work. AI plays a crucial role in taking education and research to the next level, transforming the way these organisations work and make decisions. AI can improve the traditional work of educational institutions and transform the understanding of their complexity. However, challenges remain in various aspects, whether related to the nature of our work or to the AI technologies themselves. This is still a great opportunity for every HEI. Therefore, this paper aims to analyse the potential of AI to accelerate the implementation of the strategic plans of the two universities.

Keywords Artificial intelligence · Education · Strategy · Sultan Qaboos university · Qatar university

T. Al-Awadhi (✉) · A. D. Al-Balushi · M. M. Abdullah
Department of Geography, College of Arts and Social Science, Sultan Qaboos University, Muscat, Oman
e-mail: alawadhi@squ.edu.om

M. M. Abdullah
Department of Ecology and Conservation Biology, Texas A&M University, College Station, TX 77843, USA

A. Abulibdeh
Applied Geography and GIS Program, Department of Humanities, College of Arts and Sciences, Qatar University, P.O. Box: 2713, Doha, Qatar

E. Zaidan et al. (eds.), *The Future of Education Policy in the State of Qatar*, Gulf Studies 21, https://doi.org/10.1007/978-981-97-9667-0_9

9.1 Introduction

In the past two decades, there has been a notable shift in educational methodologies, accompanied by significant technological advancements that have had far-reaching impacts across multiple domains (Roll & Wylie, 2016; George & Paul, 2020; Adams et al., 2023; Slimi, 2021; Slimi & Carballido 2023, Zaidan et al., 2023). On the other hand, the COVID-19 pandemic played an important role in accelerating digital transformation in several areas, including the education system (Mnhrawi and Alreshidi, 2023; Pantelimon et al., 2021) There is a growing trend among higher education institutions (HEI) worldwide to adopt artificial intelligence (AI) and machine learning (ML) technologies as a means to revolutionise their educational environment. The majority of literature reviews focus on the remarkable prospects that universities can leverage through the implementation of AI and ML technologies, alongside the considerable obstacles they encounter in this undertaking (Arogundade, 2023; Grassini, 2023).

AI has become a field of research, and the resulting innovations and developments that stem from it have culminated in computers, machines, and other artefacts with human-like intelligence, characterised by cognitive abilities, learning, adaptability, and decision-making ability. A study by (Kuleto et al., 2021) found that AI has been widely adopted and used in various forms in the education sector, especially by HEIs. Another study by George & Wooden, 2023 discussed the strategy transformation of AI in HEIs. AI first took the form of computers and computer-related technologies, then transformed into web-based and intelligent online education systems, and finally with the use of embedded computer systems, along with other technologies, the use of humanoid robots and web-based chatbots to perform instructor tasks and functions independently or with instructors (Bozkurt, 2023; Bozkurt et al., 2023; Timms, 2016). It is not so much the availability of technical capabilities in the classroom that is important as enabling the teacher to have these capabilities, as the student is more familiar than the teacher with AI applications by virtue of sitting in front of these applications for long periods of time, and dealing with them more broadly and deeply. The teacher must accompany the student in dealing with these applications, from a position one or two steps ahead of the student. Future jobs will be offered based on how well an individual possesses technical skills (Grassini, 2023).

Due to technological advancements and the threat posed by any pandemic like COVID-19, higher education institutions need to adopt new teaching methods to teach their fundamentals. AI tools have become a transformative force in higher education (Abulibdeh et al., 2024), offering universities a variety of opportunities to improve their operations and educational goals. Therefore, a significant acceleration in the use of technology has been observed in most educational institutions across the world. It is necessary to focus on several issues (Chen et al., 2020): (1) building infrastructure to enable educational institutions (universities) to carry out their tasks using AI; (2) preparing society to accept the idea of using AI; and (3) reviewing the strategies that universities have implemented for AI. Hence, the utilisation of advanced technologies holds promise in aiding universities in the attainment of their strategic objectives, encompassing enhancements in educational quality and streamlining administrative operations. Nevertheless, despite the considerable potential that AI holds, it also presents a number of challenges and potential threats that universities must thoroughly evaluate in order to effectively leverage its advantages.

Consequently, these tools can be used to improve learning experiences by offering customised adaptive learning platforms to meet the students' individual needs. This not only improves the outcomes of education, but also helps institutions to achieve the goals of success and the involvement of students. In addition, AI can enable universities to make data-driven decisions, such as identifying students at risk of dropping out of college, or using predictive analytics to evaluate the effects of different teaching methods. Facilitating evidence-based decision-making (George & Wooden, 2023; Hannan & Liu, 2021; Siirtola & Röning, 2019). AI contributes to the achievement of strategic objectives related to academic excellence and institutional effectiveness. In addition, AI-driven automation streamlines administrative processes, from admission to funding distribution, frees staff up to concentrate on more strategic tasks, and ultimately supports universities to pursue operational efficiency and cost reduction (Grassini, 2023).

Qatar University (QU) and Sultan Qaboos University (SQU) serve as higher education institutions in the countries of Qatar and Oman, respectively. Both of these universities, similar to numerous universities around the world, establish strategic objectives aimed at enhancing their educational and research programmes, fostering innovation, and sustaining a competitive edge (Zaidan et al., 2024). AI tools have the potential to make significant positive contributions to QU and SQU in the pursuit of strategic objectives related to education, research, innovation, and competitiveness. However, they also pose potential risks and challenges. The aim of this study is to examine the impacts, potential benefits, and challenges of integrating AI tools into higher education, and how these factors align with the strategic objectives of universities such as QU and SQU. The study reflects a comprehensive examination of the role of AI in reshaping the higher education landscape and the strategies institutions can adopt to effectively harness its potential. This study aims to answer the following research questions: To what extent do AI tools in higher education contribute to the improvement of the quality of education and student success, and how do these contributions align with the strategic goals of universities like QU and SQU? What specific challenges and potential threats are associated with the integration of AI tools in higher education, and how do universities navigate these challenges to effectively harness the benefits of AI technology in their operations and educational objectives? And in what ways does the adoption of AI-driven automation in administrative processes within universities impact operational efficiency, cost reduction, and the overall functioning of academic institutions? How does this align with universities' strategic goals for administrative efficiency and resource optimisation?

These research questions address the impact, benefits, challenges, and operational implications of AI tools in higher education, focusing on their alignment with the strategic objectives of universities such as QU and SQU. They could provide a comprehensive framework for studying the role of AI in reshaping higher education in general, and how institutions could maximise its potential while addressing issues which are likely to arise more effectively.

9.2 AI Potential in Teaching & Learning

AI was first used in the form of computers and computer-related technologies, and then evolved into implementing web-based and online-based intelligent education systems. Recently, AI-embedded computer systems, among other technologies, have also led to the use of humanoid robots and web-based chatbots, which take on teachers' tasks and function independently or together with teachers. One such area where AI has been used and is making a big impact is the education sector. The innovations and developments that have culminated in the development and use of AI have given the education sector and especially its more academic institutions the opportunity to adopt and utilise AI (Chen et al., 2020; Zhai et al., 2021; Slimi & Carballido 2023).

AI is currently being used in teaching and learning in a variety of ways, from simple applications such as chatbots that can provide students with some form of academic support, to self-learning algorithms that meet the needs and requirements of each student. It has the ability to provide a wide range of benefits for education (Abulibdeh et al., 2024). The most important thing is the ability to customise the learning experience for each student (Grassini, 2023). Using AI, faculty can analyse a student's performance, and create personalised lesson plans and assessments that correspond to each student's individual strengths and weaknesses. In addition, it can handle faculties' administrative work such as monitoring grades and tracking attendance and absences, saving them time to focus on other aspects, be it improving the quality of education, scientific research, or community service (Markauskaite et al., 2022; Zhai et al., 2021).

As a result of the increasing number of students in higher education, the variety of courses, the different teaching methods, the increasing competitiveness between universities with a lack of highly qualified teaching staff, and a lack of experience, it is necessary to look for alternatives and think creatively. With modern technology, AI could offer the appropriate alternative to compete with and potentially replace the human element in the future. With the increasing load of teaching and need for community services, faculty are struggling to perform their heavy lifting such as teaching small groups at a specific time and place while AI is available to students at any time and place. Educational experiences depend on the experience of members of faculty, and the scope of their activities designed to develop their skills, while AI depends on continuity and sustainability in development, in addition to dependence on open data sources. Emotions both positive and negative are also present in the teaching and learning process of faculties, while AI only comes with passion. The need to constantly upskill employees in education incurs additional costs, while AI merely requires constant technical support.

The roles of AI in the field of education can be classified into three primary categories (Alqahtani et al., 2023; Chen et al., 2020; Gao, 2021; Grassini, 2023; Hwang et al., 2020; Kuleto et al., 2021; Markauskaite et al., 2022): First, improved personalised learning: AI can tailor educational experiences to individual students' needs, adjusting content, pace, and resources to accommodate different learning styles and abilities. It can also analyse student data to identify patterns that indicate struggling students, enabling early intervention and support systems. AI has improved teaching methods in several ways. One important way is the development and use of intelligent tutoring systems (ITS). These systems use AI to provide students with personalised and adaptable instruction. Immersive technologies such as VR and 3D can be integrated. By analysing students' skills and needs, ITS can tailor the content and teaching approach to each learner, ensuring an optimised learning experience. These personalised lessons help students improve their understanding and retention of material. AI tools can help create accessible content and support students with disabilities, especially those who are visually impaired.

Second, the implementation of AI technology has significantly enhanced the efficiency and effectiveness of administrative functions within the realm of education through the automation of various administrative processes and tasks. Expanding the use of AI in various educational environments will help faculties become more reliant on technology; otherwise they will lag significantly behind the students. (The previous sentence is unclear, how will they "lag behind the students"? Editor) AI could be used as a tool to complete educational administrative tasks, such as setting assignments, and providing feedback to higher administration levels. It is also used to analyse large amounts of data to identify patterns and insights that can inform the development of new educational strategies and policies. AI can support and automate some administrative tasks like enrolment, scheduling, as well as grading and evaluation, plagiarism detection, and delivering feedback on students' written work, and it can free up time for faculty and staff to focus on more strategic activities. These advancements have improved instructional quality, engagement, and the overall learning experience for students (Gao, 2021).

Lastly, for research purposes, academics may have more time to concentrate on research and innovation. AI technologies can assist in data analysis, simulations, and other research-related tasks, potentially accelerating the pace of discovery. AI associated with other technologies like ML or Deep Learning (DL) can process large volumes of data to provide valuable insights for scholarly research and innovation. This gives faculty members more time to complete their academic research and further develop their research and teaching skills. There may also be more time available to provide academic advice to students, and even to support the local communities.

9.3 AI Challenges & Concerns in Teaching & Learning

AI has numerous advantages for scientific research and education; nevertheless there are still numerous challenges and limitations. With the increasing availability of online education and online educational resources, the biggest challenge is ensuring that all students have access to AI-powered tools and platforms, regardless of their socio-economic status or location. Additionally, there are concerns about the possibility that AI could perpetuate existing bias and discrimination in education.

In addition, faculties are concerned that AI tools may replace human interaction in the future, and thus it may affect the quality of classroom teaching. Ensuring that AI is used in a way that complements rather than replaces human teachers will be significantly important in the upcoming years. While there are many benefits to using AI in education, there are still ethical considerations that need to be addressed. In addition, there are concerns about the impact of AI on student privacy and data security, noting ChatGPT's ability to generate meaningful responses to questions from assessments and exams. It is often not possible to attribute such responses to a specific source—which makes it difficult to detect plagiarism (Khalil & Erkan, 2023; Qin et al., 2023). Another concern is the possible relocation of jobs in the education sector as technology advances. The automation of many administrative tasks may mean fewer jobs available for teachers and support staff.

Numerous scientific investigations (Grassini, 2023; Hwang et al., 2020; Markauskaite et al., 2022; Slimi, 2023) hypothesise that AI is likely to encounter a range of challenges in the forthcoming period, which can be categorised into three principal domains.

1. Privacy and security:

   The management of sensitive student data necessitates the implementation of strong security protocols in order to safeguard against potential breaches by hackers and unauthorised individuals.

   Universities are obligated to conform to legal and regulatory frameworks that govern the utilisation, availability, and confidentiality of data.

2. Ethical considerations in academic research:

   Ethical considerations play a crucial role in the development and implementation of AI algorithms, as they serve to uphold principles of fairness, transparency, and accountability. These considerations are of utmost importance in order to mitigate the risk of bias or discriminatory outcomes.

   The issue of excessive dependence on technology necessitates the avoidance of a potential disconnect between students and teachers.

3. The challenges associated with implementation:

   Resistance to the adoption of technology in various fields and workplaces is often encountered, as individuals who adhere to conventional systems and methods tend to exhibit apprehension or fear. There is a need for heightened endeavours to enhance the dissemination of knowledge regarding the advantageous outcomes of AI implementations in the domains of education and scientific research. This pertains to both the aspects of fostering motivation and exerting persuasive influence to encourage active engagement and involvement. Furthermore, the proficient utilisation of AI technologies necessitates additional training.

The process of integrating AI into pre-existing educational systems can present considerable complexity and necessitate substantial investments in both technological resources and training initiatives.

The implementation of AI technologies can incur substantial initial expenses in terms of hardware, software, and human resource allocation.

One potential drawback of utilising pre-packaged AI solutions in higher education institutions is the absence of customisation and adaptability. These off-the-shelf solutions may not consistently align with the unique requirements and objectives of each individual institute.

Maintenance and scalability are crucial aspects in ensuring the long-term effectiveness and currency of AI systems. It is imperative to continuously engage in maintenance activities and implement updates to sustain optimal performance.

9.4 QU and SQU Strategic Plans

The higher education systems in Qatar and Oman exhibit a significant degree of modernity. QU was established in 1977 as the inaugural public university in Qatar. Similarly, SQU was founded in 1986, commencing its academic activities with five colleges: Medicine, Engineering, Science, Agriculture, and Education. The College of Arts at SQU was inaugurated one year subsequent to the university's establishment. The governments of Oman and Qatar have demonstrated significant commitment towards the development of the education sector, with particular emphasis on higher education. The primary objective of the initiative was to offer extensive financial assistance in order to enhance the standards of education and promote scientific research. According to the Shanghai University Ranking list (QS), QU and SQU are prominent institutions among Arab universities, consistently positioned at the forefront in various university rankings.

The institution of higher education, particularly the university, is not solely regarded as a centre for instruction and knowledge acquisition. A university can be broadly defined as a social institution that encompasses the functions of education, research, and community engagement. By virtue of these three elements, the university exhibits a considerable capacity for the storage, dissemination, and generation of knowledge. Over the course of the past five decades, there have been substantial transformations observed in various aspects of human existence on a global scale. The aforementioned alterations have had a substantial impact on various establishments, including HEIs such as universities. The alterations and modifications have necessitated universities adapting their tasks, roles, and functions in alignment with global shifts. Therefore, it is imperative for universities in the present era to promptly adapt to these transformations, in order to enhance their competitive standing and align with the numerous advancements and objectives set forth by their respective governments. Both of these universities have collaborated to implement long-term strategies aimed at enhancing their competitive capabilities in teaching and learning, scientific research, and community service, at both regional and international levels. SQU has implemented a novel strategic plan that spans the period until 2040, aligning itself with the objectives outlined in Oman Vision 2040. Similarly, QU has also adopted a distinct strategic plan that extends until 2023, in accordance with the goals set forth in Qatar Vision 2030.

9.4.1 QU Strategic Plan

QU Strategic Plan is a comprehensive framework that aligns its mission, vision, and core values with its role as the national institution of higher education in Qatar. These principles guide the university's efforts in education, research, and societal engagement, positioning it as a key contributor to Qatar's future growth and development while upholding the highest standards of academic excellence, integrity, and diversity.

The mission, vision, and core values of Qatar University's strategic plan play a pivotal role in shaping the institution's identity, objectives, and its commitment to the advancement of education and research in Qatar. Qatar University's mission statement underscores its role as the foremost national institution of higher education in Qatar. The university's primary purpose is to provide high-quality undergraduate and graduate programmes that prepare competent graduates who will play a crucial role in shaping the future of Qatar. This mission embodies a commitment to educating the next generation of leaders and professionals who will contribute to the country's progress and development. The emphasis on addressing local and regional challenges in research and knowledge creation underscores the university's responsibility towards society, and its aspiration to actively address the needs and aspirations of the nation. The vision statement articulates Qatar University's aspiration to be regionally recognised for distinctive excellence in education and research. It highlights the university's aim to become an institution of choice for both students and scholars. Additionally, the vision emphasises the university's role as a catalyst for the sustainable socio-economic development of Qatar. This vision sets a clear direction for the university's future, emphasising not only academic excellence but also its broader societal and economic impact.

Qatar University's core values serve as guiding principles that underpin its operations and decision-making processes. The commitment to excellence reflects the university's dedication to maintaining the highest standards of quality and professionalism in all its endeavours. It ensures that Qatar University continually strives for the best possible outcomes in education and research. Integrity is a fundamental value with an emphasis on honesty, fairness, transparency, responsibility, and accountability. It establishes a culture of trust and ethical behaviour within the university community. Furthermore, QU emphasises innovation in the educational process. The value of innovation encourages creativity and the exploration of novel ideas. By promoting free and open inquiry and the pursuit of knowledge, QU inspires its students, faculty, and staff to develop innovative solutions that can have a significant impact on society.

9.4.2 SQU Strategic Plan

The strategic plan outlined for SQU, in alignment with the National Strategy for Education 2040, represents a comprehensive and forward-thinking approach to higher education and its role in shaping the future of Oman. One of the fundamental principles of this strategic plan is its alignment with the national education strategy, which envisions the future of Oman. The plan reflects a commitment to play a pivotal role in translating these national aspirations into tangible actions and outcomes. It underscores the importance of HEIs in driving the socio-economic progress of the country.

The key goals of the strategic plan include producing quality graduates capable of pursuing successful careers and providing solutions to societal needs. This focus on educational excellence not only benefits individual students but also contributes to the development of a skilled workforce that can drive national progress. Another goal is conducting research with international recognition and impact on the socio-economic development of the country which reflects a commitment to knowledge creation and innovation. Research outcomes that directly benefit Oman's development are central to this goal. Furthermore, the plan's emphasis on transforming knowledge into value through a vibrant innovation system highlights the importance of research and education in driving economic growth. This goal recognises the potential of academic institutions to act as catalysts for innovation and entrepreneurship. The development of a long-term strategic plan by SQU reflects a commitment to excellence, continuous improvement, and aligning its mission with the broader national development objectives. This ambitious planning process is a testament to the institution's dedication to advancing higher education, research, and community service.

The university's vision, as formulated, reflects a forward-looking approach that balances its national leadership role with international recognition for innovation, research quality, and strategic partnerships. The vision is comprehensive, focusing not only on education and research but also on the broader goals of community service and global recognition. The identification of five strategic focus areas—undergraduate teaching and learning, postgraduate teaching and learning, research, creativity and innovation, and community engagement—demonstrates a well-rounded approach to the university's role. These areas encompass the core functions of a comprehensive university, emphasising the importance of teaching, research, innovation, and societal impact. The alignment of the university's vision with the national 2040 vision for education signifies a commitment to contribute to the broader national development agenda. The university's mission, values, and vision are clearly tied to the national goals of producing a skilled, adaptable human workforce, and preserving cultural identity. Furthermore, the development of strategic goals, strategies, and initiatives for each domain and enabler demonstrates a comprehensive and actionable plan. This approach provides clarity and a structured path to achieving the long-term vision. SQU's approach to long-term strategic planning is characterised by inclusivity, international benchmarking, a comprehensive vision, and a well-structured framework for implementation. This strategic plan not only reflects the institution's commitment to excellence but also its dedication to contributing to the development of Oman by aligning its mission and values with national goals and priorities.

9.5 The Integration of AI Tools in Universities' Strategies

The integration of AI tools in universities represents a significant shift in how higher education institutions operate and manage their resources, and has had a profound impact on their strategies, both positively and negatively. These effects can be observed in various aspects of university operations and management. Based on the review of previous research, two schools of thought have emerged: using higher education to support AI or vice versa. The first philosophy is not discussed in this article; rather, the focus is on how AI can be used to benefit educational institutions, and support their strategies and long-term goals. The three main pillars of the two universities' strategic plans are research and teaching, community service by promoting creativity and innovation, and scientific research. There are great similarities in both universities' strategies in many of their characteristics and features, in terms of vision, mission, objectives, and elements of success. To ensure the success of the two strategies, the two universities have adopted a number of applicable factors (SQU, 2016; and QU, 2020), such as: introducing a common governance and management framework to enhance academic excellence; striving for financial sustainability and diversification; providing a supportive, high-quality research and educational environment.; supporting innovation and the use of modern technologies in the field of education and scientific research; creating a stimulating environment for all stakeholders, including students and faculty members; encouraging competition, equality, justice and transparency in the selection of faculty members; fostering a high quality and supportive research and teaching environment for all stakeholders; applying productivity monitoring, using quantitative measurement methods; and continuously supporting capacity building for all stakeholders.

There are several overlapping areas which can identified between AI capabilities and universities' strategies, as shown in Fig. 9.1.

With the widespread developments in various technological areas, the fourth industrial revolution, and the use of AI and ML, there is a great opportunity for the two universities to implement the vision and objectives of their strategic ambitious plans. Figure 9.2 illustrates the framework of AI integration into universities' strategy plans.

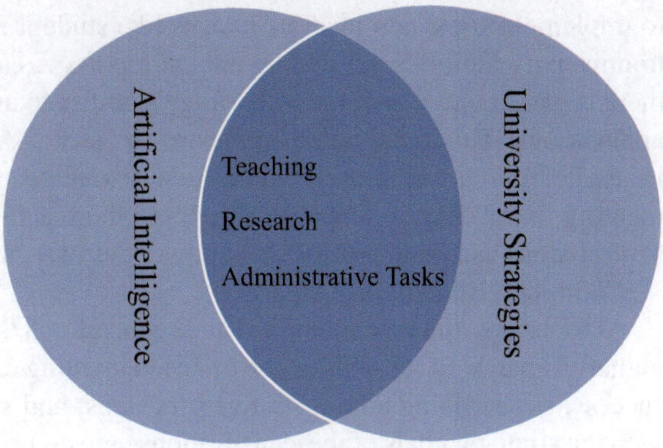

Fig. 9.1 Overlapping between university strategy and AI

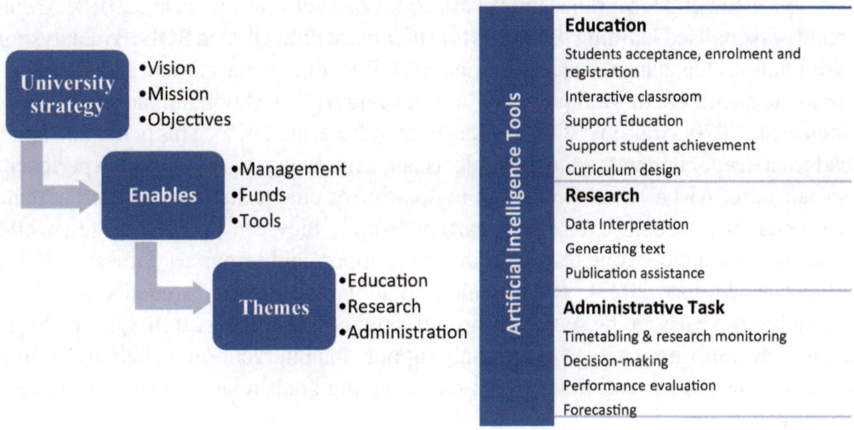

Fig. 9.2 AI and university's strategy integration

9.5.1 Positive Impacts of Integrating AI Tools on University Strategies

Strengthening University Faculties and Improving Administrative Processes

AI tools assist higher level management at any university to avoid biassed decision-making, easily extract analytics from data, and make strategic choices quickly, ensuring greater capabilities in the future (Manju & Anilkumar, 2020; George & Wooden, 2023; Markauskaite et al., 2022; Zhai et al., 2021; Zawacki-Richter et al., 2019). The question that always arises is whether AI can develop and implement tasks in the strategy drawn up by the institution, especially higher education institutions. Based on previous studies, most researchers believe that in short the answer is "no". At the same time, they see that AI and advanced analysis tools provide great benefit to those responsible for implementing strategies. AI provides a suitable and sustainable environment for achieving strategies. For example, it provides tools to implement some administrative tasks like student enrolment, timetabling, access to university facilities, producing annual reports …etc. It should also be noted that there is uncertainty about the technology's ability to understand context in advanced applications. This reality also applies to the use of AI in strategy development, as its applications are employed with difficulty to match the knowledge of those implementing it, but they are capable of helping them achieve it. Nevertheless, there are several areas can be identified to empower the two universities using AI.

Enhanced Educational Delivery

AI tools have the potential to enhance educational delivery at both QU and SQU in numerous ways. By providing personalised learning experiences, promoting student success, streamlining administrative processes, and supporting research, AI aligns with the strategic goals of these institutions, and contributes to improved educational outcomes and overall excellence (George & Wooden, 2023; Adams et al., 2023; Gaševi´c et al., 2023; Markauskaite et al., 2022; Zawacki-Richter et al., 2019). AI can create personalised learning pathways for students at both QU and SQU. By analysing individual student data, these institutions can tailor educational content and strategies to cater to each student's unique needs and learning styles (Almohammadi et al., 2017; Chen et al., 2020; Grassini, 2023; Ocaña-Fernandez et al., 2019). This personalisation leads to a deeper understanding of subjects and a more engaging learning experience. Furthermore, AI-based adaptive learning platforms enable students to learn at their own pace. If a student excels in a particular area, they can progress faster, while those who need more time receive additional support and resources (Grassini, 2023; Whalen & Mouza, 2023). Additionally, AI tools can identify students at risk of falling behind early in the semester through predictive analytics. Both QU and SQU can use this information to provide timely support and interventions to help struggling students. Early intervention is crucial for achieving goals related to student success and engagement.

Since the two universities are national universities, the number of students is expected to increase significantly in the coming years, and they also have ethical responsibilities to the local community and the labour market. Therefore, the use of AI is one of the solutions that can be relied upon greatly in the future. AI tools provide practical solutions for students, whether in providing support in important scientific subjects or obtaining assistance in greater understanding of the required work. AI can also create a supportive educational environment for students and faculty members, which is one of the strategic requirements of the two universities. AI tools provide students and academics with quick access to data, and also facilitate the transformation of teaching and learning based on self-education. This technology will provide faculty members with additional tools to explain scientific material through simple approaches. It will also provide tools for self-evaluation of the course and sending feedback automatically.

Empowered Faculty Members as Research Scholars

QU and SQU aspire to excellence in the field of scientific research, as it shown clearly in their strategy plans. Both universities look to research as the leading path to achieve excellence and take a leading role at regional and international levels. They are optimistic about using scientific research to bring incredible change and have therefore strengthened the mechanisms and enablers that help scientific research flourish. SQU and QU's strategies suggest that scientific research is the cornerstone that will lead to the remaining pillars for development and growth. According to previous studies (Alqahtani et al., 2023; Chen et al., 2020; Hannan & Liu, 2021) the contribution of AI in the field of scientific research can be specified, according to previous studies, into three main tracks: text generation, data interpretation, and accelerating publication.

Faculty members at most universities, including SQU and QU, struggle to balance teaching, scholarly research, administrative duties, and community service. Several previous studies have suggested (Alqahtani et al., 2023; Grassini, 2023; Pantelimon et al., 2021; Weigel et al., 2022) that AI can play both a positive and negative role in generating or paraphrasing texts. AI is capable of producing high-quality scientific texts without human emotions, and also enables the synthesis of research papers and published articles. Many respected researchers suffer from the inability to write outstanding scientific articles to publish in a respectable scientific journal, even if the research results are publishable and applicable. AI can also save researchers time and effort in several ways. It is therefore clear that an academic member of staff will have more time to improve his/her skills which in the long-term reflect well on the university's reputation. AI can recommend focusing and conducting specific research that is consistent with scientific research trends, and compatible with the requirements of society and the local market. AI can also enable researchers to ensure that summaries, previous studies, analyses, conclusions, and recommendations are linked to the results. These tools can also ensure compliance with journal-specific formatting and citation style requirements, saving authors time.

AI technologies have revolutionised data mining and analysis for various disciplines (OECD, 2023; Weigel et al., 2022). These technologies facilitate the management and analysis of large amounts of data, and contribute to understanding society complexity and environmental systems. It also helps in recognising patterns and interrelationships between data. These technologies also accelerate scientific discoveries, and provide valuable information for identifying and testing hypotheses, due to their ability to analyse large amounts of data. At the same time, some researchers argue that analyses performed using AI lack quality, reliability, accuracy, and transparency. AI tools are becoming increasingly important for analysing large amounts of data to understand data patterns and identify processing methods. This positive point will strengthen the hand of QU and SQU, to increase the annual publications of their members. The process of reviewing and summarising previous studies is another advantage of using AI in research. AI technologies have an excellent ability to process large numbers of studies quickly and efficiently, saving researchers time and allowing them to focus on other aspects of their projects. AI systems can search for keywords and concepts and find articles that human reviewers might ignore. They can also identify patterns and trends and highlight new research directions or gaps in the literature. Furthermore, AI systems can assess the quality of the scientific literature by analysing study design, sample size, and statistical methods, which helps researchers assess the reliability and relevance of a study for their review.

AI can conduct the scientific review process of scientific articles authoritatively and impartially. It is able to detect plagiarism, and also detect the falsification of data and results. This saves time, and has a positive effect on the rapid processing of studies, and accelerated publication. It will help authors receive accurate and targeted comments, which will help them improve their research on specific points for later publication. AI can also focus on processing initial queries, organising submissions, and sending reminders, allowing editors to focus on improving content quality, strengthening relationships, and developing editorial strategies. It can assist with language editing and proofreading, allowing editors to efficiently create or review content and find manuscript errors. For example, it can personalise the publishing experience for authors and reviewers by analysing the submission history and providing personalised guidance on manuscript preparation and evaluation, thereby improving submission quality and reviewer feedback. All these advantages can be utilised to implement the research aspect of SQU and QU, which then may improve the quality of research in both universities.

9.5.2 Negative Impacts of Integrating AI Tools on University Strategies

AI has many features and characteristics which may assist QU and SQU to achieve their strategy plan, or at least accelerate the implementation of its goals. However, AI also has disadvantages (Alqahtani et al., 2023; Heilinger et al., 2023; OECD, 2023) which may affect HEIs negatively. The main concern about using AI in all aspects of higher education is the absence of human sensitivity. Resistance from teachers is another concern about using AI. Some faculty members feel threats to their job security. AI in the future could affect the need for some administration jobs; for example, a unit dealing with admission and registration may not exist in the future in the structure of most universities. Everything from applying to university to course registration and timetabling will be done by AI Over time, students and academics may lose their passion for teaching and learning and rely heavily on AI. It is expected in the long term that this dependence can lead to the neglect of important traditional teaching methods, and a lack of development of critical thinking and problem-solving skills.

In scientific research, AI also has some drawbacks. Texts created using AI have no human meaning, making it difficult for researchers themselves to understand the context. Over time, researchers will no longer have access to current experiences in their field, and their passion for learning new things will diminish. There are issues related to plagiarism or generating false information; it can cause unfairness in competition between academics, and furthermore could make researchers ignore their own development as writers, with over-reliance on AI. There could also be an impact on ethical views, such as moralistic or religious ones, or harm could be done to certain communities or societies through ignorance of their beliefs. Moreover, complete dependence on texts generated by AI can limit the creative process and inspire bland writing styles. In addition, these texts lack transparency and therefore the possibility of rejection by the reviewers or editors. Using AI for data interpretation totally depends on the quality and quantity of available data, and this is a kind of bias against the financial capabilities of the institutions conducting the research. It is a scientific injustice. AI may lack understanding of the context of some research areas, leading to misunderstandings. It is also expected that this will reduce the opportunities for researchers to learn everything new in the field of scientific research and scientific trends and to learn from other researchers' mistakes in order to correct them in their own research.

9.6 Conclusion

The integration of AI tools in universities represents a significant transformation in higher education institutions' operations and strategies. This shift has both positive and negative impacts on various aspects of university management and functioning. The strategic plans of two universities, QU and SQU are built on pillars such as research and teaching, scientific research, and community service. These universities share common characteristics and features in their strategies, aligning their visions, missions, objectives, and success elements. To ensure the success of their strategies, they adopt various factors, including common governance frameworks, financial sustainability, support for innovation and technology, and capacity building for stakeholders.

The two universities must strike a balance between harnessing the benefits of AI to advance their strategic plans, while addressing the associated challenges. A thoughtful and strategic approach to AI integration is essential, to ensure that it enhances the overall educational and research experience, rather than merely replacing critical human elements and creativity. By carefully managing these positive and negative impacts, QU and SQU and other universities can effectively leverage AI tools to achieve their strategic objectives and maintain their commitment to excellence in higher education and research.

Bibliography

Abulibdeh, A., Zaidan, E., & Abulibdeh, R. (2024). Navigating the confluence of artificial intelligence and education for sustainable development in the era of industry 4.0: Challenges, opportunities, and ethical dimensions. *Journal of Cleaner Production, 437*, 140527.rences

Adams, C., Pente, P, Lemermeyer, G., & Rockwell. G. (2023). Ethical principles for artificial intelligence in K-12 education. Computers and Education: *Artificial Intelligence 4*: 100131. https://doi.org/10.1016/j.caeai.2023.100131

Almohammadi, K., Hagras, H., Alghazzawi, D., & Aldabbagh, G. (2017). A survey of artificial intelligence techniques employed for adaptive educational systems within E-learning platforms.

Journal of Artificial Intelligence and Soft Computing Research, 7(1), 47–64. https://doi.org/10.1515/jaiscr-2017-0004

Alqahtani, T., Badreldin, H. A., Alrashed, M., Alshaya, A. I., Alghamdi, S. S., Shuroug, K. S., Alowais, S. A., Alshaya, O. A., Rahman, I., Al Yami, M. S., & Albekairy, A. M. (2023). The emergent role of artificial intelligence, natural learning processing, and large language models in higher education and research. *Research in Social and Administrative Pharmacy, 19*(8), 1236–1242. https://doi.org/10.1016/j.sapharm.2023.05.016

Arogundade, O. R. (2023). Structuring knowledge bases with AI and machine learning. *Information and Knowledge Management, 13*(1), 31–39.

Bozkurt, A., Xiao, J., Lambert, S., Pazurek, A., Crompton, H., Koseoglu, S., Farrow, R., Bond, M., Nerantzi, C., Honeychurch, S., Bali, M., Dron, J., Mir, K., Stewart, B., Costello, E., Mason, J., Stracke, C., Romero-Hall, E., Koutropoulos, A., Jandrić, P. (2023). Speculative futures on ChatGPT and Generative Artificial Intelligence (AI): a collective reflection from the educational landscape. *Asian Journal of Distance Education, 18* (1), 53–130. https://www.asianjde.com/ojs/index.php/AsianJDE/article/view/709

Bozkurt, A. (2023). Generative artificial intelligence (AI) powered conversational educational agents: The inevitable paradigm shift. *Asian Journal of Distance Education 2023, 18.* http://www.asianjde.com/ojs/index.php/AsianJDE/article/view/718

Chen, L., Chen, P., & Lin, Z. (2020). Artificial intelligence in education: a review. *IEEE Access, 8*, 75264–75278. https://doi.org/10.1109/ACCESS.2020.2988510

Gao, S. (2021). Geospatial artificial intelligence (GeoAI). *Geography, 4*(2005), 1–16. https://doi.org/10.1093/obo/9780199874002-0228

Gaševi´c, D., Siemens G. , & Sadiq S.. 2023. Empowering learners for the age of artificial intelligence. *Computers and Education: Artificial Intelligence 4*: 10–13. https://doi.org/10.1016/j.caeai.2023.100130

George, B., & Paul, J. (2020). *Digital transformation in business and society: Theory and cases.* Springer International Publishing. https://doi.org/10.1007/978-3-030-08277-2

George, B., & Wooden, O. (2023). Managing the strategic transformation of higher education through artificial intelligence. *Administrative Sciences, 13*, 196. https://doi.org/10.3390/admsci13090196

Grassini, S. (2023). Shaping the future of education: exploring the potential and consequences of AI and ChatGPT in educational settings. *Education in Science, 2023*(13), 692. https://doi.org/10.3390/educsci13070692

Hannan, E., & Liu, S. (2021). AI: New source of competitiveness in higher education. *Competitiveness Review: An International Business Journal, 33*, 265–279.

Heilinger, J. C., Kempt, H., & Nagel, S. (2023). Beware of sustainable AI! Uses and abuses of a worthy goal. *AI Ethics, 26*, 1–12. https://doi.org/10.1007/s43681-023-00259-8

Hwang, G. J., Xie, H., Wah, B. W., & Gašević, D. (2020). Vision, challenges, roles and research issues of artificial intelligence in education. *Computers and Education: Artificial Intelligence, 1*, 1–5. https://doi.org/10.1016/j.caeai.2020.100001

Khalil, M., & Erkan E. (2023). "Will ChatGPT get you caught? rethinking of plagiarism detection." *Lecture Notes in Computer Science* 14040 LNCS: 475–87. https://doi.org/10.1007/978-3-031-34411-4_32

Kuleto, V., Ilić, M., Dumangiu, M., Ranković, M., Martins, OMD., Păun, D ., & Mihoreanu, L. (2021). Exploring opportunities and challenges of artificial intelligence and machine learning in higher education institutions. *Sustainability, 13* (18), https://doi.org/10.3390/su131810424

Manju, G., & Anilkumar, K. S. (2020). Rule-based cognitive modelling for multimodal intelligent tutoring systems. *International Journal of Psychosocial Rehabilitation, 24*, 17–26.

Markauskaite, L., Marrone, R., Poquet, O., Knight, S., Martinez-Maldonado, R., Howard, S, Tondeur, J., De Laat, M., Shum, S.B., Gašević, D., Siemens, G. (2022). Rethinking the entwinement between artificial intelligence and human learning: What capabilities do learners need for a world with AI?, *Computers and Education: Artificial Intelligence, V. 3*, https://doi.org/10.1016/j.caeai.2022.100056

Mnhrawi, S. A., & D.N.T.A. & Alreshidi, H.A. (2023). A systemic approach for implementing AI methods in education during COVID-19 pandemic: Higher education in Saudi Arabia. *World Journal of Engineering, 20*(5), 808–814. https://doi.org/10.1108/WJE-11-2021-0623

Ocaña-Fernandez, Y., Valenzuela-Fernandez, L., & Garro-Aburto, L. (2019). Artificial intelligence and its implications in higher education. *Propósitos y Representaciones, 7* (2), 536–568. https://doi.org/10.20511/pyr2019.v7n2.274

OECD (2023), Artificial Intelligence in Science: *Challenges, Opportunities and the Future of Research*, OECD Publishing: Paris, https://doi.org/10.1787/a8d820bd-en

Pantelimon, F.-V., Bologa, R., Toma, A., & Posedaru, B.-S. (2021). The evolution of AI-driven educational systems during the COVID-19 pandemic. *Sustainability, 13*, 13501. https://doi.org/10.3390/su132313501

Qatar University. (2018). *Qatar University strategy 2018–2023*. Qatar, Unpublished Report.

Qin, C., Zhang, A., Zhang, Z., Chen, J., Yasunaga, M., & Yang, D. (2023). *Is ChatGPT a General-Purpose Natural Language Processing Task Solver?* EMNLP 2023 https://arxiv.org/abs/2302.06476

Roll, I., & Wylie, R. (2016). Evolution and revolution in artificial intelligence in education. *International Journal of Artificial Intelligence in Education, 26*, 582–599. https://doi.org/10.1007/s40593-016-0110-3

Siirtola, P., Röning, J. (2019). incremental learning to personalize human activity recognition models: the importance of human AI collaboration, *Sensors, 19* (23), 5151; https://doi.org/10.3390/s19235151

Slimi, Z. (2021). *The impact of AI implementation in higher education on educational process future: a systematic review.* https://doi.org/10.21203/rs.3.rs-1081043/v1

Slimi, Z. & Carballido, B. V. (2023). Navigating the ethical challenges of artificial intelligence in higher education: An analysis of seven global AI ethics policies. *TEM Journal, 12* (2).

Slimi, Z., & Carballido, B. V. (2023). Systematic review: AI's impact on higher education—learning, teaching, and career opportunities. *TEM Journal, 12*(3), 1627–1637. https://doi.org/10.18421/TEM123-44

Sultan Qaboos University (SQU). (2016). *SQU Strategic Plan 2016–2040*. Oman, Unpublished Report.

Timms, M. J. (2016). Letting artificial intelligence in education out of the box: educational robots and smart classrooms. *International Journal of Artificial Intelligence in Education, 26*(2), 701–712. https://doi.org/10.1007/s40593-016-0095-y

Weigel, A., Caldas, C., Meyer, A., & Morris, S. A. (2022). The impact of AI on research. *Cell, 185*(15), 2621–2622. https://doi.org/10.1016/j.cell.2022.06.024

Whalen, J.; Mouza, C. (2023), ChatGPT: challenges, opportunities, and implications for teacher education.

Zaidan, E., Belkhiria, E., & Wazen, C. (2023). Universities of the Future as Catalysts for Change: Using the Sustainable Development Goals to Reframe Sustainability–Qatar University as a Case Study. In The *Sustainable University of the Future: Reimagining Higher Education and Research* (pp. 1–23).

Zaidan, E., Momani, R., & Al-Saidi, M. (2024). Entrepreneurial universities and integrated sustainability for the knowledge-based economy: Self-perception and some structural challenges in the Gulf region. Humanities and Social Sciences Communications, 11(1), 1–14. Cham: Springer International Publishing. *Contemporary Issues in Technology and Teacher Education, 23*, 1–23.

Zawacki-Richter, O., Marín, V. I., Bond, M., & Gouverneur, F. (2019). (2019) Systematic review of research on artificial intelligence applications in higher education – where are the educators? *International Journal of Educational Technology in Higher Education, 16*, 39. https://doi.org/10.1186/s41239-019-0171-0

Zhai, X., Chu, X., Chai, C. S., Jong, M. S. Y., Istenic, A., Spector, M., et al. (2021). A review of artificial intelligence (AI) in Education from 2010 to 2020. *Complexity, 2021*, 8812542. https://doi.org/10.1155/2021/8812542

Chapter 10
Nurturing the Home Learning Environment: Evidence from Qatar Reads Programme

Ozcan Ozturk, Logan Cochrane, and Reem Al-Hababi

Abstract This chapter examines the effectiveness of a family reading programme, Qatar Reads, implemented within a specific socio-cultural context, focusing on how to promote children's literacy, linguistic, and emotional development through home learning environments. In addition to the reading activities of the Qatar Reads programme, this study investigates the role of home libraries and parent–child activities in this process. Employing data collected through a survey, and building on prior research, the study finds that the Qatar Reads programme has a significant, positive impact on children's holistic development. Additionally, the findings highlight the importance of home libraries and parent–child reading activities, as pivotal components driving children's development. Despite the limitations mentioned in the chapter, these findings have implications for policymakers, educators, and families seeking to create a nurturing home environment for their children's long-term development.

Keywords Reading · Literacy · Education · Sustainability · Middle East · Qatar

10.1 Introduction

The home is where the journey of exploration, discovery, and learning begins for every child. Within these familiar walls, each room becomes a staging post for learning for young minds; the kitchen becomes a lab for discovery, the living room decorated with bookshelves evolves into a classroom for shared learning with family, and the bedroom becomes a realm of imagination during the reading of bedtime stories. Such learning environments—enriched with books, cultural conversations, and interactive activities—ignite children's curiosity, and nurture their lifelong love

O. Ozturk (✉) · L. Cochrane
College of Public Policy, Hamad Bin Khalifa University, Doha, Qatar
e-mail: oozturk@hbku.edu.qa

R. Al-Hababi
Gulf Studies, Qatar University, Doha, Qatar

E. Zaidan et al. (eds.), *The Future of Education Policy in the State of Qatar*, Gulf Studies 21, https://doi.org/10.1007/978-981-97-9667-0_10

for learning. This is the essence of the Home Learning Environment (HLE) which profoundly influences children's cognitive, linguistic, and emotional development. While the HLE is deeply embedded within the family set-up, its ripple effects go far beyond it, touching upon global aspirations and ambitions captured in the Millennium Development Goals (MDG) and the United Nations' Sustainable Development Goals (SDGs).

The 2015 target within the Millennium Development Goals (MDG) was to achieve universal primary education, during which time progress occurred in advancing access to primary education around the world. According to the World Bank (2022), when the MDGs were launched in 2000, the net percentage of enrolment for primary school was 83%, and in 2015 this had reached 89% (World Bank, 2022). While some countries made significant progress, MDG 2, that "children everywhere, boys and girls alike, will be able to complete a full course of primary schooling" was not met on the global level. As of 2018, there are an estimated 59 million primary school-aged children who are out of school (UNESCO, 2022). The Sustainable Development Goals (SDGs), starting from 2015, with an end date of 2030, set a more ambitious goal, in SDG 4 to "ensure inclusive and equitable quality education and promote lifelong learning opportunities for all." SDG 4 has ten targets and twelve indicators, which are beyond the scope of this paper to cover in detail; instead, we will focus on the three that are most relevant to childhood literacy.

The SDG target 4.1 expands MDG 2 by including free, universal secondary school education, and it also emphasises the quality of learning ("relevant and effective learning outcomes"). Quality of education is assessed in indicator 4.1.1 which reports on proficiency in reading and maths (in addition to enrolment). Target 4.2 further expands on MDG 2, as it includes universal access to early childhood development, care, and pre-primary education. The indicator for this is quite broad, and intersects with other SDGs, as 4.2.1 assesses the "proportion of children aged 24–59 months who are developmentally on track in health, learning and psychosocial well-being," along with participation in pre-primary learning. Target 4.5 reiterates the requirement of inclusive education, by focusing on the elimination of discrimination in education (this is further emphasised in Target 4.a in terms of inclusive educational facilities), which is assessed by analysing the parity indices for a range of forms of social differentiation. Inclusive education is explicitly a focal objective of the SDGs, with the overarching goal of "leaving no one behind" as a cross-cutting commitment. While these targets are an important indicator regarding the status of education, there are also socio-cultural norms that are required to support the educational objectives beyond what goes on in formal educational institutions. The Qatar Reads programme assessed in this chapter explores one such initiative, seeking to cultivate a culture of reading outside school.

10.1.1 Qatar

The State of Qatar is a relatively small state, geographically (less than 12,000 km^2) and demographically (~3 million residents). However, like the small states of Macao, Luxembourg, and Singapore, Qatar has a high GDP per capita, enabling the government to invest significantly to support rapid economic and social development, including in the realm of education (see: Cochrane & Al-Hababi, 2023; Hinnawi et al., 2024; Sellami et al., 2024). Qatar National Vision 2030 (QNV) launched in 2008 provides a framework for the national development strategies of Qatar in four pillars (economic, social, human and environmental). Education takes an important role in the human development pillar, which targets "development of all Qatar's people to enable them to sustain a prosperous society" and "to meet the needs of this generation without compromising the needs of future generations." To this end, several education and training initiatives have been designed with the overarching goals of building knowledge and skills, and fostering a capable and motivated workforce. Some of these initiatives are specific to K-12 education, including projects that improve student language, literacy, maths and science skills at all K-12 levels.

In 2021, the State of Qatar submitted a voluntary national report on its progress on the SDGs; however, SDG 4 was not reported on (PSA, 2021). Utilising data from the World Bank Data Bank (2022), insight into some educational indicators is available: gender parity for primary and secondary enrolment is 1.00 (2010 data), lower secondary completion rate is 93.6% (2020 data) but 90% for female students and 97.6% for male students (2020 data), and literacy amongst youth aged 15 to 24 is 95% (2017 data; World Bank, 2022). The upper primary completion rate disparity reported by the World Bank is interesting, because the reverse gendered trend is seen for completion of secondary school and for educational outcomes (for additional details, see Cochrane et al., 2022, 2023). It is also noteworthy that the aggregate data from the World Bank Data Bank does not reflect the complexity of the education system in Qatar, wherein there are 332 private schools (63% of enrolled students) and 207 government schools (37% of enrolled students), as of the end of academic year 2019–2020 (MOEHE 2022; Amin & Cochrane, 2023). The former largely (although not exclusively) have English as their primary language of instruction, while the latter Arabic. Due to the heterogenous nature of the education system, any analysis of educational outcomes needs to take into account the unique educational context, lest the aggregation of these figures make important differences invisible.

In this paper, we examine the multifaceted nature of the HLE, exploring its components, examining its influences, and looking at its role in achieving children's cognitive, emotional and literacy development. Our analysis is guided by the examination of a home reading programme, Qatar Reads, which serves as a good illustration, showing the power of the HLE. Through the lens of the HLE, we aim to demonstrate the path towards achieving the SDGs, showcasing the way that fostering education within the home today sets the stage for global progress tomorrow. The results of this study may guide policy to support education so that these transformations move beyond individual households and to the societal level. The remainder of the paper is structured as follows. Section 10.2 reviews the literature relevant to the programme as well as the context it operates within, while Sect. 10.3 outlines the methodology and data used. Section 10.4 provides results and Sect. 10.5 provides discussion, policy recommendations, and concluding remarks.

10.2 Literature Review

Reading literacy is a key determinant of children's academic achievement, and even success in future life (e.g. Denton, West & Walston, 2003; Snow et al., 1998). Literacy acquisition starts from a very young age—before school entry—thus, the home environment and parental activities play a significant role in enabling literacy skills (e.g. Mirasol & Topacio, 2021). Early or emergent literacy involves the development of "concepts about reading through the informal and spontaneous learning experiences of the children with their parents" (Mirasol & Topacio, 2021, p. 779). Such concepts and beliefs developed in the early years of childhood "guide" their later thinking and behaviour. In order for these skills to be fostered by parents within the home, before starting schooling and throughout childhood, parental perception of literacy skills, appropriate skills levels, approaches to engaging children with reading literacy, and beyond, are influenced by parents' own perceptions. As a result, understanding these perceptions is critical to ensure that children have appropriate and quality support (and where this may be insufficient, services which can support children and families accordingly).

Parents influence children's reading attitudes and perceptions through various means, including how often they read for their own pleasure, such as through role modelling these activities, how often they read aloud to their children, and the perceptions about reading they convey to their children (Baker & Scher, 2002; Johnson et al., 2010). Positive reading attitudes and perceptions of parents provide a positive reading environment for children (e.g. Cochrane et al., 2023). In addition to explicit and direct efforts of parents to enhance their children's literacy skills, when parents read for various purposes and in different contexts, children develop their literacy skills "incidentally," and they also develop a positive attitude and perception towards reading (Stainthorp & Hughes, 2000). Parents' perceptions were found to be significantly associated with the frequency of home-based reading activities (Vanbecelaere et al., 2021). Parents who have positive perceptions towards reading are more likely to do reading-related activities with their children more often (Vanbecelaere et al., 2021). Parents' awareness of their role in encouraging their children to read can be reflected in the home-based reading activities they establish, their ability to convey the sense that reading is pleasurable and valuable, and the overall supportive reading environment they provide (Johnson et al., 2010). This includes, but is not limited to household traits, such as having a home library (e.g. Cochrane et al., 2022). Children raised in literacy-rich home environments are more likely to become motivated readers (Boulhrir, 2017). Moreover, children who grow up in families that view reading as a pleasurable and valuable activity tend to perform better in measures designed to assess literacy, compared to those from homes where reading is undervalued as a skill (Baker et al., 1996).

Perceptions about literacy and reading, as well as the activities that follow therefrom, are not the only influencing factors. Socioeconomic status (SES) also influences parents' and children's attitudes to reading and perceptions of it, and, consequently, children's academic achievement and reading performance. Some studies, for instance, found that low-income families are at a higher risk of literacy-related issues for various reasons, including lack of awareness of the value of reading, lack of time to engage in home-based literacy activities, and financial barriers due to the prioritisation of essential needs over the purchase of books, or engagement in activities outside of the household (Mirasol & Topacio, 2021). The presence of comparatively fewer books, less common incidence of home libraries, and inability to engage in external literacy-enhancing activities that the children from low-income families experience, contribute to less developed literacy skills by the time they enter kindergarten (Chall et al., 1990, p. 16; Mirasol & Topacio, 2021). Others, however, suggest that access to books is a more important factor for academic achievement and reading development and that such gaps could be tackled by providing children with access to books and reading programmes (e.g. McQuillan, 1998, p. 69; Achterman, 2008; Evans et al., 2010, pp. 186–187; Family and Community Engagement and Afterschool Alliance, 2013). In comparison to parental perception, some research suggests that the perception of reading is more significant than access to books. For example, when provided with similar sources—books—middle-income families seemed to take more advantage of them than low-income families (Mirasol & Topacio, 2021). Since a family's SES is usually measured considering parents' education and occupation, in addition to their home possessions or settings (e.g. the number of books and availability of a designated study/reading area for the children; OECD 2011, p. 48), a second element that influences children's—and parents'– reading attitudes and perceptions is parents' educational level.

Several studies that have been conducted in different countries and socio-cultural contexts suggest that parents' educational levels correlate with children's recreational reading (e.g. Altun et al., 2022; Bracken & Fischel, 2008; Lee & Yeo, 2014; Smith, 2010). Parental educational attainment is associated with how often parents read, how often they read to their children, and the number of books and other reading materials in their homes (Altun et al., 2022; Bracken & Fischel, 2008; Lee & Yeo, 2014; Smith, 2010). According to Smith (2010), parents with higher levels of education perceived reading as a valuable activity, and so did their children. A similar conclusion has been provided by Gökbulut and Yeniasir (2018), that children of well-educated parents were more aware of the importance of reading. In terms of studies that have been carried out in Qatar, Chiu (2018) found that children of parents with high SES—i.e. education and financial status—read more frequently, and had more positive reading attitudes and better reading performance.

The available literature highlights a third influencing element of SES on children's literacy skills, which is household traits (e.g. having a home library, or access to literacy-enhancing technologies and programmes). A home library is considered a component of the home literacy environment, and is assumed to encourage the valuing of reading (Kirby & Hogan, 2008). Children's reading is found to be positively related to the size of the home library. Bergan et al.'s (2016) study suggested that the number of books in the home could act as a proxy measure of how much reading is valued by the family, and that it relates to how often parents read to their children. O'Brien et al. (2020) mention that the number of books in the home is one of the indicators of parental literacy involvement, and parents' efforts to engage their children more in home-based reading activities, which could imply the value they place in reading and to what extent they convey it to their children.

The aforementioned findings on the impact of SES components on children's reading attitudes and perceptions should be assessed with due consideration of contexts in which education and income variables are inconsistent. As noted above, Qatar has the unique traits of being rapidly developing, having high-income levels, and unique demographic trends (of a minority citizen population), which are associated with high levels of inequality. Amongst the challenges the country is working to address is lower than expected educational outcomes relative to countries with similar GDP levels and a similar level of government investment in education (e.g. Cochrane et al., 2023). Moreover, the existing literature concerning children's literacy skills in Qatar indicates that parents face challenges supporting their children in this regard, due to a lack of interest and time (Ihmeideh & Al-Maadadi, 2020). Interventions that aim to enhance parents' involvement in children's reading in Qatar are limited (Ihmeideh & Al-Maadadi, 2020) and so are the studies that investigate their impact on children's attitude to reading. Thus, the importance of working towards a better understanding of the role of parental perception in Qatar. Some studies suggest that the influence of interventions such as family literacy programmes can go beyond children's reading performance to their attitudes and perceptions. Neyer et al. (2018), for example, concluded that families that participated in a home-based literacy programme strongly associate reading with learning and academic achievement, and that the programme helped them to develop reading habits and activities.

10.3 Methodology and Data

Qatar Reads is a reading programme initiated by the Qatar Foundation, with the primary goal of enriching the reading culture within Qatar. The programme commenced in 2019, inspired by the National Reading Campaign in 2016. The programme targets children aged 3 to 13, with each participating household receiving a monthly delivery of books directly to their homes. As of 2021, the programme has made a significant impact, with over 370 families participating, benefitting more than 900 children. Qatar has a unique demographic context, where the majority of the country's residents are expatriates. The programme initially targeted Qatari Nationals, but extended its reach to non-Qataris, with over 85% of participants being Qatari Nationals.

This research draws upon data collected from a survey conducted among Qatar Reads participant households. The survey was developed based on a review of the previous surveys. The questionnaire was translated into Arabic, and ethical approval was acquired from the Institutional Review Board. The survey was sent out to the families after piloting and refining. A total of 65 parents completed the survey on behalf of 106 children enrolled in the programme, with a nearly equal gender distribution.

The study uses secondary literature, available statistics, and survey data for analysis. The literature review involved searches on Scopus and Google Scholar for evidence related to children's reading habits and household characteristics, both within Qatar and the Gulf Cooperation Council states, along with a global evidence sample. Additionally, data from authoritative sources such as the Planning and Statistics Authority and the Ministry of Education and Higher Education were referenced.

The analysis of survey data employs a combination of descriptive and inferential methods, including numerical summaries, charts, as well as correlation tests to explore associations between variables. The study uses the framework depicted as Fig. 10.1 to address the research questions. The figure shows the factors contributing to children's literacy development. Socioeconomic factors, such as household income, access to technology, and parental education levels, as well as socio-cultural dynamics including cultural values, language spoken at home, and family approach to learning can contribute to the home learning environment, and can play a significant role in shaping learning resources and family involvement in the learning process.

Different components of the HLE may affect children's literacy outcomes in various ways. This study's focus is to analyse the impacts of the HLE on children's literacy development, assessed through parents' perceptions. In particular, we analysed the correlation between HLE and parental perceptions of children's language,

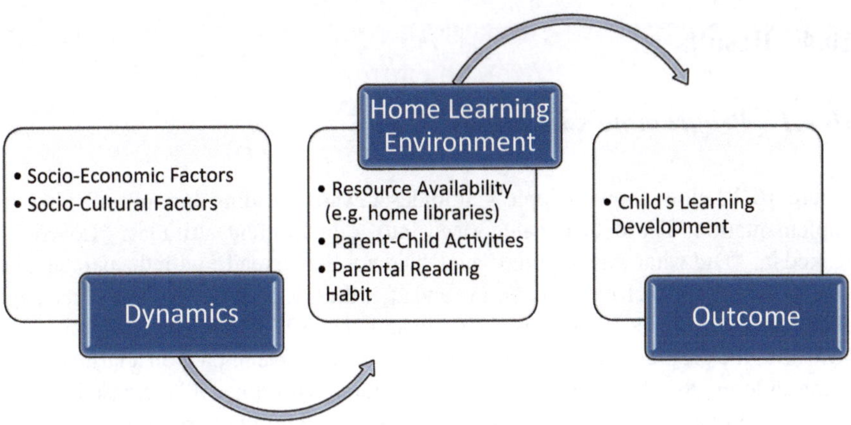

Fig. 10.1 Conceptual model of the contributions of the home learning environment to children' literacy and language development (Authors' own depiction)

literacy, and emotional development through the Qatar Reads programme. Following Bergan et al. (2016) and O'Brien et al. (2020), we used the number of children's books in the home library as an indicator of access to books. For reading activities, we used parent–child joint reading activities and parents' reading habits as role models.

On the other hand, for the outcome variable, we used the following questions as a proxy to test parents' perceptions of children's development.

I believe that QR has increased my child's language and emotional development

I believe my child enjoys reading books to himself/herself

I believe my child enjoys reading books to other people (siblings, friends, or parents)

I believe my child is a motivated reader

Where parents were asked to rank their perception of literacy development on a Likert scale of 5 (*Strongly disagree, Disagree, Neutral, Agree, Strongly agree*), we created a composite variable using the average score of these questions, and employed a combination of descriptive and inferential methods, using the survey data described above.

It is important to acknowledge the limitations of this study. The data is derived from a self-selected sub-population, namely, participants on a voluntary reading programme, potentially introducing self-selection bias. Consequently, the findings may not be generalisable to the broader population. The research relies on recall data rather than baseline data, limiting the ability to assess changes brought about by the Qatar Reads programme. The study was conducted in Arabic and English, overlooking the linguistic diversity within Qatar, necessitating future research to encompass other linguistic groups. Furthermore, the respondent sample includes a relatively high proportion of non-citizen residents, highlighting the need for more focused research on specific population groups within Qatar.

10.4 Results

10.4.1 Programme's Influence

Figure 10.2 below shows people's attitudes towards reading before and after its implementation. The data is categorised into four different attitudes: "Loved it," "Liked it," "Did what was required," and "Did not like to read," with the percentages of respondents for each category before and after the QR programme. The percentage of children who "loved it" increased significantly from 27% before the programme to 48% after the programme, suggesting that it had a positive impact on a large portion of the children. Similarly, the percentage of children who "liked it" increased, though not as notably as in the previous category, from 34% before the programme to 39% after it, indicating a moderate improvement in children's positive attitudes towards reading. On the other hand, the "Did what was required" category saw a significant decrease from 28% before the programme to 12% afterwards, which also indicates

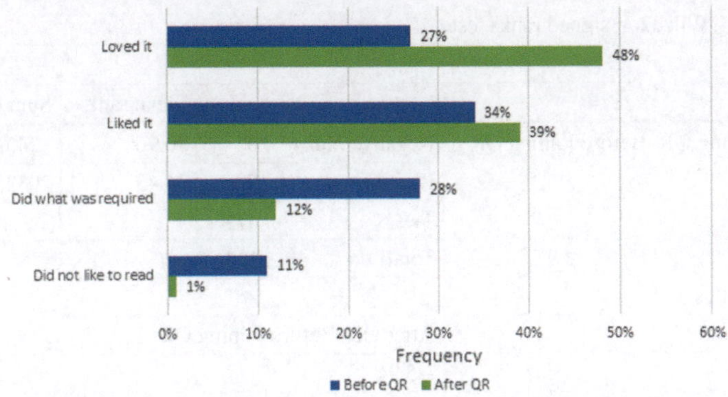

Fig. 10.2 Children's attitude towards reading before and after Qatar reads. *Source* Logan et al. (2022)

a shift towards more positive attitudes. Likewise, the "Did not like to read" category experienced a significant decrease from 11% before the programme to only 1% after it, indicating that the programme had a highly positive effect on reducing the number of people who disliked reading.

Additionally, a statistical test was conducted to assess if the positive differences between before and after the programme are statistically significant. To determine an appropriate statistical method, normality tests (Kolmogorov–Smirnov and Shapiro–Wilk) were conducted. The results suggest that the data are not normally distributed, which justifies the use of non-parametric methods for assessment, as they are robust to non-normal data. Amongst the non-parametric tests, the Related-samples Wilcoxon signed-rank test is an appropriate method, since the measures are correlated (i.e. the same measures were taken from the same subjects before and after the Qatar Reads programme). Hence, the Wilcoxon method is employed to assess the changes, which can be hypothesised as follows:

Ho: The median of differences between after joining Qatar Reads and before joining Qatar Reads equals 0.

H1: The median of differences between after joining Qatar Reads and before joining Qatar Reads is not equal to 0.

The results presented in Table 10.1 suggest that null hypotheses are strongly rejected in favour of the alternative hypothesis at a 1% significance level, suggesting that the children's attitudes towards reading significantly improved after joining the QR programme.

In order to assess the gender dimension of the changes, we conducted a gender analysis of the results in Fig. 10.2. The results shown in Fig. 10.3 indicate that girls made slightly greater progress in all categories compared to boys. However, we conducted a Chi-squared correlation test to determine whether these observed differences are strong enough for the association between gender and children's reading habits to be statistically significant. The analysis results indicate that there is no statistically significant association. Thus, we can conclude that Qatar Reads

Table 10.1 Wilcoxon signed ranks test

Ranks				
		N	Mean rank	Sum of ranks
After joining QR–Before joining QR	Negative ranks	8	30.96	245.5
	Positive ranks	59	34.45	2032.5
	Ties	47		
	Total	114		
Test statistics				
	After QR–Before joining QR			
Z	−5.94			
Asymp. Sig. (2-tailed)	<0.001			

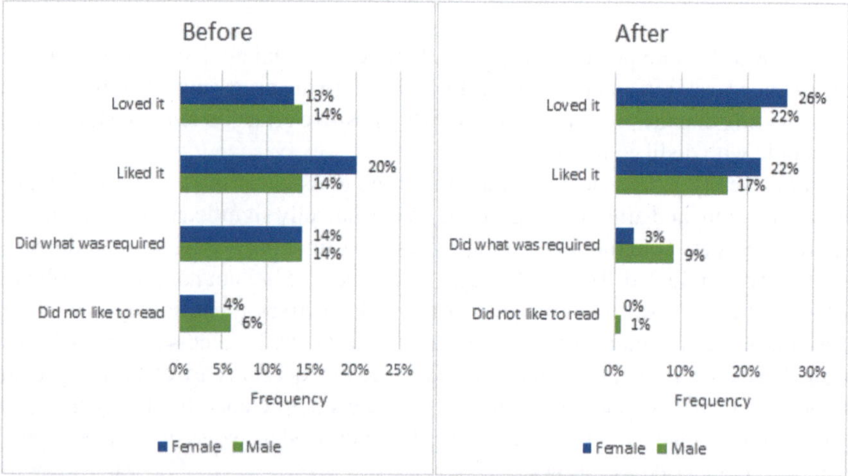

Fig. 10.3 Children's attitude towards reading before Qatar Reads, based on gender. *Source* Logan et al. (2022)

has enabled a significant positive change in the reading habits of children, and these changes do not differ significantly by gender. This is an important finding, as it suggests that there are no accessibility or content barriers related to gender, as was identified in earlier intervention evaluations. The results suggest that parents support their children's reading regardless of their gender.

Overall, Figs. 10.1 and 10.2, and the statistical test show that the Qatar Reads programme had a substantial positive effect on children's attitudes towards reading. More subjects reported loving or liking the programme, while fewer felt they were merely doing what was required or disliked reading altogether. Additionally, there was no gendered difference with regards to attitudes towards reading.

10.4.2 Home Libraries and Children's Language and Emotional Development

As depicted in Fig. 10.1, HLE is multifaceted. Having reading resources at home is one of its most important components. This is especially true for the Qatar Reads programme, as enabling the building up of home libraries is its main goal. Several studies have shown that having home libraries is positively correlated with children's reading achievement. What follows presents an analysis of home libraries with regards to different variables, and its impact on children's learning development. The first figure in Fig. 10.4 shows the availability of books within the household reported by citizens and expatriates/non-citizen residents. Qatari nationals generally had a higher percentage in the "10–20" and "20–50" categories, while expatriates had a higher percentage in the "50+" category.

The second figure shows how educational attainment affects building home libraries. Those with higher levels of educational attainment were on the higher end of book holdings before joining Qatar Reads (in the 20–50 or 50+ categories), while amongst those with a high school diploma, the larger number of book holdings was in the 10–20 range.

Next, we tested if and how home libraries affect children's language and emotional development. To do so, we conducted a correlation test between two variables; having home libraries and parents' perception of their children's language and emotional development. The null and alternative hypotheses are as follows:

Ho: There is no relationship between having children's library at home and parents' perception of children's language and emotional development.

Ha: Having home libraries more likely will increase parents' perception of children's language and emotional development

Since the data is categorical, the Chi-squared association test is the appropriate test to use. Hence, we used the Chi-squared test. The results are presented in Table 10.2. The null hypothesis is rejected in favour of the alternative hypothesis, as the corresponding p-value is smaller than 0.05. That is, there is a statistically significant

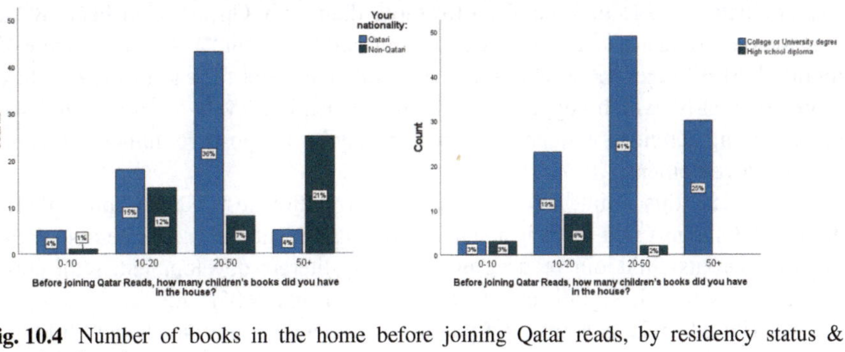

Fig. 10.4 Number of books in the home before joining Qatar reads, by residency status & educational attainment

Table 10.2 Correlation test between access to book and children's language and emotional development

	Value	Degree of freedom	Asymptotic significance
Pearson chi-square	26.2	12	0.010
Likelihood ration	17.71	12	0.125
Linear-by-linear association	0.07	1	0.781
Phi	0.48		0.010
Cramer's V	0.28		0.010
N of valid cases	114		

positive correlation between having child home libraries and parents' perception of children's language and emotional development. As for the strength of the correlation reported in Table 10.1, Cramer's V coefficient is the appropriate indicator to report, since the variables have more than two categories. The value of 0.28 indicates a moderate correlation, and suggests that 28 per cent of children's linguistic and emotional development is attributable to the availability of reading resources at home.

10.4.3 Joint Reading Activities and Parental Perception of Child's Literacy Development

Parental involvement in a child's learning is another significant dimension of the home literacy environment. Several studies have supported the connection between parent–child literacy and reading activities and children's literacy development. Figure 10.5 examines the reading habits of college graduates and high school graduates, in terms of how often they read to their children weekly. For college graduates, 44% reported reading "2–4" hours per week to/with their children. A total of 21% of university graduates reported "5–10" hours reading to their children. The same percentage of respondents reported "0–1" hour of reading, while only 3% of university graduates read to their children for more than 10 h. On the other hand, 8% of high school graduates stated they read to their children from "2–4" hours, while 4% reported "0–1" hour per week. University graduates tend to read to their children more frequently, which suggests that higher education levels are associated with more reading activities with children, which may have a positive influence on child literacy development.

To validate this point, we conducted a correlation test. Results presented in Table 10.3 suggest that there is a statistically significant and moderate association between parents-child joint reading and children's literacy development, as the corresponding asymptotic significance value is smaller than 0.05. This further supports the descriptive analysis above that parent–child reading activities foster children's literacy development.

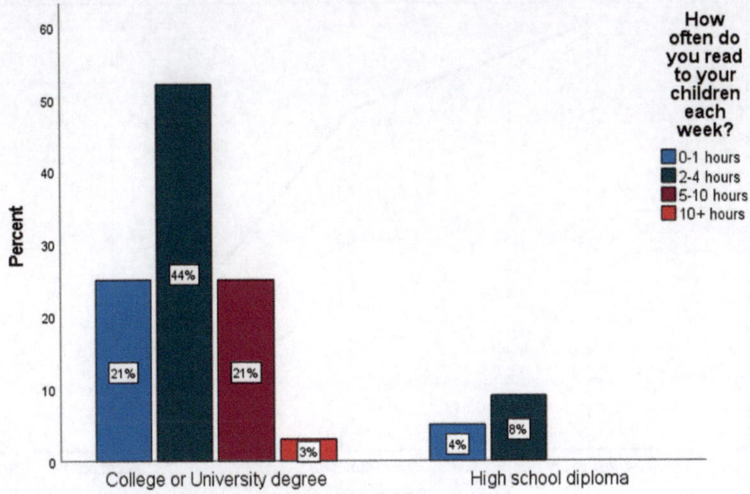

Fig. 10.5 Parental reading habits and reading to children

Table 10.3 Correlation test between parents-child joint reading and children's literacy development

	Value	Degree of freedom	Asymptotic significance
Pearson Chi-square	33.059	9	0.001
Likelihood ration	38.696	9	0.001
Linear-by-linear association	14.258	1	0.001
Phi	0.539		0.001
Cramer's V	0.311		0.001
N of valid cases	114		

However, the positive impact of parental involvement is stronger on children below the age of ten (in primary school) compared to children above the age of ten (secondary), which suggests that children tend to read on their own as they get older, and parents' time spent with them gets shorter compared to time spent with children in early years (Fig. 10.6). Figure 10.7 shows that parents' perception of development decreases for the children in the secondary stage. These results suggest that the QR programme should be tailored differently for children in secondary years.

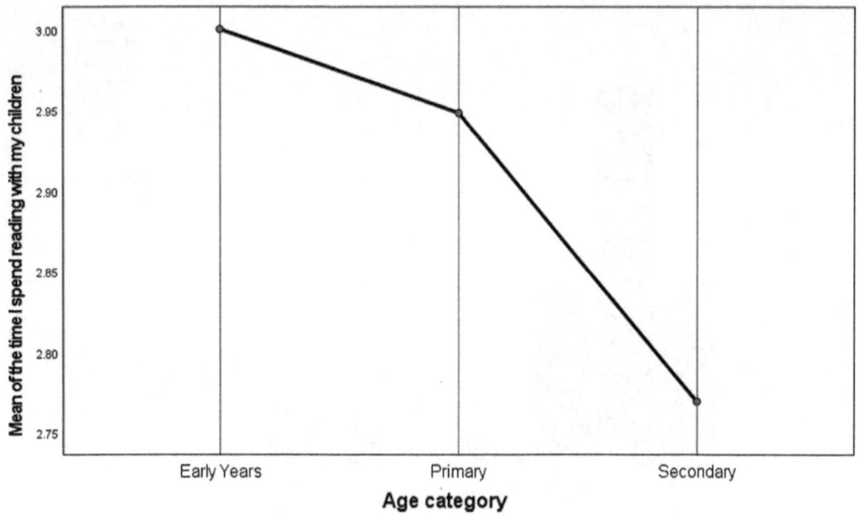

Fig. 10.6 Parental reading involvement with different age categories

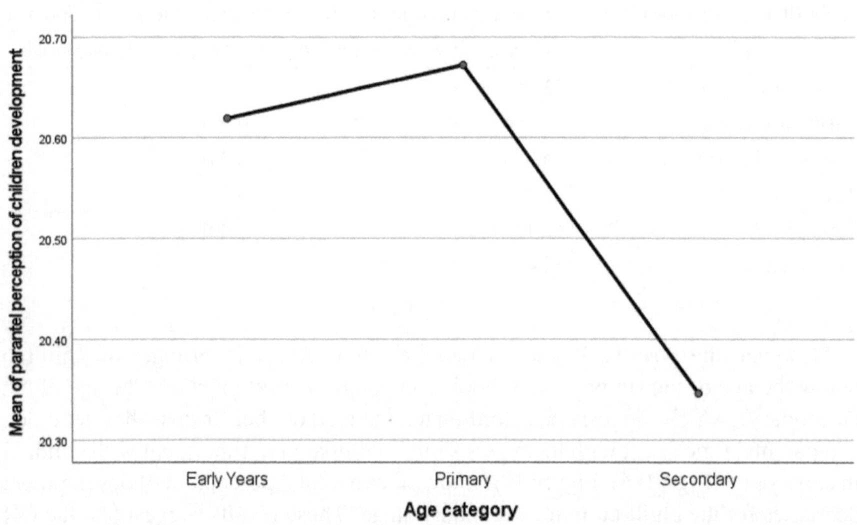

Fig. 10.7 Parental perception of children's literacy skill development based on age category

10.5 Discussion, Policy Implications and Concluding Remarks

In this chapter, we report on a comprehensive assessment of the Qatar Reads programme, which aims to advance children's literacy, linguistic proficiency, and emotional development through the enrichment of home learning environments. Our analysis of survey data has revealed the multifaceted benefits of the programme, while also pinpointing areas where its efficacy and effectiveness could be enhanced for different age groups. Among the observed outcomes, our findings indicate the significant, positive overall impact of the Qatar Reads programme on children's holistic development. This suggests that the programme has not only improved children's attitudes towards reading but also contributed to a reading-centric culture within the home. Results also indicate that these positive outcomes are consistent for both boys and girls, suggesting that the programme's content is thoughtfully designed to cater to the needs and preferences of both boys and girls. Furthermore, results reveal the pivotal roles of home libraries and parent–child activities in promoting the cognitive and emotional development of children. These findings are in alignment with the literature on the positive roles of having home libraries and conducting parent–child activities. Nonetheless, a nuance emerges when it comes to parent–child activities, as the positive effect on the parents' perception of children's development appears to decrease as children grow older, suggesting that the secondary stage may require a distinct approach to parent involvement.

These findings have direct implications for policymakers, educators, and families. First, policymakers, when formulating educational policies, should recognise the importance of such family-oriented programmes, and provide continuous support to enable them to flourish. The Ministry of Education should also explore avenues for expanding and diversifying customised tertiary educational programmes, to enhance educational achievement at a broader population level. Second, the results highlight the value of establishing and maintaining home libraries and the need to encourage families to do so. Third, there should be an emphasis on promoting parent–child joint reading activities, as they have proven to be effective in promoting children's development and strengthening the bond between children and parents. However, it is worth noting that tailoring such programmes to different age groups is crucial, especially for children in the secondary stage, where the impact of parental involvement appears to diminish, suggesting a need for more independent reading approaches.

One of the key limitations of this chapter revolves around the relatively small sample size, and the fact that the sub-population group analysed voluntarily participated in the reading programme, which creates a self-selection bias within the respondent group, presenting limitations regarding the extent to which our findings can be generalised. Another limitation of this research is its reliance on recall data rather than having access to baseline data collected at the time of programme enrollment. This absence of baseline data could potentially introduce bias into the assessment of the programme's impact. Nonetheless, it is worth highlighting that for a context where evidence is scarce, this research significantly contributes to an under-researched area, while offering new avenues for further research.

Bibliography

Achterman, D. L. (2008). *Haves, Halves, and Have-Nots: School Libraries and Student Achievement in California* [Doctoral dissertation, University of North Texas]. https://digital.library.unt.edu/ark:/67531/metadc9800/m2/1/high_res_d/dissertation.pdf

Altun, D., Erden, F. T., & Hartman, D. K. (2022). Preliterate young children's reading attitudes: connections to the home literacy environment and maternal factors. *Early Childhood Education Journal, 50*, 567–578.

Amin, H. & Cochrane, L. (2023). The development of the education system in Qatar: assessing the intended and unintended impacts of privatization policy shifts. *British Journal of Middle Eastern Studies,* 1–21.

Baker, L., & Scher, D. (2002). Beginning readers' motivation for reading in relation to parental beliefs and home reading experiences. *Reading Psychology, 23*, 239–269.

Baker, L., Sonnenschein, S., Serpell, R., Scher, D., Fernandez-Fein, S., Munsterman, K., Hill, S., Goddard-Truitt, V., & Danseco, E. (1996). Early literacy at home: children's experiences and parents' perspectives. *The Reading Teacher, 50*(1), 70–72.

Bergen, E., Zuijen, T., Bishop, D., & Jong, P. (2016). Why are home literacy environment and children's reading skills associated? What parental skills reveal. *Reading Research Quarterly, 52*(2), 147–160.

Boulhrir, T. (2017). Using binary logistic regression to predict long-term effects of early-age home literacy environment on reading motivation. *American Journal of Educational Research, 5*(8), 858–862.

Bracken, S., & Fischel, J. (2008). Family reading behaviour and early literacy skills in preschool children from low-income backgrounds. *Early Education and Development, 19*(1), 45–67.

Chall, J. S., Jacobs, V. A., & Baldwin, L. E. (1990). Literacy and Language among Low-Income Children. In *The Reading Crisis: Why Poor Children Fall Behind* (pp. 1–19). Harvard University Press. https://doi.org/10.2307/j.ctvjhzs0b.5

Chiu, M. (2018). Qatar family, school, and child effects on reading. *International Journal of Comparative Education and Development, 20*(2), 113–127.

Cochrane, L., Ozturk, O., Al-Hababi, R., Khataee, H., Al-Malki, F., & Nourin, H. (2023). *The Impact of Household Traits and Parental Educational Attainment on Reading Habits in Qatar*.

Cochrane, L. & Al-Hababi, R. (2023). *Sustainable Qatar: Social, Political and Environmental Perspectives* (p. 380). Springer Nature.

Cochrane, L., Ozturk, O., Khataee, H., Al-Hababi, R., Al-Malki, F., & Nourin, H. (2022). Fostering a reading culture in Qatar: Evidence from Qatar reads. *Development Studies Research, 9*, 82–94.

De Baryshe, B. D., & Binder, J. C. (1994). Development of an instrument for measuring parental beliefs about reading aloud to young children. *Perceptual and Motor Skills, 78*(3), 1303–1311.

Denton, K., West, J., & Walston, J. (2003). *Reading: Young Children's Achievement and Classroom Experiences* (Report No. NCES-2003-070). National Center for Education Statistics. https://eric.ed.gov/?id=ED480265

Evans, M. D. R., Kelley, J., Sikora, J., & Treiman, D. J. (2010). Family scholarly culture and educational success: books and schooling in 27 nations. *Research in Social Stratification and Mobility, 28*(2), 171–197.

Family and Community Engagement and Afterschool Alliance. (2013). *The Life-Enhancing Benefits of Reading in Out-of-School Programs Issue Brief*. Family and Community Engagement and Afterschool Alliance. http://afterschoolalliance.org/documents/Afterschool-Literacy-Brief.pdf

Gökbulut, B., & Yeniasır, M. (2018). Analysis of children's interest in books and their reading levels depending on the education status of family. *Quality & Quantity, 52*(S1), 235–245.

Hinnawi, R., Sever-Mehmetoğlu, S. D., Ozturk, O., & Tok, E. (2024). Towards sustainability in education: lessons from the FEE eco-schools program in Qatar. *Education for Sustainable Development Cluster Research Grant Hosted at the College of Public Policy, HBKU*. https://doi.org/10.5281/zenodo.10807808

Ihmeideh, F., & Al-Maadadi, F. (2020). The effect of family literacy programs on the development of children's early literacy in kindergarten settings. *Children and Youth Services Review, 118.*

Johnson, E., Bornman, J., & Alant, E. (2010). Parents' perceptions of home reading activities: comparing children with and without learning disability. *Perspectives in Education, 28*(1), 34–43.

Kirby, J. & Hogan, B. (2008). Family literacy environment and early literacy development. *Exceptionality Education International, 18* (3).

Lee, M., & Yeo, K. (2014). Influence of home literacy environment on children reading attitude. *Journal of Education and Practice, 5*(8), 119–138.

McQuillan, J. (1998). Real crises, real solutions. In *The Literacy Crisis False Claims, Real Solutions* (pp. 67–86). Heinemann.

Mirasol, R., & Topacio, K. (2021). Reading perceptions, needs, and practices among parents of an urban poor community in the Philippines. *Reading Psychology, 42*(7), 777–787.

Neyer, S., Szumlas, G., & Vaughn, L. (2018). Beyond the numbers: social and emotional benefits of participation in the imagination library home-based literacy programme. *Journal of Early Childhood Literacy, 21*(1), 60–81.

O'Brien, B. A., Ng, S. C., & Arshad, N. A. (2020). The structure of home literacy environment and its relation to emergent English literacy skills in the multilingual context of Singapore. *Early Childhood Research Quarterly, 53*, 441–452.

Organisation of Economic Co-operation and Development. (2010). *PISA 2009 at a Glance.* OECD Publishing. https://doi.org/10.1787/9789264095298-en

PSA. (2021). Qatar Voluntary National Review 2021: Report on the Implementation of the 2030 Agenda for Sustainable Development. Planning and Statistics Authority.

Ministry of Education and Higher Education (MOEHE). (2022). *Public Education Sector*. https://www.edu.gov.qa/en/Pages/pubschoolsdefault.aspx?ItemID=48

Sellami, I., Sever, S., Ozturk, O., Zaman, A., Amin, H., Tok, E. (2024). *Changemakers for Sustainability: Nurturing Skills, Values, and Citizenship through Makerspaces.* Discover Education (Forthcoming).

Smith, N. (2010). *A Study of Middle Grades Students' Reading Interests, Habits, and Achievement.* https://doi.org/10.17615/bfwq-2689

Snow, C., Burns, S., & Griffin, P. (1998). *Preventing Reading Difficulties in Young Children.* National Academy Press.

Stainthorp, R., & Hughes, D. (2000). Family Literacy Activities in the Homes of Successful Young Readers. *Journal of Research in Reading, 23*(1), 41–54.

UNESCO. (2022). Out-Of-School Children and Youth: Out-of-School Children and Youth. | UNESCO UIS

Vanbecelaere, S., Matsuyama, K., Reynvoet, B. & Depaepe, F. (2021). The role of the home learning environment on early cognitive and non-cognitive outcomes in math and reading. *Frontiers in Education, 6*.

World Bank. (2022). Data Bank: School enrollment, primary (% net)—World | Data (worldbank.org).

Part II
Inclusive Education

Chapter 11
The Expanding Private School Market in Qatar: Exploring the Consequences

Nouf Alkaabi

Abstract Privatisation plays a significant role in the education system in Qatar, as approximately 334 private schools are functional in the country, serving around 40.2% of Qatari students and 72% of non-Qatari students. This rapid growth of private education in Qatar has sparked apprehensions regarding the potential implications and ramifications of the private school market in the country. Considering this perspective, the present chapter consolidates the existing scholarly research to illustrate the effects and difficulties associated with private educational institutions on accessibility, educational outcomes, and the preservation of cultural heritage in Qatar. Several suggestions are proposed for the prospective trajectory of private education.

Keywords Private schools · Privatisation · Vouchers · Qatar

11.1 Introduction

Over the last twenty years, Qatar has made substantial efforts to improve its primary and secondary education systems, as well as its higher education sector, significantly changing its educational landscape. This alteration is occurring at a pace and scale that demands reassessment. More specifically, with 334 private schools that accommodate 40.2% of the Qatari student population and 72% of non-Qataris in 2022 (MEHE 2022), it is evident that the system has a significant privatisation component (PPC, 2011).

The argument over public versus private schools has generated considerable attention within the academic community, leading to diverse findings regarding the influence of school type on promoting equal opportunities and enhancing academic outcomes (Galey, 2015; OECD, 2012). Gershberg (2005) suggests that the global

---

N. Alkaabi (✉)
Director of Planning, Quality, and Innovation, Civil Service and Government Development Bureau, Doha, Qatar
e-mail: alkaabi.nouf84@gmail.com

E. Zaidan et al. (eds.), *The Future of Education Policy in the State of Qatar*, Gulf Studies 21, https://doi.org/10.1007/978-981-97-9667-0_11

discussion around the role of government in service provision has emphasised innovative concepts such as competition, free markets, decentralisation, and privatisation. Primarily, due to the rapid growth in demand for education, it is believed that the funding of education systems solely by the government is neither realistic nor feasible (Rizvi, 2016). However, it can be argued that the intentions of government and for-profit private schools are not aligned, and that "the notion of education as a public good will disappear as for-profit companies dominate the sector" (Ridge et al. 2016, p. 55).

With the growing interest in the privatisation of education in Qatar, this chapter focuses on the private school options available for students. It examines the implications for achieving equal access to quality education for all. Through a comprehensive survey of literature about the Qatari context and the experiences of some other countries, the chapter provides details about the growth of the private sector over the years, the funding of private schools, and how they are governed. It also synthesies findings from the literature, to demonstrate the impact and challenges of private schools concerning access, school performance, private sector governance, and the preservation of cultural identity in Qatar.

11.2 Context

Qatar is a small Arabian Gulf country with a population of less than three million. As the world's second-largest natural gas exporter, the country's economy highly depends on oil, gas, and an expatriate workforce (PSA, 2018). The country's economic growth resulted in an influx of high- and low-skilled expatriate workers, creating a ratio of one national to nine foreign expatriates (Wittrock et al., 2017). Like other GCC countries, Qatar relies on inexpensive migrant labour for its development programmes in infrastructure and energy (Babar, 2015). It is estimated that petroleum-rich Arabian countries are the third-largest importers of migrant workers in contemporary economies (ESCWA, 2007). Over the last decade, the country's population more than doubled, reaching 2,846 thousand in 2020, compared to 1,147 thousand in 2010; the population grew 7.1 times from the population census in 1986 (PSA, 2021). It has been estimated that Qataris constitute a little under 15% of the population in Qatar, making them by far in the minority; this has led to a significant decrease in the ratio of Qataris to non-Qataris (World Population Review, 2023).

Due to the limited availability of skilled Qatari workers, the government sought a proficient expatriate labour force to support the substantial developmental endeavours and projects. (Cheema, 2015). Today, the country consists of a highly diverse and mobile population, due to development projects in the petroleum sector, and sports initiatives such as the 2022 World Cup. This has resulted in an imbalanced population structure in age and gender groups that lacks homogeneity between the different expatriate groups ideologically, culturally, economically, and socially (PSA, 2021). Consequently, there are tremendous implications for service provision, primarily in the health and education sectors. Internationalisation, rapid infrastructure development, and expansion of public services were necessary to tackle the country's influx of expatriates (PPC, 2011).

11.3 Private Education in Qatar

Like other countries in the region, Qatar invested the income generated from oil and gas exports in developing the country's infrastructure and education system, and caring for the well-being of its citizens (PPC, 2011). After the discovery of oil, Qatar established its inaugural government school in 1947, providing structured education exclusively to male students. Subsequently, a second educational institution was founded in 1951 (MOE 2005). In 1956, the Ministry of Education (MOE) (Wizarat Al-Maarif) was established, to offer free public education to all children in Qatar, both boys and girls. To attract students, the MOE provided several incentives that included monthly financial rewards, free food, clothes, and transportation, among other things (MOE 2005). At that time, the government aimed to eliminate illiteracy among citizens (Nasser, 2017). Since its establishment, the MOE has witnessed consistent growth, and gradually developed its policies and services. In 1957, basic public education was generalised for all Qataris and residents of Qatar, and the number of schools increased gradually. At the beginning of 1959/1960, the preparatory and secondary stages were introduced, and educational programmes were developed to include vocational, religious, and industrial education. In 1963, more schools opened, and enrolment rates were increasing (PPC, 2011).

The period from 1980 to 1990 was characterised by an enhancement of the quantitative and qualitative aspects of education by developing school infrastructure, administration, and national curricula (MOE 2005). The government continued establishing the needed infrastructure and providing schools with human resources (Brewer et al., 2007). Since Arabic was the official language in education, it prevented many non-Arab expatriate families from enroling their children in government schools. This prompted the state to allow expatriate communities to open schools to meet their cultural needs in addition to Arabic and foreign private schools (PPC, 2011). Community schools were the first form of private school in the country. In the 1960s, community private schools were established, to cater to the needs of different groups of expatriates. They were mainly pre-primary daycare centres to help both Qatari and expatriate parents care for their children while they were working. The numbers grew in time, and more private schools opened. The MOE was responsible for licencing and supervising these schools. Also, it provided them with textbooks and healthcare. According to the MOE's 1969 statistical report, there were 14 community schools, 11 Arabic schools, and three foreign schools, two of which were Pakistani (MOE, 1970).

Private schools began to attract non-Qatari students before 1985. As for Qatari students, they began to go to private schools at the beginning of the nineties (MOE, 2005). At the time, government schools accommodated almost all the Qatari students, and about two-thirds of non-Qatari students (PPC, 2011). The private sector grew moderately until 2000, when the number of private schools escalated. During the 1970s, 1980s, and 1990s, the country attempted many reforms to improve its essential education provision, and to decrease illiteracy rates (Nasser, 2017). However, there were growing concerns regarding the quality of instruction and insufficient classroom capacity in government schools, due to the population increases (Al-Banai & Nasser, 2015; Al-Kubaisi, 1999).

In the early 2000s, the Qatari government attempted a large-scale reform called Education for a New Era (EFNE) that significantly impacted the country's education sector. The reform was one of the country's attempts towards building a world-class education system that supports its economic and social development and meets its ambitious aspirations towards competing in the global economy. The reform called for radical shifts in vision and approach, resulting in tremendous educational landscape changes. Among these changes was adopting choice and variety as guiding principles for the new system. The RAND Corporation, the contracted designers of the reform, attempted to create diverse schooling options for parents, and allow them to choose according to the quality of the school's performance. It was believed that these principles would encourage competition and reinforce school accountability (Brewer et al., 2007).

Since then, Qatar's education has grown exponentially in size and diversity, and has a more diverse student population in government and private education (Ari & Koc, 2022). According to the Ministry of Education and Higher Education (MEHE) historical statistics, the number of private schools did not exceed 20 from 1964 until 1980, and the number of government schools was still increasing rapidly. The number of private schools rose in the mid-80 s and nearly doubled between the 90 s and 2000. After the EFNE initiative and the introduction of vouchers, the number of private schools increased to reach 334, compared to 209 government schools in 2022 (MEHE 2022).

It has been argued that the growth in the education sector over the last decade was primarily attributed to the increase in the expatriate student population, which grew four times in both government and private schools compared to Qatari citizens (Amin & Cochrane, 2023). Moreover, many Arab and foreign expatriates prefer to enrol their children in international private schools (Ari & Koc, 2022). On the other hand, the capacity constraints in government schools in the face of continued growth in population made many Qataris and other Arab expats opt for private schools that follow British or American curricula. Today, the country has four different types of schools: government schools, Arabic private schools, international private schools, and community schools. International schools follow different curricula, such as American, British, French, and German ones, while Arabic private schools implement the Qatari National Curriculum (MEHE 2023b). There are 22 community schools in Qatar that residents in the country established under the supervision of their embassies. These schools are expected to follow similar licencing and supervision procedures as other private schools (Regulating Private Schools Law No. 23 2015). Many residents are keen to enrol their children in community schools so that they can continue their education in their original countries upon the expiration of their contracts, or the occurrence of any political circumstances which might force them to leave (Al-Bidairi, 2019).

There are 30 different curricula in private schools, to promote variety and choice in the system. Although Qataris make up only 30% of the student population in private schools, private schools accommodate dozens of nationalities, due to the openness of the country to the world which has more than 85 nationalities living there (Al-Jabarai, 2021). Analysis of student data in the last ten years marks a 91% increase in the total number of students in private schools compared to a 39% increase in the total number of students in government schools, and an 87% increase in the number of non-Qatari students as opposed to a 37% increase in the number of Qatari students. According to the most recent statistics published by MEHE, 334 private schools admit 209,088 students compared to 209 government schools with 129,249 students (MEHE, 2022). In 2022, Qataris made up 50.1% of the total student population in government schools, and the percentage of Qataris who chose government schools dropped from 72.6% in 2013 to 60% in 2022; this signals a growing interest in private schools among younger Qatari parents (MEHE, 2022).

11.4 Governance of Private Schools

In the past decade, the education private sector witnessed the issuance of two major documents that primarily addressed the governance of the sector. First, Law (23) of 2015 for Regulating Private Schools was issued in November by an Emir's decree. According to the law, private schools aim to provide diverse and quality educational opportunities for Qataris and children of communities residing in Qatar (Regulating Private Schools Law No. 23, 2015). By law, the MEHE regulates private schools and is responsible for developing policies and monitoring a school's adherence to the regulations. It regulates tuition fees, student registration, teacher recruitment, and curriculum quality in private schools. The MEHE also licences private schools, after receiving and studying applications from investors. It aims to create a competitive environment among educational service providers, and promote the development of a transparent and accountable private sector through rigorous licencing regulations. By law, investors cannot open a school without the MEHE's approval (MEHE, 2021). The law appealed to many parents and educators, as it addressed their concerns regarding the quality of private schools, and granted more judicial power to the MEHE to hold schools accountable and sanction them with penalties (Raslan, 2015).

The second shift was the issuance of the MEHE's new structure. In January of 2016, the Supreme Education Council (SEC) was abolished, and a revised organisational structure for the MEHE was announced in the Emir's Decree No. (9) of 2016 (Emir's Decree No. 9 2016). The new structure included a separate sector with multiple departments, to govern private schools and address the growth of the private school market. Prior to that, the SEC had a single department within the Education Institute, to licence private schools and manage their affairs. According to the new structure, private schools were monitored by three newly established departments, using rules and regulations that differ from what is applied in government schools. The revised structure included a dedicated department for licencing and relicensing schools, another for managing private schools' affairs, and a separate department for managing private learning centres.

The MEHE offers support to private schools through its different departments; it also supervises and monitors the quality of private schools (Al-Jabarai, 2021). For instance, all private schools are expected to adhere to the MEHE's safety and security standards, including fire regulations, emergency procedures, and periodic checks for all equipment used in this regard, and are sanctioned when not doing so (MEHE, 2021). In addition, all private schools are expected to obtain the MEHE approval of their curriculum and education plans. Schools may not change the approved educational curriculum except after obtaining authorisation from the Schools Licensing Department (MEHE, 2021). Islamic studies, Arabic language, and Qatari History are mandatory subjects for Qatari and Arab students in private schools (MEHE, 2023b). The MEHE is responsible for establishing the needed instruction time for each of these subjects; Arabic language and Islamic Studies are required to be taught in Arabic, while Qatari History could be taught in the school's preferred language. Private schools are required to allocate four instruction hours per week for the Arabic language, two hours per week for Islamic Studies, and an open number of hours for Qatari History. Also, private schools are expected to teach any additional subject mandated by the MEHE (MEHE, 2021). Despite having the policies in place, there were growing concerns among parents regarding the schools' adherence to teaching the compulsory subjects, which provoked the MEHE to revise and improve its inspection practices in 2021. In response, it was reported that many private schools modified their annual plans, and hired more staff accordingly. Moreover, private schools were encouraged to organise and participate in extra curricular activities that promote the Arabic language, Islamic values, and Qatari culture (Al-Sharq, 2021).

To further ensure quality in all schools, the School Evaluation Department mandates all private schools to apply for National Evaluation after 2 years of operation, to be eligible for National Accreditation. Based on the outcomes of their evaluation, private schools can renew their accreditation every three or five years. Low-performing schools are evaluated on a yearly basis, until they demonstrate evident improvement. Schools that fail to acquire accreditation are denied vouchers (Raslan, 2016). This policy did not appeal to many investors; there have been several complaints and requests to appeal against the evaluation outcomes from private school owners. They argued that the school evaluation teams tend to be lenient with some schools and strict with others; and the application of the policy was inconsistent and marred with shortcomings (Salah, 2023).

Al-Ansari and Zahirovic (2021) describe the private sector as diverse and highly inconsistent, because of the variety of curricula and standards implemented. Dr. Abdulla Al-Kubaisi, the first appointed President of Qatar University, supported that viewpoint, and argued that the introduction of the vouchers in Qatar and the growth of the education market have resulted in governance "dualism" between government and private schools. As trust in government schools has diminished, and more Qataris are choosing private schools, the government's influence over education is at stake, as private schools identify with Western culture and follow their original systems, which would eventually harm the homogeneity and cohesion of the society (Al-Afifi, 2020).

11.5 The Funding of Private Schools

How and to what extent governments fund private schools matters tremendously, primarily in achieving equitable access to education for all children. The government's financial involvement in private schools helps remove barriers to education, primarily for students from low socioeconomic status (OECD, 2012). The funding of private schools takes two forms: a school is government-dependent if it receives more than 50% of its core funding from the government through student vouchers or for payment of its teaching staff. On the other hand, an independent private school receives less than 50% of government financial support for student vouchers or teacher salaries (OECD, 2012). There are two types of voucher systems for students: universal voucher systems that target all students, and targeted voucher systems for segments of students such as disadvantaged students. It has been argued that targeted vouchers can address equity challenges, while universal vouchers lead to increased stratification (OECD, 2010b).

In the Gulf region, Qatar and UAE have the highest percentage of private schools (Ridge et al., 2016). The private education sector in both countries adopts free-market and open labour models that allow employers to decide who to hire, and how much they pay their employees. Prior research suggests that Qatar has become an attractive country for international school operators, due to its booming education market and diverse population (Sawalhi & Tamimi, 2023). MEHE officials supported this argument in a recent press statement, explaining that Qatari businesspeople are attracted to investment opportunities in education, and the MEHE receives many applications for private school licences every year (Al-Jabarai, 2021).

While government schools are funded solely by the government, private schools depend on student tuition fees, vouchers, and grants from private donors (Cheema, 2015). It has been estimated that the annual fees of private schools in Qatar average around 7,340 USD in elementary school, 7,936 USD in middle school, and 9,288 USD in secondary school. Fees vary according to school type; for instance, international schools charge higher fees than Arabic and community private schools (GASERC 2022). In a report issued by the Organization of Economic and Cooperation Development (OECD) in 2012, it is estimated that private schools in Qatar, Dubai, and Jordan receive financial funds of 1% or less from the government in comparison to Finland, the Netherlands, and Hong Kong, where such schools receive 90% of their funds from the government. The above was supported by a recent comparative study conducted by the Gulf Arab States Educational Research Center (GASERC) in 2022. The study reported that 71% of private schools in the GCC region receive no funds from local institutions, depend primarily on the income generated from student tuition fees, and usually follow a centralised process to obtain approvals from ministries of education for fee increases.

Operators are not allowed to receive funds or donations from any party before obtaining written approval from the MEHE. In addition, school operators are expected to submit a list of all tuition fees and expenses collected from students at the beginning of the academic year. They are not allowed to increase these fees or costs without the approval of the MEHE; otherwise, the school is obligated to refund the amounts collected from the students (MEHE, 2021). Ridge et al. (2015, 2016) argue that due to the limitations on fee increases in Qatar, many private schools charge nominal fees for transportation, field trips, and materials, to name a few. It has been argued that almost 25% of private schools in the GCC region fail to cover their expenses relying solely on students' fees (GASERC 2022).

In a press statement from the Director of Private School's Licensing Department in Qatar, he revealed that the MEHE limited fee increases for private schools in the academic year 2022–2023, and that several necessary controls have been put in place. Schools cannot apply for any fee increase; only schools that suffer financial losses, or which transfer to a new school building are exempted. The statement indicated that approvals were limited to schools that submitted sufficient evidence and a strong rationale to justify the needed increase (Abdulrahman, 2022).

Recently, the government has implemented several initiatives to support private school owners financially. For instance, in Qatar, the MEHE exempts private school operators from electricity/ water fees and customs duties, to support investors. Moreover, in 2016, the government offered 12 plots of land for private vendors to establish private schools (MEHE 2021). Such initiatives have succeeded in attracting some Qatari investors interested in establishing private schools.

11.6 The Use of Educational Vouchers in Qatar

Educational Vouchers were introduced in 2008 to provide various quality schooling options that cater to Qatari parents' needs and ambitions. The vouchers were believed to have been set in place to enable the country to promote the principles of choice and variety, and encourage competition among schools. Consequently, the system would incentivise high-performing schools, and attract investors to open branches of outstanding schools nationwide (Badoor, 2012). When the vouchers were first introduced, schools were required to follow accredited national or international curricular standards, obtain accreditation from a recognised national or international body, and teach the three compulsory subjects: Islamic studies, Arabic language, and Qatari history (PPC 2011). Today, private schools must receive the National Accreditation from the School Evaluation Office to be eligible for the vouchers (MEHE 2023b). The children of Qatari parents, whether working in the public or private sector, are qualified to receive a voucher of 28,000 QRs, to help them pay their children's tuition fees in private schools; a total of 120 private schools benefited from the vouchers in the academic year 2021–2022 (MEHE 2022).

Amin and Cochrane (2023) examined the growth in private schools after the introduction of vouchers to subsidise private schools for Qataris. Their research confirmed that introducing vouchers accelerated the privatisation of the sector. According to the latest statistics, more than 24,293 Qatari students benefited from the vouchers in 2021/2022 (MEHE 2022). Although the private sector has expanded exponentially over the last decade, government schools achieved a slow-paced increase, despite the continued demand for vouchers among Qataris. Today, the system has witnessed a 110% increase in the Qatari student population in private schools, compared to a 12% increase in government schools (Amin & Cochrane, 2023).

11.7 Outstanding Private Schools in Qatar

To raise the quality and variety of private school provision in the EFNE reform, the SEC launched the 'Outstanding Schools' initiative in 2007 to open branches of exceptional international schools locally (PPC, 2011). The initiative's objective was to cater to the educational requirements of both Qataris and expatriates, facilitating access to high-quality international education. Additionally, it aimed to serve as a benchmark for other schools in the nation to emulate in terms of exemplary practices.

In 2009, three schools were opened: DeBakey High School for Health Professions from the United States of America, London International School, and Sherborne also from the United Kingdom (PPC, 2011). The Qatar Finland International School and S.E.K Spanish School were later established as part of the 'Outstanding Schools' programme (Sawalhi & Tamimi, 2023). According to law (23) of 2015, the MEHE can establish branches of outstanding schools in Qatar and provide them with financial support, including a free school building, furniture, and financial loans (Regulating Private Schools Law No. 23 2015). Despite being financially supported by the MEHE, some schools would operate as for-profit institutions (Zayed, 2015). According to Sawalhi and Tamimi (2023), the five outstanding schools have strayed from the initial intent behind establishing them, due to structural reforms within the MEHE and constant management changes. They argue that, although these schools still operate today, as they have since the inauguration of the initiative, the programme "has slowly been buried" (Sawalhi & Tamimi, 2023, p. 61).

The above section provides a factual review of the growth of the private sector in Qatar. Also, it sheds light on some initiatives the MEHE has undertaken to support the sector's growth and richness.

11.8 Implications of the Growing Private Sector

The K-12 education system in Qatar is described as diverse and unique for its provision structure and inconsistent outcomes (Al-Ansari & Zahirovic, 2021). The growing number of for-profit private schools has several implications for the system's performance regarding access, quality, cultural identity, and outcomes.

11.8.1 Access

It has been argued that the growth of for-profit private schools "threaten(s) access to education for certain groups at the lower end of the socioeconomic spectrum" (Ridge et al., 2016, p. 50), especially when private schools operate independently and are granted autonomy regarding their admission policy. The literature affirms that when private schools are granted total freedom over admission of their students, they tend to select high achievers from socioeconomically advantaged families, intensifying the inequity in education systems (OECD, 2010a). Additionally, private schools attract more advantaged students because most of their students come from privileged families (OECD, 2012). Elacqua's research confirms that privatising education in Chile resulted in a highly segregated school system because of selective admission and student sorting; and government schools served the needs of low-income families and had a more diverse student population than private schools (Elacqua, 2012). According to the Private School's Student Registration Circular issued in 2021, private schools in Qatar are granted autonomy in terms of the student admission policy while they are obliged to abide by other MEHE guidelines for grade entry age and the academic year calendar (MEHE 2021b). Regarding this issue, Ridge et al. (2016) write that government schools in Qatar and the UAE are "far more student, teacher, and community-centred" compared to for-profit private schools, which focus on profit generation (p. 53). Supporting this claim, GASERC's study has revealed that in the event of failure or lateness in paying tuition fees, private schools would withhold students' test results, deprive them from taking their exams, or transfer them to another school. There have been numerous instances in the past where parents expressed their rage and dissatisfaction to the MEHE about private schools withholding their children's report cards for financial or other unclear reasons. They believed that banning access to students' results was unacceptable, as it embarrassed their children in front of their classmates (The Peninsula, 2021).

On the other hand, the capacity constraints in government schools in the face of continued growth in population made many Qataris and non-Qataris opt for private schools. While all Qataris are granted free access to government schools, expatriates, primarily Arabs, who wish to enrol their children in government schools are restricted due to limited availability of places (Al-Ansari & Zahirovic, 2021). To address the issue of over-crowded schools and increasing demand, the MEHE issued the government school admission policy, which admits GCC students and children of parents who work in the government sector (MEHE 2023a). Hence, private schools are the only option for some segments of the student population. For that reason, many private schools suffer from enormous pressure from parents to lower fees (Ridge et al., 2015). In recent interviews with parents, they expressed their frustration with the financial burdens imposed by some community private schools in Qatar, and the academic complications that affect their children's learning (Al-Bidairi, 2019). Ridge et al. (2015, 2016) argue that Westerners earn much higher wages than Arab and Asian expatriates in Qatar, and are likelier to receive educational subsidies than other expatriates. This leaves other families with limited options in terms of school choice, as many of them cannot afford to enrol their children in even low-fee private schools that tend to be of low quality. Elacqua (2012) discusses the impacts of private schools' tuition fees on choice in Chile, and reports that the cost of education is the leading factor for the segregation between public and private schools.

Other factors determine Qatari students' tendency to enrol in private schools, the first of which is choice. In the past, most Qatari and Arab families used to go to government schools, while foreign expatriates enrolled their children in private schools (PPC, 2011). However, the variety of schooling options for Qataris and other Arabs made many of them opt for private schools following British or American curricula. Other factors attract more advantaged students to privately managed schools, including admission criteria, policy decisions, academic performance, the student population's profile, and the school environment; consequently, the variety of private schools makes public schools the last resort for students of high socioeconomic status (OECD, 2012). The recent preference of many Qataris and Arab communities to educate their children in English has made them admit their children to international schools with good reputations (PPC, 2011). However, the availability of several diverse options could confuse parents and jeopardise the reputation of government schools. According to a survey conducted by the Social and Economic Survey Research Institute (SESRI), Qatari parents who opt for private schools tend to struggle to choose the right school for their children, and are more likely to change schools, compared to parents in favour of government schools (Benmansour, 2017).

11.8.2 Quality of Performance

According to Ammermuller (2005), education systems with a large proportion of private schools tend to have more inequality in outcomes. The argument is supported by Cheema (2015), who investigated the difference in the performance of government and private schools in the Program for International Student Assessment (PISA) between 2006 and 2012. The study's statistical analysis reveals that the sector suffered a significant literacy performance gap in favour of private schools in maths, reading, and science in three consecutive assessment cycles. The study also reveals that non-Qataris performed better than Qatari students who scored lower than first- and second-generation immigrants in both government and private schools. He argues that the fact the gaps have not shrunk over the past three cycles puts the country's government schools in a detrimental position. However, in 2018, it was reported that 60% of private schools performed below expectation in the National Evaluation, and they were subject to close inspections due to their failure in meeting the MEHE requirements (Raslan, 2018). This indicates that many private schools continue to struggle to meet the MEHE's quality measures, as current inspections do not identify these shortcomings early on. On the other hand, the report reveals that government schools achieved remarkable progress compared to private schools in evaluations (Raslan, 2018); the above might signal the absence of a clear definition of quality that is consistent across all schools in Qatar. This has led to confusion among many Qatari parents who struggle to choose between government schools, to maintain Islamic and national values, or private schools for the quality of the school environment and curriculum (Benmansour, 2017).

Another example of inconsistencies is that Qatari students in government schools receive more hours of Arabic and Islamic Studies instruction than Qataris in private schools. Although the MEHE obliges all private schools to teach the national curriculum in these subjects, the quality and hours of instruction fluctuate between schools (Al-Ansari & Zahirovic, 2021). This intensifies inconsistency issues even among Qatari students.

11.8.3 Cultural Values

Hillman and Ocampo Eibenschutz (Al-Ansari & Zahirovic, 2018), p.11) argue that English is considered a superior language in Qatar, "linked to an identity of access to cultural capital and success". Their study reveals that many Qatari parents pay a lot of money for IELTS and TOEFL courses, besides their children's private school fees, because they believe that being taught in English as the medium of instruction in private schools would increase the chances of their children being accepted into prestigious universities inside Qatar or abroad. However, a study of the lived experiences of Qatari private school graduates has revealed that studying in an international school "may create tension, stress, and difficulty for some Qatari adolescents as they transition between the expectations and realities of the westernized school environment and that of an Islamic home life" (Graham et al., 2019, p. 345). Similarly, Amin and Cochrane (2023, p. 19) argue that the move towards international private schools has led to a linguistic and cultural divide because of the "socio-cultural values, and worldviews embedded within imported international curricula".

Qatari parents seem to acknowledge the negative influence of some international private schools on their children's language and cultural values. In a recent survey, Qatari parents reported that government schools preserved their children's values and national identity more than private international schools (Benmansour, 2017). This issue is attributed to the low instruction hours for Islamic Studies, Arabic language, and Qatari History in private schools. According to GASERC's study (2022), it is estimated that an average of 1.1 h is allocated for teaching Qatari History and

2.4 h for Islamic studies in international private schools. The study also warned of the potential threats to social cohesion and national identity from private schools' hidden curriculum and peer influence. Similarly, Al-Kubaisi warned of the harms of the Qatari families' negligence of teaching children their mother language. He stressed that the Arabic language is the intellectual and cultural vessel that preserves the identity of Qatari society. Also, he appreciated recent government decisions that officially shifted the language of instruction to Arabic in all government schools, and necessitated the provision of Arabic language programmes in most of Qatar's university colleges (Al-Afifi, 2020). In essence, recent initiatives implemented by the MEHE to monitor the preservation of the Arabic language and Qatari identity in private schools are promising, and need to be continued.

On another note, there were some growing concerns over the alienation of students of similar socioeconomic backgrounds created by some community schools. On that issue, in a press interview, Dr. Majidi Atef, social sciences professor in Qatar University, spoke about the negative consequences of creating two realities for students whereby they follow their home country's curriculum and simultaneously engage in homogenous experiences away from the reality of Qatari society, which creates a kind of turmoil in the student's culture, and may lead to some psychological complexes (Al-Bidairi, 2019).

11.8.4 Governance of Private Schools

In Qatar, private schools have to go through strict regulations and continuous demands from regulating bodies that limit their ability to make class size decisions, impose fee increases, and alter building standards (Ridge et al., 2016). Prior studies into private schools in Qatar have found that several for-profit schools perform in 'survival mode' by focusing on covering expenses and adhering to regulations, compared to government schools, which tend to focus more on improving the learning experience of students, and contributing to the community (Ridge et al., 2016). A similar finding was obtained from analysing parents' views on the low quality of instructional practices in some community schools in Qatar. They complained that some private investors reduced teacher salaries to cover expenses, which jeopardised the quality of teaching and learning in these schools (Al-Bidairi, 2019).

Moreover, operators of the outstanding school branches in Qatar have raised the same issues. Sawalhi and Tamimi (2023) reported that due to a lack of support from the MEHE for the programme, these schools are focusing primarily on generating income and adhering to stricter regulations than other private schools. In addition, these schools face several challenges that include the need for more autonomy, the rigidity of MEHE procedures, and the difficulty in communicating with the MEHE and parents (Sawalhi & Tamimi, 2023). Similar challenges were found to exist in GASERC's study (2022), which indicated that private schools in GCC countries suffer from ineffective governance and the absence of a clear vision of the role of the private sector in addressing the government need for cost-efficient schooling options.

The study points out that "the absence of a contemporary vision of the status of private education and its role in society has led to improvisation in developing educational policies" (p. 439).

On a positive note, many educators believe that Law No. 23. Of 2015 for Regulating Private Schools signalled a turning point in the governance of private schools in Qatar, as it covered all aspects needed to hold schools accountable for performance and ensure the provision of quality education, regardless of investors' interest in generating profit. Supporting that, Ridge et al. (2015) report that private school inspections in Qatar have improved in quality and accountability. The continuous revision and tweaking of policies over the past few years has resulted in improved monitoring and guidance for private schools.

For instance, the ministry requires all private schools to follow their recruitment criteria, and the appointment of teachers and school principals have to be approved by the Private Schools Affairs Department (MEHE 2021). These new regulations have prevented favouritism and manipulation in recruitment, as one community school principal explained in a published press interview (Al-Bidairi, 2019). Moreover, the MEHE revised its monitoring guidelines for inspecting the quality of teaching compulsory subjects in private schools through classroom observations, and reviews of academic plans in 2021 (Al-Sharq, 2021). And to enhance cohesion among private schools, the MEHE issued the Educator's Ethical Code of Conduct in 2022, which included clear guidelines and expectations of school owners, principals, administrative and teaching staff, parents and students. The document focused on six principles: national identity, discipline, professionalism, sincerity, transparency and justice. According to officials in the MEHE, the document supports Qatari sought principles and emphasises the importance of respecting Islamic values and Qatari culture. It also intends to promote positive behaviour among schoolchildren, and the spirit of responsibility (QNA, 2022).

Yet the absence of clear policies to support low-quality schools means that inspections were only carried out for reporting and accountability purposes, and poor-performing schools that serve the needs of disadvantaged students were left untouched. The above argument was supported by the voice of frustrated parents, who confirmed that some community schools suffer from a lack of supervision, and urged the ministry to send inspection teams to visit schools and attend student classes, to evaluate the school's performance and provide needed support (Al-Bidairi, 2019).

11.9 Discussion and Recommendations

Due to the increasing demand for education, governments all over the world have adopted privatisation policies to address cost limitations, while providing access to education for all. This new reality has shifted the focus of policymakers towards the extent they need to privatise and how they can regulate the new private actors in the system (Rizvi, 2016). Based on the total number of public and private schools, the trend towards privatisation of education appears evident in the State of Qatar (PPC,

2011). Undoubtedly, the private sector has been of tremendous support to the government in addressing the growing demand for education, especially from the foreign expatriate population in the country. Yet the findings of this study have revealed considerable limitations in the current provision model that calls for immediate action to address them.

It has been argued that it is likely that social and economic inequality increases when governments adopt a market-driven model in education (Ridge et al., 2016). Consequently, stratified education systems based on socioeconomic factors affect social cohesion in those societies (OECD, 2012). To support that, Elacqua's research revealed that three decades of "unrestricted choice" for parents had intensified inequities and stratification between public and private schools in Chilean education (Elacqua, 2012, p.2). Moreover, the privatisation movement has distracted the public from solving the real problems, and turned them into consumers who care only for their own well-being (Ravitch, 2020). On that issue, Rizvi (2016) argues that privatisation has redefined the notion of 'the public good' in governments, and made them more transactional in dealing with citizens; he explains:

Citizens are rendered as investors and consumers, and not as members of a polity who share certain common traditions, spaces, and experiences. What is more, knowledge and education are valued and desired almost exclusively for their contribution to the processes of capital formation and enhancement.

The findings of the present study push the discussion in three directions. First, the extent to which public funding covers schooling costs for students matters tremendously. More public financial involvement in privately managed schools would ease that burden, and enable more disadvantaged students to attend privately managed schools (OECD, 2012). The findings emphasise the urgent need to review the school funding model implemented in Qatar, and assess the efficacy of the public/private partnership in achieving equitable access for all children. Given the diversity of the student population in the country, government-funded private schools would offer more accommodating and efficient opportunities, in comparison to government schools. All children are to be educated at the expense of the government, and when public funds are not directed to support all students, it usually generates equity challenges in the short and long run. It has been argued that when governments provide funding for private schools, this tends to increase choice and competition in any system (OECD, 2012).

Secondly, investing more in improving the quantity and quality of government schools is considered a favourable policy option for the country's future, as it would result in improved coherency and harmony in the system's outcomes (Al-Ansari & Zahirovic, 2021). Not neglecting the role of private schools in catering to the diverse needs of the country's heterogeneous population, the transition of more Qataris to private schools would intensify the neglect in the quality of government schools, and make them the last resort for Qataris. Ravitch (2020) argues that when the public are not satisfied with public schools, governments have the responsibility to improve them and not give public funds to entrepreneurs to build new schools. New improvement initiatives that address the quality concerns in government schools need to be put in place. Moreover, a slow phasing out of the vouchers for Qataris is a recommended policy initiative, to increase the proportion of Qatari and Arab students in government schools, and change the views of parents about government schools (Al-Ansari & Zahirovic, 2021).

The system could also enhance equitable access by redirecting the educational vouchers to serve the needs of disadvantaged students from resident communities, and provide them with cost-effective learning opportunities provided by private schools. Elacqua (2012) suggests that reconsidering voucher policies to cater to the needs of underserved students, and incentivising schools to admit them could mitigate inequities in education. In addition, the government could support nationals by establishing non-profit private schools that implement context relevant curricula as a significant policy option for "alleviating some of the existing inequalities" and offering alternatives for Arab communities who cannot access government schools due to capacity constraints (Ridge et al., 2016, p.55).

Finally, while variety can promote choice for parents, rigorous controls need to be enforced on all schools to eliminate stratification, and ensure quality opportunities for children. The study reveals that the variety of schooling options is one of the merits of the Qatari education model; yet it makes the development of coherent quality standards and learning objectives across different systems hard to accomplish. Not to mention the toll which such variety takes on graduating Qatari students with a wide variety of knowledge, skills, and views of the world. This calls for serious discussions regarding what is the real aim behind adopting choice as a key principle to reform the country's education, and to what extent it is aligned with the government's vision for the Qatari society. The development of a national framework detailing expected knowledge, skills, and attitudes for each grade to be implemented in all schools is a suitable policy recommendation to address the current situation. The success of this recommendation requires the implementation of supportive accountability measures, such as centralised assessments and quality checks across all schools.

The future improvements call for comprehensive private sector reform through equitable legislation, an improved funding model and accountability-enhancing policies that ensure consistent quality education for all children in Qatar.

11.9.1 Conclusion

The current study attempts to provide a realistic analysis of the current education provision model in the Qatari context, while highlighting its various components and limitations. The private/public partnership in education has been a highly debated topic in the literature for many years; however, limited studies have discussed its implications on GCC countries which have witnessed rapid growth of the private education market in the last two decades. This study intends to provoke new discussion of the ways of looking at the role of the private sector in education, and how this partnership can be redirected to achieve equitable access to quality, and coherent opportunities for all children in Qatar.

Bibliography

Abdulrahman, A. (2022). Hamad Al-Ghali, Director of the Private Schools Licensing Department for Al-Sharq: There will be no increase in private school fees next year. *Al-Sharq.* https://shorturl.at/0trtk

Al-Afifi, A. (2020). Dr. Abdullah Al-Kubaisi: The "educational voucher" system is a problem that has caused "education dualism.". Al-Arab https://alarab.qa/article/28/12/2020/%D8%AF-%D8%B9%D8%A8%D8%AF%D8%A7%D9%84%D9%84%D9%87-%D8%A7%D9%84%D9%83%D8%A8%D9%8A%D8%B3%D9%8A-%D9%86%D8%B8%D8%A7%D9%85-%D8%A7%D9%84%D9%82%D8%B3%D8%A7%D8%A6%D9%85-%D8%A7%D9%84%D8%AA%D8%B9%D9%84%D9%8A%D9%85%D9%8A%D8%A9-%D9%88%D8%B1%D8%B7%D8%A9-%D8%B2%D8%B1%D8%B9%D8%AA%D9%87%D8%A7-%D8%A8%D8%AE%D8%A8%D8%AB-%D9%85%D8%A4%D8%B3%D8%B3%D8%A9-%D8%B1%D9%86%D8%AF

Al-Ansari, M., & Zahirovic, I. (2021). Contemporary national identity in Qatar: strong foundations and growing challenges. *Contemporary Qatar: Examining State and Society* (pp. 195–210). Singapore.

Al-Banai, N. and Nasser, R. (2015). The educational reform in Qatar: Challenges and successes. In *Proceedings of INTCESS15–2nd International Conference on Education and Social Sciences,* 2–4 February 2015 Istanbul. Istanbul: INTCESS15, pp. 678–683.

Al-Bidairi, A. (2019). Community schools suffer from administrative and academic complications and high fees. *Lusail.* https://tinyurl.com/nhh2rejt

Al-Jabarai, M. (2021). 30 curricula in private schools. *Al-Watan.* https://tinyurl.com/yf5ykp3b

Al-Kubaisi, A. (1999). *Attempts to reform the educational system in the State of Qatar.* https://qspace.qu.edu.qa/handle/10576/9302

Al-Sharq, (2021). Developing plans for teaching compulsory subjects in private schools. *Al-Sharq* https://tinyurl.com/5n7yrkuu

Amin, H., & Cochrane, L. (2023). The development of the education system in Qatar: assessing the intended and unintended impacts of privatization policy shifts. *British Journal of Middle Eastern Studies,* 1–21. https://doi.org/10.1080/13530194.2023.2198688

Ammermüller, A. (2005). Educational opportunities and the role of institutions. *ZEW-Centre for European Economic Research Discussion Paper,* (05–044).

Ari, I., & Koc, M. (2022). *Education Quality in Qatar: Challenges and Recommendations for Improvement* (SSRN Scholarly Paper 4015810). https://doi.org/10.2139/ssrn.4015810

Babar, Z. (2015). Population, power, and distributional politics in Qatar. *Journal of Arabian Studies, 5*(2), 138–155.

Badoor, A. (2012). 28 thousand riyals for the educational vouchers. R*aya.* https://tinyurl.com/4psm7ypk

Benmansour, N. A. (2017). Education in Qatar: the complex reality of citizen satisfaction. *SESRI Policy Brief*. Doha: SESRI. https://qspace.qu.edu.qa/bitstream/handle/10576/6383/Education%20in%20Qatar%20EN.pdf?sequence=2&isAllowed=y

Brewer, D.J., Augustine, C.H., Zellman, G.L., Ryan, G.W., Goldman, C.A. & Ryan, G. (2007). *Education for a New Era: Design and Implementation of K12 Education Reform in Qatar*. Rand-Qatar Policy Institute.

Cheema, J. R. (2015). The private–public literacy divide amid educational reform in Qatar: What does PISA tell us? *International Review of Education, 61,* 173–189.

Economic and Social Commission for Western Asia. (ESCWA). (2007). *International Migration and Development in the Arab Region: Challenges and Opportunities*. ESCWA.

Elacqua, G. (2012). The impact of school choice and public policy on segregation: Evidence from Chile. *International Journal of Educational Development, 32*(3), 444–453.

Galey, S. (2015). Education Politics and Policy: Emerging Institutions, Interests, and Ideas: Galey: Research in Education Policy. *Policy Studies Journal Yearbook, 43.* https://doi.org/10.1111/psj.12100

Gershberg, A. I. (2005). *Towards an education decentralization strategy for Turkey: Guideposts from international experience: Policy note for the Turkey education sector study*. World Bank Institute.

Graham, J. M., Saudelli, M. G., & Sheppard-LeMoine, D. (2019). The lived experiences of Qatari adolescents attending international secondary school. *Mental Health, Religion & Culture, 22*(4), 344–356.

Hillman, S., & Ocampo Eibenschutz, E. (2018). English, super-diversity, and identity in the State of Qatar. *World Englishes, 37*(2), 228–247.

Ministry of Education. (MEHE). (2005). *Annual Statistics Report 2004–2005*. Ministry of Education.

Ministry of Education and Higher Education. (MEHE). (2021a). Private Schools and Kindergarten's Licensing Guidebook. Doha: MEHE. Available from: https://elr.edu.gov.qa/apps/ELR/Arabic/Documents/LicseningHandbook-Ar-NewData.pdf

Ministry of Education and Higher Education. (MEHE). (2023a). *Frequently Asked Questions*. https://www.edu.gov.qa/ar/

Ministry of Education and Higher Education. (MEHE). (2023b). *Private Education Sector*. https://www.edu.gov.qa/en/pages/pvschoolsdefault.aspx

Ministry of Development Planning and Statistics, (2008). *Qatar National Vision 2030*. MDPS.

Ministry of Education and Higher Education. (MEHE). (2021b). Private School's Student Registration Circular. MEHE

Ministry of Education and Higher Education. (MEHE). (2022). *Annual Statistics Report 2021–2022*. Doha: MEHE. https://www.edu.gov.qa/ar/

Ministry of Education. (MOE). (1970). *Annual statistics report 1969–1970*. Doha: Ministry of Education.

Nasser, R. (2017). Qatar's educational reform past and future: Challenges in teacher development. *Open Review of Educational Research, 4*(1), 1–19.

OECD. (2010a). *PISA 2009 results: what makes a school successful?* OECD.

OECD. (2010b). *PISA 2009 results: Overcoming social background: Equity in learning opportunities and outcomes* (Vol. II). OECD.

OECD (2011), *"Private Schools: Who benefits?"* PISA in Focus, No. 7, OECD Publishing.

OECD. (2012). *Public and private schools: How management and funding relate to their socio-economic profile*. OECD.

Organizational Structure of the MEHE Emiri Decree No. 9. (2016). https://tinyurl.com/3uwep7hk

Permanent Population Committee. (PPC). (2011). *Education development in the State of Qatar 1980–2008*. Permanent Population Committee: Doha. ppc.gov.qa/Admin/ReportsAndStudies/Population_Studies_Issue11_Education_in_Qatar_1980–2008_AR.pdf

Planning and Statistics Authority, (2008). *Qatar National Vision 2030*. PSA.

Planning and Statistics Authority. (PSA). (2018). *National Development Strategy 2018–2022*. PSA.

Planning and Statistics Authority. (PSA). (2021). *Main results of the general census of population, housing and establishments 2020*. PSA: Doha. https://www.psa.gov.qa/en/statistics/StatisticalReleases/General/Census/Census_2020_Res_Summary_En.pdf

Qatar News Agency. (2022). *The Ministry of Education issues an Ethical Code of Conduct for private schools and kindergartens*. QNA. https://tinyurl.com/bdekz3r4

Raslan, M. (2015). Private school curricula are under control. *Al-Raya*. https://tinyurl.com/7nw25pry

Raslan, M. (2016). A new system for evaluating private schools. *Al- Raya*. https://tinyurl.com/5xxhfytm

Raslan, M. (2018). 60% of private schools performed poorly. *Al-Raya*. https://tinyurl.com/4s9vu5d5

Ravitch, D. (2020). *Slaying Goliath: The passionate resistance to privatization and the fight to save America's public schools*. Vintage.

Regulating Private Schools Law No. 23. (2015) https://tinyurl.com/57wvsyjb

Ridge, N.Y., Kippels, S.M., Shami, S. (2015). Private education in the United Arab Emirates and Qatar: implications and challenges. Sheikh Saud bin Saqr Al Qasimi Foundation for Policy Research. https://doi.org/10.18502/aqf.0113

Ridge, N. Y., Shami, S., Kippels, S. M. (2016). Private education in the absence of a public option: The cases of the United Arab Emirates and Qatar. *Forum for International Research in Education 3* (2), pp. 41–60.

Rizvi, F. (2016). Privatization in education: Trends and consequences (Education Research and Foresight Working Paper No. 18). UNESCO. https://unesdoc.unesco.org/ark:/48223/pf0000246485

Salah, I. (2023). The Administrative Court Cancels a decision not to grant the school national accreditation. *Al-Raya.* https://tinyurl.com/267zy4ne

Sawalhi, R., & Tamimi, A. (2023). Leading outstanding international schools in Qatar: Lessons learned. *Leadership and Policy in Schools, 22*(1), 60–72.

The Peninsula. (2021). Parents complain against private schools for withholding academic reports. *The Peninsula.* https://thepeninsulaqatar.com/article/15/11/2021/parents-complain-against-private-schools-for-withholding-academic-reports

The Gulf Arab States Educational Research Center. (GASERC). (2022) *The Reality of Private Education in the GCC*. GASERC.

Wittrock, J., Kimmel, L., Hunscher, B., & Le, K. T. (2017). Proxy reporting in education surveys: Factors influencing accurate reporting in the 2012 Qatar Education Study. *International Journal of Social Research Methodology, 20*(6), 737–748.

World Population Review. (2023). *Qatar Population 2023*. https://worldpopulationreview.com/countries/qatar-population

Zayed, W. (2015). Shura Council approves a draft law regulating private schools. *Al-Sharq*. https://tinyurl.com/2p6cmv4f

Chapter 12
Multilingual Students in Qatar's Education City International Branch Campuses: Advancing a Social Justice-Centred Policy Agenda

Sara Hillman

Abstract This chapter presents new insights and understandings based on a qualitative meta-synthesis of studies exploring the challenges faced by multilingual students in Qatar's Education City international branch campuses (IBCs). These challenges include emotional, academic, and language barriers, sociocultural hurdles, and issues pertaining to students' linguistic identities and sense of belonging in IBCs. The synthesis culminates in strategic pathways forward and policy recommendations, aimed at enhancing diversity, inclusion, and equity in Qatar's Education City IBCs. Key recommendations include promoting linguistically responsive instruction, shifting from monolingual to multilingual approaches, and embracing more local epistemologies, ethics, and pedagogies. The social justice-centred policy agenda presented in this chapter is not only applicable to the specific context of Education City IBCs, but could be extended to English-medium and transnational higher education settings across the Arab Gulf states and more globally.

12.1 Introduction

As part of the dramatic increase in "internationalisation at home" higher education initiatives and education abroad (Jackson, 2020), many international branch campuses (IBCs) have been established during the past two decades, particularly in the Arab states of the Gulf region and East Asia (Graham & Eslami, 2019). According to the Cross-Border Education Research Team (2020), U.S. universities are the largest exporter of IBCs, with 86 IBCs around the world, and six of these American IBCs can be found in a demarcated zone called Education City within the State of Qatar.

S. Hillman
Texas A&M University at Qatar, Doha, Qatar

S. Hillman (✉)
Hamad Bin Khalifa University, College of Humanities and Social Sciences, Doha, Qatar
e-mail: shillman@hbku.edu.qa

E. Zaidan et al. (eds.), *The Future of Education Policy in the State of Qatar*, Gulf Studies 21, https://doi.org/10.1007/978-981-97-9667-0_12

Qatar has welcomed IBCs onto its shores, in partnership with the Qatar Foundation (QF), creating a dynamic educational ecosystem within Education City. IBCs have been advertised as a "Western education in a Middle Eastern setting" (Hillman et al., 2021), and students obtain the same degree as they would if they were studying on the university's parent campus. IBCs generally adopt the policies and curriculum of the parent campus, and often recruit their leadership and faculty from their respective parent universities as well, with the goal of offering students at the IBC an equivalent educational experience to that on the home campus (Graham et al., 2021).

A prominent feature of IBCs is their reliance on English-medium instruction (EMI). EMI, as conceptualised by Macaro et al. (2018), refers to the utilisation of the English language for teaching academic subjects in countries where English is not the first language of the majority of the population. IBCs in Education City must teach in English per their contractual obligations with QF, aligning with QF's vision of preparing its citizens for a rapidly evolving global landscape (Kane, 2014). Recent EMI research has focused on how EMI extends its influence beyond the classroom, affecting entire university ecosystems and local communities (Hillman, 2023a). This makes EMI a phenomenon that intertwines with sociopolitical, economic, ideological, and power dynamics. Hillman et al. (2023) underscore how EMI often triggers intense debates concerning internationalisation, Westernisation, Englishisation, linguistic imperialism, global competitiveness, and asymmetrical power relationships. Specifically related to the context of higher education in Qatar, Vora (2019) argues that the Education City IBCs have led to new societal divides, and also accentuated existing ones among Qataris:

> Having an Arabic and English option for higher education in Doha in the form of the local Qatar University and the more globally oriented Education City has created new divisions among Qataris, or intensified preexisting ones: as students of the same generation graduate in higher numbers from both places, often from within the same family, they increasingly live in different linguistic and social worlds (p. 45).

Recognising these tensions, this chapter presents new insights and understandings about the challenges faced by multilingual students in Qatar's Education City IBCs. This chapter unfolds in five distinct sections. Following this introduction, the second section provides some contextualisation of Qatar's Education City IBCs. The third section explores some of the challenges faced by multilingual students in the IBCs, based on a meta-synthesis of previous research. The fourth section of the chapter provides some strategic pathways forward for the Education City IBCs, promoting a social justice-centred policy agenda. These recommendations emanate from the studies highlighted in the third section, and offer strategies to enhance diversity, inclusion, and equity. Finally, the fifth section offers some brief policy implications. From promoting linguistically responsive instruction to embracing more local epistemologies, ethics, and pedagogies, these policy implications aim to promote a more inclusive and equitable educational experience for students in Qatar's Education City IBCs.

12.2 Education City International Branch Campuses

Recognising the potential for economic and social development in Qatar through higher education, the late Emir Sheikh Hamad Bin Khalifa Al Thani and his wife Sheikha Moza Bint Nasser established the Qatar Foundation for Education, Science, and Community Development (QF) in 1995. QF is a state-led nonprofit organisation, and its flagship initiative, Education City, was launched in 1997. Education City serves as a designated hub for hosting six American IBCs (referred to by QF as "partner universities"). These IBCs are fully funded by QF, and acknowledged for their academic specialisations and degrees.

IBCs play a pivotal role in Qatarisation, a national strategy aimed at increasing the participation of Qatari nationals in different sectors of the workforce. Qatari nationals currently constitute half of the entire Education City student population (Qataris Make Up Half of Education City Student Population, 2023). In order for Qatar to build the capacity of its nationals through education, QF aims to keep increasing the overall percentage of Qatari students enrolled in IBCs, and has established key performance indicators as part of their agreements with the IBCs to meet specific targets regarding the admission of Qatari students. These IBCs participate in advancing Qatarisation through various admission, curriculum, and outreach programmes (Al-Saleh, 2022).

Education City is able to offer an "international degree, regional culture" (Hillman et al., 2021), so that Qatari students can study degree programmes in English at a branch of a U.S. university alongside international peers, while still residing in Qatar with their families. The IBCs admit many of the top Qatari secondary school graduates, who stay in Qatar for higher education. Notably, 75% of Qatari students in QF's partner universities are women, aligning with one of Sheikha Moza's original objectives for Education City: to provide talented young Qatari women with a viable alternative to studying abroad, often impractical due to cultural and religious constraints (Crist & Powell, 2017). Additionally, Qatari national students are sponsored by government ministries or state-owned companies such as Qatar Energy and Qatar Airways, thus virtually ensuring their employment in Qatar upon graduation from the IBCs.

While QF has created a dynamic educational landscape, fostering cross-cultural interactions and knowledge exchange, attending Education City IBCs is not without its challenges for students. Walsh (2019) interviewed executives and directors in the leadership of the IBCs about the challenges during the early years of operations, and noted that English language proficiency was a significant issue at the branch campuses. Many students arrived with lower language skills than expected, highlighting a gap between IBC expectations and the students' abilities. This shortfall was reported to pose a considerable challenge to both students and faculty. One of Walsh's Qatari participants stated, "students had to spend too much time taking English, and some have been very angry. The issues and challenges needed to be rectified. There is a need for flexibility regarding culture and language" (p. 282). However, Walsh reports that "both Qatar and IBC leadership stressed that this problem was primarily resolved after the first two years of operations" (p. 282). While the language barriers may not be as big a hurdle as they were in the early days of the IBCs, a number of studies continue to report challenges when it comes to teaching discipline-specific content in English in the IBCs, particularly addressing multilingual students' language development needs in the realms of academic English writing and reading (Graham & Eslami, 2020; Hillman et al., 2019; Kane, 2014; Mitchell et al., 2021; Pessoa et al., 2014; Pessoa et al., 2018). At Carnegie Mellon University in Qatar (CMU-Q), Pessoa et al. (2014) describe how they have worked on creating curriculum for scaffolding student writing development since "their [students'] written English lags behind their speaking abilities, even after attending a year-long foundation program housed in Education city, called the Academic Bridge Program" (p. 130).

Moreover, while prospective IBC students may satisfy most of the admission requirements, achieving the necessary English proficiency scores on standardised tests can be a barrier for multilingual students gaining admission to Education City IBCs. Standardised English tests often act as gatekeepers, restricting access to the IBCs and contributing to various divisions and inequities. To address this barrier, Weill Cornell Medicine-Qatar (WCM-Q) has a Foundation Programme, which integrates a year of English instruction with study skills aimed at enhancing critical reasoning skills (Hajjar and Gotto, 2013). Likewise, at Texas A&M University at Qatar (TAMU-Q), there is a programme to provisionally accept students who score low on standardised tests, and they often need to complete a year of remedial coursework before they can transition to fully admitted status (Hillman, 2022c).

Beyond language barriers, many students are confronted with different epistemological frameworks, ethical paradigms, and value systems in Education City IBCs. For example, public gender segregation is still a customary practice in Qatari society, and is considered an integral part of the social structure (Theodoropoulou & Ahmed, 2018), but this is not a practice which is valued or promoted within the IBCs. Education City was actually the "first large-scale co-educational institution of higher education in the country" (Al-Saleh, 2022, p. 8), and many Qatari parents initially hesitated to send their children there. As Crist (2015) points out, common concerns about Education City IBCs include the potential erosion of cultural values, norms, and language attributed to foreign institutions, either directly or indirectly. A minority of students coming from gender-segregated public schools have to become accustomed to learning in a coeducational environment (Rudd, 2018), but even local students who have come from coeducational schools often report that they practised de facto segregation in their school contexts, leading to discomfort when required to participate in mixed-gender group work and projects. Additionally, the values embedded within the home campuses, often mirroring the sociopolitical climate and historical context of the United States or a specific US state, can occasionally clash with the values of students in IBCs (Al-Saleh & Vora, 2020). Thus, this cultural dissonance can present an additional layer of complexity for IBC students to navigate (Hillman, 2023b).

In order to form new insights and understandings about the challenges faced by multilingual students in Qatar's Education City IBCs, the next section synthesises data from previous studies that investigated obstacles faced by students in Education City IBCs. Three of the most dominant themes are presented: language barriers, sociocultural hurdles, and challenges pertaining to linguistic identities and belonging. In order to amplify the voices of students, and illustrate the themes that emerged from the data, a number of verbatim quotes from students are presented and interpreted within each of the presented themes.

12.3 Challenges Encountered by Multilingual Students in Qatar's Education City International Branch Campuses

12.3.1 Language Barriers

A number of studies indicate that multilingual students in Qatar feel ill-prepared for the IBCs in terms of the linguistic environment, and struggle emotionally and academically at the beginning of their college journey in adapting to an English-medium American curriculum (e.g., Alkuwari, 2019; Kane, 2014; Hillman and Ocampo Eibenschutz, 2018; Pessoa et al., 2014). Students largely attribute their struggles to what they view as inadequate preparation in their pre-collegiate education, irrespective of whether they attended public or private schools. A second-year premedical student's account in Kane's (2014) study is an example of this. The student highlighted the emotional toll the transition from an Arabic-medium school to an English-medium IBC took, to the point where the student remained silent throughout their whole first semester:

> Coming from an Arabic-medium school, the whole thing is so, just so different. The first week in Cornell was so tough. Around half of the class were from Arabic-medium schools. They are the ones who found it difficult. …In biology, where there was a lot of written stuff, that was very difficult at the beginning. I didn't do it, but I saw people translating the first month, the first semester, literally translating from English to Arabic…At the beginning I was not interacting at all. I was silent for the whole first semester. I remember in the English class the professor called me [in] because I wasn't interactive. I should discuss more. Coming from an Arabic-medium school, you really need two to three years to adjust. The way the curriculum prepares you in Arabic school is just nothing. In English it's nothing. You get zero [preparation] really (p. 97).

Similarly, Hillman and Ocampo Eibenschutz (2018), who interviewed 85 Qatari students in foundation English courses at TAMU-Q, commented that many of their participants had gone through several shifts between English and Arabic as the medium of instruction during their pre-university education and "they expressed great frustration in these abrupt shifts" and felt that their education "did not prepare them and instill confidence for an EMI higher education experience nor an Arabic-medium one" (p. 237). Students who attended private and English-medium schools also felt ill-prepared. An alumna who was interviewed in Alkuwari's (2019) dissertation study of Qatari students' experiences in IBCs commented that Qatari students had to study more than non-Qatari students: "It was more challenging…I feel our schooling wasn't that okay. Even though I was in an international school, it wasn't the right international school" (p. 222). This student acknowledged that not all private schools in Qatar offer the same level of preparation for an American university system. Furthermore, Hillman (2023a, 2023b) found that some Qatari students grappled with feelings of inadequacy about being placed in foundation English courses, and perceived institutional bias from the IBCs based on their nationality:

> I was dismayed that I was placed in Foundation English II and pre-calculus. All of my Qatari friends were also taking the same classes. Then when I went to my first mandatory advising, my advisor said that I would graduate in five years. At some point I felt like a failure and dumb and I felt like they made me feel like this. And I was not the only one in this situation. Every Qatari guy in my batch was delayed a year. And actually only one is graduating in five years and the rest including me in six years. I felt like I was being treated differently because I was a Qatari student and that it was discrimination (n.p.).

Parallel themes concerning the emotional and academic ramifications of language barriers are echoed in research conducted by Bendriss and Bendriss (2015), in which they surveyed 75 students who had been enrolled in the foundation and first year of the undergraduate programme at WCM-Q, and found that 72% of the students encountered the need to translate scientific terminology into Arabic. They also found that 61% of students refrained from posing questions in classes due to shyness or embarrassment that they did not understand terminology. Thus, the reliance on translation hindered their comprehension and engagement in coursework, and students needed more support in acquiring specialised vocabulary. Likewise, Weber (2018) reported that limited vocabulary and difficulties with nuance and connotation were challenges faced by Arabic-speaking medical students at WCM-Q. Kane (2014) also addresses this barrier, presenting the experience of a male first-year medical student who felt socially isolated due to language barriers: "I struggled socially, yes. I did in the first year actually because of the language barrier. I was like…I was like hibernating, just sitting in the corner of the class trying to absorb this new language. I wasn't very social, I only socialised with Arab students at the beginning…" (p. 97). This isolation can hinder the overall college experience and affect students' confidence.

In addition to difficulties comprehending materials and participating, studies also point to students' challenges with academic reading and writing in the IBCs. Pessoa et al. (2014) observed that some of their students' reading comprehension difficulties stemmed from "limited vocabulary knowledge, background knowledge, and reading stamina" (p. 127). They also reported that students had "difficulties understanding the genre expectations and style of English academic writing" (p. 127). This was attributed to students' limited exposure to reading and writing, and initial difficulties in understanding the expectations and style of academic writing in American universities. One participant Dima described how in her high school "we only wrote 100–250 words in English class. They [the teachers] give you the topic. The students write paragraphs for each topic and memorise each paragraph without thinking" (p. 137). This quote illustrates how students like Dima may be used to relying on rote memorisation rather than critical thinking skills for writing. Dima also struggled with vocabulary as a barrier: "This is my problem in reading and writing, my vocabulary. I don't understand. I have to translate. If I don't understand the reading, then

I won't be able to write" (p. 137). This struggle with vocabulary not only affected students' reading comprehension but also hindered their ability to express themselves effectively in writing.

12.3.2 Sociocultural Hurdles

In addition to facing the language barrier, students in Education City IBCs may face a shift in social norms and values, creating internal tensions and anxieties, as they navigate the environment. For example, the coeducational environment of Education City IBCs can be a profound and challenging experience for many students (Golkowska, 2016; Rudd, 2018; Vora, 2019). A vivid example of this is portrayed through the experience of Hessa, a student interviewed in Alkuwari's (2019) study. Hessa recalled the shock and discomfort she felt when interacting with male peers at an IBC: "I was shocked. It is all new for me that I am sitting talking to a man, a boy at that time…it made me feel stressed" (p. 209). Her narrative provides a window into the immediate nature of this transition, starkly juxtaposing a segregated upbringing with the mixed-gender environment of the IBCs. It is posited by Alkuwari that many Qatari students like Hessa are unaccustomed to such interactions, and the lack of gradual exposure or guidance exacerbates their anxiety. In fact, such feelings are so pervasive that many students revert to self-segregation in mixed-gender classrooms, a form of behaviour observed and commented on by Small (2022). In this instructor's classes, Qatari males and females, despite being present in balanced numbers, would often sit separately, and this self-segregation was not merely physical. When working in mixed-gender groups, Small reports that students exhibited palpable nervousness. Female students, in particular, were apprehensive about the potential ridicule they might face from their male counterparts, especially if they made a mistake, fearing damage to their reputation:

> Although my classrooms contained generally balanced numbers of Qatari males and females, they would self-segregate and sit on opposite sides of the room. Although they would work in gender-mixed teams when I asked them to do so, many would be nervous and uncomfortable. When I asked Qatari female students about their hesitation, they said that if they were to say something "dumb" in class, the male students would make fun of them and they did not want their reputations to be marked by being judged as "stupid" or "silly" (Small, 2022, p. 32).

It is clear from previous studies that despite students' hesitations, IBC educators largely promoted mixed-gender group work anyway, viewing it as a pedagogical necessity and as preparation for future professional interactions (Vora, 2019). As Vora (2019) points out, although professors see value in normalising "work-related heterosociality," students often feel disadvantaged in such settings. Alkuwari (2019) also touches upon this tension, discussing the deliberate strategies of IBC faculties to mix genders in assignments and group projects. While such practices were aimed at prepping students for future workforce environments, they sometimes overlooked the students' comfort and cultural sensitivities. Students interviewed by Alkuwari recognised the coeducational nature of the institutions, but they also yearned for a more culturally sensitive model than faculty-imposed interactions, models that aligned more with their values and did not make them feel uncomfortable. Golkowska (2017) also notes, however, that rather than attempting to challenge or deconstruct the existing social systems and spaces, many Qatari students opt for autonomy through navigating these systems and spaces. The division between genders continues to be respected and observed by students, and students modify their verbal and non-verbal behaviours to comply with the principles of modesty.

Alkuwari (2019) also discusses how Qatari students found themselves torn between the possibilities to deviate from societal norms, and the strong influence of their upbringing and cultural background while attending IBCs. The quote from his participant Muneera below highlights the conscious effort required to navigate different social norms. Muneera emphasises the internal struggle that some Qatari students may face when confronted with opportunities to challenge established norms, while inside the IBC environment:

> It doesn't matter if you agree or disagree with them [social norms], the social norms exist, and you have to abide by them. And you have to adhere to them regardless of your age, regardless of your background, and regardless of your education. And it is a conscious choice and conscious effort to do that for the local [Qatari]. You have everything you would need [in the IBC] to facilitate any act that you would do to break those norms. The university setting, the events happening at the university, you can participate and you can break the social norms, nothing stopping you except your conscience...(pp. 213–214).

Muneera also referred to the IBC as "American soil" and commented how "once you walk in the university it is unlike other universities. So, when you go, you are on American soil, so you don't have to abide by the same [Qatari] rules" (Alkuwari, 2019, p. 219).

Students' experiences in common areas and student lounges in IBCs in Alkuwari's (2019) study were also reflective of the challenges they faced in terms of navigating gender norms in IBCs. Shaheen's account sheds light on the complexities of such communal spaces:

> Some guys [male students] when they have three hours' gap between classes, they would take a nap [in common areas in the IBC] for an hour or an hour and a half...expat girls have no issue about sleeping there [common areas] on the floor. But it is impossible for the Qatari girls to lay down on the floor and sleep. We [Qatari student club] cared about them. It was personal to us as Qataris. We created this organisation to have the Qatari voice heard in the university (p. 216).

Male students could comfortably nap in common areas during gaps between classes, while female Qatari students often found it impossible to relax, due to societal expectations. In the case of Shaheen's IBC, the Qatari student club was eventually successful in obtaining a female-only lounge area (Alkuwari, 2019), but this has been a challenge in other IBCs, such as TAMU-Q, due to policies on the home campus in Texas, which view gender-segregated spaces as discriminatory or in violation of laws promoting gender equality. Qatari females at TAMU-Q, complained in Hillman et al.'s (2017) study that the student lounges tended to be male-dominated areas, and that they did not feel comfortable hanging out in them:

> I feel like because we don't have a place to stay, like after we finish our classes, we just go home. We don't stay here. Like boys stay here, like until 8:00 or 9:00 p.m. They finish their studies here; they do everything here. But for me personally, I just leave after my class is finished (p. 7)

There is a very small, makeshift lounge exclusively for female use, but a consequence of not having a larger female-only space was that the females felt like they missed out on informal networking and extracurricular activities. Similarly, Vora's (2019) interview with Maryam and Rania, two sisters attending TAMU-Q, underscored how students had to navigate parental influence and cultural expectations that conflicted with the academic culture of the IBC. The two sisters described how they wanted to stay later than 8 pm on campus to study, but their parents would not let them. Likewise, Hessa, in Alkuwari's (2019) study stated "There are a lot of things I wanted to go to at the university like they show movies sometimes, but I didn't go, maybe because it was late. Girls can't be late (p. 233)." While some female Qatari students may have more autonomy in making decisions about their study habits, others are bound by parental restrictions. This can lead to a sense of missing out on opportunities for academic and personal growth in IBCs.

12.3.3 Challenges Pertaining to Linguistic Identities and Belonging

Education City IBCs also introduce challenges surrounding linguistic identities and belonging for many multilingual students. The majority of students who attend the IBCs are bilingual Arabic-English speakers, and they have to navigate the interplay between Arabic and English language realms, which can deeply influence their sense of self and their sense of acceptance within their communities. As one student expressed in Kane's (2014) study, speaking English can sometimes feel like assuming a different persona—an IBC persona—that is distinct from their home self:

> Especially in college, most of the time you're speaking English, and once you go back home, you don't speak English. You don't ever speak English; you always speak in Arabic. It feels different, [like] when I'm talking in English, that's not really me, that's just the Cornell E'temaad, but once I'm back home, that's normal E'temaad. Yeah, sometimes this happens. The thing is nowadays that I'm talking in English a lot in college, sometimes when I go back home I kind of forget a few words in Arabic. Like I actually slip an English word in, and it sounds really wrong (p. 103).

In the English-medium setting of the IBCs, students reported that spontaneous translanguaging between Arabic and English (i.e., mixing and shuttling between the two languages) happens more frequently, but they recognised the importance of reverting to Arabic only in certain contexts to avoid appearing arrogant or detached from their cultural roots (Kane 2014; Hillman, 2022b). Hillman (2022b) notes that students often faced criticisms from peers and family for mixing English with Arabic outside the context of the IBC:

> Here [in the university] it's easier to me to speak in English. But like I was sitting in my cousin's and I spoke in English. They're like, 'we know that you're at Texas A&M, but you don't have to talk in English.' They say that I'm bragging…like you went to Texas A&M, like you've changed a lot. It makes me feel ashamed (Sara) (p. 224).

Some students also struggled with Arabic language loss, mostly in terms of reading and writing *al-fusHa*, also referred to as Modern Standard Arabic (MSA). The experiences of students in Pessoa and Rajakumar's (2011) study, reveal the unintended consequences of language transitions. While pursuing an education in English, students often neglected their Arabic language skills, leading to difficulties in spelling and writing in Arabic, and feelings of embarrassment or shame:

> The Arabic language is being neglected. I forgot how to write papers in Arabic. My spelling became much worse; I don't know how to spell words anymore; I asked my cousins who are 10 for the spelling; it's embarrassing. I haven't written [in Arabic] in two years. We should learn in English but we should not forget our Arabic (p. 162).

Similarly, Virginia Commonwealth University in Qatar (VCU-Q) students in Yyelland and Paine's (2009) study emphasised feeling self-conscious about not knowing specific design terms in Arabic or appearing too Westernised:

> I find it very, very hard to deal with my Qatari or Arabic clients when I am speaking about design because I don't know what the terms are in Arabic, but I try to make it a point to speak in Arabic, and it has made me very self-conscious about not being Westernised and preserving my culture and heritage and traditions (p. 127).

The importance of maintaining proficiency in both languages became evident for many students in their academic and personal lives, but it was a struggle for many students to maintain their Arabic, and many did not get the opportunity to develop professional skills in Arabic that they may need for the workplace. Hillman (2023b) reports that "While most Qatari national students can converse fluently in Qatari Arabic dialect, they do not develop professional skills for presenting and writing in MSA [Modern Standard Arabic] at many of the IBCs and private universities" (p. 65). Multilingual students in Education City IBCs must grapple with their linguistic and cultural identities and contend with societal perceptions and expectations. The delicate balance between mastering enough English to succeed in academic life and retaining Arabic is often fraught with insecurities and sometimes alienation from various communities.

12.4 Pathways Forward

In considering the different linguistic, sociocultural, identity, and belonging challenges, previous studies have presented a number of pathways forward, to reduce the barriers for multilingual students in Education City IBCs and enhance diversity, inclusion, and equity in IBCs. These include promoting linguistically responsive instruction, shifting from monolingual to multilingual approaches, and embracing more local epistemologies, ethics, and pedagogies in IBCs.

12.4.1 *Promoting Linguistically Responsive Instruction*

Offering more robust language support within the IBC curriculum can help multilingual students achieve academic success, without necessarily prolonging their path to graduation. One way that IBCs can be more linguistically inclusive for multilingual students is through the promotion of linguistically responsive instruction (LRI) or "a pedagogical approach encompassing the knowledge base, dispositions, and practices required to support content learning and advanced academic language development in discipline-specific classes typically designed for presumed monolingual student in contexts dominated by monolingual norms" (Gallagher & Haan, 2019, n.p.). An important part of LRI is that content instructors "must display a willingness to assume responsibility for teaching language" (Coady et al., 2011, p. 235). LRI strategies might include using more extra-linguistic supports like visual tools, and supplementing and modifying written texts and oral language in ways that support multilingual students. For example, a Qatari petroleum engineering professor at TAMU-Q has developed a YouTube series called Petrohow, which has short, engaging videos in English about petroleum engineering topics (Petrohow). The professor is also developing parallel videos in the Arabic language. Hillman (2021) has proposed establishing multidisciplinary faculty learning communities on LRI in Qatar's IBCs in order for faculty to share the responsibility of teaching discipline-specific academic language, and to support multilingual students. Similarly, Pessoa et al. (2018) and Mitchell et al. (2021) have created a curriculum for scaffolding student writing development, recognising that students may have diverse linguistic backgrounds. They have scaffolded a sequence of assignments to help students develop genre awareness and prepare them to write in the disciplines, especially analytical and argumentative texts. Instead of viewing multilingual students as having linguistic deficiencies, or blaming the school systems that they come from, IBC instructors can help students to acknowledge their existing language skills and potential, and emphasise that these multilingual skills can be a foundation upon which to further enhance their language proficiencies.

12.4.2 *Shifting from Monolingual to Multilingual Approaches*

When it comes to promoting diversity, there is a risk of overlooking the multilingual and multicultural diversity that exists in IBCs, if they function around an English monolingual ideology (De Costa et al., 2021). While multilingual approaches are a subset of linguistically responsive instruction, monolingual assumptions and constructs (i.e., English-only) in IBC curriculum and course design need to be challenged, acknowledging the multilingual and multicultural nature of the student body. This may involve more translation, translanguaging, and other multilingual approaches in IBCs. Reynolds (as reported in Hillman et al., 2022) describes how he encourages his students to incorporate sources in languages other than English for writing assignments in a first-year writing course at CMU-Q. He discusses how students often have a deeper knowledge of the geopolitical context they are writing about when they access sources in languages they know other than English. The use of translanguaging to facilitate content and language learning should also be explored, allowing students to draw on their multilingual skills (Eslami & Graham, 2019; Hillman et al., 2019). IBC instructors are encouraged to implement practices that align with English as *a* medium of instruction rather than a monolingual EMI (Graham et al., 2021). This may come through multilingual approaches to assignments (Hillman, 2023a; Hillman et al., 2022) and consideration for the role of other languages beyond just Arabic and English (Hillman, 2022a; Hillman et al., 2019). Similarly, Nebel (2017) states that instructors "need to rethink and unthink the monolingual assumptions and constructs that dominate U.S.-based writing studies, whether at international branch campuses or locally operated extensions of American (or other institutions)" (p. 37). IBC instructors should design curricula that leverage students' unique linguistic and cultural advantages, and allow students to contribute their insights and expertise in subjects where they possess a linguistic or cultural advantage. Small (2022) refers to this as a kind of *micropraxis* and offers this example:

> A colleague and I engaged in micropraxis in our IBC classrooms by designing curricula where students had the advantage. For example, my colleague chose to teach poetry written in Arabic by Palestinian authors. Although the book included English translation, that process often muddled important aspects of the poetry. Her students could understand the original language and the political contexts of the work while she had to rely on them to help her (re)consider the lesser quality of the translation (p. 170).

By not dismissing the valuable linguistic and cultural resources that students bring to the classroom, IBCs can help promote equity or a more level playing field (De Costa et al., 2021).

Translanguaging should also be embraced as a sign of ethical caring (Graham and Eslami 2020) and as a valuable resource for communication and creativity (Reynolds, 2019). Collaboration between languages should also be emphasised, rather than competition. Reynolds (2019) calls for encouraging students to sustain and develop their proficiency in multiple languages, fostering multilingual discourse and communication skills:

> Students should sustain the development languages they understand well; speakers of majority languages should learn minoritised languages of their communities as well as languages that enable them to operate across boundaries, negotiating linguistic difference will be taught as a core communication skill, multilingual discourse will be visible in mass media and a common source for creativity, all students will have the opportunity and encouragement to learn and develop sustainable resources in more than one languages (p. 73).

Hillman (2023a) illustrates a pedagogical intervention that bridges emotions, creativity, and translingual pedagogy with IBC students, and allows them to explore their emotions regarding their relationship with English and Arabic (and other languages).

Furthermore, more courses should be offered in Arabic language and culture at IBCs for both native and non-native speakers of Arabic. While Georgetown University in Qatar (GU-Q) has a well-developed Arabic programme, and Northwestern University in Qatar (NU-Q) has announced their commitment to multilingual approaches, students in engineering, medicine, business, and the art and design fields need more options. These could be electives or additional Arabic modules added to existing English courses. Kane (2014) reported back in 2014 that more transnational medical training was needed in order for English-medium students to adjust to Arabic clinical settings effectively, and as of 2021, WCM-Q faculty were working on developing an Arabic for medical purposes course. Likewise, discussions are in place to offer Arabic modules in technical writing to engineering students at TAMU-Q. This would be similar to an Arabic for professional purposes course that has already been offered at CMU-Q. Qatar Foundation has also announced a "commitment to preserving and celebrating the richness of the Arabic language and promoting its enduring relevance to the world" (Qatar Foundation, 2020, para. 3) through the launching of initiatives such as TEDinArabic, a platform for TED talks in Arabic, and other flagship initiatives for QF students. As argued by Hillman et al. (2021), IBCs are multilingual institutions and "need language policies and practices that better reflect the reality of students' (and faculty and staff's) diverse linguistic and cultural worlds and prepare them to work in multilingual and multimodal contexts" (p. 249).

12.4.3 Embrace More Local Epistemologies, Ethics, and Pedagogies

Another pathway forward is to embrace more local epistemologies, ethics, and pedagogies within IBC curriculum and institutional cultures. In this way, IBCs can create a more culturally inclusive campus climate, curriculum, and instructional practices. This starts first with open dialogue, and creating platforms where students can freely discuss their challenges and expectations, providing feedback on what they believe could make their academic journey smoother. Rudd's (2018) study of students' responses to an imported honour code at TAMU-Q, calls for rhetorically listening to students in order to better understand their experiences and perspectives, and to ascertain their linguistic and cultural needs. One might question whether IBCs truly hear and value the voices of students, especially Qatari students, or inadvertently marginalise them. Alkuwari (2019) observes that often solutions are presented without the direct involvement of students, and are predominantly rooted in Western perspectives. One example of this may be instructors who force students to participate in mixed-gender groups, versus allowing more flexible group dynamics.

Belongingness also plays a pivotal role in students' academic achievement and emotional well-being (Wong, 2023). Walsh's (2019) findings highlighted Qatari participants' apprehensions about potential cultural and religious erosion in IBCs.

The participants emphasised the importance of IBCs engaging with Qatari families and the community to build trust and acceptance of change, yet simultaneously preserving cultural and religious aspects within Education City. Alkuwari (2019) similarly identified the significant influence of family in a students' choice of higher education and "the tensions faced by Qatari students as they negotiate being a 'home' student on an international campus" (p. 14). As evidenced in Akuwari's study, familial commitments often take precedence over university engagements. For example, seven participants in Akuwari's study gave priority to scheduling time with family over university activities if they conflicted as to time, day or season (p. 174).

Alkuwari (2019) proposes a framework focusing on inclusivity for Qatari students, particularly regarding their participation in extracurricular events (though he also acknowledges that it is important to recognise individual differences of Qatari students and not bracket them all together). Akuwari suggests that subsequent studies could probe into tailoring IBCs to the local ethos without compromising their international reputation and credentials, and how there is a nuanced balance between localising and globalising that needs further exploration. A deeper collaboration with Qatari families and communities could be pivotal, however, to ensuring that Qatari students feel fully integrated into the educational experience. This might come through service learning, community projects, cultural events, or extension programmes for family members.

Furthermore, course design and pedagogy should resonate with local viewpoints, stressing the value of the cultural and linguistic backgrounds and needs of students (Reynolds, 2021). Embedding regional histories, thought processes, and ethics can be instrumental in bridging cultural gaps, as it not only makes the content more relevant for students but also celebrates and respects the local cultures. Students from diverse backgrounds can see their cultures reflected and valued in their education. Importantly, students need not only to perform well academically but also to feel a sense of belonging and support within their campus communities.

12.5 Policy Implications

Based on these pathways forward, the following policy implications can be derived for helping IBCs embrace a more social justice-centred framework:

12.6 Provide Comprehensive Linguistic Support:

Implement linguistically responsive instruction (LRI) in all courses to ensure inclusive pedagogy tailored to diverse multilingual and multicultural students. Brief training programmes should be established for faculty that focus on LRI, and encourage the integration of visual tools, translations, and scaffolding in written texts and oral language, to aid multilingual students. There should also be incentives for instructors of different disciplines to collaborate with language and writing specialists to help scaffold writing assignments for students.

12.7 Promote Multilingual Curricular Approaches:

IBCs should re-evaluate their curriculum to move away from a purely English-centric approach and encourage multilingual approaches, especially in courses with significant regional relevance. There can be more courses or modules that utilise Arabic alongside English, and this also offers rich opportunities for multiversity collaborations across Education City. Instructors should also explore options for more multilingual approaches to assignments, allowing students to draw from their full linguistic repertoires. Embracing diversity comes through embracing multilingual diversity.

12.8 Strengthen Local Connections and Inclusivity:

IBCs must actively integrate local epistemologies, ethics, and pedagogies into their curriculum and institutional culture. They should create more platforms and forums for students to voice their challenges, expectations, and cultural experiences. Faculty should also undergo training to better understand the sociocultural dynamics of the student body, thereby enabling them to be more empathetic and better equipped to handle diverse classroom environments. IBCs should also collaborate more with local communities and families to ensure that curricula and extracurricular activities respect and reflect Qatari traditions and values. This can strengthen local connection and inclusivity.

12.9 Emphasise Student Well-Being and Belongingness:

Beyond academic performance, IBCs should prioritise students' emotional well-being and sense of belonging within the campus community. Courses and campus activities should foster inclusivity and ensure students feel integrated and valued, and should represent the interests and identities of the diverse student body. There should be safe spaces on campus where students can share their experiences and concerns. Mentorship programmes that pair new students with upper-level students or alumni in the community can also help guide students.

12.10 Conclusion

In conclusion, a synthesis of previous research has shown that some of the challenges multilingual students in situations such as Education City IBCs face are language barriers, sociocultural hurdles, and challenges pertaining to linguistic identities and belonging. As the percentage of Qatari students increases (Qataris make up half of Education City student population, 2023), it is imperative that Education City IBCs restructure their approach, so as to advance a more social justice-centred agenda. This agenda, when implemented with care and consideration for the unique cultural and linguistic landscape of Qatar, can help IBCs better serve their students and enhance diversity, inclusion, and equity in IBC educational landscapes. By adopting comprehensive linguistic support, implementing multilingual approaches, incorporating more local ethos, and prioritising students' well-being and sense of belonging, IBCs can create an environment where all students, irrespective of their background, can thrive academically, emotionally, and socially.

Bibliography

Al Saleh, D. & Vora, N. (2020). Contestations of imperial citizenship: Student protest and organizing in Qatar's Education City. *International Journal of Middle East Studies, 52* (4), 733–739. 10/1017S00200743820001026

Alkuwari, M. S. (2019). International branch campuses in Qatar: Qatari students' experience of campus life [Dissertation, University of Edinburgh]. *Edinburgh Research Archive*. https://doi.org/10.7488/era/435

Al-Saleh, D. (2022). Who will man the rigs when we go? transnational demographic fever dreams between Qatar and Texas. *Environment and Planning C: Politics and Space*, 1–17. https://doi.org/10.1177/23996544211063205

Bendriss, G. & Bendriss, R. (2015). Raising awareness among health educators and educational institutions of the need to overcome the language barrier among non-native English speaking students of health sciences in Qatar: a needs analysis and solutions. *Journal of Local and Global Health Science, 1*. https://doi.org/10.5339/jlghs.2015.1

Coady, M., Harper, C. A., & de Jong, E. J. (2011). From preservice to practice: Mainstream elementary teacher beliefs of preparation and efficacy with English language learners in the state of Florida. *Bilingual Research Journal, 34*, 223–239. https://doi.org/10.1080/15235882.2011.597823

Crist, J., & Powell, J. J. W. (2017). Higher education systems and institutions, Qatar: In J.C. Shin, & P. Teixeira (Eds.), *Encyclopedia of international higher education systems and institutions* (pp. 1–5). https://doi.org/10.1007/978-94-017-9553-1_484-1

Crist, J. T. (2015). Innovation in a small state: Qatar and the IBC cluster model of higher education. *The Muslim World, 105*(1), 93–115.

Cross-Border Education Research Team. (2020). *C-BERT international campus listing*. http://cbert.org/resources-data/intl-campus/.

De Costa, P. I., Green-Eneix, C., & Li, W. (2021). Embracing diversity, inclusion, equity and access in EMI-TNHE: Towards a social justice-centered reframing of English language teaching. *RELC, 52*(2), 227–235. https://doi.org/10.1177/00336882211018540

Education City in Qatar. (2022). *Qatar Foundation*. https://www.qf.org.qa/education/education-city

Eslami, Z. R. & Graham, K. M. (2019). Arabic-English bilingualism as a social justice issue in Qatar higher education. *Global Neighbors*. TESOL International Association.

Gallagher, C. E., & Haan, J. E. (2019). Call for abstracts: Special issue on linguistically responsive instruction in higher education. *TESOL Quarterly*. https://onlinelibrary.wiley.com/page/journal/15457249/homepage/november2019callforabstracts.html

Golkowska, K. U. (2016). Negotiating academic identity on a North American branch campus. *Arab World English Journal, 7* (4).

Golkowska, K. (2017). Qatari women navigating gendered space. *Social Sciences, 6* (123). https://doi.org/10.3390/socsci6040123

Graham, K.M., & Eslami, Z.R. (2020). Translanguaging as an act of ethical caring in the U.S. international branch campus. In C.C. Lin & C. Zaccarini (Eds.), *Internationalization in action: Leveraging diversity and inclusion in globalized classrooms* (pp. 9–26). Peter Lang Publishing.

Graham, K. M., & Eslami, Z. R. (2019). Attitudes toward EMI in East Asia and the Gulf: A systematic review. *Language Problems and Language Planning, 43*(1), 8–31. https://doi.org/10.1075/lplp.00030.gra

Graham, K. M., Eslami, Z. R., & Hillman, S. (2021). From english as the medium to english as *a* medium: perspectives of EMI students in Qatar. *System, 44*(2), 229–252. https://doi.org/10.1016/j.system.2021.102508

Hajjar, D. P., & Gotto, A. M., Jr. (2013). Launching of an American medical college in the Middle East: Educational challenges in a multicultural environment. *International Journal of Higher Education, 2*(2), 67–75. https://doi.org/10.5430/ijhe.v2n2p67

Hillman, S., Salama, G., Ocampo Eibenschutz, E., Awadh, S., & El-Said, L. (2017). Being female and an engineering student in Qatar: Successes, challenges, and recommendations. In *ASEE Annual Conference and Exposition Proceedings*, June 25–28, Columbus, Ohio. https://www.asee.org/public/conferences/78/papers/18122/view

Hillman, S. (2022c). Negotiating between cultural and academic values: Undergraduate education choices and divisions in the State of Qatar. In A. Ibrahim & O. Barnawi (Eds.) *Past, present, and future of higher education in the Arabian Gulf region: Critical comparative perspectives* (Chapter 10). Routledge/Taylor Francis.

Hillman, S. (2021). Linguistically responsive instruction in international branch campuses: Beliefs and practices of liberal arts and STEM instructors. *TESOL Quarterly, 55*(4), 1162–1189. https://doi.org/10.1002/tesq.3078

Hillman, S. (2022a). Intersecting scapes, global flows, and new identity options: Female Qatari students' informal learning of Turkish language. In S. R. Mehta & F. Al-Rashdi (Eds.), *Language, culture, and identity in the Arab World* (pp. 134–152). Routledge/Taylor Francis.

Hillman, S. (2022b). Navigating identity and belonging as international branch campus students: The role of linguistic shame. In S. Hopkyns & W. Zoghbor (Eds.), *Linguistic identities in the Gulf: Waves of change* (pp. 215–230). Routledge/Taylor Francis.

Hillman, S. (2023a). 'I feel unbalanced': Bridging emotions, creativity, and translingual practice with EMI students. *The Language Learning Journal*. https://doi.org/10.1080/09571736.2023.2239861

Hillman, S. (2023b). Language educational policies and ideologies in Qatar: Is resistance to English-medium instruction the right resistance? In M. Wyatt & G. El Gamal (Eds.), *English as a medium of instruction (EMI) in the Arabian Gulf* (pp. 55–69). Routledge/Taylor Francis.

Hillman, S., Graham, K. M., & Eslami, Z. R. (2019). Teachers' translanguaging ideologies and practices at an international branch campus in Qatar. *English Teaching and Learning, 43*(1), 41–63. https://doi.org/10.1007/s42321-018-0015-3

Hillman, S., Graham, K. M., & Eslami, Z. R. (2021). EMI and the international branch campus: Examining language ideologies, policies, and practices. *Australian Review of Applied Linguistics, 44*(2), 229–252. https://doi.org/10.1075/aral.20093.hil

Hillman, S., Li, W., Green-Eneix, C., & De Costa, P. I. (2023). The emotional landscape of English medium instruction (EMI) in higher education. *Linguistics and Education*. https://doi.org/10.1016/j.linged.2023.101178

Hillman, S., & Ocampo Eibenschutz, E. (2018). English, super-diversity, and identity in the state of Qatar. *World Englishes, 37*(2), 228–247. https://doi.org/10.1111/weng.12312

Hillman, S., Reynolds, D., & Elsheikh, A. (2022). Expanding communicative repertoires through plurilingual pedagogies in IBC classrooms in Qatar. In D. Coelho & T. Steinhagen (Eds.), *Plurilingual pedagogy in the Arabian Peninsula: Transforming and empowering students and teachers* (pp. 114–130). Routledge/Taylor Francis.

Jackson, J. (2020). *Introducing language and intercultural communication.* Routledge.

Kane, T. (2014). Whose lingua franca? The politics of language in transnational medical education. *The Journal of General Education, 63*(23), 94–112. https://doi.org/10.1353/jge.2014.0015

Macaro, E., Curle, S., Pun, J., An, J., & Dearden, J. (2018). A systematic review of English medium instruction in higher education. *Language Teaching, 51*(1), 36–76. https://doi.org/10.1017/S0261444817000350

Mitchell, T. D., Pessoa, S., Gómez-Laich, M. P., & Maune, M. (2021). Degrees of reasoning: Student uptake of a language-focused approach to scaffolding patterns of logical reasoning in the case analysis genre. *TESOL Quarterly, 55* (4). https://doi.org/10.1002/tesq.3085

Nebel, A. (2017). Linguistic superdiversity and English-medium higher education in Qatar. In L. Arnold, A. Nebel, & L. Ronesi (Eds.), *Emerging writing research from the Middle East-North Africa region* (pp. 27–40). The WAC Clearinghouse and University Press of Colorado.

Pessoa, S., Miller, R. T., & Kaufer, D. (2014). Students' challenges and development in the transition to academic writing at an English-medium university in Qatar. *IRAL, 52*(2), 127–156. https://doi.org/10.1515/iral-2014-0006

Pessoa, S., Mitchell, T. D., & Miller, R. T. (2018). Scaffolding literacy at a branch campus of an American university in the Middle East: Interdisciplinary collaborations. In M. Rajakumar (Ed.), *Western higher education in global contexts* (pp. 133–147). Lexington Books/Rowman & Littlefield.

Pessoa, S., & Rajakumar, M. (2011). The impact of English-medium higher education: The case of Qatar. In A. Al-Issa & L. S. Dahan (Eds.), *Global English and Arabic: Issues of language, culture, and identity* (pp. 153–178). Peter Lang.

Petrohow [YouTube Channel]. Retrieved September 26, 2023 from https://www.youtube.com/@petrohow

Qatar Foundation. (2020). *QF and TED launch TEDinArabic—A global platform for showcasing ideas in the Arabic language.* https://www.qf.org.qa/stories/qf-and-ted-launch-tedinarabic-a-global-platform-for-showcasing-ideas

Qataris make up half of Education City student population. (2023). *Gulf Times.* https://www.gulf-times.com/article/667978/qatar/qataris-make-up-half-of-education-city-student-population

Reynolds, D. (2019). Language policy in globalized contexts. *WISE, Qatar Foundation.* Retrieved from www.wise-qatar.org/app/uploads/2019/08/language-policy-in-globalized-contexts.pdf

Reynolds, D. (2021). Reflections of (dis)location: What is "intercultural communication" in transnational higher education English-medium instruction? *RELC, 52*(2), 253–269. https://doi.org/10.1177/00336882211017554

Rudd, M. (2018). "It makes us even angrier than we already are": Listening rhetorically to students' responses to an honor code imported to a transnational university in the Middle East. *Journal of Global Literacies, Technologies, and Emerging Pedagogies, 4*(3), 655–674.

Small, N. (2022). *A rhetoric of becoming: USAmerican women in Qatar*. Parlor Press.

Theodoropoulou, I., & Ahmed, I. (2019). Ethnographing gender roles and power in intercultural communication in Qatar. *Journal of Arabian Studies, 8*(1), 141–160. https://doi.org/10.1080/21534764.2018.1533697

Vora, N. (2019). *Teach for Arabia: American universities, liberalism, and transnational Qatar.* Stanford University Press.

Walsh, P. (2019). Establishment of an American branch-campus model of higher education: Qatar's early goals, rationales, and challenges. *Athens Journal of Education, 6* (4), 271–290. https://doi.org/10.30958/aje.6-4-2

Weber, A. (2018). English writing assessment and the Arabic speaker: a qualitative longitudinal retrospective on Arabic-speaking medical students in Qatar. In A. Abdelhamid & H. Abouabdelkader, *Assessing Writing in the 21st Century Arabic World: Revealing the Unknown.* Palgrave Macmillan.

Wong, B. (2023). Exploring the spatial belonging of students in higher education. *Studies in Higher Education.* https://doi.org/10.1080/03075079.2023.2243285

Yyelland, B., & Paine, P. (2009). Qatari women and ten years of American design school: Student perceptions of a cross-cultural educational experience. *International Journal of the Humanities, 7*(9), 119–129.

Chapter 13
Learning Ecosystems for Equitable Education and Lifelong Learning

Rosie Clayton and Atif Shafique

Abstract This chapter presents an overview of 5-years of research and development work conducted by WISE/Qatar Foundation into learning ecosystems. This work has included one of the first international studies into the concept of learning ecosystems, setting out a definition which has since been adopted by other researchers and practitioners around the world, and the identification of key characteristics and impacts. Which has supported global growth in interest in learning ecosystems at both practice and policy levels. Recently, WISE has focused on understanding the potential of this concept in Qatar, in particular looking at how learning ecosystems could drive efforts towards more equitable and inclusive education, and the promotion of lifelong learning. Our work makes the case for recognising all forms of learning and connecting the learning experiences that young people have inside and outside school, enabling young people in Qatar today to develop the core knowledge, skills, and capabilities that they will need to navigate an ever-changing world.

13.1 Introduction

The learning experiences of a young person occur across multiple, varied learning environments. Humans are born to learn, and continue to learn throughout their lives, through every interaction and experience. These experiences shape our knowledge, skills, and character, our self-efficacy, our sense of belonging and well-being, and they begin within our families, communities, and neighbourhoods.

Learning continues through the more structured settings of K-12, higher and tertiary education, and in the workplace. Through more informal extracurricular activities, such as sports and the arts, through visits to museums, and other cultural institutions and faith settings. Through outdoor learning in the natural environment and parks, and from exposure to social media and other cultural influences. The broader networks of support, experiences, and influences, in addition to school itself, play a seminal role in shaping who we become, and how we fulfil our aspirations and ambitions.

R. Clayton (✉) · A. Shafique
WISE Research Fellows, Qatar Foundation, Doha, Qatar

E. Zaidan et al. (eds.), *The Future of Education Policy in the State of Qatar*, Gulf Studies 21, https://doi.org/10.1007/978-981-97-9667-0_13

Not every young person, however, has the opportunity to access a breadth of learning experiences, within and outside of school, through family and in their community. It is increasingly well understood that the equity gap in education starts with gaps in lived experience, and that this often correlates with the neighbourhoods where young people live, and the amenities and resources available to them. Informal learning experiences outside of school that build knowledge, skills, networks, and socio-cultural capital, are important in helping young people discover and explore their passions, ignite their motivation to learn, understand and navigate the places and communities where they live, and develop their capacities to contribute to an ever-changing world.

In recent years, the concept of 'learning ecosystems' in education has come to encapsulate this wider understanding of the variety of critical influences on child and youth development, and the support given for ongoing growth and positive life outcomes. Formal education settings such as schools and colleges are essential foundations for learning, and remain the primary focus of public education policy.

But interest is now turning to the diversity of organisations and stakeholders outside of the formal education system that support child and youth development. And how learning ecosystems—which bring together diverse combinations of providers (schools, cultural organisations, businesses, and community organisations, as well as government agencies) to create new learning opportunities and pathways to success—can be cultivated, and institutionalised through policy. To address the complex and enduring challenges around equity and lifelong learning.

In Qatar, there has been a growing impetus to understand the potential role and value of learning ecosystems in expanding opportunities for youth learning, as well as addressing equity, inclusion, and social cohesion. It is also increasingly well recognised that transformational change which improves learning outcomes for all youth cannot be achieved through smart policies or well-designed programmes alone. Change requires a collective effort among community members, and multi-stakeholder collaborations that are deeply rooted in their local contexts.

WISE's 2019 report, *Local Learning Systems: Emerging Models*, set out the following definition of a learning ecosystem, which has since been adopted by practitioners and policy-makers around the world (Hannon et al., 2019)

Definition: *Learning ecosystems comprise purposeful connections between diverse combinations of providers (schools, cultural organisations, businesses, community organisations as well as government agencies) to create new learning opportunities and pathways to success*

Characteristics of- learning ecosystems:

- They diversify learning resources and pathways for learners
- They activate and share resources for learning in new ways from diverse sources
- They are dynamic in composition and porous around the edges
- They are supported by helpful infrastructure
- They comprise formal and informal learning institutions, both traditional and new entrants
- They have distributed governance
- They are learner driven or have learner agency at their heart
- They make an attempt to meet twenty-first-century challenges in some way, beyond academic attainment alone

---

13.2 Learning Ecosystems in Qatar—the Context

Qatar National Vision 2030 (QNV 2030), launched in 2008, sets out an ambitious vision for the country's human, social, economic, and environment development, with education at the core of these four pillars. The aim of QNV 2030 is to transform Qatar into "an advanced society capable of sustaining its development and providing a high standard of living for its people." (Qatar National Vision, States of Qatar).

It addresses five major challenges facing Qatar:

- Modernisation and preservation of traditions.
- The needs of the current generation and of future generations.
- Managed growth and uncontrolled expansion.
- The size and quality of the expatriate labour force, and the selected path of development.
- Economic growth, social development, and environmental management.

Education across all phases of life is considered a central aspect of social and economic development in Qatar. There have been significant investments in educational institutions, including innovative new 'demonstrator' schools, the development of Education City offering a thriving higher education environment with both home-grown and international universities, and informal learning institutions such as the cultural institutions under the aegis of Qatar Museums. Plus many community-oriented initiatives which are part of the Qatar Foundation, such as Qatar National Library.

The evolution of Qatar's education system is a story of significant national progress, but also ongoing structural challenges that have held back its economic and human capital potential.

Qatar is ranked 29th on the World Economic Forum's Global Competitiveness Index (2019), which measures how well countries perform on a range of factors that contribute to economic productivity. Qatar is ranked the eighth best country for Information and Communication Technology (ICT) adoption, and leads the world in the percentage of the adult population that uses the internet. It is also recognised as among the best in the world for investing in the skills of its current and future workforce, including through high-quality higher education and vocational training, skilled graduates, the availability of skilled workers, critical thinking in teaching, and good pupil-to-teacher ratios. As noted in the World Bank report *Maximising the Contribution of Education in Diversifying Qatar's Economy* (2017), one of its strongest assets for engaging in education reform is the existence in Qatar of many schools that are functioning at a level comparable to that of the most successful schools worldwide, and of established models of successful innovation.

Moreover, Qatar has made significant and rapid strides in driving up participation in learning. For example, from 2006 to 2012, the proportion of children not in learning at the primary level almost halved from 82% to 42%, while the rate of secondary schoolchildren not in learning declined from 73 to 52%. This, in addition to wider government measures, has contributed to a very high employment rate and zero youth unemployment, an indicator on which Qatar notably exceeds other high-income countries, according to the World Bank's Human Capital Index.

Nevertheless, significant challenges remain. Despite the progress made, Qatar still has a high percentage of people not meeting basic learning levels, especially in comparison to other high-income countries. Its overall performance on human capital, according to the World Bank, is higher than the average for the Middle East and North Africa region, but lower than the average for high-income countries. It struggles in particular with 'learning poverty', with 35% of ten-year-olds not able to read and understand a simple text by the end of primary school. This is significantly higher than the region's average for its income group (14%). It ranks 109th in the education subset indicator within the United Nations Human Development Index (2020), which is defined as an average of mean years of schooling (of adults) and expected years of schooling (of children).

Surveys from Qatar's Ministry of Education also show clear attainment gaps where data is available, with males significantly lagging behind females, and substantial gaps between government and private secondary school attainment. This incidence of discrepancies between government and private schools is supported by Principals' and teachers' perceptions about their students' motivation to learn, the quality of extracurricular activities, and the education resources which households can provide to support learning. The overall picture, therefore, suggests that there are significant challenges for Qatar to address, but that efforts to tackle them can build on a foundation of progress and investment in increasingly high-quality institutions.

In comparison to other countries around the world, Qatar's primary challenge is not necessarily about financial resource and investments. Rather in how available resources are effectively directed and utilised across the entire system, particularly in addressing inequalities in access to high-quality learning, and opportunities to learn inside and outside of the classroom. It reinforces the importance of building local leadership across the education ecosystem; and increasing communities' and residents' access to and participation in learning.

WISE's 2021 research report *Developing Local Learning Ecosystems in Qatar to Advance Equity, Inclusion, and Social Cohesion* provided an in-depth study of the educational landscape through the lens of a learning ecosystem. It looked at the variety of actors and organisations involved in education and learning provision, and the ecology of support systems that help young people to thrive inside and outside of formal education, and to develop crucial knowledge, skills, and attributes for success in life and work. The role of the informal learning sector in particular was identified as crucial to the development of knowledge, skills, and social capital in Qatar, through rich contextualised real-world learning experiences outside of school, in the workplace and within the community.

In 2023, Sever and Tok (2023) undertook a mapping exercise of the full range of learning ecosystem institutions and initiatives in Qatar. See Fig. 13.1 below for a visual representation.

The conversation about the value of learning ecosystems therefore sits against a backdrop of significant national growth and change, and is orientated around how

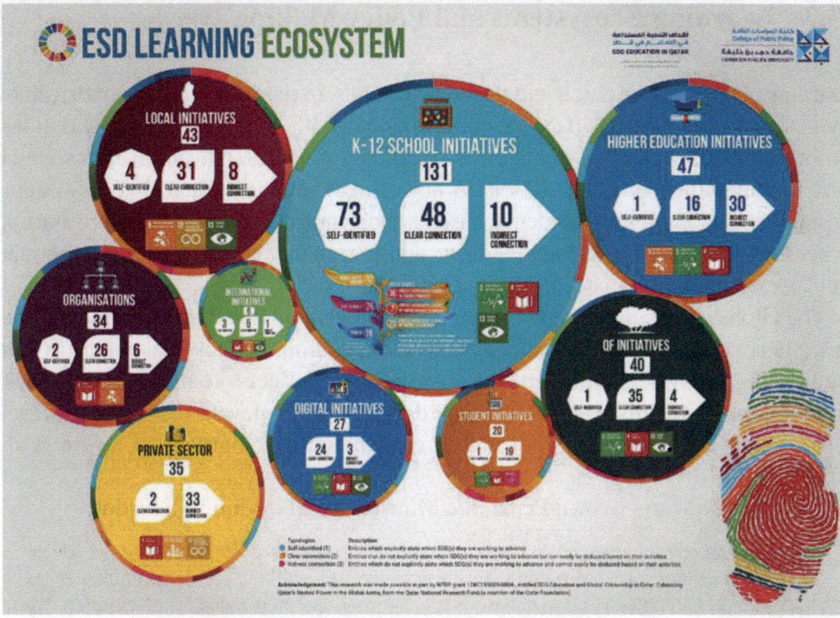

Fig. 13.1 Education for Sustainable Development Learning Ecosystem

learning ecosystems can help maximise national learning assets, support the development and spread of high-quality education practices and out of school learning opportunities, and create stronger collaborations between organisations built on trust, and a common commitment to benefitting more young people.

It is important to consider how learning ecosystems can form part of the national development story, and to realise the evolution of relationships, practices, and processes in line with the aspirations of the Qatar National Vision. Recent initiatives reflect a strong appetite among local stakeholders in Qatar for the idea of learning ecosystems. These include the WISE Doha Learning Days, the establishment of UNESCO Learning Cities in Qatar, and exploration of the idea of Cities of Learning at the WISE Summit in 2019.

Such initiatives have enabled a better understanding of existing innovative practices for greater equity, and of the forms of local collaboration and partnership already under way. They further suggest the way that the growth of more structured and orchestrated learning ecosystems could better connect organisations, build cultures of trust, and provide spaces for experimentation in education provision and lifelong learning.

13.3 Learning Ecosystems and Policy Making

Learning ecosystem research and development has, to date, typically been driven by civil society organisations (CSOs), education and skills providers, and Foundations. However, in the wake of a number of significant international publications which have provided the theoretical rationale, powerful exemplars of learning ecosystems in practice, and a growing evidence base on their impact, policy-makers are starting to see the potential of this approach to address challenges around equitable and inclusive education.

The Global Education Leaders Partnership, Dream a Dream and Learning Planet Institute conducted a study in 2022, which brought global policy-makers firmly into the learning ecosystems dialogue, to identify their distinct role in the development of learning ecosystems, and some of the different policy levers that can be utilised across complex systems to support their growth and development. (Clayton et al., 2022).

As noted by Maria Brown Perez, the Minister of Education in Ecuador:

> "Families and communities are at the centre of the education ecosystem. The formal education system cannot solve everything. Teachers cannot solve all the problems or challenges children face. Our objective is to make our education system more equal for all children. We need inclusion and equal opportunities for all—and need to accelerate policies that bring us closer to these objectives. We are pursuing three pillars—cognitive, social-emotional skills, attitudes and behaviours—to guide education. We believe change has to be drastic, not just tweaking."

WISE's 2019 report *Local Learning Ecosystems: Emerging Models* reviewed the existing literature, and noted that writers exploring the concept and potential of learning ecosystems are driven by a shared sense of the predicament facing conventional education systems.

The report highlights three interrelated issues:

1. The exhaustion of the existing educational paradigm, on its own terms (especially with regards to equity and access);
2. The need for a shift in the purpose and outcomes of education, in the context of rapid, fundamental change (beyond foundational competencies and the accumulation of qualifications);
3. The need for new organisational models to deliver this shift.

This final point, the need for new organisational models—that are, in practice, collaborative structural formations between different actors and institutions—which deliver shifts in purpose, practice, and outcomes, is where the concept of learning ecosystems really starts to take root.

WISE has since focused on applying this thinking to the educational context of Qatar, in particular looking at how the local learning ecosystem can be strengthened through better knowledge-sharing and collaboration, to distribute learning opportunities more equitably and drive lifelong learning. In Qatar, the challenges inherent in balancing a transformational agenda for societal and economic modernisation, with a commitment to the local context, has created a complex policy environment.

On the one hand, Qatar has invested heavily in cultivating new and innovative institutions and world-leading models of practice, as mentioned above. On the other hand, decision-makers are hesitant about tampering with a centralised, top-down educational model because of the experience of past reforms such as Education for a New Era (EFNE). Such reforms did not account for the importance of local context, capacity, and knowledge. A policy framework that prioritises learning ecosystems offers a way of balancing these complex priorities. Learning ecosystems help to cultivate institutions and practices for twenty-first-century learning, driven by principles of equity, learner agency, local autonomy, and pedagogical innovation. At the same time, learning ecosystem models are iterative and collaborative—putting the views of educators, students, and parents, and the wider community at the heart of governance, and working with the grain of local culture and expertise.

13.4 Equity and Learning Ecosystems—Policy Implications

WISE's 2021 research report outlined the need to develop policies and incentives, to develop capacity for partnership working and innovation, both within schools and at a systemic level. A systematic effort is also needed to codify innovative practices, and improve coordination capacity, e.g. via an intermediary entity or broker which 'orchestrates' the learning ecosystem. The role of NGOs and civil society organisations (CSOs) should be further explored to complement governmental policies and efforts. Two particular opportunity areas are set out below.

13.4.1 Innovative Practices, Pedagogies, and Teaching and Learning Strategies

There is a diversity of innovative practices that promote equity, inclusion, and cohesion across the education system, including in instructional practices and methods, and many innovative schools and new types of learning spaces (such as mobile labs, and satellite learning hubs). Innovation, however, appears concentrated in a small number of schools, networks, and communities of practice.

A wide breadth of learning opportunities are also available through the informal learning sector and extracurricular activities. However, private and government schools do not take advantage of these opportunities equally. Generally, extracurricular activities are less prevalent in government schools. The promotion of non-cognitive skills and social-emotional learning appears more developed and more widespread among private schools.

Some challenges and gaps can be addressed through new policy and practice adjustments at a national level. Policies to support the recognition of the value of existing informal learning across the system and across civil society, and to provide incentives for those opportunities to be mainstreamed across the student population in Qatar, can be a good pathway. This can be done by enabling more grassroots level collaboration (see below).

13.4.2 Collaboration and Partnership Working

There are significant opportunities for partnerships and collaboration; however, there is a lack of knowledge across the system about collaboration opportunities and how to access them, with weak knowledge-sharing and a poor communications infrastructure for disseminating information and promoting learning opportunities at a system level.

Typically, small-scale collaborations arise related to specific topics or for a specific purpose, for example between schools where there are curriculum similarities. But there are low levels of collaboration between different types of schools and education providers. Launching and sustaining collaborations is challenging in Qatar, with government schools particularly constrained.

There is also a need to grow professional confidence around collaboration, for example through setting clear standards and expectations, and investing in leadership development. Higher capacity for partnership working is needed, both centrally and within schools, with the necessary incentives and structures in place at a system level. A lack of coordination capacity to broker relationships and partnerships hinders collaborative efforts. There is also a lack of implementation support—for example, supporting teachers to engage and connect with informal learning providers.

These challenges require a multi-layered 'whole system' response to improve equity and inclusion, and help the country achieve its national development goals. This is where the growth of 'learning ecosystems', that build alignment and purposeful collaboration across institutions and sectors, could add significant value.

A first step towards raising awareness around different learning opportunities, and creating an environment for multi-sectoral collaboration in education is through learning festivals—sometimes called innovation festivals—as highlighted by another study conducted by WISE and Remake Learning. These are "community-oriented, time-bound events that bring together multiple stakeholders around a purpose or theme, such as the celebration of learning or the promotion of science, technology, engineering, arts, and mathematics (STEAM), with the aim of providing participants with creative and immersive learning experiences."

13.5 Towards a National Policy Framework

WISE's most recent work has focussed on three critical enablers of better collaboration and partnership working, to support the diffusion of innovative practices for equitable learning, and the growth of coordinated learning ecosystems to sustain this.

1. **The effective communication of learning opportunities** to schools and directly to students and parents. This includes both formal and informal learning opportunities that promote twenty-first-century skills.
2. **Creating incentives for stakeholders to engage in such opportunities.** This includes demonstrating the value proposition, quality and impact of such learning, and exploring specific forms of incentives, such as teacher accreditation, student portfolio building, and credits.
3. **Providing support for the implementation of opportunities in schools**, for example direct teacher support, external partners delivering sessions in schools, changing timetabling, and the practical support needed at a school and classroom level, to enable learners to participate meaningfully in innovative learning programmes.

Through extensive consultation with local multi-sectoral stakeholders in Qatar, a number of useful insights were highlighted, which are particularly relevant for decision-makers, budget holders, educational leaders, and policy-makers.

Key insight 1: Knowing and communicating the value proposition and impact of informal learning can boost engagement. If it is clear to educators that such opportunities are valuable, and that they positively impact their students' learning and respond to their own needs as teachers, then collaboration and engagement is more likely. A clear value proposition is also more likely to encourage students and their parents to participate.

Key insight 2: A learning ecosystems approach can help stakeholders find effective ways to promote and embed new skills and competencies (including industry-specific skills and programmes) into school curricula and core teaching practices. This has been a major goal of a plethora of strategies and initiatives across Qatar, and is something that existing digital platforms can also contribute to.

Key insight 3: Legitimacy and endorsement from key stakeholders is a critical enabling factor. The policy environment needs to be conducive not only to the growth of new types of learning models and approaches, but also to the types of organisations that can support the delivery of new forms of learning, particularly in the informal learning sector. Leadership and buy-in are important, not only from institutions, but also from young people, for student-centred approaches.

Key insight 4: Any digital platforms that are used to support learning ecosystems need to be usable, practical and should enable educators to use them seamlessly, as part of their everyday teaching experience. This means having an understanding of how teachers currently seek support/collaboration, and what the practical opportunities are to better facilitate this and to improve the 'connecting infrastructure'.

Building on these insights, further research has suggested four specific policy recommendations which should be adopted, to improve equitable education and strengthen the learning ecosystem infrastructure. These recommendations have received widespread stakeholder support, but require careful consideration when it comes to practical implementation.

Recommendation 1—Policymakers introduce new 'learner entitlements'—providing every young person with guaranteed access to careers/out of school experiences/informal learning

Stakeholders consulted were supportive of the idea of learner entitlements (or learner opportunities) and discussed the fact that it would be helpful to have a definitive centralised list of opportunities available on an annual basis, and a clear pathway for getting approvals for learners to participate in these opportunities

A shared platform between private and public schools and nurseries/providers was seen as essential in making something like this work, so that opportunities can be promoted and viewed in one place

Various existing initiatives around volunteering were highlighted, which prove motivational to learners, and could be better harnessed. Requiring non-academic stakeholders to offer learning across the school curriculum could also help realise this recommendation.

It was felt that quality assurance structures for learning opportunities should also be developed, with the need for thinking about how best to align them to school extracurricular/curricular activities—so young people can start to build a portfolio of experiences from an early age.

Questions were raised about how learning experiences are categorised, for example, whether a framework would be needed for this, to help educators and learning experience providers promote and understand what is on offer

It is also important for parents to understand the value of informal learning and out of school learning, and for their expectations around the knowledge and skills gained to be improved.

Funding entitlements for schools are essential for the effective implementation of this recommendation

Practice Example: Qatar Foundation's Pre-University Education Experience and Placement Offer

Qatar Foundation's Pre-University Education has a comprehensive industry engagement scheme, with a dedicated member of staff for developing partnerships and placement opportunities. Students have access to real-world learning opportunities within the school curriculum, through extracurricular activities, and through an internship programme. Employers and research institutes involved include QEERI, ExxonMobil, QBRI, QBB, QU, and Doha Bank, offering placements across design and manufacturing, environmental engineering, data analysis and statistics, green energy, retail, and operations.

Recommendation 2—Policymakers work with industry, learning providers, and civil society organisations, to establish a process of validating non-formal and informal learning and skills—so that these experiences are valued alongside formal qualifications

Having recommended a policy around learning entitlements and opportunities, the second recommendation focuses on validating these opportunities and experiences, and generating a clear value proposition.

Stakeholders discussed developing a cross sectoral common language around skills, and a clear process and framework for the assessment or evidencing of skills, which is understood by all stakeholders across the learning ecosystem. This process would likely raise the quality of opportunities on offer, where competency attainment and progression is clearly defined.

Stakeholder recommendation: *A shared language around skills, building expectations that could be defined across tiers of student ages and abilities*

Digital systems including a digital platform to help the Ministry of Education understand who is accessing the opportunities, the utilisation of opportunities, and track engagement could support this policy.

Digital credentialing of informal learning should be piloted and coordinated with the Ministry of Education, learning from promising practices elsewhere in the world. It was seen as important to guard against the 'informal learning entitlement policy' becoming akin to the existing volunteering hours. Achievement should be based on learning outcomes and skills gained, not hours of participation, and this is something that digital credentialing can help with, also in importing rigour and quality assurance to the process.

Practice Example: RSA Badge Standard

The Royal Society of Arts, Manufacturers, and Commerce in the UK has developed a 'Standard' for recognising and validating informal learning, in partnership with City & Guilds, an awarding body for vocational and skills-based learning. (RSA 2019)

This framework sets out a process for quality assuring different modes of learning, and providing a digital record of:

- the contents of the learning
- the context of the learning
- the contribution of the learner

"*In the nineteenth century, the RSA invented the modern exam. Now we're championing a twenty-first century innovation to validate the skills and capabilities that exams miss. We want to make all learning visible.*" (RSA 2019)

Recommendation 3—Policymakers make investments in strengthening the core infrastructure of learning ecosystems in Qatar's schools, through funding new learning ecosystem coordinator and brokerage roles within schools, to organise and coordinate learning experiences, and facilitate collaborations with other schools and informal learning providers

This recommendation was seen as essential to achieving the overarching strategic goal of more equitable education, and facilitating greater access to out of school learning experiences and innovative opportunities. Stakeholders discussed whether each school should have a dedicated role, focused on partnership development. With a remit of both identifying new informal learning opportunities and sharing best practice with other schools. This should be a paid role, with a specific budget attached

The infrastructure for developing partnerships, particularly between the public sector and private sector, is significantly lacking in Qatar, and all high-performing systems around the world place a strong emphasis on brokerage and collaboration. Existing digital platforms could be better leveraged to promote non-formal learning opportunities and to foster collaboration

Practice Example: New Roles in Schools

When thinking about coordination and brokerage across a learning ecosystem, some key roles and functions are needed., including:

- the Partnership Broker
- the Enabler (someone from the Ministry of Education who can create the conditions for educators and other providers to engage in learning ecosystem activity)
- the Advocate (actors within institutions that champion learning ecosystems/coordinate opportunities within their institution)
- the Facilitator (this could be someone from tech/civil society/the community that helps support and provide the infrastructure for learning ecosystems approaches. For example, a technology provider like Microsoft could tailor and expand access to their online platforms and courses, and create APIs for a connective infrastructure of learning)

Recommendation 4—Policymakers support the expansion of professional learning opportunities for teachers to build their knowledge of industry and informal learning partners. The provision of teacher internships, building knowledge, skills, and capabilities in creating relevant curricula

Teacher professional learning and development has been highlighted consistently as an area for development and improvement, and stakeholders discussed making mandated professional learning much more adaptive, relevant, and meaningful for educators. So they are motivated to seek out CPD, and it is not just seen as a box-ticking exercise.

Experiential modules should start early in a teacher's career, and the idea of a teacher credit profile which builds up over time and incentivises lifelong learning was suggested. Peer-to-peer professional learning for educators was discussed, as well as consideration of ways to normalise innovative pedagogies and practices which speak to the Government's strategic goal, in all schools.

Being mindful of teacher workload is an important consideration. As is learning not just for teachers but for parents, caregivers, and for those involved in community learning too.

Practice Example: Remake Learning

Remake Learning (remakelearning.org) is a peer network based in the city of Pittsburgh in the USA, comprised of over 1,200 teachers, artists, librarians, designers, learning scientists, out of school educators, technologists, school administrators, and more. The network brings educators and innovators together, to create learning opportunities that are engaging, relevant, and equitable for students. In the 15 years since its inception, Remake Learning has hosted over 220 events, transformed over 400 learning spaces into STEM labs, media hubs, makerspaces, and collaborative classrooms, and brought over 150,000 children, youth, and families together at Remake Learning Days festivals in Pennsylvania, and over 20 other regions around the world.

13.6 Conclusion and Summary of Key Points

Unlocking greater equity for all learners requires a 'whole system' response—with more structured forms of collaboration and ecosystem coordination.

Through WISE's research work, consultation and ongoing dialogue with local partners, it is clear that the concept of learning ecosystems resonates with local experiences, and there is a clear value in supporting their growth and evolution in Qatar. With a focus on improving the sharing of knowledge and practices through more structured and better supported forms of collaboration, and growing institutional capacities and skills for innovation and partnership working.

Learning ecosystems could add significant value, in helping to overcome a number of the contextual conditions which foster inequalities, and that make innovation and collaboration particularly challenging in Qatar. Including inequalities in access to knowledge, opportunities, and resources, particularly between government and private schools; helping the informal learning sector work more closely with schools through better designed and coordinated activities; strengthening local leadership capacity and capabilities; better aligning incentives; and fostering a wider sense of collective responsibility across the public and private sectors. There is a pressing need for improved system coordination and brokerage, with the addition of intermediary organisations currently largely missing from the national infrastructure.

It is time to bring this idea firmly into the national policy dialogue on the future of education, skills, and lifelong learning. The four recommendations outlined in Sect. 13.5 provide entry points for policy-makers interested in considering practical strategies that could be adopted at a national level, to augment the community level activity already under way. Reform is an adaptive process where it is important to balance the requirements of a transformation agenda with the grain of local culture, lived experiences, and expertise.

At the core, it is important to recognise that the development and growth of learning ecosystems is an evolutionary journey towards deepening relationships, practices, and shared intent between different actors, institutions and people. Relationships which often start small through collaboration on areas of mutual interest. The ambition therefore is to structure and grow these relationships in small but meaningful ways towards more purposeful, intentional, and wide-ranging collaborations. Deeper integrations which actively start to address some of the entrenched systemic challenges around inequitable access to opportunity, and skills gaps, which no one single organisation or entity can solve alone.

In Qatar, all the 'ingredients' exist to make this work: it just needs a willingness across the system, governmental support, and effective coordination and governance.

Bibliography

Clayton R. et al. (2021) *Developing Local Learning Ecosystems in Qatar to Advance Equity, Inclusion and Social Cohesion.* https://www.wise-qatar.org/developing-local-ecosystems-qatarreport/

Clayton, R., Hannon, V. & Stephenson, M. (2023) Exploring Local Learning Ecosystems in the Global South: Pathways to thriving for every child. https://www.learning-planet.org/action-group/exploring-local-learning-ecosystems-in-the-global-south

Clayton R, & Shafique A. (2023) *Strengthening Learning Ecosystems in Qatar*. https://www.wise-qatar.org//app/uploads/2023/11/2023-wise-report-strengthening-learning-ecosystems-in-qatar-en.pdf

Digital Badging: Validating the capabilities exams fail to recognise (2019). Retrieved from: https://www.thersa.org/design-for-life-our-mission/infrastructure/digital-badging

Hannon V. et al. (2019) *Local Learning Ecosystems: Emerging Models*. https://www.wise-qatar.org/app/uploads/2019/05/wise_report-rr.1.2019-web.pdf

Sever S.D. & Tok, E. (2023). Education for Sustainable Development Learning Ecosystem in Qatar. National Priorities Research Program Grant [NPRP12C-0804–190009] from the Qatar National Research Fund (a member of The Qatar Foundation)

State of Qatar Ministry of Development Planning and Statistics (2015) Realising Qatar National Vision 2030. The Right to Development. Qatar's Fourth National Human Development Report

State of Qatar Ministry of Education and Higher Education (2018) Education in the Schools of Qatar: Annual Report For The Academic Year 2017–2018.

State of Qatar, Qatar National Vision 2030. Retrieved from: https://www.gco.gov.qa/en/about-qatar/national-vision2030/

World Bank. (2017). Maximising The Contribution of Education in Diversifying Qatar's Economy. Qatar Education and Training Sector Report

World Bank. (2020). Qatar: Human Capital Index 2020. Retrieved from: https://databank.worldbank.org/data/download/hci/HCI_2pager_QAT.pdf?cid=GGH_e_hcpexternal_en_ext
World Economic Forum. (2019). The Global Competitiveness Report 2019. Retrieved from: http://www3.weforum.org/docs/WEF_TheGlobalCompetitivenessReport2019.pdf.
Ziegler, L. et al. (2023) Building Skills, Strengthening Communities: the Impact of Learning Festivals. Wise.

Chapter 14
Undergraduate Qatari Students' University Choice in Qatar

Mounia Zidani and Mohammad S. Alkuwari

Abstract The decision-making process of Qatari undergraduate students when choosing a university is not well understood. This chapter addresses the existing research gap by synthesising findings from two distinct qualitative inquiries in 2019 and 2023. These investigations aimed to reveal the determinants that impact students' decision-making processes. The study looked at students from both a local university and multiple international branch campuses. The results found significant similarities but also major differences between the two types of institutions. These findings have implications for policymakers, administrators, faculty, researchers, and students in the Arab world, particularly within the Gulf region and specifically Qatar.

Keywords University choice · Higher education · Higher education system · Qatar · GCC · Qualitative research

14.1 Introduction

The exponential growth of Qatar's higher education (HE) in the last two decades has led to an increased choice of universities for secondary students to pursue their tertiary education (Crist & Powell, 2017; MacLeod & Abou-El-Kheir, 2017). From four universities up until the 1990s, thirty have been created since the 2000s, offering a wide range of options from STEM to non-STEM programmes, governmental to private institutions, and local to international systems, with most universities offering English as the medium of instruction. These options are significant for the country's sustainable development (Ahmed, 2019; Baghdady, 2016) as they allow students to gain the necessary knowledge and skills from different disciplines and institutions, resulting in what are hoped to be quality graduates who will become the workforce

M. Zidani (✉)
The Office of Graduate Studies, Qatar University, Doha, Qatar
e-mail: mounia.zp@qu.edu.qa

M. S. Alkuwari
College of Education, Qatar University, Doha, Qatar

E. Zaidan et al. (eds.), *The Future of Education Policy in the State of Qatar*, Gulf Studies 21, https://doi.org/10.1007/978-981-97-9667-0_14

and future leaders ready to serve the country's progress and advancement across industries and sectors.

University choice plays a significant role in shaping students' future academic and professional careers. Worldwide, factors influencing college and university choices for students have been extensively discussed by a significant number of authors in the literature. Some of these factors include family influence, cost, location, the reputation of the institution, ranking of the institution, the availability of desired programmes, the availability of resources, the flexibility of the university's schedule, and financial aid (Hai et al., 2023; Hossler & Stage, 1992; Wijayanto & Sumarwan, 2016; Wilson & Adelson, 2012). Salami (2013) argues that inappropriate career decisions seen in students' choice of university may negatively impact the entire society.

Considering the crucial role of selecting a university for students' futures but also for building the national capacity and contributing to the knowledge economy for Qatar's society, this chapter presents the factors influencing Qatari students' decisions. In other words, why did students select the university they attended? Hence, it is essential to question university choice among Qatari students in Qatar. While the above research provides valuable insights internationally, the current research on the topic remains limited within the Arab world.

This chapter intends to begin filling this gap by offering a viewpoint regarding Qatari undergraduate students' university choices. The chapter presents qualitative data and selected quotes from two qualitative research studies conducted in 2019 and 2023, using semi-structured interviews with students from a local university (LU) and students studying at Qatar's International Branch Campuses (IBC). Both of these qualitative studies explored undergraduate Qatari students' university choices.[1] The qualitative research approach offers insight into the perspectives and voices of students. This approach enabled the findings to provide transferability to regions with similar contexts (Cohen et al., 2001). In addition, the chapter aims to contribute to the existing literature by emphasising implications for policymakers and by suggesting recommendations. The study findings have significant relevance to higher education administrators, faculty, researchers, and students in the Arab world in general, and the Gulf region specifically.

14.2 Higher Education Institutions (HEIs) in Qatar

Qatar's economic development in the past few decades has led to the development of all facets of society, placing education as the driver of its transformation (Hassen, 2022). Qatar has witnessed exponential growth in the tertiary education sector over

[1] IBC Participants' testimonies quoted in the chapter are from: Al-Kuwari, M. S. (2019). International branch campuses in Qatar: Qatari students' experience of campus life. University of Edinburgh. Those of LU Participants are from Zidani, M. First Year Students' Engagement in the Gulf Region: The Case of Qatar. Université de Lorraine. Ph.D. thesis currently under revision.

the past twenty years. Numerous new institutions offer undergraduate and postgraduate programmes, funded by the public and/or private sectors. The current Higher Education Institution (HEI) landscape encompasses a wide variety of universities that reflect the country's commitment to providing diverse university types and programmes. Public and military universities represent 47% of available options. In comparison, QF's IBC and private HEI hold a significant place in Qatar, representing the majority of institutions, with 52% (MOEHE 2021). Qatar University encompasses the majority of the programmes offered in the country, with a total of 50 bachelors, 47 masters, and seven doctorate programmes (Qatar University, 2023), with a student body of over 23,000 (Planning & Statistics Authority, 2021). Today, the country has a total of 348 programmes, including certificates (4%), diplomas (12%), bachelor (40%), graduate certificates (2%), graduate diplomas (3%), masters (30%), and doctorates (9%) (MOEHE 2021).

The variety of universities, programmes, and disciplines is essential, considering the diverse profile of students in Qatar. In fact, the population of university students is primarily composed of Qataris, followed by the children of expatriates, who vary in terms of their duration of stay in the country representing the country's workforce, ranging from short to long term, depending on their parents' personal and professional opportunities and choices. Based on the most recent official statistics published by the Planning and Statistics Authority (2021), Qatari students constitute approximately 45% of the student body in private higher education institutions, while this proportion accounts for 69% at Qatar University, the largest national university.

Students who are long-term expats consider Qatar as their "home." These long-term expats are commonly (but not only) from Arab and Asian countries. These "third-culture kids" represent an important category for the country. Born and raised in the country, they embrace or are at least able to navigate within the context of the country. Students who fall into the category of short- to medium-term expats typically stay a few years and may choose to pursue their education journey in Qatar. Lastly, the final group of students is international students. Those are individuals applying from abroad, and who decide to study in Qatar.

To conclude, the landscape of universities' student bodies is characterised by diversity; this complex blend of students requires the country to offer different options for them to pursue their tertiary education, including different systems, disciplines, and different language instructions to meet the diverse needs. Alkuwari (2019) points out a wide range of diversity within the Qataris. They do have some commonalities. However, they also come with differences. For example, those differences stem from many layers of their upbringing, such as family values, school types, and life choices. Therefore, learning about Qatar's diverse higher education profile and the Qataris' interest in those different offerings can help shape future higher education decisions.

14.3 Factors Influencing Undergraduate Qatari Students' Choice of University

The choice of university is an important step for secondary education graduates. Selecting a university means choosing one's university experience, including the quality of programmes, faculty, and, consequently, a future career. Choosing a university is important for students to attain successful occupational trajectories, primarily through acquiring a university degree. This subsequently facilitates access to improved employment circumstances (Kasalak, 2021; Retamosa et al., 2019). In addition, choosing a university is a crucial decision for students, as it will undoubtedly shape their university experience (Grainger, 2015), including how they engage academically with their programmes and faculty, and socially with their peers, and also how they experience campus life (Briggs et al., 2012; Wut et al., 2022). Research has demonstrated that selecting a university that aligns with a student's academic and career goals increases the likelihood of achieving success in both their university experience as well as in their chosen career path (Hai et al., 2023; Stipanovic et al., 2017).

In recent years, several studies have been conducted in the GCC region to better understand the factors influencing students' university selection process. According to the findings by Al-Ali Mustafa et al. (2018) and Hillman (2022), the quality of education, cultural values, the cost of education, demographics (gender, nationality, parental education, and parental occupation), academic reputation, university reputation, student characteristics, medium of instruction, and coeducational environment are considered as the most critical factors in the university selection process for secondary students in Qatar. This is supported by the research conducted by Echchabi and Al-Hajri (2018), which demonstrates that students in Oman prioritise the university's reputation and academic quality as primary determining factors in their decision-making process. Furthermore, students' completion time and faculty qualifications also emerge as important factors in university selection (Gajanova et al., 2020; Korcsmáros et al., 2021). Additionally, in the United Arab Emirates, Rababah's (2016) research reveals that family and peer influence are the two most important factors in the decision-making process for university choice.

Therefore, it is relevant to try to understand the factors that influence student choice. As already mentioned, this chapter sheds light on two significant institutions: a LU and a collection of IBCs. From examining those two, public and private, this chapter will attempt to help the reader understand how the participants are making their university choices. In what follows, we summarise the findings from semi-structured interviews with over 50 students sharing their experiences at the institutions mentioned above. Results indicate that the main factors influencing Qatari undergraduate students in selecting a LU include the admission process, the free education, the institution's reputation, quality of education, acquiring a degree from their own country (including pride, comfortability, practicality, and family support), gender segregation, family influence, and it being a second choice. In contrast, the main factors influencing Qatari undergraduate students to select one of the IBCs are their early exposure to higher education, perception of the higher standard of Western education, experiencing the influence of the environment, the university's location, the application processes, the IBC's reputation, personal life aspirations after university choice and family influence. The findings offer an opportunity to shed light on Qatari students' university choices and compare the findings regarding the two different HE systems.

14.4 LU and IBC Student Choices

The studies above have revealed that LU and IBC students have both similar and distinctive motives for choosing a university. Table 14.1 illustrates the students' reasons for choosing a university according to both studies. The table provides a comparison of their reasons.

14.4.1 Application Process and University Admission

Regarding LU students, there was a sense that access to and studying at an LU has become easier than in previous years. This was evident in their observations that the admission process has become more streamlined and less stringent and that the LU provides a wide range of Arabic-language programmes. For example, Mohamed (male, LU) addresses the issue of access by comparing it favourably with previous years, stating, "Because the university at that time had a less developed system, and there were many unnecessary requirements to access it." Maha (female, LU) echoed this thought, stating that IBCs have more complex and burdensome procedures than

Table 14.1 Student reasons for university choice

LU students	Students reasons for choice	IBC students
✓	Application process and university admission	✓
✓	Institution reputation	✓
✓	University location	✓
✓	Family influence and support	✓
✓	Environmental reasons	✓
	Perception of western education	✓
✓	A second choice	
	Early exposure to higher education	✓
✓	Free education	
	Personal aspirations	✓

Source Authors

this LU. She states, "…other universities are much more complicated than here… with tuition and admission procedures." Furthermore, the students think this LU "does not put too much pressure on students compared to universities (IBCs)… and with English too" (Fatma, female, LU).

Regarding IBC students and the university admission process, the application process also seems to influence the participants' decisions. The participants shared different reasons related to the application process regarding competitiveness, the timing of the admission process, or how the IBC staff have treated them positively. One of the participants compared the main campus in the USA and Qatar and favoured the Qatar IBC even more than its main campus. Ruqaya (female, IBC) discusses competitiveness: "…Out of all Qataris, out of all applicants, they only take 70 or 50. So think about it, I will be the chosen one, you know! So I wanted to be the chosen one [laughs]."

Ruqaya's quotation compares the main campus to its branch campus in Qatar, favouring the branch campus in Qatar because, in comparison, it showcases the fact that the IBC is highly selective compared to the main campus of the same university. This competitiveness is seen as valuable by Ruqaya. However, other participants valued the fact that the IBC staff treated them as students who desired to join their university. Another reason was their early acceptance notification compared to other universities. Once they received their admission letter, they seemed to not spend more time on active applications with other university options.

14.4.2 Institution Reputation

The findings for LU students indicate that the choice of the LU is also motivated by its reputation. According to them, the university "has a good reputation, it's well-known in Qatar" (Abdullah, male, LU). This LU is perceived as "the best university in Doha" (Mayassa, female, LU), people usually refer to Qatar by mentioning Doha, its capital, which means this LU is the best in the country, not only the city of Doha. Some students, like Aljohara (female, LU), chose this university specifically for the quality of their programme: "I chose this university because it has a good reputation in this specialty." The reputation of this higher education institution is a significant criterion in their decision to pursue their studies there. This finding is consistent with previous research suggesting that a university's reputation significantly impacts students' decision-making when choosing a higher education institution (Helm, 2007; Kurniawan et al., 2021).

The research findings also demonstrate that obtaining a respected degree is an important criterion for students, hence choosing this LU. According to Ahmed (male, LU), obtaining his "university degree from this university is not like from other universities." It can lead to "working in a good position in the country, like in a ministry." For others, obtaining a degree from this LU is seen as a "more highly recognised degree" that is even perceived as "a real degree" (Rowdha, female, LU) or a "solid degree" from an "accredited university" (Abdullah, male, LU). For Aljohara (female, LU), the quality of education offered by this university is crucial, even if the chosen specialty is available elsewhere: "But I chose this university because of its good reputation in this speciality, it's also available at (IBCs) but not at the same level. I asked specialists before applying for the university." The quality of education and obtaining a solid and recognised degree in the country are another crucial criterion these students consider.

For IBC students, the reputation of the institution that partnered with the IBC played an important part, regardless of the IBC they wanted to join. Abdulrazaq (IBC) stressed,

> (The institution), in general, is an environment that hosts special students. It is rare to find a student below average in (the institution). All the students are high performers. Classes are smaller; Qatar shows interest in (the institution), and all resources are channeled to it. (The institution's) graduates are well-known in the job market. So, they get access to better job opportunities compared to others. So, that's what made me choose (this institution).

Students like Abdulrazaq saw the high-profile attention and investment in the IBCs; therefore, they trusted the quality and value that IBCs bring through the institution they partnered with, which they saw as contributing positively to their career after graduation.

Furthermore, the reputation of the IBC that the students can have a solid study plan that will allow them to graduate on time is the opposite of the reputation of one of the local universities among students. Ruqaya (female, IBC) shared what she heard from her friends at the local university: "I have been hearing of a lot of problems with girls…" she later continues, "As they don't graduate fast enough, or they hold them back. They changed their study plan, so I said why put myself in such a situation when I can be in a prestigious university..." She refers to the IBC as a prestigious university. Ruqaya did not want to choose the local university because of what she heard from her friends about that university having unclear academic plans, which can be a reason for their graduation delay. Therefore, she is choosing a university that she believes to be prestigious with a reputation for solid plans to allow her to graduate on time, without putting herself in a difficult situation.

14.4.3 University Location

LU students indicated that location was important in choosing a university. For some students, this choice is a civic matter intertwined with pride. Studying at this LU in their country is a civic and obvious decision: "Because it's my country's university, it's recognised, and any Qatari would choose it first! And also, the speciality I want is here, so why go elsewhere?" (Fay, female, LU). While studying abroad with a scholarship is also an option for these students, the data analysis highlights a strong preference for studying in their own country. Abdulaziz (male, LU) explains that studying in his own country is a natural choice for him, a "normal" choice. He acknowledges the rapid development of the LU, reinforcing his decision: "It is my country's university first… that is normal. Secondly, the university has developed quickly regarding technology and education in every aspect. Ten years ago, I used to drive my aunt to the university… so I can see the difference. Honestly, that's why I chose it." Choosing this university is a rational choice rooted in their belonging, leaving out other options for alternative higher education institutions.

For IBC students, the location of IBCs also played an essential role in recruiting Qatari students. Eight participants expressed their decision to enrol in an IBC as being motivated, encouraged, supported, or even provoked by the presence of a family member who either currently or formerly attended the same IBC. Being in Qatar for Qataris can be seen as an advantage for the ones who want an American education but prefer to stay in Qatar or cannot study abroad. Aisha (female, IBC): "My sister and mom are around. If I go abroad, that means I will be there on my own." In Aisha's example, she explains that one of the factors that influenced her choice was the location, to be close to her sister and mother. The selection of the location might be motivated by a variety of factors, yet familial connections and influence emerged as a prominent consideration.

14.4.4 Family Support and Influence

Mohamed (male, LU) believes that studying in Qatar was a good decision because he feels comfortable there. He clearly expresses the opinion that he prefers to be among his family and friends and sees no point in studying abroad unless he's forced to: "I think it was a good opportunity to study here because I thought I would be comfortable… it's better to be here with my family and friends. What will I do abroad… I'll go if I'm forced to…." Another reason for staying at home is the fear of failure. This factor led some to choose security by enrolling at the LU: "I was afraid things would go wrong, and I'd come back with nothing" (Saif, male, LU). Fahad (male, LU) approaches the practical aspect of this decision by pointing out the expensive cost of studying abroad, and the responsibility of being the head of the family, which makes his staying in Qatar easier.

Moreover, he mentions the presence of foreign universities in the country offering similar study options: "Because it will cost a lot first, and I'm the householder, and even if I graduate, I'll work in Qatar, so why bother studying abroad? Besides, we have foreign universities on our soil … so why go abroad?". Also, Aljohara (female, LU) shares this preference and explains that the idea of studying outside of Qatar was not feasible due to her emotional attachment to her family: "Outside of Doha wasn't an option for me because I don't like being far from my family." Similarly, Maha (female, LU) illustrates the importance of family considerations and emotional security in her choice of staying in Qatar. She mentions that despite being offered sponsorship to study in the UK, she preferred to decline the opportunity, and took her father's concerns into account, to avoid being alone in a foreign country where she would lack support: "Because the sponsor said they would send me to the UK, but my father was worried about me, he didn't say no, but I still refused because I considered my father's fears, but also because I didn't want to be alone; it's tough, especially for a girl, to be alone when she needs help, when she cries... she needs support, a girl is a girl, she needs someone by her side."

For those who had and/or currently have a family member at this university, the choice was motivated by the positive influences and support they could get from them: "Honestly, I wanted to study elsewhere, but my cousin is here, and she helps me a lot. When I applied, I saw that it was a good university." (Moza, female, LU). She also expresses the positive influence of her eldest sister: "Yes, it helps me a lot, my big sister who will be graduating at the end of the semester convinced me that this university is very easy, thank God." In the case of Mubarak (male, LU), both his parents graduated from the LU, and he currently has a sister who is studying there. This created a favourable environment for him: "My whole family went to this university, so the environment helped me... my parents graduated from there, and I have a sister in education there now." Aljohara (female, LU) also demonstrates the positive influence of her family, particularly her father, who also went to the same university and advises her a lot: "I heard a lot of things from my surroundings, but I thought that everyone had a different experience; I was very influenced by my father, he told me about his experiences at university and told me that I just had to enjoy myself and have fun, and so far there are a few teachers who remember him and still ask me if I'm his daughter." Family members' successful experience at the LU inspires and influences students positively, giving them an optimistic and encouraging perspective before beginning their college journey.

IBC students expressed the opinion that family influences their decisions and plays a role in selecting a university, as the conversation Ali had with his father discussing what would be a valuable education option shows that students value parents' opinions about their education (Ali's conversation with his father will be described later). One of the male participants was unable to study, sharing the fact that he became the only child in the family after they lost his brother, so his mother decided that he should study in Qatar. There are also some examples of students making their own decisions based on their understanding of their situation, their parents' preferences, and other family ties, like the story of Nawal. Nawal (female, IBC) explains: "My parents weren't really in favour. I am quite close to my father, so when I saw he wasn't very keen on me living abroad [laughs] I just decide to weigh my options and honestly, they weren't bad options." She explained that the options of the IBC universities are good compared with the other options abroad. By weighing her family preferences and other family ties as she described, she made her own decision based on those factors. The following quote illustrates the influence of the family. Ruqaya (female, IBC): "He [her father] encourages us to study abroad. Even my sister didn't want to study abroad. However, he told her you are going to study abroad. I mean, usually, it is the opposite: the daughter wants to study abroad, and the father is not keen." The previous examples illustrate different ways families' influence and support contribute to shaping students' choice of university.

14.4.5 The Influence of the Environment

The influence of the university's campus and its varying experiences shape the decisions of LU and IBC students. However, students view the environment from a significantly different perspective. For LU students, the findings indicated a reluctance to belong to a coeducational environment, reflecting a strong adherence to the cultural norm of gender segregation, which is still widely observed in Qatar, especially in most government institutions. For example, Mayassa (female, LU) articulates the importance of holding on to traditional cultural norms by firmly expressing her refusal and considering it unacceptable: "Frankly, no, because it's something unacceptable to us."

Other LU students enunciate similar thoughts. Reem (female, LU) also clearly expresses this: "Because it's not a mixed environment," meaning a gender-segregated environment where both genders have their own learning spaces. On the other hand, Lolwa (female, LU) nuances these opinions by highlighting the presence of female students, implying the approbation of the application of gender segregation: "Because it has a high percentage of girls." Shamsa (female, LU) articulates her thoughts about it and mentions reasons for comfort: "I don't think I'll be comfortable: if I'm not able to deal with girls, how can I deal with boys?[laughs]...we're more comfortable like that." This evidence clearly reflects a preference by some students for a single-gender environment, hence choosing this LU.

IBC students wanted to explore a different environment, while others wanted to continue with English language as the teaching medium. The first example had spent most of his life being a Qatari in Qatar, and wanted to try something different, but did not want to or could not study abroad. Mubarak (male, IBC) shared this while discussing one of his motives: "To experience an American culture at least." In Mubarak's conversation, he suggested that even though he could not study abroad (in the United States of America) as he wanted, the opportunity to study in one of the IBCs would provide him with some experience of the American culture. Therefore, the IBC, an American university in Qatar, was an option for him and others.

On the other hand, more than one IBC participant who studied their K-12 education in an English language medium did not want to study in a local university that teaches in the Arabic language, because they felt their English language was stronger to help them perform better academically. Therefore, they wanted to continue in a similar teaching environment. The general thought of having Qatari teachers with an Arabic-language level that can help them through their academic journey might not be the case for everyone. This can be due to the K-12 schooling systems they had attended, their upbringing, and more.

14.4.6 Additional Reasons

14.4.6.1 LU Students Second Choice

As previously depicted in Table 14.1, distinct areas of interest were observed among the various student groups. For some LU students, this LU was a second choice. The data collected highlighted the fact that for other students, this LU was not their initial preference but a second choice. Ahmed (male, LU) states, "Initially, I applied to the army, but I was rejected," which led him to explore other options. The cost of tuition caused Rashid (male, LU) to join the LU after he spent two years at an IBC. He explains that he was forced to pursue his second choice (LU) because he could not finish his studies at an IBC due to incurring a debt of half a million Qatari riyal that he had to repay to his sponsor unless he obtained a degree from another university. These qualitative data reveal that the choice of university was limited or hindered by external factors, such as rejections.

14.4.6.2 LU Students and Free Education

For some students, the choice of university is mainly motivated by financial considerations. For example, Alya (female, LU) would have preferred to study in Malaysia, but could not due to financial constraints. She has a job and thinks she can afford tuition fees, but knows she would also need more for living expenses, making her decision harder: "Honestly, I didn't expect to finish my studies here… I intended to study in Malaysia, but I didn't have a choice. I thought I could pay the tuition fees because I work, but I'll need daily living expenses, and I don't want to ask my family for money.“ Ali (male, LU) also chose his current university for financial reasons: "Because if I got into another university, it would cost me financially….“ Although Qatari students benefit from government support and have access to free education (Stasz et al., 2007), financial considerations are a concern when selecting a university, and often force them to choose the free education offered by this LU.

14.4.6.3 IBC Students' Early Exposure to Higher Education

Early exposure to higher education showcases the power of student outreach and recruitment efforts. Hessa (female, IBC): "…it was very early in my life, so I didn't know what computer science was, and I didn't know how they studied for it. But I was so excited, I got their brochure and hugged it like this [embracing her arms to her chest] for the love of it. Alhamdulillah [Thanks to Allah], after a struggle, I got accepted."

The above quotation demonstrates the participant's excitement about the university brochure, she hugged it with excitement without knowing much about the university. It sounded like some students' choices can happen from an initial interaction. In the case of Hessa, she made hers at that early stage.

14.4.6.4 IBC Students and Their Perceptions of Western Education

The next reason that influenced the participants' choices was their perception of Western education in general. Some participants saw for themselves, or through their families, saw Western education as a higher quality option. Ali, one of the participants, was having a conversation with his father when he considered studying abroad. Ali (male, IBC) related what his father told him about seeking quality education in the following quote: "If you are going to study abroad, go for something worth it, a valuable university." Ali explained that his father meant that if you are going to study abroad, go for a Western education and do not consider Arab/Middle Eastern universities for reasons of quality. Ali chose to study at a Western university in one of the IBCs in Qatar, a Western university "perceived" as having a better-quality education (see Al-Kuwari 2019 for insight into this perspective).

14.4.6.5 Personal Life Aspirations

Future aspirations also influenced the programme of study. Ameena (female, IBC) shared her opinion that in specific programmes like medicine, the longer years of study could make her life harder: "For example, marriage, even the work is hard, and I feel it doesn't fit the Qatari female." She later elaborates that after many years of studying, she will be just starting her journey as a professional: "I would only be an unspecialised doctor. General medicine. If I want to be a major, I would need more years, so I spend my life in studying. I do have other goals; my life is not only about study. I have a goal to open a business. I don't want to be imprisoned with studying." She chose not to go into a medical-related programme for two different reasons. However, for the same general cause: that long years of studying and irregular working hours would negatively impact her social life and plans, such as getting married or opening a business.

14.5 Discussion

This chapter sheds light on Qatari undergraduate students' university choice in Qatar, between a local university and IBCs in the country. The results demonstrate that there are similarities in how students choose their university from both entities. For instance, the application and admission process was cited on both sides. The LU has improved its admission process, making it more straightforward and, therefore, more accessible for students than before. In contrast, IBC students reported they appreciated the positive admission application process. Next, the university's reputation for providing quality education and offering recognised degrees also influences students' decisions on both sides. Also, while students from the LU perceive their university as less intense and pressurising than IBC universities, the participants from the IBC university saw the university as being less problematic in terms of the study plan clarity, which would help them to graduate on time.

Moreover, IBC students from different school systems, single and mixed gender learning environments, with Arabic or English as a medium of instruction, expressed the desire to experience a different environment, try different cultural settings, or stay in the same English language teaching environment. At the same time, LU students prefer to continue to adhere to a gender segregation environment. Interestingly, all the students interviewed from the LU had come from public schools that adhere to single-gender education. This suggests they preferred a continuation of this environment.

This also implies that the LU may be perceived as a safe haven for some and a challenging place for others, depending on their prior educational experience, upbringing, and more. This demonstrates the benefit of providing different options to Qatar's diverse student population. Finally, family influence and support are also factors affecting both sides. Students always consider their family's opinions in their decisions, even when they are indirect. Maha's narrative from the LU is similar to that of Nawal from an IBC and illustrates their family influence. Both made their own decisions, considering different factors, including their family preferences for them to stay in the country, even though their families accepted the idea of them studying abroad. LU students revealed other factors. Specifically, financial considerations came into play for some students who found it easier or at least cheaper to pursue their tertiary education there rather than studying elsewhere.

Additionally, a sense of pride served as motivation to choose it. Others prioritise comfort and practicality when making their choice. The analysis also highlights that some students selected the local university in the country because they could not achieve their first option. Furthermore, the IBC is often selected by students with early exposure during their secondary education. Students also considered location advantages in addition to personal career aspirations. This study reveals that students are primarily influenced by the application and admission process, institution reputation and desire for quality education, family support and influence, and highlighting their lack of external information and support. The analysis shows limited avenues to find support for the decision-making process, demonstrating a gap in this area. This is, therefore, an opportunity for policymakers and practitioners to develop and offer a wide range of support in different shapes and forms to facilitate students' decision-making. This chapter stresses the importance and need to develop planning and policy areas to support their choice.

Qatar has made significant strides to support the country's rapid economic growth through education. The establishment of a diverse range of universities in the past two decades is strong evidence of the government's will to offer quality education to serve Qatari students, so as to build the national capacity's workforce, and to be the driver of the nation's transformation with the needed knowledge and competencies, as well as to meet the needs of the diverse expatriates in Qatar. Today, the new generation is more than ever offered various in-country education options. With the increased university options, students' university choices become increasingly complex, while it remains a crucial aspect that will influence their future life. Like any in life, this decision should not be taken without a thorough decision-making process.

Nevertheless, instead, it must be an informed one that stems from a deep and solid self-reflection process. Without the support of a robust and comprehensive university choice programme, and dedicated and qualified professionals, students may choose a university that might not be the best fit, potentially leading to a negative student experience, and ultimately to student dropout or an unhappy career path. The Ministry of Education and Higher Education (MOEHE) can lead in implementing policy and planning decisions to avoid such situations. The recommendations below are very concrete and require decision-makers and policymakers, here meaning the MOEHE, to take the necessary actions and devote further human, technical, and financial resources to create an ecosystem that supports student decision-making.

14.6 Policy Implications

14.6.1 Connecting Students via Outreach Mediums

University visibility plays a vital role, and the power of student outreach and recruitment initiatives increases visibility for prospective students. There are many ways to reach out to students, physically and virtually. It is equally important to create an outreach medium for the students to use as a resource, to help them during their informed decision-making process. However, with the changing student way of thinking and approachability, it is highly recommended to keep on learning how to approach helping future students to be connected and engaged, to help them choose their university. This could be achieved through the creation by the MOEHE of an interactive student-friendly medium to reach out to the younger generations, i.e., an app, that could encompass a list of all the universities in Qatar, including filter options that the community could use to identify universities based on their language of instruction, degree level, discipline, programmes, acquired accreditation, ranking, tuition fees, financial aid, and scholarships. In addition to that, this medium could provide contact information, admission deadlines, and requirements for entry. Streamlining information regarding the available options in Qatar will help raise awareness of the different options in the country, allowing students to evaluate their options, and avoiding situations where students make a choice thinking that it is the only one, like Ahlem (female, LU), who indicates going to this particular LU because there are no other universities in Qatar: "*because I live in Doha and there's no other university.*" It is interesting to find a student unaware of 33 other institutions, showing an apparent lack of awareness (See Fig. 14.1 for an overview of all recommendations).

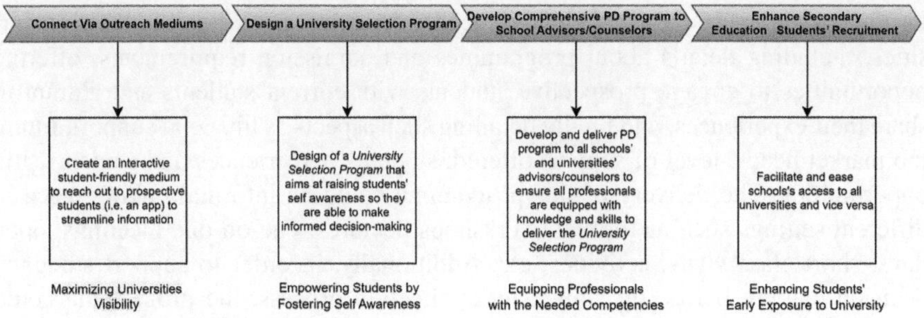

Fig. 14.1 Summary of recommendations

One practical approach to assisting students in making informed decisions about their choice of university and major is the development of a medium that offers access to various tests, such as career quizzes, interest assessments, and aptitude and personality tests. This medium should also include a comprehensive table containing information about each university in Qatar, encompassing the information listed above. Additionally, incorporating a section for student reviews of the universities and instant communication platforms for students to seek guidance and clarification could be advantageous. To ensure the medium meets the needs of secondary students, it is recommended for a thorough study to be conducted among this population to gather insights and tailor the medium accordingly.

14.6.2 Establishing a University Selection Programme

A university selection programme could help students learn about a university path that fits their aspirations. Students sometimes use what they have access to, such as the Ministry of Education and Higher Education approved list of universities or some ranking websites such as The World University Rankings and QS World University ranking, as seen in the following quote from: Ali (male, IBC), "I believe the ranking. And if it's on the Supreme Education Council[2] list of universities." The development of a programme can help students learn beyond a simple search online, by equipping them with tools to factor in more complex information related to the university choice, such as finding ways to learn about facts that matter to the student experience in their university choice.

This programme entails exploring different areas during school time in a dedicated period and subsequently creating the space to allow students to gain the needed information, self-reflect, receive support, and ultimately make informed decision-making. The different areas may include identifying personal and professional interests, setting professional goals, discussing career paths, raising personality self-awareness

[2] Currently Ministry of Education and Higher Education.

through psychometric tests, identifying academic ability, exploring available universities, including details about programmes and admission requirements, offering opportunities to engage prospective students with current students and alumni to share their experiences, and finally detailing such aspects as financial support, future job market needs, level of support offered, students' experiences and students' life opportunities. The delivery of this programme could be informal, taking place in different settings such as through workshops, lectures, one-on-one meetings, open days, short class visits, activities, etc. Additionally, in order to support students' decision-making before the university application process, the programme could also offer support throughout the admission process. The analysis shows that the application process influenced the participants' decision; hence, it is important to facilitate the availability of the process to all students, especially vulnerable ones, commonly known as *at risk students,* such as first generation students that have different sociocultural capital. These students often face challenges when applying to universities, so that offering support throughout the application process would increase their chances of university access. This programme could be offered across Qatar to all students, starting, for example, at preparatory education school, and placing greater emphasis on it during later years, for example, during secondary education. This career-advising programme would be aimed at empowering students by fostering self-awareness, providing them with the necessary tools to make an informed decision and increase university access, also promoting inclusiveness.

14.6.3 Offering Professional Development for School Advisors and Counsellors

The human connection and the ability to tailor the guidance to a particular student to help with their university choice can be helpful contributions to the students' access to the needed information to guide them. Hence, competent and qualified school advisors and counsellors are vital at this critical time. To ensure the quality support and delivery of the programme mentioned above, a comprehensive professional development programme is needed among all schools and universities' advisors/counsellors, to ensure all professionals are equipped with skills such as teaching, listening, questioning, and coaching.

Any career guidance and counselling programme should be specifically designed to assist individuals in making well-informed decisions regarding their education and career choices. The objective is to improve an individual's self-awareness abilities and offer them the chance to participate in educational and professional exploration and career planning. These programmes must encompass a thoughtful series of activities and experiences that are designed to address particular competencies, including self-evaluation, decision-making, establishing objectives, and career development. Qualified individuals should lead these programmes to ensure their effective implementation.

Furthermore, effective management is crucial to support comprehensive career guidance programmes, including the provision of adequate facilities, materials, resources, and opportunities for advisors and counsellors to consistently update their professional knowledge and skills. Several methods should be considered to deliver the programme effectively. Lastly, an evaluation mechanism should be in place that is based on the results of process and outcome to drive improvement (US Department of Education, 2014). In addition, the professionals should be able to deliver and facilitate the programme mentioned above, enabling students to make informed decisions. The professional development programme and the university selection programme could be developed, offered, and supervised by the MOEHE as the country's leading body of education and higher education, and promoted in all public and private schools.

14.6.4 Enhance Secondary Education Student Recruitment

University outreach and recruitment are critical to students' choice and access to university. So much can be done to reach out to the mass prospective students. For example, universities visiting students in their secondary schools is an effective approach that can help students, such as seen in Hessa's example above, about her early introduction to higher education. Currently, many universities do implement this strategy. However, a limited number of schools identified as suitable for them are called, "feeder schools" leaving many secondary education students without the same opportunity. University access and interaction with all secondary education schools would allow maximum recruitment opportunities to promote universities to students nationwide.

Similarly to the suggestions mentioned earlier, the MOEHE could facilitate streamlining the contact information of all schools to ease access to all universities. Furthermore, secondary education student recruitment could also be done through virtual and physical university visits, organised by the schools, allowing students to visualise and get inspired to go there. Finally, the MOEHE could organise a university forum, where all universities in Qatar could be represented in one location, to promote, introduce their university, and answer students' queries.

14.6.5 Conclusion

This chapter focuses on Qatari students' university choices, as they represent the dominant population in both types of institutions. Moreover, both share the same objective: to recruit Qataris as their main targeted student body, so as to achieve Qatar National Vision 2030 by supporting the human development of Qatari nationals (General Secretariat for Development Planning 2008).

It is crucial to keep learning about the different factors influencing students' choices in relation to their higher education. More investigations are needed to understand how students choose their university, to make recommendations to policymakers and practitioners, and to provide all students in the country with the needed support to make an informed decision that suits their aspirations. The above are only a few examples among many other options that can be created by involving the stakeholders. Expanding the study to other institutions in Qatar and other student populations, including non-Qataris, both expats and international students, is suggested. This will help gain a deeper understanding of the needs of all students, and provide better access to information that will help informed decision-making, for a positive university experience and student future.

Bibliography

Ahmed, F. (2019). Foreign universities in Qatar: A critical review of policy and sustainability issues. *Global Economic Observer, 7*(1), 50–59.

Al-Ali Mustafa, S., Sellami, A. L., Elmaghraby, E. A. A., & Al-Qassass, H. B. (2018). Determinants of college and university choice for high-school students in Qatar. *International Journal of Higher Education, 7*(3), 1. https://doi.org/10.5430/ijhe.v7n3p1

Alkuwari, M. S. (2019). International branch campuses in Qatar: Qatari students' experience of campus life. *University of Edinburgh.* https://doi.org/10.7488/era/435

Baghdady, A. (2016). Transforming Qatar's PSE: Achievements and concessions. In: M. E. Tok, L. Alkhater & L. Pal (eds.) *Policy-Making in a Transformative State.* Palgrave Macmillan: London. https://doi.org/10.1057/978-1-137-46639-6_5

Briggs, A. R. J., Clark, J., & Hall, I. (2012). Building bridges: Understanding student transition to university. *Quality in Higher Education, 18*(1), 3–21. https://doi.org/10.1080/13538322.2011.614468

Cohen, L., Manion, L., & Morrison, K. (2001). *Research methods in education* (8th ed.). Routledge Publishers.

Crist, J.T., Powell, J.J.W. (2017). Higher Education Systems and Institutions, Qatar. In: Shin, J., Teixeira, P. (eds) *Encyclopedia of International Higher Education Systems and Institutions.* Springer, Dordrecht. https://doi.org/10.1007/978-94-017-9553-1_484-1

Echchabi, A., & Al-Hajri, S. (2018). Factors influencing students' selection of universities: the case of Oman. *Journal of Education Research and Evaluation, 2*, 83–88. https://doi.org/10.23887/jere.v2i2.13694

Gajanova, L., Kliestikova, J. & Nadanyiova, M. (2020) Building and managing the value of higher education, *INTED2020 Proceedings*, pp. 2578–2582.

Grainger, J. (2015). *There's just something about the place" – an exploration of students' relationships with their university.* (Unpublished Ph.D.thesis). University of Huddersfield.

Hai, N. C., Thanh, N. H., Chau, T. M., Sang, T. V., & Dong, V. H. (2023). Factors Affecting the Decision to Choose a University of High School Students: a Study in Giang Province, Vietnam. *International Journal of Evaluation and Research in Education (IJERE), 12* (1), Article 1. https://doi.org/10.11591/ijere.v12i1.22971

Hassen, T. B. (2022). A Transformative state in the wake of COVID-19: what is needed to enable innovation, entrepreneurship, and education in Qatar? *Sustainability, 14* (13). https://doi.org/10.3390/su14137953

Helm, S. (2007). One reputation or many of corporate reputation. *Corporate Communications: An International Journal, 12*(3), 238–254.

Hillman, S. (2022). Negotiating between cultural and academic values. In A. Ibrahim & O. Z. Barnawi, (eds.) *The past, present, and future of higher education in the Arabian Gulf Region* (1st ed.) (pp. 185–205). UK: Routledge. https://doi.org/10.4324/9781003049609-13.

Hossler, D., & Stage, F. K. (1992). Family and high school experience influences on the post-secondary educational plans of ninth-grade students. *American Educational Research Journal, 29*(2), 425–451. https://doi.org/10.3102/00028312029002425

Kasalak, G. (2021). Does university choice decision matter for school engagement and student alienation? *Waikato Journal of Education, 26* (2), 135–148. https://doi.org/10.15663/wje.v26i2.732

Korcsmáros, E., Kosár, S. T., & Csinger, B. (2021). *Higher Education Institution in the 21st century — Key Factors for Prospective Students* (J. Maci, P. Maresova, K. Firlej, & I. Soukal, Eds.; pp. 378–385). https://doi.org/10.36689/uhk/hed/2021-01-038

Kurniawan, Y., Retnowati, E., Darmawan, D., Wisnujati, N. S., & Hardianingsih, A. (2021). The influence of economic income level and knowledge on students' consumption level. *Journal of Science, Technology and Society, 2*(1), 25–34.

MacLeod, P., & Abou-El-Kheir, A. (2017). Qatar's English education policy in K-12 and higher education: Rapid development, radical reform and transition to a new way forward. In R. Kirkpatrick (Ed.), *English language education policy in the Middle East and North Africa* (pp. 171–197). Springer.

Ministry of Education and Higher Education. (MOEHE) (2021). *Higher Education in Qatar*. https://www.edu.gov.qa/en/Pages/higheredudefault.aspx

Planning and Statistics Authority. (2021). *Chapter IV education statistics*. https://www.psa.gov.qa/en/statistics1/pages/topicslisting.aspx?parent=Social&child=Education

Qatar University. (2023). *About*. https://www.qu.edu.qa/about

Rababah, A. (2016). Factors influencing the students' choice of accounting as a major: The Case of X University in United Arab Emirates. *International Business Research, 9*(10), 25. https://doi.org/10.5539/ibr.v9n10p25

Retamosa, M., Millán, Á., & García, J. A. (2019). Thinking about going to university? Segmenting undergraduates. *5th International Conference on Higher Education Advances (HEAd'19)*. Fifth International Conference on Higher Education Advances. https://doi.org/10.4995/HEAD19.2019.9208

Salami, O. (2013). The factors determining the choice of career among secondary school students. *International Journal of Engineering Science, 2*, 33–44.

Stasz, C., Eide, E., Martorell, F., Goldman, C. A., & Constant, L. (2007). *Post-secondary education in Qatar: Employer demand, student choice, and options for policy, 644*. Rand Corporation.

Stipanovic, N., Stringfield, S., & Witherell, E. (2017). The influence of a career pathways model and career counseling on students' career and academic self-efficacy. *Peabody Journal of Education, 92*(2), 209–221. https://doi.org/10.1080/0161956X.2017.1302217

US Department of Education. (2014). Career *guidance and counseling programs*. [FAQs]. US Department of Education (ED). https://www2.ed.gov/about/offices/list/ovae/pi/cte/cgcp.html

Wijayanto, H., & Sumarwan, U. (2016). Analysis of the factors influencing Bogor senior high school student choice in choosing Bogor Agricultural University (Indonesia) for further study. *Journal of Education and e-Learning Research, 3*(3), 87–97.

Wilson, H. E., & Adelson, J. L. (2012). College Choices of academically talented secondary students. *Journal of Advanced Academics, 23*(1), 32–52. https://doi.org/10.1177/1932202X11430269

Wut, T.-M., Xu, J. (Bill), & Lee, S. W. (2022). Does university ranking matter? Choosing a university in the digital era. *Education Sciences, 12* (4), Article 4. https://doi.org/10.3390/educsci12040229

Chapter 15
Gender Inequality in Higher Education Institutions in the GCC States: Current Status and Ways Forward

Tasneem Amatullah and Martina Dickson

Abstract This chapter presents the current landscape of women in higher education in the Gulf Cooperation Council (GCC) and discusses the potential implications for policies and practices in higher education related to gender equality from the literature. A specific focus is placed on the status of female student populations, with an emphasis on the unique challenges faced by mothers pursuing higher education. Furthermore, the chapter makes an examination of the position of female academics in GCC within a global perspective, while also addressing potential consequences. A thorough investigation of female leadership roles in the GCC is also conducted. Finally, we explore the potential implications for policies and practices in higher education related to gender issues at the governmental, institutional, and societal levels. We propose a framework (Mott, H. (2022). Gender equality in higher education: maximising impacts. British Council.) for gender equality in higher education. We leave the readers with this framework to adapt and implement as relevant to their own context.

15.1 Introduction

The educational systems of the GCC (Gulf Cooperation Council) have undergone immense transformational development since the commercial discovery of oil in the late 1950s and early 1960s (OPEC n.d.). Initially, education was often based upon religious *madrasahs,* and focused on boys' education. However, as the economic revenue in the GCC countries grew, leaders in the region began to invest heavily in primary education systems, then secondary, and finally tertiary and higher education systems. By the 1970s, most countries in the GCC were opening their first national universities, while still often sending their citizens abroad for postgraduate study. In this chapter, we begin by laying out the background and context of the chapter, outlining the current status of gender in higher education across the Gulf. We then

T. Amatullah (✉) · M. Dickson
Emirates College for Advanced Education, Abu Dhabi, UAE
e-mail: tasneem.amatullah@ecae.ac.ae

E. Zaidan et al. (eds.), *The Future of Education Policy in the State of Qatar*, Gulf Studies 21, https://doi.org/10.1007/978-981-97-9667-0_15

look in detail at the status of female student bodies in the Gulf and pay particular attention to the status of females in higher education who are also mothers—a significant demographic of undergraduate student bodies in the GCC. Later in the chapter, we explore the status of female academics in the GCC within a global context and consider the possible impacts of this along with an exploration of the situation of female leaders in the GCC. In the final section of the chapter, we consider the possible policy and practice implications of gender-related issues in higher education, with particular consideration of gender empowerment possibilities, with a view to broadening the accessibility and career opportunities offered by the sector.

Higher Education Institutions (HEIs) significantly shape and transform societies, economies, and even individual lives. The HEIs have evolved globally over several decades, driven by societal changes, technological advancements, educational innovations, and diverse pedagogical approaches. Likewise, in the context of Middle Eastern and Gulf Cooperation Council (GCC) countries (Bahrain, Kuwait, Oman, Qatar, Saudi Arabia, and the United Arab Emirates), higher education has been central to transforming the educational landscape, with the introduction of the American model of education in the nineteenth century (Bridi & Al Hosani, 2022). With the urge to move from an oil-based economy to a knowledge-based economy, this part of the world has been striving for excellence in education, with substantial investments in establishing world-class universities and research centres, and revolutionising the entire education system. Additionally, with international branch campuses from the US, UK, and France, the GCC nations are becoming an emerging hub of higher education (Alam et al., 2013; Bridi, 2020; Wilkins et al., 2012). In fact, the national visions of each of these GCC nations clearly align with each other, and aim to advance education in all realms from preschool to higher education, including STEM education, technological advancements, and research and development to name a few areas, in order for their countries to become sustained knowledge-based economies (see, e.g., UAE Centennial Vision 2071, Qatar National Vision 2030, Oman 2040 Vision).

On one hand, there have been rapid advancements in the educational landscape, and on the other, there are great strides to bridge the gender gap and to include more women in the labour market. The 2022 Global Gender Gap Report (GGGR) notes that overall gender parity rose from 67.9% in 2021 to 68.1% in 2022 (p. 9). Covering 146 countries, the Global Gender Gap Index framework focuses on four pillars: economic participation and opportunity, educational attainment, health and survival, and political empowerment. It is noted that no country has fully achieved gender parity, with Iceland at 90.8%, Finland at 86%, and Norway at 84.5% in the top three. The Table 15.1 showcases the gender gap index rankings of the GCC nations whose percentages range from 60.9% to 71.6%, while there is an evident discrepancy in ranking between the UAE and the other GCC nations.

The GGGR groups the Middle East and North African countries as the MENA region to record the regional performance on the gender gap. The MENA region has reached greater parity in educational attainment and health and survival, with 96.2% and 96.4%, respectively; however, the numbers are strikingly low for economic participation and political empowerment at 46% and 15.1%, respectively. Another significant finding in this report is the gender gaps in tertiary education, lifelong

Table 15.1 Gender gap index: GCC countries

Rank	Country	Percentage
68	United Arab Emirates	71.6
127	Saudi Arabia	63.6
130	Kuwait	63.2
131	Bahrain	63.2
137	Qatar	61.7
139	Oman	60.9

Note Extracted from Global Gender Gap Report 2022

learning, and skills prioritisation. The last five years have seen increased enrolment and graduation rates for women worldwide, and they over-represent their gender in the fields of education, health, and welfare as compared to men. Nevertheless, women are under-represented specifically in the STEM, IT and engineering, and manufacturing fields (GGGR 2022).

So, what has been the social status of women in these nations historically, and how has it evolved? The position of women was transformed as the gender dynamics evolved along with globalisation, the national need to diversify the workforce, and to keep pace with the economic development of the country. Samier (2015) argues that the representation of women in the dominant literature does not portray the true picture, which lacks understanding of the traditional cultures and rights of women in this part of the world. In her words,

> The tendency in the literature is to present cultural and religious values as 'resistances' to modernisation, for example, in Maestri (2011) or 'hindrances' in Ramakrishnan and Ilias (2011), displaying a 'Western' bias in dealing with traditional cultures in other parts of the world, and limiting the discussion of education to that of 'schooling' and Western models. However, in many Muslim countries, like the UAE, and contrary to much Western literature as Raghavan and Levine (2012) argue, there has been considerable reform in women's rights (Samier, 2015, pp. 243–244).

Contrary to the dominant literature, women from the Arabian Gulf were not confined to their homes historically, but rather they participated in agricultural work, trade, and investing their inherited money in businesses (Maestri, 2011). Furthermore, Al Oraimi (2011) notes that women were "social agents", and their responsibility and accountability on family matters were based on "natural capacities" rather than on gender (p. 80). Özoral (2022) writes in length about the social status of women in Saudi Arabia, the UAE, and Qatar, highlighting the fact that there is a need for more opportunities for women to contribute to the workforce, given the social, political, and cultural barriers that women face (pp. 1155–1162). He argues that while progress has been made, significant gender disparities persist, especially in attaining senior leadership roles; for example, women's under-representation in the political sphere continues to be low. Furthermore, women continue to face "contradiction and uneasiness created by deep-rooted traditions" on an individual and social level (Özoral, 2022, p. 1162).

Abdulkadir and Müller (2020), elaborating more on the barriers, explain the "oil curse" thesis relevant to this region. Borrowing from Ross (2008), Abdulkadir and Müller explain that "there is a causal link between oil production, or the structure of economy more generally and the lack of gender equality" in this region (2020, p. 10). According to this thesis, having abundant natural resources, oil, in this case, reduces women's participation in the workforce. Ross (2008) goes on to say that "women [...] are underrepresented in the workforce and in government because of oil–not Islam" (p. 107). Ross further acknowledges that there is a patriarchal influence in Muslim-majority countries which is more linked to the nature of their economies which, in turn, affects female participation in the workforce, in short "petroleum perpetuates patriarchy" (2008, p. 120). Other scholars have a different narrative; for example, Al-Sayegh (2010) reminds us of the global economic depression of 1929 that led to the decline in the world economy and not the discovery of oil. Specifically, in the case of the UAE, it is noted that since the country's formation in 1971, there has been a gradual change in women's roles and their contribution to the society. Likewise, in the context of Qatar, Amatullah (2018) shares counter-narratives on female leaders in the higher education context, bringing to light the positive stories of leadership experiences of female leaders who acknowledge the support from the country, their spouses, their parents, and their supervisor, to help them grow in their academic as well as professorial journeys. As with any nation, these leaders also shared some challenges that are experienced by women across the globe.

Now, the worry is that although there has been a rapid increase in women participating in the workforce, the puzzling fact is that women still primarily hold lower status jobs in most sectors, and do not reach senior leadership positions, both in academia and the political sector (Özoral, 2022, pp. 1159–1160). The GGGR's longitudinal data confirms that women's share in ministerial positions globally has nearly doubled from 2006 to 2022, rising from 9.9% to 16.1%; yet if we see the big picture, this is just 16 per cent as compared to 84% of men's share. Furthermore, narrowing down our focus to the MENA region, it is reported that female representation in the political sphere has actually witnessed a decline as compared to that in the early 1970s. Reiterating a similar trend, the British Council's report titled Gender Equality in Higher Education: Maximising Impact by Helen Mott (2022), identifies the significance of HEIs to promote and sustain gender equality; nevertheless, HEIs also perpetuate gender inequalities as "they also reproduce discrimination against women, often 'by default rather than design'" (p. 5). A current concern is that these institutions need to acknowledge and address the structural and practical barriers that hinder women's growth and advancement to senior leadership roles. In summary, the status of women in higher education in the GCC has been shaped by the interplay of factors such as oil and culture for some scholars, and others point towards the economic depression of 1929. It is notable that scholars also argue that it is not the influence of Islam as a religion that hinders women's participation in the workforce. In what follows, we further analyse the existing research in this field, to present a holistic picture of the current status of women in higher education, by synthesising the literature on the status of female students, female academics, and female leaders in higher education, while also shedding light on research into improving the status of females in academia.

15.2 The Current Status and the Existing Research

15.2.1 Female Students in Higher Education

Malik and Courtney (2011) suggest (somewhat ironically) that a key way of promoting the gender balance and women's participation in higher education is through the cumulative knock-on effect benefits of women's participation in higher education, including increased economic independence and often the subsequent heightening of family standing it results in, leading indirectly to greater empowerment. This being true, countries in the GCC (Gulf Cooperation Council) are ahead of many of their global counterparts. The GCC is made up of six countries (The United Arab Emirates, Saudi Arabia, Qatar, Oman, Kuwait, and Bahrain), and these have, on the whole, made great strides in women's educational reform since the 1960–1970s and the formation of the cooperation. In terms of the Global Gender Gap Index, progress in gender representation in both the labour market and education has been high in the GCC countries, the latter of which is of most relevance to this chapter. All countries in the GCC now have gender parity in primary education, and most in secondary. This is a startling fact given the low levels of female literacy in the 1960s in most of the GCC countries, and as the earliest schools tended to be religious schools for boys with no provision for girls' schooling until later. It is a testimony to the economic and political will of these countries to empower and provide women with the opportunities which have led to today's status for women in education, leading to a situation where the number of women enrolled in tertiary education is higher than that of men in all of the GCC countries (GGGR 2022). It is well understood, and acknowledged by many of the region's leaders, that women have a vital role to play in the developing knowledge economy in their countries; see, for example, the UAE's National Strategy for Higher Education[1], Qatar's National Vision 2030[2], Oman's Vision 2040[3], etc. There is a gender imbalance in faculty overall in higher education institutions in the UAE (38% females versus 62% males) (Dickson & Al Harthi, 2023). The GCC as a whole has a gender disparity in student enrolment in higher education (in women's favour) with the exception of Saudi Arabia. In the UAE, for example, 66.35% of females are enrolled in tertiary education, compared to 46.4% of males. In Qatar, close to 59% of females are enrolled in tertiary education, but only a tiny 8% of males are students in the same sector (GGGR, 2022).

[1] https://www.moe.gov.ae/En/MediaCenter/News/Pages/h2030.aspx

[2] https://www.gco.gov.qa/wp-content/uploads/2016/09/GCO-QNV-English.pdf

[3] https://www.mof.gov.om/pdf/vision_documents_en.pdf

Female students are much more likely than male students to be carers as well as students, and when they are student-parents, carry the lion's share of the childcare commitments (Dickson, 2020). It is, therefore, worth presenting here the status of female student mothers. Worldwide, students who combine caring with studies are documented as having a more challenging experience, as they juggle multiple responsibilities. This is true in all global contexts, and studies in regions such as Europe and the US outline the multiple demands on students with caring responsibilities, which leads to a constant juggling act of managing and coordinating childcare, trying to keep up with college academic work, and also perhaps feeling at times a sense of not fitting in with non-parent peers (e.g. Moreau & Kerner, 2015; Walkup, 2006), particularly during years of taking undergraduate degrees. It may be extremely isolating for students trying to juggle their commitments while attempting to forge social relationships with peers having different home backgrounds and responsibilities. Mahaffey et al. (2015) describe the student-parent demographic as "hidden" students who are often not properly supported by university services, and who are vulnerable to institutional policies not designed to equip the non-traditional student with the tools they need to succeed. In the UAE, an in-depth qualitative study of students and student mothers showed that motherhood in undergraduate study was a relatively common occurrence for cultural reasons (Sherif, 2013). Perhaps because it was not uncommon, student mothers found that networks of other mothers and peers were often very valuable for the support they provided, ranging from the practical to the emotional (Dickson & Tennant, 2019). This did not mean of course that the juggling act was any less challenging, and with a very short maternity leave granted to students, many returned to their studies with a newborn having not fully recovered physically from the birth, and often in an emotional turmoil as a result.

Within institutions, much as there are challenges facing students as outlined in this chapter so far, it is also true that there is evidence to suggest that many student mothers exhibit great resilience and tenacity, exhibiting advanced personal skills of time management and focus, which results in higher performance than non-parent students (e.g. Reed et al., 2021; Thomas et al., 2012). This may be because the juggling between academic work and childcare responsibilities may require students to work more intensely in shorter periods of time. Salee (2015) describes ways in which students who have supportive spouses can often seem to have work-life balances which are indeed more optimal than those of non-parents, while Peterson (2016) also notes that efficient study strategies are often a hallmark of student-parents, being more able to self-advocate.

15.2.2 Female Academics

In many countries in the Gulf, expatriate males outnumber nationals, but male expatriates also outnumber female expatriates in higher education. There are fewer females in senior leadership positions than males, and the number of women employed in higher education at lower academic ranks (such as lecturer and assistant professor) are far greater than for males in the same ranks (Dickson & Al Harthi, 2023). A very small proportion of women serve in academia in the highest academic rank (professor)—in the UAE, for example, the study by Dickson & Al Harthi (2023) indicated that this was just 5%. Although this is particularly pronounced in this statistic, the representation of females in higher ranking positions or in positions of leadership is lower than males globally. The gender gap in both enrolment and in positions of leadership exist all over the world, including in Pakistan (Batool et al., 2013).

Why does this happen? There are various hypotheses as to why this gender imbalance occurs. Much research has indicated the cause to be related to the often-inequitable burden of caring which women undertake in the home—domestic chores, childcare, caring for elderly parents, etc. Others have written of the "academic housekeeping" which often occurs in academic institutions (e.g., Aiston & Fo, 2021). Women are often more likely to volunteer themselves, or to feel unable to refuse tasks and lower-level administrative roles, even though taking these on while in a lower academic rank effectively keeps them in those ranks. Positions related to departmental organisation, pastoral/student services, etc., often carry with them long and unspecified job descriptions, taking women away from the vital work of academic research. Women are also more likely to prioritise their teaching responsibilities over their research, which can impact later opportunities for eligibility for promotion applications. Judson et al. (2019) highlight that there are also differences in the types of service work that men and women do, with men performing more visible and quantifiable leadership roles often related to research, and women performing more invisible, non-promotable tasks such as pastoral care or supporting and training colleagues. Women must often decide between caring for their families and their careers, due to the masculinised ideals of competition, the unrelenting nature of academic work, and expectations for a major academic of being geographically mobile (Ivancheva et al., 2019).

There is also some research to suggest that some women are less likely to put themselves forward for job interviews and that men often display more confidence in their own abilities, leading them to apply for jobs which they may not quite be qualified for, which women are less likely to do. Women are also less likely to appear as the first author of publication from group research work, or to be named as first author in requests for research grant funds (Ballenger, 2010; West et al., 2013), all of which have a direct impact on promotion opportunities. A key consequence, of course, of women being more likely to have a lower position in higher education is the consequent salary reduction, which induces its own power dynamic.

Shepherd (2017) deftly argues against this notion, and instead strongly challenges the "notion of women's missing agency—characterized by a lack of confidence or ambition and a tendency to opt out of applying for the top jobs—as an explanation for their continued underrepresentation" (p. 82). Instead, Shepherd argues, a deep acknowledgement of the structures which cause the gender imbalance and later "leaky pipeline" to occur is more appropriate as explanatory tools. These include constructs of conservatism and extreme career capital, both of which require adaptations of systems and institutions—rather than women themselves making personal adaptations. Returning to the earlier point about the major share of domestic housekeeping being undertaken by women, there have been some suggestions that alleviating the pressure of domestic chores in a woman's home would indirectly affect women's work effectiveness/productivity. However, UAE based studies have shown that this is not the case, and women do not experience feeling less pressure around the "thinking" which goes into home and childcare.

The fact that there is such a high proportion of female student bodies in the Gulf, particularly in undergraduate ones (WEF Report) is striking, since this is not nearly represented in their faculty body. This may have implications for future female academics who may not see high enough numbers of male faculty members (Herrmann et al., 2016) to imagine this as a possible career for them and for them to consider a position in higher education. Aiston and Jung (2016) describe the seeming contradiction between a global increase in student numbers (the GCC countries are a prime example), yet a failure of women to "progress through the academic hierarchy in significant numbers and enter senior leadership positions" (p. 17).

15.2.3 Female Leaders

There has been a gradual increase globally in women's share of senior and leadership roles in the last five years, with a significant increase in the hiring of women, especially since 2016 (GGGR, 2022 p. 36). Moreover, there is also a notable tendency that the hiring of women has increased in the sectors where women were already overrepresented, and the same was the case for men. With regard to the GCC nations, we see that the indicator for advancement to leadership roles ranges from 4.65 to 5.7 out of 7 (in a 1–7 ranking with 7 being the highest and 1 the lowest). There is no reported data on firms with female majority ownership, female top managers, and the share of workers in informal work, while the gender pay gap and women's membership on boards is generated only for the OECD countries (GGGR, 2022, pp. 218–349).

Globally, while there have been major developments to promote gender equality in leadership in higher education, a report by the Royal Society of Chemistry (2018) highlights the fact that "It is resoundingly clear from our findings that there is still a lack of progress in developing and retaining women in leadership positions in the chemical sciences" (p. 3). This void of women in leadership positions in the STEM field speaks volumes. The report also points to some notable deficiencies such as "women's progress and retention, the pipeline of women in higher education, gender pay inequality, and lower impact publishing" (2018 p. 3). The same is reiterated in another report by UNESCO (2021) that there are persistent forms of gender-based inequality in higher education, despite the encouraging statistics on female enrolment as discussed earlier. The UNESCO report sheds light on the "second glass ceiling", referring to the markedly low number of women in senior management positions in higher education. In the Latin American region, only 18% of the public universities have female leaders, while it is 15% in European universities (p. 31). The report further stresses that "The number of the world's top universities that are led by women has increased for the first time in three years but still accounts for less than one-fifth of leading institutions" (p. 32). This dearth of women in leadership positions in higher education "reflects a failure to maximize female talent" (p. 32). The research stresses that the more female executives there are, the better the organisation performs (Noland et al., 2016).

There are several barriers that hinder women from rising in academic and leadership roles, with executive and full professor positions being male dominated, worldwide (Liu, 2013; Mayuzumi, 2015; Yousaf & Schmiede, 2017). In the context of the GCC, research has repeatedly documented that cultural barriers, patriarchy, and gender norms prevent women from taking on senior leadership positions, and the same is true in the political sector (Madsen, 2010; Metcalfe, 2008; Romanowski & Al-Hassan, 2013). The structural barriers such as those concerning recruitment, promotion, affiliation to masculine traits of leadership and reinforcement of cultural stereotypes, specifically gender stereotypes, play a major role in obstructing women's rise to senior leadership positions. On the other hand, Prager's (2020) research offers a glimpse from the Emirati context, where young women share a greater responsibility for taking up future leadership roles "to combat not only the socio-cultural factors but also the pressure of keeping up appearances and appropriate behaviors within cultural boundaries" (p. 69).

15.3 Improving the Status of Females in Academia

It falls to institutions and governments then to question the roots of these female empowerment challenges, and to question why there remains such a lack of women in educational leadership positions, in particular.

Gender imbalance in positions of leadership for women in higher education is attributed in part to a lack of empowerment (Alotaibi, 2020), though the true rationale for it is much more likely to be complex and multi-levelled. Suggesting that the disparity is due to a lack of female empowerment, leading to women being less likely to put themselves forward for promotional opportunities, etc., belies the complexities in the presentation of opportunities in the first place; there may be a dearth of opportunities for women. There may also be systematic barriers in place, in practice, preventing women from being qualified to apply for leadership positions, and making it challenging for them to gain the experience and/or qualifications needed for eligibility. There may also be external factors contributing to women's own decisions not to apply for leadership positions, including personal decisions to focus on other aspects of their personal and professional lives. Where existing leadership positions are held predominantly by males, it can sometimes become a perpetuating cycle to continually bring in a similar demographic of leaders, particularly since current leaders are often involved in the selection of new ones. For all of these reasons, the gender imbalance in positions of leadership in higher education persists. There have been various reforms introduced in the GCC which aim to improve this situation, in connection with the national visions referenced earlier. It is yet to be seen whether this will be effective in addressing the relatively low numbers of women in leadership. The study by Alotaibi (2020) found that, similarly to other studies, a number of barriers—personal, cultural, and organisational—contribute to the current status of women. A qualitative focus group study of female academics carried out by Ashencaen Crabtree and Shiel (2019) emphasises ways in which gendered career pathways hinder career progression, with the female participants reporting often feeling pushed towards administrative paths (of the academic housekeeping kind) and teaching, while more emphasis is placed upon the "masculine" research routes, which, in turn, affects career progression. This is supported by the recent research of Hakiem (2022) in the context of Saudi Arabia.

Naz and Ashraf (2020) write about the suppression of women within the patriarchal society, and similarly to other academics, cite higher education as a remedy for both oppression and suppression, from any external or internal sources, and how participation in higher education can be directly correlated with female empowerment. The quandaries faced by female leaders in higher education include elements of political allegiances, prioritising, and the navigation of constantly changing domains, which require both experience and instinct to navigate (Pasque & Nicholson, 2023).

15.4 Practice and Policy Implications for Women's Empowerment

UNESCO-IESALC put forth some notable recommendations to bridge the gender parity in leadership, specifically in higher education, that we could learn from (Galán-Muros et al., 2023). Drawing from these recommendations and aligned with our research on this topic, we propose some practices and policy implications for women's empowerment that are applicable in the GCC context. We propose our recommendations on three different levels: governmental level, institutional level, and societal level.

15.4.1 Governmental Level

The governments need to revisit their policies related to gender on a larger scale, to align it better with higher education policies regarding equal pay, work-life balance, etc. Furthermore, they need to raise awareness on a national level about gender inequality in academia, to create a more equitable and sustained culture of gender equality in higher education institutions (Galán-Muros et al., 2023, p. 7).

15.4.2 Institutional Level

At the institutional level, institutions need to carry out a critical analysis of the current hiring and promotion policies that focus not just on the quantitative outcomes but also focus on the qualitative performance indicators, to better understand female contribution. Institutions further need to create more opportunities for women to reach and maintain leadership roles, by providing more funding, mentoring, and networking opportunities that will be supportive of their growth (Galán-Muros et al., 2023, p. 1). Moreover, institutions need to proactively engage in raising awareness and advocacy for gender equality and women's empowerment through meetings and events. Faculty with gender-specific speciality must be hired within the faculty team and encouraged to research in this area to increase awareness (Mott, 2022). HEIs also need to build partnerships with relevant stakeholders, to promote collaboration in raising awareness for gender equality and women's empowerment, while also promoting gender-transformative pedagogy in teaching, and preparing future teachers within their institutions.

15.4.3 Societal Level

At a societal level, the challenge of gendered norms can begin as early as kindergarten. Research has indicated that children start to form gendered ideas and conceptions of roles in working life, from as young as three (Sinno & Killen, 2009), and so there is a role for schools to support the exposure to a wide diversity of roles for girls, boys, men, and women. Governing educational bodies would also do well to undertake audits of resources used in education: media, imagery, role models, etc. Positive displays of gender diversity in pictures accompanying texts can have an impact on students' ideas of roles accessible to different genders (Biemmi, 2015), which is in itself an empowering move. Children begin to recognise gendered cues even prior to kindergarten, and the play involved in early childhood care settings can be deeply gendered as a result of educators' own personal experiences and conscious or sub-conscious gendered stances (Chapman, 2016). At primary school, children can participate in hands-on workshops and activities which empower both genders to recognise disparity and inequality and understand how these can be rectified.

This could then be extended into upper levels of education, exposing both girls and boys to diversity training on university courses, and gender diversity within student and teacher roles on these courses. This would help young people to be less constrained in their university choices (and therefore later their career choices) and empowered by role models and seeing possibilities (again in both genders) in a variety of fields (Porter & Serra, 2020).

In summary, we recommend the HEIs to advocate for and holistically implement the framework designed by Mott (2022) for women's and girls' empowerment that connects stakeholders from different levels: governmental, institutional, and societal levels, for sustained gender equality. Mott (2022) presents five inter-related aspects that would drive women's empowerment. The five aspects as shown in Fig. 15.1 are (1) increased awareness and agency of individual women and girls (individual power); (2) fairer access to resources and opportunities; (3) dialogue, collaboration, and collective action; (4) a supporting legal and policy environment; and (5) changes in attitudes, beliefs, practices, and discriminatory social norms (Mott, 2022, p. 73). The author further presents a detailed explanation of the challenges and possibilities to achieve these outcomes (see Mott, 2022, pp. 73–77), but what we share here is our interpretation of the framework that will serve the purposes of this study. The framework includes four quadrants that are connected with the primary aspect, focusing on more collaboration and transparent communication between different stakeholders. Furthermore, change needs to happen simultaneously at individual and institutional levels, and in both formal and informal ways. Finally, the outermost arrows remind us that gender equality and women's empowerment is a process which is interconnected and that it will come to fruition when there is increased awareness about gender equality, and changes in attitudes, beliefs, and practices, to combat discriminatory social norms. Additionally, there needs to be fair access to resources and opportunities, irrespective of gender, along with consistent support and reform in the legal and policy space. We leave the readers with this framework to adapt and implement as is relevant to their own context.

15.5 Conclusion

In this chapter, we have synthesised various key elements in consideration of women's empowerment and striving for equality in higher education. Women's representation in key areas of higher education, including in leadership, has been discussed and contextualised within gender gap data from the GCC. The framework of empowerment has been described, which emphasises the importance of increasing agency among individual females, working towards access to opportunity, which is built upon principles of social justice and fairness, and other changes in beliefs and practices. These adaptations cannot be achieved simply by individuals, but within a broader support system of peers/colleagues, educators, administrators, and at wider levels of governmental support, too. All countries in the GCC have much to gain from this endeavour; the future economies of the country will in fact be highly dependent upon achieving successful gender parity within educational systems. It will also reduce the GCC countries' long-term and unsustainable dependency on the skills of an expatriate workforce. We have emphasised the lifelong aspect of exertions to both empower women and to directly impact on their career paths, but we have stressed that this must begin from education in the very earliest years. This is not merely a girls' or women's problem, since the whole of society stands to gain from increasing equity and fairness for both genders in educational opportunities which are forward thinking, rooted in lifelong learning stances, and linked to SDGs (Sustainable Development Goals).

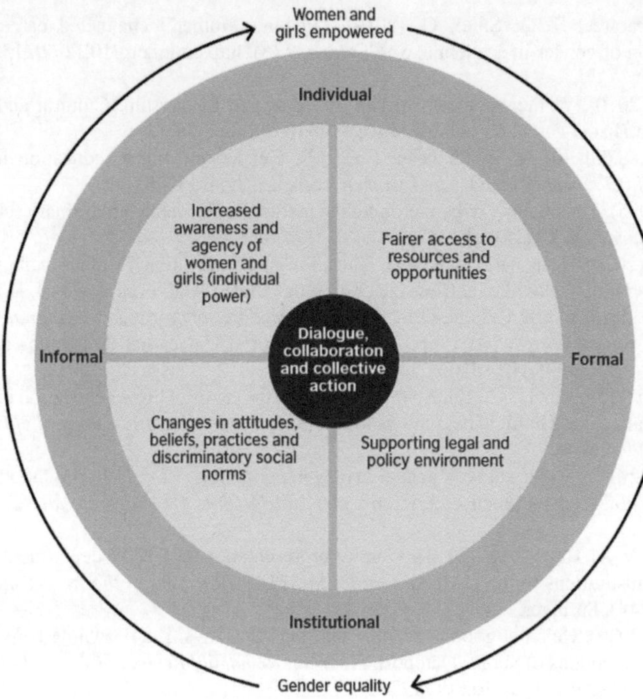

Fig. 15.1 The British Council's theory of change for women and girls' empowerment. *Note*. The framework is borrowed from Mott (2022). *Gender equality in higher education: maximising impacts*. British Council Report

Bibliography

Abdulkadir, R., & Müller, H. (2020). The politics of women empowerment: Female leaders in the UAE. *Hawwa, 18*(1), 8–30. https://doi.org/10.1163/15692086-12341368

Aiston, S. J., & Jung, J. (2016). Women academics and research productivity: an international comparison. In *Globalised re/gendering of the academy and leadership* (pp. 17–32). Routledge.

Aiston, S. J., & Fo, C. K. (2021). The silence/ing of academic women. *Gender and Education, 33*(2), 138–155.

Al Oraimi, S. Z. (2011). The concept of gender in Emirati culture: an analytical study of the role of the State in redefining gender and social roles. *Museum International, 63* (3–4), 78–91. ISSN 1350–0775.

Alam, F., Alam, Q., Chowdhury, H., & Steiner, T. (2013). Transnational education: Benefits, threats and challenges. *Procedia Engineering, 56*, 870–874. https://doi.org/10.1016/j.proeng.2013.03.209

Alotaibi, F. T. (2020). Saudi women and leadership: Empowering women as leaders in higher education institutions.

Al-Sayegh, F. (2010). *Women and Economic Changes in the Arab Gulf: The Case of the United Arab Emirates, 10*(2), 17–39. https://doi.org/10.1111/j.1949-3606.2001.tb00419.x

Amatullah, T. (2018). *Female Leadership Narratives in Higher Education in Qatar in the light of Islamic Leadership framework.* (Electronic Thesis or Dissertation). https://etd.ohiolink.edu/

Ashencaen Crabtree, S., & Shiel, C. (2019). "Playing Mother": channeled careers and the construction of gender in academia. *SAGE open, 9* (3). https://doi.org/10.1177/2158244019876285

Ballenger, J. (2010). Women's Access to Higher Education Leadership: Cultural and Structural Barriers. In *Forum on public policy online, 5.* Oxford Round Table.

Batool, S. Q., Sajid, M. A., & Shaheen, I. (2013). Gender and higher education in Pakistan. *International Journal of Gender and Women's Studies, 1*(1), 15–28.

Biemmi, I. (2015). The imagery of gender in Italian textbooks. Research into primary school books. *Foro de Educación, 13* (18), 15–35.

Bridi, R. M. (2020). Transnational higher education and international branch campuses in the Gulf Cooperation Council countries: the case of the United Arab Emirates. In J. Domenech, P. Merello, E. de la Poza, & R. Peña-Ortiz (Eds.), *Proceedings of the 6th International Conference on Higher Education Advances* (pp. 371–381). Editorial Universitat Politècnica de València. https://doi.org/10.4995/HEAd20.2020.11063

Bridi, R. M., & Al Hosani, N. (2022). An examination of the status of higher education in the United Arab Emirates: From humble beginnings to future challenges. *Eurasian Journal of Educational Research, 99*, 26–44.

Chapman, R. (2016). A case study of gendered play in preschools: How early childhood educators' perceptions of gender influence children's play. *Early Child Development and Care, 186*(8), 1271–1284.

Dickson, M., & Al Harthi, M. (2023). Gender representation in STEM departments in higher education institutions in the UAE. In *Gender in STEM Education in the Arab Gulf Countries* (pp. 221–244). Springer.

Dickson, M. (2020). He's not good at sensing that look that says, I'm drowning here! Academic mothers' perceptions of spousal support. *Marriage & Family Review, 56*(3), 241–263. https://doi.org/10.1080/01494929.2020.1712302

Dickson, M., & Tennant, L. (2019). Emirati university student mothers post-childbirth: Support systems in the home. *Journal of Women and Gender in Higher Education, 12*(1), 88–105.

Galán-Muros, V., Bouckaert, M., Roser, J. (2023). *The Representation of Women in Academia and Higher Education Management Positions.* UNESCO-IESALC Policy Brief Series.

Global Gender Gap Report (GGGR). (2022). *World Economic Forum Global Gender Gap Insight Report*. https://www3.weforum.org/docs/WEF_GGGR_2022.pdf

Hakiem, R. A. A. D. (2022). 'I can't feel like an academic': Gender inequality in Saudi Arabia's Higher Education system. *Higher Education.* https://doi.org/10.1007/s10734-022-00914-9

Herrmann, S. D., Adelman, R. M., Bodford, J. E., Graudejus, O., Okun, M. A., & Kwan, V. S. (2016). The effects of a female role model on academic performance and persistence of women in STEM courses. *Basic and Applied Social Psychology, 38*(5), 258–268.

Ivancheva, M., Lynch, K., & Keating, K. (2019). Precarity, gender and care in the neoliberal academy. *Gender, Work & Organization, 26*(4), 448–462. https://doi.org/10.1111/gwao.12350

Judson, E., Ross, L., & Glassmeyer, K. (2019). How research, teaching, and leadership roles are recommended to male and female engineering faculty differently. *Research in Higher Education, 60*(7), 1025–1047. https://doi.org/10.1007/s11162-018-09542-8

Liu, S. (2013). A few good women at the top: The China Case. *Business Horizons, 56*, 483–490.

Madsen, S. R. (2010). Leadership development in the United Arab Emirates: The transformational learning experiences of women. *Journal of Leadership and Organizational Studies, 17*(1), 100–110.

Maestri, E. (2011). Gulf Arab Women between Cultural Resistance and Knowledge Economy. In A. Ramakrishnan & M. Ilias (Eds.), *Society and Change in the Contemporary Gulf* (pp. 74–91). New Century Publications.

Mahaffey, B. A., Hungerford, G., & Sill, S. (2015). College student mother needs at regional campuses: An exploratory study. *AURO Journal, 21*, 105–115.

Malik, S., & Courtney, K. (2011). Higher education and women's empowerment in Pakistan. *Gender and Education, 23*(1), 29–45.

Mayuzumi, K. (2015). Navigating orientalism: asian women faculty in the Canadian Academy. *Race Ethnicity and Education, 18*(2), 277–296.

Metcalfe, B. D. (2008). Women, management and globalization in the Middle East. *Journal of Business Ethics, 83*(1), 85–100.

Moreau, M. P., & Kerner, C. (2015). Care in academia: An exploration of student parents' experiences. *British Journal of Sociology of Education, 36*(2), 215–233.

Mott, H. (2022). *Gender equality in higher education: Maximising impacts.* British Council.

Naz, A., & Ashraf, F. (2020). The relationship between higher education and women empowerment in Pakistan. *UMT Education Review, 3*(2), 65–84.

Noland, M., Moran, T., & Kotschwar, B. (2016). *Is gender diversity profitable? Evidence from a global survey.* Peterson Institute for International Economics.

Organization of the Petroleum Exporting Countries (OPEC). (n.d.). https://www.opec.org/opec_web/en/about_us/170.htm

Özoral, B., (2022). The effect of globalization in the gulf countries and the changing roles of women: a comparative study on Saudi Arabia, the United Arab Emirates and Qatar. *Visionary Journal, 13* (36), 1150–1169. ISSN: 1308-9552.

Pasque, P. A., & Nicholson, S. E. (Eds.). (2023). *Empowering women in higher education and student affairs: theory, research, narratives, and practice from feminist perspectives.* Taylor & Francis.

Peterson, S. (2016). Community college student-parents: Priorities for persistence. *Community College Journal of Research and Practice, 40*(5), 370–384. https://doi.org/10.1080/10668926.2015.1065210

Porter, C., & Serra, D. (2020). Gender differences in the choice of major: The importance of female role models. *American Economic Journal: Applied Economics, 12*(3), 226–254.

Prager, L. (2020). Emirati women leaders in the cultural sector: From 'state feminism' to empowerment?". *Hawwa, 18*(1), 51–74.

Ramakrishnan, A., & Ilias, M. (2011). Introduction. In A. Ramakrishnan & M. Ilias (Eds.), *Society and Change in the Contemporary Gulf* (pp. 1–9). New Century Publications.

Reed, S., Grosz, M., Kurlaender, M., & Cooper, S. (2021). A portrait of student-parents in the California community colleges. *UC Davis Wheelhouse Center for Community College Leadership and Research, 6*(2), 1–14.

Romanowski, M. H., & Al-Hassan, F. S. (2013). Arab Middle Eastern women in Qatar and their perspectives on the barriers to leadership: Incorporating transformative learning theory to improve leadership skills. *Near and Middle Eastern Journal of Research in Education, 1*, 1–13. https://doi.org/10.5339/nmejre.2013.3

Ross, M. L. (2008). Oil, Islam and women. *The American Political Science Review, 102* (1), 107–123. https://www.jstor.org/stable/27644501

Royal Society of Chemistry. (2018). *Diversity landscape of the chemical sciences.* Retrieved from https://www.rsc.org/globalassets/02-about-us/our-strategy/inclusion-diversity/cm-044-17_a4-diversity-landscape-of-the-chemical-sciences-report_web-2.pdf

Sallee, M. W. (2015). Adding academics to the work/family puzzle: Graduate student-parents in higher education and student affairs. *Journal of Student Affairs Research and Practice, 52*(4), 401–413. https://doi.org/10.1080/19496591.2015.1083438

Samier, E. (2015). Emirati women's higher educational leadership formation under globalization: Culture, religion, politics, and the dialectics of modernization. *Gender and Education, 27*(3), 239–254. https://doi.org/10.1080/09540253.2015.1028901

Shepherd, S. (2017). Why are there so few female leaders in higher education: A case of structure or agency? *Management in Education, 31*(2), 82–87.

Sherif, S. (2013). Macroeconomic policy, localization and reducing unemployment: The crucial human resource issues for the UAE. *Competitiveness Review, 23*(2), 158–164. https://doi.org/10.1108/10595421311305352

Sinno, S. M., & Killen, M. (2009). Moms at work and dads at home: Children's evaluations of parental roles. *Applied Developmental Science, 13*(1), 16–29. https://doi.org/10.1080/10888690802606735

Thomas, J., Raynor, M., & Al-Marzooqi, A. (2012). Marital status and gender as predictors of undergraduate academic performance: a United Arab Emirates context. *Learning and Teaching in Higher Education: Gulf Perspectives, 9* (2). Retrieved from http://lthe.zu.ac.ae/index.php/lthehome/article/viewDownloadInterstitial/80/34

UNESCO-IESALC (2021). *Women in higher education: has the female advantage put an end to gender inequalities?* https://unesdoc.unesco.org/ark:/48223/pf0000377182

Walkup, V. (2006). To publish or not to publish: A discussion of reporting and disseminating research into the experiences of students who are mothers returning to higher education. In *Paper presented at the 36th Annual Standing Conference on University Teaching and Research in the Education of Adults, July 4–6, Trinity and All Saints College, Leeds, UK*.

West, J. D., Jacquet, J., King, M. M., Correll, S. J., & Bergstrom, C. T. (2013). The role of gender in scholarly authorship. *PLoS ONE, 8*(7), e66212.

Wilkins, S., Balakrishnan, M. S., & Huisman, J. (2012). Student choice in higher education: Motivations for choosing to study at an international branch campus. *Journal of Studies in International Education, 16*(5), 413–433. https://doi.org/10.1177/1028315311429002

Yousaf, R., & Schmiede, R. (2017). Barriers to women's representation in academic excellence and positions of power. *Asian Journal of German and European Studies, 2*(2), 1–13. https://doi.org/10.1186/s40856-017-0013-6

Chapter 16
The Journey Towards Inclusive Education in Qatar

Nouf Alkaabi and Najlaa Al-Maadid

Abstract This chapter critically evaluates the journey of Qatar towards implementing inclusive education for children with disabilities, taking into account how this affects its adherence to the UN Convention on the Rights of Persons with Disabilities. In this context, the report outlines the legal framework, regulations, and practices that are in place in support of inclusive education and identifies the barriers that remain challenging in terms of implementation. The article most importantly highlights Qatar's legislative strides and policy innovations designed to foster educational equity, focusing on the evolution from specialised schooling to integrated educational frameworks. The narrative excellently captures the huge mileage the nation has covered to expand access and improve educational practice for children with disabilities, through the establishment of special institutions and coordinating national strategy. Despite making significant strides, the chapter also describes several of the continuing challenges that remain, such as programme limitations for students, diagnostic difficulties, and lack of support structures. Second, it has shown the complexity involved in transitioning from partial to fully inclusive educational systems and the need for sustained commitment, research, and policy refinement as long as progress is occurring. In conclusion, the chapter acknowledges Qatar's leadership in promoting inclusivity and calls for continued effort in addressing gaps and ensuring that education is made accessible and equitable to all children.

16.1 Introduction

Qatar has always been a supporter of children's right to education, especially children with disabilities (Al-Atteyah & Lazarus, 2007; Clayton, Amaral & Shafique, 2021; MoFA, 2023). In 2008, Qatar was one of the first countries at the regional and global level to sign and ratify the 2006 Convention on the Rights of Persons with Disabilities (CRPD) (MoFA, 2023). The country's ratification of the convention signals its serious

N. Alkaabi (✉) · N. Al-Maadid
Director of Planning, Quality, and Innovation, Civil Service and Government Development Bureau, Doha, Qatar
e-mail: alkaabi.nouf84@gmail.com

E. Zaidan et al. (eds.), *The Future of Education Policy in the State of Qatar*, Gulf Studies 21, https://doi.org/10.1007/978-981-97-9667-0_16

attempt to join the global community in their efforts to protect the rights of individuals with disabilities and eliminate forms of discrimination against them. However, domestic laws have long sought to promote global ideals such as equity and equality. For instance, in 2001, the Qatari government issued the Compulsory Education Law, which obligates all parents to enrol their children at age six in schools for at least nine years of schooling, or until they are 18 years of age. According to the law, all children are eligible for free access to education, regardless of their nationality, gender, and ability; the law only exempts children with severe disabilities or injuries, providing there is a submission of valid medical reports to units authorised by the Ministry of Public Health (MoPH) (Law No. 25 of Compulsory Education, 2001). Following that, the government issued Law No. 2 of 2004 in respect of People with Special Needs, which aimed to regulate the rights of individuals with disabilities include the right to education, rehabilitation, medical care, work, accommodation, participation in sports, and access to public facilities (Law No. 2 of People with Special Needs, 2004). According to Al-Hendawi et al. (2023), the issuance of the previous laws in Qatar emphasised two major concepts that aligned with global ideals, (1) that all children must have access to free education and (2) all children with disabilities have the right to be educated.

Regionally, there is growing evidence of modern views towards the inclusion of children with disabilities in education and other facets of life, led by Qatar, Kuwait, and Bahrain (Weber & City, 2012). Yet Al-Hendawi et al. (2023) argue that in the GCC and other Arab countries, the inclusion of children with disabilities in education is fairly new, and countries need time to develop infrastructures and needed resources to realise some of the highly demanding global standards. In their view, the achievement of global ideals such as equity and inclusion sought by the United Nations Educational, Scientific and Cultural Organization (UNESCO) and United Nations International Children's Emergency Fund (UNICEF) has been a long-standing challenge faced by most countries, even ones with advanced inclusion practices (Al-Hendawi et al., 2023). However, over the past few years, Qatar, like other GCC countries, has made tremendous efforts to develop inclusive educational opportunities for all children, which is noteworthy. Qatar's dedication to the implementation of internationally sought inclusive strategies, influenced by the UN's CRPD, has put the country at the forefront of change in the region, as it expanded and diversified its offerings to children with disabilities over the past few years (Guldberg et al., 2017). The government has implemented many initiatives to improve the provision of educational opportunities and diagnostic services for children with disabilities (Clayton, Amaral & Shafique, 2021).

This chapter sheds light on the strategic initiatives carried out by the Qatari government in the inclusion of children with disabilities in education. It also examines the limitations of the current programmes and identifies opportunities for future growth.

16.2 Early Attempts at Inclusion in Qatari Schools

The inclusion of children with disabilities in education was first witnessed in 1975, when the Qatari Ministry of Education set up a separate classroom in one government school for boys with a hearing disability (Al-Said, 2002). This was followed by the establishment of a separate school for the deaf in 1981 for boys, followed by another one in 1982 for girls. In 1984, the ministry established a specialised school for children with mental disabilities. The ministry's efforts continued towards the provision of inclusive education for children with disabilities, when it established a unit in 1992 for learning and emotional support for students with learning disabilities and psychological problems (Al-Atteyah & Lazarus, 2007). In 1992, the Qatar Association of Rehabilitation of Special Needs (QARSN) was formed to provide social, psychological, medical, financial, educational, and rehabilitative care for the children with disabilities registered with it, whether Qataris or residents of Qatar, regardless of their gender, nationality, religion, social, or financial level (MoFA, 2011). The association supported the establishment of care centres that catered to the varying needs of individuals with disabilities. It also played a significant role in engaging individuals with disabilities in the wider community, and educating the public of their rights (QARSN, n.d.)

In 1998, prompted by the inspiration of Her Highness Sheikha Moza Bint Nasser, Al-Noor Institute for the Blind was established, to educate the blind and visually impaired. The institute offers its services to individuals, starting from the age of three until 21. The institute supplies early intervention programmes and provides home care for younger children to prepare for school once they are old enough. The institute offers its students the needed education and training and also facilitates their employment after graduation. It includes comprehensive care services for its students, that incorporate educational, cultural, and medical services (MoFA, 2011).

The first private school for students with disabilities which offered instruction in English was opened in 1996. Students with visual disabilities were sent to other Gulf countries to get their education at the expense of the government, or kept at home. In 1998, a new school for the blind was opened (Al-Atteyah & Lazarus, 2007). According to Al-Atteyah & Lazarus (2007), before 2001, the Ministry had adopted a specialised schools model, where children with disabilities would get their education in segregated schools; and there was minimal evidence of mainstream inclusion at the time.

16.3 The Inclusion of Children with Disabilities in Schools: A New Vision

A new vision for the inclusion of children with disabilities was developed, and specialised entities were established, to achieve the country's commitment to international ideals during the Education for a New Era (EfNE) reform in 2002. The inclusion of children with disabilities in mainstream education was attributed to the newer policies developed by the reform designers. The Supreme Education Council (SEC), the governing body behind the reform and controlling schools at the time, was responsible for developing the Additional Educational Support Needs (AESN) policies and initiatives that were carried out by the SEC's AESN department, which was established in 2007 in the Education Institute (MoFA, 2011). The department developed the Additional Educational Support Policy and its supporting documents, which provided guidance to schools regarding the integration practices of students with disabilities in independent or government schools. The policy was divided into five sections: the first section focused on students with learning difficulties, the second targeted students with disabilities, the third focused on mutual aspects between students with learning difficulties and students with disabilities, the fourth focused on positive behaviour in schools, and the fifth included guidelines for working with parents to communicate information and receive the necessary assistance from the Education Institute (SEC, 2009). It has been argued that the policies of the SEC helped to officially promote inclusive practices at the system level and support all students' learning and participation in education (Hamid & Mohamed, 2021). In addition, the department provided schools with guidance and learning materials to support the learning of students with disabilities in government schools (SEC, 2009). According to the newly introduced policies, classroom teachers and administrative staff should implement the essential educational strategies to enhance the learning of all children, regardless of their abilities and needs (Hamid & Mohamed, 2021).

Part of the SEC's strategy was to develop inclusive schools, accommodating physical spaces for receiving students with a disability, provided in all schools, with teachers who were trained, administrative supporting staff assigned new roles, and the national curriculum modified to accommodate the learning needs of students with disabilities (CRPD, 2015). The SEC coordinates efforts with other specialised institutions such as Shafallah to assess students and provide them with the needed support. It also collaborates with Shafallah and Al-Noor Institute to offer several training sessions for teachers and school principals on how to effectively integrate students with disabilities in schools (MoFA, 2011). According to Al-Atteyah's study (2012) which evaluated the experience of students with disabilities in mainstream schools, it was revealed that government schools had supportive environments for inclusion, as evident in teachers' awareness and the improved services offered to the students and their parents. In addition, schools had improved physical facilities, accommodating learning spaces, and modified curriculum materials. Since then, efforts towards enhancing the opportunities and services provided to students with disabilities have continued.

Following the launch of Qatar's National Vision in 2008, the National Development Strategy 2011–2016 was the first official guiding document that identified the goals and targets for the education and training sector. According to the strategy, the Qatari government aimed to continue its reform efforts, by focusing on core policy themes that include inclusiveness, equity, equality, and mobility. Among its many high-level goals, the strategy allocated a separate goal to provide a "specialized system to ensure high-quality education for students with disabilities" (p.135). The analysis done prior to the strategy development revealed that the growth of the student population resulted in an increased demand for specialised education and training for students with disabilities. Also, teachers needed further training to support those students, and experts were needed to help students with moderate and severe disabilities (GSDP, 2011). The strategy urged the government to increase the capacity of specialised institutions for students with disabilities, to lower the pressure on mainstream schools. It also included initiatives to support the capacity development of teachers and specialists working with children with disabilities (GSDP, 2011).

Following the launch of the first strategy, the SEC started several programmes that aimed to support the learning of students with disabilities. One of the programmes of the strategy aimed to provide parents of children with disability with improved assessments, to help them choose the right programmes or schools, through the establishment of Roaa Center. The centre was opened in 2015 to provide clinical and evaluative services for students with disabilities and to enable them to participate in education alongside their peers to the maximum extent possible. The centre provides medical diagnosis and detailed assessments of students' comprehension skills, interdisciplinary evaluations of functional and performance abilities, screening and early identification of functional and learning disabilities of preschoolers, comprehensive behaviour assessments, consultations, and follow-up services. Also, the centre issues referrals for specific disciplines in Hamad Medical Corporation (HMC) if the answers to the children's needs are not available in schools, or if they are in need of assistive devices or further diagnostic tests (HMC, n.d.1).

The centre has several facilities that include a well-resourced library with specialised books, speech and language diagnosis rooms, an occupational therapy room, and a professional development room to provide training for special education teachers. Also, the centre aims to monitor the process of children screening and to generate reports for parents (Zard, 2017). Since its establishment, the centre has evaluated children with disabilities and transferred them to appropriate schools, where they could get additional educational support and services. According to the former director of the centre, it has supported parents and teachers by providing them with needed consultation and guidance (Al-Sharq, 2017). However, there have been several occasions when parents complained about the long waiting lists and the duration of the assessment at Roaa Center. In addition, parents questioned the validity of the evaluation methods and the quality of the performance of the specialists working there (Zard, 2017).

16.4 Mainstream or Specialised Schools

While the first NDS 2011–2016 focused on establishing the foundational systems for inclusion in education, the second NDS 2018–2022 emphasised the urgency of equitable access of all children to education, regardless of gender, age, and (dis)abilities. Through the improved provision of several options for children with disabilities, students are either enrolled within government schools or in specialised schools, depending on the services needed and the severity of the disability. In mainstream government schools, educational programmes are designed, based on individual educational plans for each student. Within the regular government school setting, services are designed to support students' needs through remedial plans. Schools also provide support to students in resource rooms, and special arrangements and facilities for student evaluations and classroom activities (MEHE, n.d.). All government schools adopted a three level system for integrating students with disabilities in schools; the level of support varied based on the student's circumstances. They offered students additional support through remedial sessions and assistive technology. Also, schools were required to form internal committees to monitor the learning of students and offer support to their parents (MoFA, 2011). According to Al-Atteyah (2012), government schools accommodated students with disabilities with learning spaces that included spacious classrooms, suitable lighting, easily accessed classrooms, and washrooms. On the other hand, some government schools were lacking fully equipped learning resource centres, enough spaces for movement or light, and audio alarms.

Since the launch of the second NDS 2018–2022, the MEHE has focused on improving access to educational opportunities for children with disabilities, by implementing two strategic initiatives. First, the MEHE equipped 70 government schools with the needed facilities, teachers, and learning resources to admit students aged 4–18 years with mild and moderate disabilities in mainstream education. In these schools, students with disabilities learn in separate fully resourced classrooms, and get proper support when needed from specialised teachers (Al-Raya, 2021). According to the MEHE's latest statistical report for the academic year 2021–2022, there were 4,132 students integrated in mainstream schools (MEHE 2022). Second, the MEHE established Al-Hedaya Specialised Schools for students with disabilities, which provided intensive educational services and supportive programmes for students with *Autism Spectrum Disorder* (ASD) and intellectual disabilities (MEHE, n.d.).

Moreover, the MEHE implemented a programme to adapt the national curricula according to the educational goals specified in the individual educational plan for each student. This programme has been offered for students with intellectual disabilities, autism disorders, or students with physical disabilities (MEHE, n.d.). A second programme was launched to deliver the general curriculum approved by the Ministry of Education and Higher Education without any change or modification for students who have various kinds of learning difficulties. These include slow learners, students with minor and moderate hearing impairment, and students with cochlear implants. They also include students with visual impairment and cerebral palsy. The general curriculum is offered in the classroom and in resource rooms in all public schools (MEHE, n.d.).

The MEHE recognised the importance of partnering with the private sector for the provision of education for students with disabilities to address the increasing demand. According to the NDS 2011–2016, more specialised schools were needed, to cater for the increasing student population (GSDP, 2011). For that purpose, the SEC adopted a public/private partnership strategy for an inclusive provision of educational opportunities through the development of specialised government and private schools for students with disabilities (Weber & City, 2012). Specialised private schools were licensed by the MEHE, as well as private care centres. Qatar Foundation (QF) established Awsaj Academy in 1996, as a non-profit specialised school serving students with learning challenges in grades K-12. The Academy has more than 620 students. Awsaj Academy is the only school of its kind in Qatar, and one of only a handful worldwide dedicated to working with students who have demonstrated aptitude in certain areas, while facing academic challenges in others (Hukoomi, 2023). Moreover, there were 23 licensed centres that accommodate 623 students in 2021–2022 (MEHE, 2022). These centres are monitored by the MEHE and follow the MoPH regulations in their service provision.

To support the country's movement towards more inclusive education, the College of Education at Qatar University initiated a bachelor's degree and a Diploma degree in Special Education in 2015. The programmes include several courses to teach students about disability and inclusion (Hassanein et al., 2021). When the SEC was replaced by the Ministry of Education and Higher Education in 2016, the Department of Special Education and Gifted Students Care was established by the Emiri Decree No. (9) to provide educational programmes for both students with disabilities and gifted students. The department expanded in scope and was responsible for providing screening/assessment services, developing and monitoring educational plans and programmes, ensuring the quality of the services provided to AESN students, and guiding parents in how to support the learning of their children (Emiri Decree No. 9, 2016).

16.5 The Inclusion of Children with Autism in Education

Regionally, Qatar has been a pioneer in its efforts to improve the access of individuals with autism to health and education services, due to the active support of Her Highness Sheikha Moza Bint Nasser and the Qatari government (Guldberg et al., 2017). For instance, in 2007, Qatar presented a proposal to the United Nations, which was adopted unanimously on December 18 of the same year, to consider April 2 as World Autism Awareness Day (MoFA, 2023). Moreover, in April 2017, the National Autism Plan (NAP) 2017–2021 was launched by Qatar's Prime Minister to communicate the country's serious efforts to improve the lives of families with autistic children and ensure the provision of quality services to all individuals with autism in Qatari society (MoFA, 2019). The NAP had 44 policy recommendations in six dimensions for improving service provision for individuals with autism, which included diagnosis, awareness, education, intervention, and transitioning into adulthood. The strategy aimed to implement programmes to improve the diagnostic services for children with autism, and their access to appropriate interventions. It intended to address parents' concerns regarding access to educational services and provide them with the needed guidance and support for raising a child with autism. Also, the strategy aimed to increase awareness of parents and the whole community about autism, and how to integrate autistic individuals in society (Murad, 2019).

Qoronflesh et al., 2019 reported that ASD has not been recognised as a national priority in the GCC countries which has led to fragmented and inconsistent efforts in supporting children with ASD. However, they indicated that Qatar was the first in the region to launch a national strategy for autism in 2017, which demonstrated the government's serious attempts to improve services provided to children with ASD (Qoronflesh et al., 2019). It has been argued that the speed of development and approval of the NAP 2017–2021 was attributed to the attention and strong support of the country's leadership (Guldberg et al., 2017).

Following the launch of the NAP, a forum was organised by the Ministry of Public Health (MoPH) that aimed to change society's false views of children with autism and build supportive systems of early intervention and education for parents and the wider community. Relying on the experiences of local and foreign experts, the forum focused on training Qatari cadres in the health and education sectors to deal with cases of autism, changing the culture of society regarding accepting and integrating children with autism, as well as training families who deal with the emotional side of raising a child with autism, and providing them with various skills and related experiences (Murad, 2019).

The MoPH played a significant role, by leading massive improvement efforts in diagnosis, early intervention services, and spreading awareness about autism. Similarly, the education sector adopted complementary efforts in the development of inclusive educational opportunities for children with autism; the efforts were primarily led by QF and *MEHE.* Also, the World Innovation Summit for Health (WISH) has been a key player in moving forward the agenda for improving services for individuals with ASD. In 2016, WISH held the "WISH Autism Forum 2016", which brought in policymakers and experts from all over the world to exchange ideas and practices that aim to improve the lives of individuals with ASD. The forum has resulted in the development of a comprehensive framework for the health and wellbeing of people with autism. The framework identified the key players in the development of services for individuals with ASD and their roles (Qoronflesh et al., 2019).

It has been reported that regionally, Qatar and Oman are among the first states to have adopted early intervention practices, and made progress in providing primary care and services for children with autism (Qoronflesh et al., 2019). Some of these programmes include the establishment of Renad Academy and Al-Hedaya government schools, among other initiatives that are detailed below.

16.5.1 Shafallah Center

In 1999, H. H Sheikha Moza Bint Nasser established Shafallah Center for Persons with Disabilities to provide exemplary services to persons with mental disabilities and autism, aged three to sixteen years old. The centre operates to provide educational and rehabilitation services to prepare children for independent participation in society. It plays a significant role in community awareness and spreading information to society. In addition, it provides parent counselling, and support for families and individuals (Shafallah, n.d.). There are other supporting organisations that operate under the governance of the Shafallah Center that include Alnoor Institute for The Blind, Best Buddies-Qatar, and Shafallah Medical Genetics Center.

Since 2013, Shafallah Center has operated under the supervision of the Qatar Foundation for Social Work, and it seeks to set up, develop, plan, and implement programmes and strategies for special education and pre-vocational training to suit students with mild and moderate intellectual disabilities and autism. The centre provides other support services such as adaptive sports division, art, music, and assistive technology (Hukoomi, 2023). It adopts the MEHE National Curriculum and accommodates it according to the individualised plans of the students. It also incorporates additional programmes for academic, cognitive, and social development. The centre aims to contribute locally to the provision of model services for children with disabilities to support the realisation of Qatar Foundation's vision to develop the capabilities of various groups in Qatari society. The centre also has a separate unit to cater for the needs of children with severe mental disability that includes Down syndrome and brain paralysis (Shafallah, n.d.).

According to the latest statistics published by MEHE, Shafallah admits 624 students (MEHE 2022). The centre has specialised facilities that include a well-resourced library, therapy halls for developing cognitive and motor skills, and a 3-dimensional theatre digitally equipped for live broadcast and film screening.

Shafallah Center also enables its students professionally through vocational training and supports them to find suitable jobs through partnerships with other institutions (Bedaiwi, 2019). The Rehabilitation and Vocational Training Unit in the centre provides vocational training and social integration activities for students aged 16–21 to enhance their employability and independence skills. Shafallah has several partners who work closely with the Employment Unit in the centre to provide students with additional training and job opportunities. It contacts potential employers and arranges accommodating jobs for the centre graduates, according to their capabilities and desires. The Employment Unit also ensures the suitability of the work environments for its graduates and coordinates with the different employers regarding the needed on-the-job training, support, and adjustments. Both Qatar Airways and Green Houses have supported the centre in providing training and job opportunities for its graduates (Shafallah, n.d.).

16.5.2 Renad Academy

Renad Academy (RA) is a school located in Education City dedicated to children who have been diagnosed with mild to moderate autism spectrum disorder (ASD); the academy opened in 2016 and is a member of Qatar Foundation (Hukoomi, 2023). The academy's vision is to become a centre of excellence for children with autism and provide an optimum learning environment for young children, their parents, and the wider community (Clayton, Amaral & Shafique, 2021). More than 620 students are registered in Renad and its satellite locations, which are provided with an inclusive learning environment (MEHE, 2022). The academy offers parents online and personal support that helps them to navigate through the challenges of raising autistic children. Part of the academy's innovative model is its research-grounded practices and its communication campaigns, which have brought the outer community closer, and increased their awareness (Clayton, Amaral & Shafique, 2021).

Renad develops individualised learning schemes that are different for every child; these are developed based on their diagnosis and individual learning needs. The units of learning cover basic literacy and numeracy skills, behavioural skills, receptive/expressive communication, and social skills. Due to its partnership with Al-Shaqab equestrian facility, weekly equine therapy sessions are provided to children. The academy also organises field trips for the children to interact with the wider world and improve their communication skills (Renad, n.d.). It is expected that the academy will "forge a pathway for improved services for people with autism" and become a distinguished model at the regional level (Clayton, Amaral & Shafique, 2021, p.80).

16.5.3 Al-Hidaya Government Schools

In the past decade, the MEHE has made tremendous investments to establish specialised schools to support children with disabilities. There are 70 government schools that provide inclusion services for students with physical disabilities, autism, and intellectual disabilities (MEHE, n.d.). On the other hand, Al-Hidaya Schools for students with mild and moderate intellectual disabilities and students with autism offer intensive and integrated services that are provided through a multidisciplinary team approach. Al-Hidaya schools offer children different programmes that include physiotherapy, occupational therapy, behavioural therapy, and specialised academic units provided by trained teachers. Parents can register their children in Al-Hidaya schools after they go through a comprehensive diagnostic assessment at one of the approved list health centres. Medical reports are then sent from the MEHE to schools, to develop educational or behavioural plans for each student, based on their report; these reports are discussed and agreed upon with parents before implementation (Tawalbeh, 2023). Since the launch of the second NDS 2018–2022, nine branches of Al-Hidaya schools have been opened in different municipalities; the school has a total of 371 students (MEHE, 2022).

The school accommodates the National Curriculum based on every child's need, and the trained specialists follow guided instructions provided by the MEHE. Some of the schools' modern facilities include sensory simulation rooms, and assistive technology that activates the students' senses and improves their thinking and language skills (Tawalbeh, 2023). The sensory simulation rooms in Al-Hidaya were designed to engage the visual, olfactory, tactile, and auditory senses of children and develop their independence and skills (Jones, 2020). In addition, the school offers children several opportunities to improve their communication skills and community participation, by engaging them in activities that include art exhibitions to present their work (MEHE, n.d.).

16.5.4 Al-Maha Center for Children and Young People

A comprehensive assessment was carried out by Holland Bloorview Kids Rehabilitation Hospital in December 2013, of the children's facilities in Hamad Medical Center HMC, which revealed that they were "inadequate" in providing a safe and supportive environment for those children with complex and long-term healthcare needs. Also, the assessment revealed that the medical services provided to vulnerable children by HMC were isolated, due to heavy reliance on medical support and technology. Such environments were not supportive of children, as they were not family-centred and did not connect the child to the outer world. Following the assessment, HMC worked closely with consulting firms to develop a comprehensive plan for developing a centre that catered to the needs of children in an integrated and family-oriented environment (HMC, n.d.2).

In 2015, the MoPH announced the establishment of Al-Maha Hospital under the management of HMC; a full-service building to provide necessary care for children diagnosed with autism and delayed growth in the Al-Wakra area. Al-Maha Center was to focus on training cadres and deploying them in all centres and specialised institutions in Qatar (Saber, 2019). The centre adopts innovative practices and evidence-based strategies, to cater for the health and learning needs of children with complex and long-term healthcare needs. This is because children with complex medical needs rely heavily on specialised medical techniques, nursing support, and clinical care. For this reason, the environments in which these sick children are cared for combine the stimulating qualities of home, school, and the outer community, where children can learn from the world around them within a warm, comfortable, and child-friendly environment (Al-Meraghi, 2015). The centre offers early intervention programmes and a range of therapy and rehabilitation services, including a gait laboratory, life skills programme, music and art therapy classes, and a hydrotherapy pool (Al-Arab, 2020). It was officially opened in 2022; and it is the first to provide a "one-stop facility" for children with complex and long-term healthcare needs in the region (HMC, n.d.1).

The Guldberg et al. (2017) report revealed that the country's journey towards true inclusion of children with autism is in its early stages, as is evident in low levels of awareness and late detection of development delays in children, which usually happens at ages four or five by their schools. On the other hand, the report revealed that the experiences of parents are improving, as the country has committed to improving its service provision for children with autism over the past decade. The researchers reported that some of these initiatives have increased parents' awareness and acceptance of children with autism. According to a study conducted by Doha International Family Institute (DIFI) that evaluated some programmes of the NAP in 2019, it appeared that there were challenges faced by families in accessing educational services in schools and receiving proper individualised plans for their autistic children. Although the NAP called for the ensuring of access to children of education with a focus on inclusionary options, there were some limitations due to the limited availability of seats and qualified staff in some schools. In terms of awareness, the study revealed that there were false or fixed beliefs among many people about the causes of autism, its symptoms, and the behaviours of children with autism. Moreover, the study also called for more financial support for parents who cannot afford it, as many of them are struggling to pay for the highly priced programmes provided by private schools and centres (DIFI, 2019).

16.6 Insights from the Qatari Inclusion Experience

There are growing positive trends towards inclusion in education inspired by Qatar's serious efforts to integrate individuals with disabilities in society (Hamid & Mohamed, 2021). Over the last few years, the Qatari government has invested heavily to improve the provision of diagnostic services and educational opportunities for children with disabilities (Clayton, Amaral & Shafique, 2021; Bedaiwi 2019). Moreover, there has been an increase in early childhood special services, and the education of students with disabilities has become more directly driven by national education standards (Al-Hendawi, Khair and Keller 2017). It has been reported that Qatar's MEHE has adopted internationally sought inclusive strategies, influenced by the UN's CRPD, and is now at the forefront of change in the region; as it has expanded and diversified its offerings to children with disabilities over the past few years (Guldberg et al., 2017). The ratification of the UN's CRPD in 2008, and the increasing demand for specialised educational services for students with disabilities have encouraged the initiation of several national strategic programmes with a noticeable impact on society, like improved capacity, awareness, and quality of services, as explained in detail below.

16.6.1 Improved Awareness

The issuance of several laws and strategies over the past two decades confirms the country's acknowledgement of and attempt to protect the rights of individuals with disabilities. The inclusion of individuals with disabilities has been supported by the highest authorities in government, which, in turn, has increased awareness among society. The UN's Committee on the Rights of Persons with Disabilities appreciated the initiation of several campaigns to raise awareness of the rights of individuals with disabilities in Qatar (CRPD, 2015). At the beginning of the period to increase inclusion in government schools, the SEC provided several training sessions and campaigns to educate teachers and parents (MoFA, 2011). It has been reported that there was increased awareness regarding autism in Qatar, and there were promising initiatives to address the needs of autistic individuals carried out by the MEHE and QF, following the launch of NAP 2017–2022 (Guldberg et al., 2017).

Moreover, the inclusion of students with disabilities in government schools resulted in improved levels of acceptance among other students, as they felt they shared more similarities than differences in an inclusive setting, as reported by Al-Atteyah & Lazarus (2007). They argued that teachers demonstrated increased awareness and they were supportive, encouraging, and accommodating to their students' needs. Similarly, according to Khalifa et al. (2017), specialised schools keep parents informed about their children's performance, using several methods such as e-mails, phone calls, and reports. The study revealed that parents appreciated the schools' confidentiality policies, and the eagerness of the school staff to listen to and support parents. The interactions between parents and schools helped to educate the parents about their children's rights and learning progress.

Also, the offering of specialised courses in special education through QU's College of Education has further increased awareness among students in teacher training programmes. Hassanein et al. (2021) argue that students of Special Education programmes in QU demonstrated more awareness and confidence to teach children with disabilities in schools, due to the programme's focus on developing their theoretical and applied components.

On the other hand, Al-Atteyah (2012) reported that teachers found it challenging to integrate students with disabilities in regular classrooms, without providing them with the necessary facilities, supports, and materials. They believed that separate classrooms within a regular government school were the most suitable arrangement for integrating their students in education, followed by specialised centres. Moreover, experts have spoken about the limited knowledge among parents about the varying opportunities to educate children with disabilities. Also, they reported that parents' limited beliefs, and feelings of embarrassment and fear, hindered the inclusion of children with disabilities in society (Bedaiwi, 2018). Hadidi and Al-Khateen (2015) reported that teachers hold negative views towards inclusion in schools, and they had many concerns regarding teaching students with disabilities in the regular classroom. They believed that teachers' resistance and negative attitudes delayed the provision of inclusive educational opportunities, despite the massive investments to develop infrastructures and the material aspect. Despite improved awareness levels among educators and parents, the concept of true inclusion of children with disabilities in education and society remains a challenging concept to fully comprehend and apply.

16.6.2 Diversified and Improved Educational Opportunities

Qatari schools have become more accommodating and inclusive due to continuous efforts to integrate children with disabilities in education. According to the UN's report, clinics have opened to assess students' capacities and to prepare reports for parents and teachers that inform individualised educational plans. New schools had been opened with appropriate arrangements that enhanced accessibility, teachers and specialists were trained, and educational books had been translated into Braille and or voice recorded (CRPD, 2015). In addition, the new school buildings have become barrier-free and so accommodating compared to older school buildings (Al-Atteyah & Lazarus, 2007). In a press interview, Dr. Al-Meghaisib, former dean of the College of Education in QU, praised Qatar's efforts in integrating children into mainstream education over the past two decades (Bedaiwi, 2019).

In addition, the capacity of educational institutions that cater to the needs of children with disabilities has increased exponentially, which has increased the number of students who have benefited from these services. For instance, the inclusion of students with disabilities in mainstream education started in 2003 with less than 27 students in two government schools (Al-Atteyah & Lazarus, 2007). Today more than 4,000 students are integrated into 70 government schools with their peers (MEHE, 2022).

On the other hand, there have been several challenges that have hindered the effective inclusion of individuals with disabilities in education. These challenges tend to vary in degree and impact according to the student's disability, nationality, and socio-economic status. Some of the main challenges include the difficulty to access education, limitations in diagnosis, the rising cost of the services, and the inconsistent governance of the sector.

16.6.3 Limitations in Diagnosis

The experiences of parents in the pre- and post-diagnostic phases have been described as challenging by many parents who struggle after diagnosis and are in need of emotional and technical support. It has been reported that the diagnosis of children with autism is troublesome and "patchy", which usually forces many Qatari parents to take their children abroad to receive proper diagnoses (Guldberg et al., 2017, p. 29). Moreover, many parents have experienced difficulties in receiving proper diagnosis services, due to the duration and validity of the process. A study has reported that the diagnosis period needed to be shortened, and support to parents during and after the process was necessary to reduce stress among mothers and fathers (DIFI, 2019). The above information was supported by the continuous calls from parents to reassess the evaluation methods applied at Roaa Center which many of them believed were inaccurate, insufficient, and lengthy. In addition, parents complained about the difficulty in booking appointments for the evaluation, and some of them had to wait for more than a year to get assessed, which delayed their children's school entry age (Zard, 2017). Added to that, it has been argued that support given to parents of children with disabilities is insufficient and inconsistent after diagnosis, as reported by parents (Guldberg et al., 2017).

16.6.4 Segregated Approach to Inclusion

According to the UN's committee report published in 2015, one of the criticisms of the special education model applied in Qatar was the segregated approach in the provision of services, as evident in the increase in specialised schools and workplaces for individuals with disability. The committee believed that specialised solutions could create barriers to the true inclusion of persons with disability in the local community (CRPD, 2015). Similarly, Al-Kaabi (2010) argued that the educational programmes provided to students with disabilities are not considerate of many of the challenges faced by disabled children and their parents. For instance, many programmes do not enable them to engage with the outer world and acquire the skills needed to join the labour workforce. On the other hand, the employment sector itself is not aware of their rights and needs, as well. The findings of Al-Atteyah's study (2012) confirmed that teachers believed that having separate classrooms within a regular government school was the most suitable arrangement for students with disabilities, followed by specialised centres.

Moreover, it has been reported that the system suffered from inconsistent provision of opportunities provided to the children of residents in Qatar, while expatriate parents expressed their concerns regarding the unavailability of sufficient opportunities to educate and support their children (Guldberg et al., 2017). This discrepancy was also supported by the UN committee, which expressed the necessity to provide migrant and country residents with the same attention (CRPD, 2015).

16.6.5 Cost Burdens

The rising cost of educating children with disabilities has been a primary source of financial and psychological pressures on families (Fekri, 2020). It has been argued that the financial burdens on parents of children with autism are high and need to be evaluated and reduced (DIFI, 2019). Parents have often complained about the rising costs of educating their children in specialised schools, as tuition fees surpassed 95,000 Qatari Riyals a year in some schools. They have also expressed their frustration with the additional costs they had to bear for supporting services such as consultations and training offered by private centres. They argued that the private centres overpriced their services and were only concerned with generating revenue (Fekri, 2020). For instance, the tuition fees in Awsaj Academy range between 93,000 and 122,000 Qatari Riyal (Awsaj, 2023).

16.6.6 Scattered Governance

One of the obvious concerns with the current provision model is the number of institutions responsible for offering education to students with disabilities, which include MEHE, QF, Qatar Social Work, MoPH, and private schools/centres. This has resulted in duplication of efforts and inefficiencies in resource allocation. Several experts have expressed their frustration with the lack of coordination between the education, health, and social sectors in their service delivery, which has resulted in delays in integrating individuals with disabilities into education (Bedaiwi 2018). Consequently, the above situation has affected the quality of student data generating information to the public. To support that, there have been repeated concerns regarding the quality of available data concerning students with disabilities. For instance, data about students with autism are at a preliminary stage in educational institutions and need to be improved (Guldberg et al., 2017; DIFI, 2019). Similarly, the UN's report revealed that several institutions that support children with disabilities provided insufficient information (CRPD, 2015).

16.6.7 Insufficient Professional Development for Educators

Despite the increased awareness of the rights of individuals with disability in Qatar, yet specialised knowledge regarding methods of true inclusion remains limited. For instance, the Guldberg et al. (2017) study called for improved professional development for educators working with autistic students. Similarly, the findings of Al-Atteyah's (2012) study emphasised the importance of educating school principals about the needs of students with disabilities, ways to cater for them, tools to engage and support their families, and the types of extracurricular activities they can participate in. It has been argued that teachers needed more time to achieve the learning objectives in classrooms that integrate students with disabilities, who tend to be distracted or annoyed by misbehaving classmates. Teachers also felt underprepared to cater to the needs of students with disabilities, and government schools suffer from a shortage of specialised teachers (Al-Atteyah, 2012).

16.7 Discussion and Recommendations

Weber and City (2012) argue that the inclusion of students with disabilities in education is a novice concept in the Gulf region, which makes it, in their view, a "challenging arena" for policymakers and reformers in the region (p. 85). Over the past few years, tremendous efforts have been made by Qatar and other GCC countries to develop inclusive educational opportunities for all children, yet there is no consensus among GCC countries in terms of how to integrate students with disabilities in schools

(Weber & City, 2012). This varying experience of countries in the region is an indicator that Qatar, like other countries, is at the experimentation stage, and this calls for more research to evaluate the experiences of individuals with disabilities in education and learn about ways to move forward. Moreover, policymakers need to invest more resources to achieve equitable access to inclusive education, by expanding the type of services offered to students with disabilities, making sure no child is neglected or discriminated against.

The massive investments made by the Qatari government have undoubtedly resulted in improved access to education for individuals with disabilities. Having the will and the means, Qatar needs to continue its efforts in the development of services for such students and make systematic efforts to improve practices based on evidence to support the growth and development of inclusivity in the country's education (Al-Hendawi et al., 2023). The findings reveal that there are fragmented efforts to integrate individuals with disabilities in education, which calls for more cohesion and coordination between educational, medical, and social institutions locally. Also, the findings of this chapter emphasise the necessity to focus efforts in two areas: improved governance to achieve equitable access for all children with disabilities to inclusive education, and increased awareness of means and methods for inclusion in education, employment, and society as a whole.

First, in Qatar, the educational needs of individuals with disabilities are provided by several institutions which have resulted in varying degrees of quality. However, there are models of excellent practice in the system offered by some institutions like Shafallah and Renad Academy. For that reason, collaboration between the varying agencies responsible for educating children with disabilities is necessary, for knowledge transfer and overall system growth. The above situation necessitates the development of a national strategy for educating individuals with disabilities that defines the roles of each institution and areas of collaboration. Also, it calls for the adoption of standards for quality provision that are implemented consistently by all schools and care centres. This would minimise inequities in provision and increase parental knowledge of what constitutes good education for their child. Finally, Qatar cannot achieve true integration of individuals with disabilities in education without offering education and additional support services free of charge, as the cost has been a huge barrier to inclusion.

Second, regionally, it is believed that educating children with disabilities remains a challenge; and Arab countries need to consider more equitable and inclusive opportunities in mainstream education, and slowly move away from isolated education facilities. To achieve that, generous investments need to be allocated to the development of assistive technology, curriculum, and human resources working in schools (UNESCWA, 2021). Moreover, the country needs to invest heavily in research and continuous professional development for those educators teaching and supporting individuals with disabilities. It has been reported that Qatar has struggled to expand its offering, due to the limited supply of qualified human resources, and the need to intensively train the current workforce in schools (GSDP, 2011; Zard, 2017). In addition, teacher education programmes need to include compulsory intensive courses on special education and continuous learning opportunities for professionals working in schools, to learn differentiation strategies and accommodate teaching methods (Hamid & Mohamed, 2021). Along with continuous learning opportunities for educators to develop their practices following evidence-based inclusive strategies, to ensure the provision of quality services for students with disabilities in all schools (Al-Hendawi et al., 2023).

Moreover, Al-Atteyah's (2012) study revealed that there is a need to intensify the support provided to parents of students with disabilities, and opportunities for community participation. She emphasised the need for bringing the students' issues to common awareness and engaging the wider community to address them. It has been argued that despite improved awareness, many experts called for more education for parents regarding the rights of individuals with disabilities, and ways to integrate them effectively in society (Bedaiwi 2018). Similarly, Khalifa et al. (2017) called for schools to further support parents, by providing them with more information regarding support organisations that could provide additional psychological, social, physical, and financial support to families with children with disabilities.

16.7.1 Conclusion

Since its adoption of the United Nations Convention for Persons with Disabilities DRPD in 2008, Qatar has made significant progress in improving the education of students with disabilities. The Qatari experience towards inclusion of individuals with disabilities is a unique and rich one. This chapter has provided a demonstration of some of the national strategic initiatives carried out by the core institutions, such as the MEHE, QF, and MoPH. Also, it has offered a general discussion of the dividends and the challenges in achieving equitable access to inclusive educational opportunities for all.

Bibliography

Al Atteyah, A., & Lazarus, B. (2007). "Hope in the Life": The children of Qatar speak about inclusion. *Childhood Education, 83*(6), 366–369.

Al-Arab. (2020). The completion of the al maha center for the care of children with multiple and long-term health needs this year. *Al-Arab*. https://tinyurl.com/26ece66k

Al-Atteyah, A. (2012). Requirements for integrating children with special needs with their regular peers from the point of view of their teachers. *Journal of Early Childhood and Education (University of Alexandria), 10*(1), 193–290.

Al-Hendawi, M., Keller, C., & Khair, M. S. (2023). Special education in the Arab Gulf countries: An analysis of ideals and realities. *International Journal of Educational Research Open, 4*, 100217.

Al-Kaabi, I. (2010). The socio-family and care staff opinion of services and role of children with special needs in Qatar. *International Journal of Business and Social Science, 1* (3).

Al-Meraghi, M. (2015). Establishment of Al Maha Center for Children and Youth at Al Wakra Hospital. *Al-Sharq*. https://tinyurl.com/yrmtmyw4

Al-Raya. (2021). Rehabilitating 70 government schools as centers to integrate students with disabilities. *Al-Raya.* https://tinyurl.com/y5tywy4e

Al-Said, M. (2002). *Qatar's experience in the special needs field.* Doha, Qatar: Special Needs Committee, Supreme Council for Family Affairs.

Al-Sharq. (2017). A mechanism to transform cases of people with disabilities into Ruaa Center. *Al-Sharq.* https://tinyurl.com/bd7htz3z

Awsaj. (2023). *Awsaj admission fees for the academic year 2023–2024.* Fee Structure–Awsaj Academy

Bedaiwi, S. (2018). Experts: Qatar is keen to consolidate the rights of people with disabilities. *Lusail News.* https://lusailnews.net/article/politics/qatar/04/06/2018/%D8%AE%D8%A8%D8%B1%D8%A7%D8%A1-%D9%82%D8%B7%D8%B1-%D8%AD%D8%B1%D9%8A%D8%B5%D8%A9-%D8%B9%D9%84%D9%89-%D8%AA%D8%B1%D8%B3%D9%8A%D8%AE-%D8%AD%D9%82%D9%88%D9%82-%D8%B0%D9%88%D9%8A-%D8%A7%D9%84%D8%A5%D8%B9%D8%A7%D9%82%D8%A9

Clayton, R., Amaral, A., & Shafique, A. (2021). Developing local learning ecosystems in qatar to advance equity, inclusion and social cohesion. *Report of WISE.*

Committee on the Rights of Persons with Disabilities. (2015). *Committee on the rights of persons with disabilities considers initial report of Qatar*. Ohchr.org. http://www.ohchr.org/EN/NewsEvents/Pages/DisplayNews.aspx?NewsID=16353&LangID=E

Doha International Family Institute. (2019). The well-being of families living with Autism in Qatar. *DIFI*. https://doi.org/10.5339/difi_9789927137938

Fekri, N. (2020). Al-Sharq delivers the suffering of families with disabilities due to high costs. *Al-Sharq.* https://tinyurl.com/yr9fzmhd

General Secretariate for Development Planning. (2011). *The National Development Strategy 2011–2016.* GSDP. NDS Cover 2011 (psa.gov.qa)

Guldberg, K., Ashbee, E., Kossyvaki, L., Bradley, R., & Basulayyim, A. (2017). Meeting the needs of pupils with autism in Qatar: moving forward. *Report of WISE.*

Hadidi, M. S., & Al Khateeb, J. M. (2015). Special education in Arab countries: Current challenges. *International Journal of Disability, Development and Education, 62*(5), 518–530.

Hamid, M. S., & Mohamed, N. I. A., (2021). Empirical investigation into teachers' attitudes towards inclusive education: a study of future faculty of Qatari schools. *Cypriot Journal of Educational Science. 16 (2)*, 580–593. https://doi.org/10.18844/cjes.v16i2.5636

Hassanein, E. E., Alshaboul, Y. M., & Ibrahim, S. (2021). The impact of teacher preparation on preservice teachers' attitudes toward inclusive education in Qatar. *Heliyon, 7* (9).

Hamad Medical Corporation. (n.d.1) Al Maha Center for Children and Young People. HMC. Al Maha Center (hamad.qa)

Hamad Medical Corporation (n.d.2). *School Rehabilitation Services Programs for Students with Disabilities.* HMC. https://www.hamad.qa/EN/Hospitals-and-services/CDRC/About/Our%20Services/Pages/School-Rehabilitation-Services-Program.aspx

Hukoomi (n.d.). Qatar E-Learning government. https://portal.www.gov.qa/wps/portal/disabilityqa/Services/Education/!ut/p/a1/04_Sj9CPykssy0xPLMnMz0vMAfGjzOI9DVzdDRzdDbz9_f1NDByNXXy8PU1NDIINDYEKIoEKDHAARwNC-sP1o_AqcTWAKsBjRUFuhEGmo6IiAIlcQ3o!/dl5/d5/L2dBISEvZ0FBIS9nQSEh/

Jones, G. (2020). Sensory Room in Al-Hidaya Kindergarten and School. *Experia.* Sensory Room in Al-Hidaya Kindergarten and School (experia.co.uk)

Khalifa, B. M. S., Khalil, M. F., & AlKhamra, H. A. (2017). Collaboration with families of children with disabilities in qatar: A parent's perspective. *Journal of Child and Adolescent Behavior*

Compulsory Education Law No. 25. (2001). Al Meezan—Qatary Legal Portal | Legislations | Law No. 25 of 2001Mandatory Education

Ministry of Education and Higher Education (n.d.). *Special Needs and Gifted students.* MEHE. https://www.edu.gov.qa/en/pages/pubschoolsdefault.aspx?itemid=150

Ministry of Education and Higher Education. (2022). *Annual Statistics of Education in the State of Qatar.* MEHE. https://www.edu.gov.qa/Documents/HigherEdTracks/Annual%20Statistics%202020-2021_%20%D9%85%D8%B9%D8%AA%D9%85%D8%AF[2083].pdf

Ministry of Foreign Affairs. (2011). *The initial report of the State of Qatar on the implementation of the agreement International Committee for the Rights of Persons with Disabilities.* CRPD.C.QAT.1_ar.doc (live.com)

Ministry of Foreign Affairs. (2019). *The Qatari delegation to the United Nations organizes a high-level session on occasion of World Autism Awareness Day.* https://ny.mission.qa/en/media/news/detail/2019/04/17/qatar's-permanent-mission-to-un-holds-high-level-session-on-occasion-of-world-autism-awareness-day

Ministry of Foreign Affairs. (2023). *The State of Qatar organizes a high-level dialogue on the occasion of World Autism Awareness Day.* https://tinyurl.com/3kt6ne44

Murad, E. (2019). Qatar trains its cadres to deal with individuals with autism. *Al-Jazeera net.* https://tinyurl.com/3zd6mche

Organizational Structure of the MEHE Emiri Decree No. 9. (2016). https://tinyurl.com/3uwep7hk

People with Special Needs Law No. 2. (2004). Al Meezan—Qatary Legal Portal | Legislations | Law No. 2 of 2004 in respect of People with Special Needs

Qatar Association of Rehabilitation of Special Needs. (n.d.). *About the QARSN.* https://qsrn.org

Qoronflesh, M. W., Essa, M. M., Alharahsheh, S. T., Al-Farsi, Y. M., & Al-Adawi, S. (2019). Autism in the Gulf states: A regional overview. *Frontiers in Bioscience-Landmark, 24*(2), 334–346.

Renad. (n.d.). *About Renad Academy.* Home—Renad Academy

Saber, H. (2019). Al Maha is a national project serving children on the autism spectrum and people with disabilities. *Al-Sharq* https://tinyurl.com/fhrfdryf

Supreme Education Council. (2009). *Additional Educational Support Needs. A Pack of Policies, Guidance Documents and Support Materials for Schools.*

Shafallah. (n.d.). *About Shafallah center.* AboutShafallah

Tawalbeh, K. (2023). Al-Hedaya special-needs school helps students with disabilities to learn fast. *Tribune.* Al-Hedaya special-needs school helps students with disabilities to learn fast—Read Qatar Tribune on the go for unrivalled news coverage (qatar-tribune.com)

United Nation's Economic and Social Commission for Western Asia. (2021). *National strategies for individuals with disabilities in the Arab region.* UNESCWA. https://e-inclusion.unescwa.org/sites/default/files/resources/mapping-national-disablility-strategies-arab-region-arabic.pdf

Weber, A. S., & City, E. (2012). Inclusive education in the Gulf Cooperation Council. *Journal of Educational and Instructional Studies in the World, 2*(2), 85–97.

Zard, F. (2017). Ruaa raises a crisis for parents of people with special needs. *Al-Arab.* https://tinyurl.com/2s4a4h7j

Chapter 17
Public Universities and Educational Sustainability in Qatar: A Case Study of Qatar University

Cesar Wazen

Abstract This book chapter explores the transformative journey of public universities, with a focus on Qatar University, in the context of Qatar's evolving higher education landscape. It analyses the historical, social, and economic aspects that have shaped the trajectory of Qatar University, emphasising its pivotal role in fostering knowledge, community engagement, and leadership. The chapter highlights the institution's commitment to gender inclusivity, outreach initiatives, and the development of a knowledge-based economy, in alignment with Qatar National Vision 2030. Furthermore, it examines Qatar University's strategic positioning within the emerging higher education hub in Qatar, emphasising its collaborations, partnerships, and contributions to research and innovation. The chapter concludes by considering the future of public universities in the region, and their vital role in shaping the educational landscape and supporting national and global initiatives.

17.1 Introduction

In the following chapter, we highlight the pivotal role of Qatar University in shaping the educational landscape of Qatar and beyond. Established in 1977, QU has not only been a benchmark for academic excellence but also a catalyst for cultural enrichment, scientific advancement, and community engagement. QU stands out as the country's first, and for a long time, only national university. Its establishment marked a milestone in Qatar's commitment to providing higher education opportunities for its citizens and large expatriate population. Prior to the university's founding, Qatari students often had to pursue higher education abroad. The establishment of QU provided local students with access to higher education within their own country. After detailing the growth of QU and its evolution over the years into becoming a comprehensive institution, offering a wide range of programmes, by delineating the university's history, we will also tackle its unique contributions

C. Wazen (✉)
Director of International Affairs Office, Qatar University, Doha, Qatar
e-mail: cesar.wazen@qu.edu.qa

E. Zaidan et al. (eds.), *The Future of Education Policy in the State of Qatar*, Gulf Studies 21, https://doi.org/10.1007/978-981-97-9667-0_17

to research and innovation, and its profound impact on the local community. We will also examine how QU has contributed to the implementation of QNV2030, by developing the Qatari workforce, nationalising education, and fostering community engagement, all within a very active international collaboration perspective, boosted by the educational system created by the State of Qatar. By examining the Qatar University journey, we gain valuable insights into the broader theme of the "Future of Education in Qatar".

17.2 Foundation and Modernisation

Qatar University (QU) started out as a teacher training college for men and women in 1973 and remained so until its inception as a university in 1977. Although "young" in age, it has a history full of reform and transformation, making it able to adapt to and adopt the State of Qatar's ambitious soft power politics—higher education playing an essential part in this. This process led the university to quickly develop from preparing trainees for school teaching to a comprehensive university with 11 colleges, offering "the widest range of academic programs—50 Bachelors, 47 Masters,—in Qatar tailoring them to meet the needs of Qatari society", as it presents itself on its website. Along the way, QU benefited from the research funds established by the state to develop into one of the fastest rising universities in the region in research. Bolstered by its state-of-the-art Research Complex, with 18 research centres of excellence. The 500 collaborative projects in over 130 countries have led the university all the way to the top of the world in indicators related to the international research network, according to some rankings.

For a long period, QU was the only national university in the State of Qatar. It continues to be the largest and most prominent university in the country, despite the ever-changing higher education landscape in Qatar, now comprising 10 public and six military higher education institutions. Qatar Foundation universities add nine more choices for higher education to the Qatari community, and nine private institutions complete the new landscape of higher education in Qatar.

In that sense, the future of higher education in Qatar is being written as we go along, with the main strategy set as early as 2008, with the adoption of Qatar National Vision 2030, and the focus on shifting to a knowledge-based economy, where education in general and higher education in particular are essential components (Table 17.1).

17.3 Main Events Influencing the Recent History of QU

Further to the reform initiated in 2003, a set of significant developments throughout the years have influenced QU significantly. In 2008, HH Emir Hamad bin Khalifa Al Thani announced the adoption of Qatar National Vision 2030 (QNV) to guide planning decisions. It set the goal of moving from a "carbon-based" economy to a

Table 17.1 Evolution of the vision and mission of the university (*Source* Baker, D. (2024). Making Qatar University (1973–2018) an informal history, Qatar University Press)

Strategic Plan	Vision	Mission
2006–2009	Qatar University seeks to be a model national university that offers a high-quality, learning-centred education	To improve the quality of Qatar University graduates
2009–2012	Qatar University shall be a model national university in the region recognised for high-quality education and research and for being a leader of economic and social development	Qatar University (a) prepares competent graduates by providing high-quality programmes; (b) conducts research that addresses contemporary regional challenges and advances knowledge; (c) contributes actively to the needs and aspirations of society
2010–2013 (updated) 2013–2016	Unchanged	Qatar University is the national university of higher education in Qatar. It provides high-quality undergraduate programmes that prepare competent graduates, destined to shape the future of Qatar. The university community has diverse and committed faculty who teach and conduct research, which addresses relevant local and regional challenges, advances knowledge, and contributes actively to the needs and aspirations of society
2018–present	To be regionally recognised for distinctive excellence in education and research, an institution of choice for students and scholars and a catalyst for the sustainable socio-economic development of Qatar	Qatar University is the national institution of higher education in Qatar. It provides high-quality undergraduate and graduate programmes that prepare competent graduates, destined to shape the future of Qatar. The university community has diverse and committed faculty who teach and conduct research, which addresses relevant local and regional challenges, advances knowledge, and contributes actively to addressing the needs and aspirations of society

more diversified economy, and then progressively to a "knowledge-based economy". The vision placed the development of human capital—and thus education—at the forefront of government and regulatory activities.

The QNV was followed in 2011 by the National Development Strategy 2011–2016.

In 2012, the National Research Strategy set 23 goals and 76 objectives to enhance research and development critical to the development of a knowledge-based economy. The aim of the research strategy was to make Qatar an international centre for research and development excellence and innovation and to enlist Qataris in the research efforts.

The year 2012–13 witnessed a turning point in the history of Qatar University. In that period, changes in government policy profoundly influenced the university, notably the introduction of the National Research Strategy, which opened the way to major funding for research, and the Supreme Education Council's resolution that the university should revert to Arabic as the language of instruction in all fields other than Science, Engineering, and Pharmacy and should exempt students on Arabic-language programmes from Foundation requirements. In addition, a review of the university's development was initiated by the SEC. To match these changes, the university intensified its commitment to graduate studies, by launching an array of Master's programmes and its first PhD programme. It also opened or planned to open new research centres. With the cancellation of the English proficiency requirements, the university had to accommodate a significant influx in enrolment, forcing the university to hire hundreds of additional Arab-speaking faculty members and, thanks to the need to restructure what remained of the Foundation Program, to reconsider how to integrate a flood of poorly prepared students onto degree programmes, some of whom were old returning students who had not been able to complete their degree years before. The First Year Experience programme was developed to cater for this type of, student, and large new facility projects were launched for Research Affairs, Engineering, and Student Affairs, alongside additional classrooms and student housing. Significant change affected the university at all levels.

While other colleges were being inaugurated along the way, the foundation of the College of Medicine in 2015 was another turning point in the history of Qatar University, as well as within the health sector in general in Qatar. Further to the role it was to play on the national level during the pandemic in 2020, it was the first major step in the nationalisation of health sciences in the country. It also adopted a different approach to the American-based medical education already running in Education City, offering a problem-based learning European model, which was a revolutionary step in the history of the country. The university also faced some challenges in its young history that were imposed by geopolitical matters, such as the blockade on Qatar in 2017, and which extended to more than three years, and after that there followed the COVID-19 Pandemic.

While some of its scholarship holders abroad, and exchange students and faculty were affected by the blockade, the university adopted a policy that was in line with the State of Qatar policy of non-retaliation towards nationals of the blockading countries. During this diplomatic crisis, Qatar University ensured that students from nations implementing a blockade against Qatar were able to accelerate their academic requirements (like the necessary exams) so that they could evacuate prior to the blockade. Moreover, the number of students from these countries in fact increased for Fall 2017 up from 1,787 to 1,837.

The COVID-19 Pandemic provided Qatar University with the opportunity to showcase two important areas of strength: 1. The swift shift to online learning (within one week), thanks to the infrastructure it had invested in as part of its new strategy; 2. The expertise it was able to bring to the table to support the efforts of the Ministry of Health in curbing the effects of the pandemic through its qualified personnel at the newly established College of Medicine and Health cluster, as well as the rich network of international partners that it was liaising with and exchanging information with for protocols. The most striking example of the latter was the collaboration with Peking University, with a number of staff members involved as first respondents in Wuhan, months before the pandemic hit Qatar. Through the Chair of the State of Qatar for Middle East studies, established through an endowment by the State of Qatar in 2014, intervention protocols used by the Chinese team were compiled and translated into English and Arabic, and provided to QU. In those times, this time-consuming activity proved to be vital, and literally life-saving, at the national level.

More recently of course, the FIFA World Cup 2022, a game changer for the whole country and even for the Arab region was successfully hosted by the State of Qatar and again, QU was not on the sidelines. Once again, QU contributed to this national success on all levels. Its students were involved as volunteers in the organisation; its researchers were involved in investigating crowd management and developing intelligent crowd management and control systems. The systems included multiple components for crowd counting, face recognition, and abnormal event detection—which would all help to control and manage crowds at Qatar 2022. Moreover, stadium design research resulted in the development of the cooling technologies used in the stadiums, a technique that had not been attempted in open-air structures before. In terms of facilities, tackled in another part of this chapter, two of the teams participating in the tournament, including the tournament winners, chose the QU campus to host their teams. If nothing else, these contributions were proof of the state-of-the-art facilities and expertise QU provides and they represent the seal of approval for the appreciation of the excellence to be found in QU's young history.

17.4 Nationalisation of Education

The main contribution to the nationalisation of education by Qatar University remains the fact that it caters for the largest number of Qatari (and non-Qatari) students among higher education institutions in Qatar (ranging in between 20 and 30,000 students per year) representing 60% to 70% of the total number of enrolled students. In terms of graduates, QU has provided 55% to 60% of graduates in recent years. These students are all graduates of American accredited programmes, yet they have been prepared with Qatari values throughout their studies, with the university creating a set of courses in the First Year Experience Program, Core Curriculum Program, and Arabic Program ensuring that any graduate from the QU, while holding an accredited American system based degree, has sufficient knowledge of the culture, language, and values related to Qatari society.

On top of creating opportunities for nationals and residents to pursue higher education at the highest level in Doha, the government has provided funding for nationals to pursue their studies abroad, in an attempt to speed up the process of preparing the next generation of leaders, while study programmes were still not offered completely in higher education institutions in Qatar. Initially started in the 1960s as the only option for students willing to pursue higher education studies, the programme was transformed, as of 1977, into an opportunity for young Qataris to study abroad. The range of programmes being wider than at QU at that time, it also enabled them to be immersed in different cultures, which positioned them as perfect fits for possible roles in a government which plays a major role in conflict resolution globally.

Again, the national university benefited from this generous provision of scholarships to build its own capacity in teaching assistants, professors, and researchers who have graduated from the top universities worldwide and joined its faculty, to groom the next generation of leaders and cadres in their home country. As one of our former presidents put it, even if the university ends up losing these highly qualified nationals to other entities in the country, it considers it as part of its mandate to provide the nation with the next generation of leaders.

This sums up the role the university intends to play and is mandated to play on the national level: bridging the gap to help the State of Qatar fulfil its national vision as early as 2030.

The main challenge remained, however, the low attractiveness of the teaching profession, especially among young people. This worldwide issue is more acute in a country like Qatar, where critical mass is missing, and needs have to be filled in more strategic positions, especially for male students who are needed in the police or military. The university's skewed gender balance (3:1 to the favour of women) shows one of the main roles it is meant to play: to empower young women coming from traditionally conservative households to attend a world-class university and receive an accredited degree only a few miles away from home.

On a different level, and as mentioned earlier, the nationalisation of the health profession in Qatar was significantly boosted by the establishment of the College of Medicine at QU in 2015. In fact, in the whole Gulf region, the issue of nationalisation of the health profession is a matter of national concern. The countries in the Gulf usually outsource this sector or rely on expatriate professionals to fill the gap. Alongside a high turnover, this also proves to be inadequate in that profession, specifically due to the language and cultural barriers affecting the communication between patients and medical staff. Qatar University bridged the gap that, despite running successfully in Education City, especially in terms of research and advancement of the health sector in Qatar, Weill Cornell was not able to cover. Another sign of the complementarity between QU and QF-based universities is that they worked hard together in order to avoid redundancy.

As far as research is concerned, the centres of research at Qatar University operate on the basis of a set of research priorities linking any research to the Qatari society's needs. Those priorities and focus areas were regrouped under FOUR research pillars: Energy and Environment, Health and Biomedical Sciences, Information and Communication Technology, and Social Sciences and Humanities. The core research areas were selected to address the national needs of the State of Qatar, in alignment with Qatar National Development Strategy 2 (QNDS-2 2018–2023) and Qatar National Vision 2030 (QNV 2030). Many examples of important research efforts aimed at the betterment of the Qatari society and catering to its needs came out of these priorities, one of which is the Workers Welfare Index Social and Economic

Survey Research Institute (SESRI,) a unique survey-based tool that measures and monitors the impact of the Qatari government's policies on guest workers. Now in its fifth wave, the survey—which was administered in more than seven languages-—evaluates several aspects of the working and living conditions of guest workers, including safety and security at working sites and in living compounds; human rights and labour rights; finance and remittances; and treatment by their employers. In addition to assessing and tracking workers' welfare, the index identifies areas for improvement to guide policy formulation.

Another example tackles the nationalisation of food processing, as part of food security policy, especially triggered by the impact of the 2017 blockade on Qatari society and the threats it held for its population. The Agricultural Research Station, a specialised research farm and centre for agricultural research was established by QU in the north of Qatar. The station develops innovative agricultural techniques, including developing biofertilisers for the improvement of plant productivity, salt tolerance, and fruit quality; development of hydroponics and irrigation systems; and the application of integrated agricultural pest management and plant disease control. The station also develops and creates organic and sustainable crops and produces sustainable organic fertilisers to increase and preserve agricultural soil fertility. They are also constructing an on-campus Vegetable Factory Pilot Plant with the help of a Japanese company.

Qatar University, therefore, has strived to ensure a holistic approach to the nationalisation of education as a broad term, including academic, research, student life, and facilities, and has successfully played a part in the success of Qatar as a nation, to face up to the threats of geopolitical and health crises, but also to get a return on the significant investment it has placed in education, and research and development.

17.5 Research and Innovation

The Qatar National Research Fund (QNRF) was established in 2006, and given the responsibility "to monitor and direct the research activity and measure outcome impact". The QNRF developed four programmes to support research and deepen the research culture in Qatar: the National Priorities Research Program (NPRP) and the Undergraduate Research Experience Program (UREP), both of direct relevance to the university; and the Secondary School Research Experience Program and the Young Scientists Research Experience Program that the university linked to by organising its own students' outreach initiatives such as the Young Scientist Centre (QU-YSC), a scientific hub that trains and empowers the next generation in STEM subjects, through a range of programmes, making STEM accessible to high school students. Another initiative related to these programmes is the QU Student grant programme, which aims to foster the research culture and develop research capacity at Qatar University, by enhancing the research experience of our undergraduate and graduate students through different research projects.

The National Priorities Research Program (NPRP) supports research projects through grants of $20,000 to $350,000 per project, for a period of up to three years. There is an Exceptional Proposals category for larger or longer projects of special importance to national goals. The government made a commitment to provide up to 2.8% of GDP for research. As one of QU's "key performance indicators" is to "conduct research that addresses contemporary challenges and advances knowledge" through projects that are "in alignment with national research priorities", the university strengthened its research infrastructure, and made a commitment to "establish institution-wide research centers/units for excellence in research that supports research priorities". Units were created, helping researchers prepare proposals and identify sources of funding, review budgets, and ensure compliance with university and sponsor policies and procedures.

By 2014–15, QU had already developed the following centres: the Biomedical Research Center (QUBRC), the Center for Advanced Materials (CAM), Environmental Studies Center (ESC), Gas Processing Center (GPC), KINDI Lab for Computing Research, Qatar Road Safety Studies Center (ARSSC), and Qatar Mobility Innovations Center (QMIC). The Research Complex opened in 2014 and includes the Laboratory Animal Research Center (LARC) and Central Laboratories Unit (CLU), which provide analytical and technical support and equipment for the research and testing needs of researchers. The laboratories for research purposes were accredited by the Association of Laboratory Accreditation (A2LA) in 2010, and renewed in 2014; the CLU also sustains its ISO 17025 accreditation status.

In 2014, the university approved a five-year "Road Map: Advancing Research for Qatar's Future" prepared by the University Research Council, that centred on four priority areas: (1) Energy, Environment, and Resource Sustainability; (2) Social Change and Identity; (3) Population, Health, and Wellness; and (4) Information, Communication and Technologies (ICT). It set the goal of making QU a regional leader in research. Total research funding to the university rose to QR154.8 million in 2013–14, with the bulk going to Engineering (QR81 million), Arts and Science (QR32.9 million), the QU Wireless Innovations Center (QR7.9 million), and Pharmacy (QR4.3 million). This was generated by grants, sponsorship, and contracts with industrial partners, including Qatar Petroleum, RasGas, Exxon Mobil, and Qatar Shell.

Investment in facilities went even further with the purchase of a state-of-the-art research vessel, built by one of Spain's most prestigious shipyards, with the QU president stating that "Janan is not just a vessel, it launches a new area of research that is expected to attract Qatari scholars who will conduct advanced studies in Qatar's waters and those of the Gulf". A statement that summarises the national importance of this step but also the ambitious scope of its work, mainly driven by the State of Qatar realising early enough the negative effects of its gas extraction activities on marine life off-shore. The evolution in the number of grant submissions has also seen a significant rise both in scope and numbers. While dominated initially by Pharmacy, Engineering, and Arts and Sciences, the total grants submitted during the 2020–2024 cycle show a more balanced distribution across colleges (Fig. 17.1).

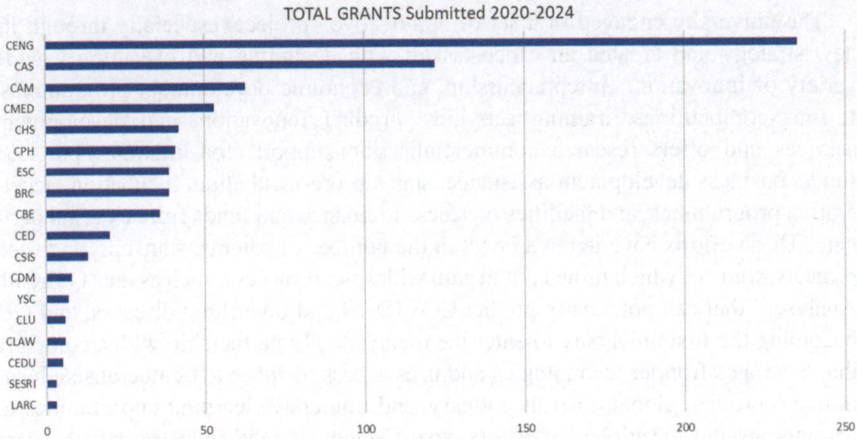

Fig. 17.1 The total grant submissions made to Qatar University over the period 2020–2024

Faculty members' success rates in securing grants from the National Priorities Research Program were highest in Law and the Social and Economic Survey Research Institute (SESRI). Under the Undergraduate Research Experience Program (UREP), the university received grants for more than 1,200 students over six years. As the following chart, Fig. 17.2 shows the number of indexed publications grew steadily, in part due to funding, and in part to scholarly collaborations. Collaboration with Education City universities and Qatar Foundation affiliated entities has grown, as summarised in the graph below in the same period (2013–2022), showing that the top 10 collaborators with QU on co-authored publications include five universities from Education City aside from QF itself and Aspetar (Source SciVal).

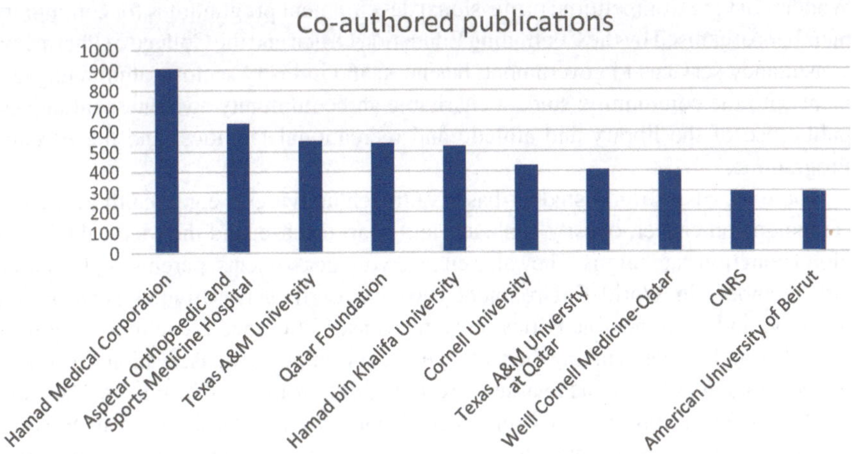

Fig. 17.2 Publications jointly between QU, other universities, and Qatar Foundation

The university engaged in a set of "innovative" projects especially through its new strategy and created an office tasked with designing and executing a wide variety of innovation, entrepreneurship, and economic development programmes. It runs competitions, training activities, product innovation, and development services and offers research commercialisation support, consultations, pre-seed funds, business development assistance, start-up pre-incubation, incubation, acceleration programmes, and facilities of access to finance and funds from external partners. These efforts have led to a boost in the number of patents, start-ups, and seed projects, some of which turned out to gain widespread success, such as the "Q Breath Analyser" that can potentially predict COVID-19 and other lung diseases, and QU becoming the first university to enter the metaverse. In partnership with a company that leverages frontier technologies and uses access to space to democratise education and provide global, interdisciplinary, and immersive learning opportunities to learners around the globe, educators, space scientists, and industry experts were brought together to deliver interdisciplinary curricula, and prepare students for the future of work.

17.6 Community Service and Gender Inclusivity

From its inception, Qatar University has evolved along with the society it serves. The first strategic plan already tackled the need for community service activities. The resulting initiatives included the formation of the QU Community Service Committee (CSC), the QU Alumni Association (QUAA), and the Center for Volunteerism and Civic Responsibility, plus assuming responsibility for managing the Time Bank Global Change Makers project. Other initiatives include high school outreach programmes such as Gasna, Al Bairaq, Life is Engineering, Computing Contest, and Wooden Bridge Competition; professional development programmes for community members organised by the Continuing Education Office and the College of Pharmacy; consultancy services to government, business, and industry sectors; alumni engagement with the community; student charitable and community awareness initiatives; public use of the library and athletic and recreational facilities; and the Parents' Programme.

There are also several student-based outreach activities—e.g., health awareness campaigns on cancer, obesity, diabetes, and heart disease, and the Medical Clinic's blood donation campaigns. They also engage younger students, parents, and community members in World Environment Day and Earth Day activities such as tree planting, and awareness activities on water conservation and recycling. They have played an active role in taking the message of environmental conservation to schools, to encourage K-12 students to assume responsibility for the environment. They also led "Let's Read" programmes for World Book Day and participated in Reading Together for Qatar, a volunteer tutoring programme at the College of Education that provides K-12 students with literacy tools.

Student participation in campus life was further enhanced by the establishment in 2010 of the QU Student Representative Board—the first such student body at QU. It provides a forum for connecting students and university administration (through elected student representatives from the colleges) in a manner that enhances the alignment of goals and common understanding. Its establishment is in line with QU's commitment to engage students in active participation in university affairs and provides an opportunity for them to experience and learn the essentials of involvement in a democratic process.

The Student Activities Department organises an annual Charity Fair, and the students in the CAS Leadership Course organise event days for the elderly, orphans, and hospital-bound young cancer sufferers and provide meals for low-skilled labourers, tea staff, janitors, and gardeners at QU (among others). QU also contributes to celebration events held on National Day (December 18), as well as the Emir-decreed National Sports Day. Its facilities, which include the Aquatic Complex, gymnasium, and stadium are venues used by various public and private institutions in Qatar. These include Shafallah Center for the Disabled, Qatar Association for Rehabilitation of Special Needs, Qatar Paralympics Committee, and the Internal Security Force (Lakhwiya) and Qatari armed forces for their routine training, especially diving.

The university has a vibrant internal intellectual life, with lectures and discussions on topics such as the war in Yemen, Qatar's international role, the global balance of power, the architecture in Doha, and environmental and ecological challenges—and these are generally open to the public, and many are attended by students from particular programmes. In general, however, the university wishes to see a stronger engagement of students in activities on campus and has made that one of its priorities in planning for the future.

With a ratio historically close to 3:1 in favour of female students, Qatar University strives to prepare its women population for the job market. While being proud to have the majority of its student body composed of females, the university is well aware of the fact that quantity does not provide for equal chances, and hence strives to make sure the female students are well equipped to achieve their potential in their professional life and to provide for gender balance at the leadership levels countrywide. For that purpose, QU has undertaken many steps to promote gender inclusivity in higher education.

Starting with research, ranging from Female Labor Force Participation to entrepreneurship, Qatar University engaged through its Social and Economic Survey Research Institute (SESRI) in studies aiming at improving female labour participation, not only at the university but in Qatar at large. Using nationally representative surveys, they are studying how micro-level factors such as age, education, and marital status affect women's participation in the workforce in Qatar (Lari et al., 2022), and how public attitudes affect this participation. The aim of this research is to inform policymakers how state-directed initiatives can encourage women to participate in the labour market, but it also calls for the enactment of policies to facilitate a rapid increase in the number of women employed in Qatar. At the level of Entrepreneurship for Women, researchers at the College of Business and Economics are working on creating a platform to promote entrepreneurship for women in Qatar (Qatar University Homepage). The project has explored ways of creating opportunities to stimulate and support Qatari women's engagement in action-oriented entrepreneurial activities, such as practical training, international collaboration, knowledge exchange, and shared learning to foster entrepreneurship among Qatari women, in order to enable them to initiate new business start-ups. A special issue of *Emergent Materials* (EMMA), a multi-disciplinary journal of Qatar University (QU) focused on women in nanotechnology, with the aim of highlighting the work of research groups led by female researchers around the world. The issue included distinctive projects in the field of nanotechnology, written and supervised by women.

Outreach efforts are also essential at QU, to attract women to specific fields, by celebrating the success of prominent ones and encouraging future generations to join these fields. Presentations with distinguished distinct alumni shedding light on their achievements, the Cyber Women initiative that aims to inspire and encourage females to engage with STEM subjects, the road show of the College of Law documenting the stories of female pioneers in law, development, international law, and diplomacy, the International Day of Women and Girls in Science celebrated annually, bringing together prominent women and honouring female researchers, are just a few examples of the significant outreach efforts aiming at empowering successful women, and providing guidance to young women engaging in their university life.

Students also play an integral part in this effort, and regularly participate and win awards related to recognising young scientists (L'Oreal Prize). The university also organises its own women's entrepreneurship week, highlighting women entrepreneur success stories, and providing a platform for students to discuss their concerns with them. The university also pays a lot of attention to the needs of its own female population, providing on-campus state-of-the-art childcare and extended nursing hours.

17.7 Knowledge Economy

Following the early days of its reform, and preceding the announcement of QNV 2030, Qatar University placed itself as "the principal reservoir of knowledge and expertise" in Qatar, so as to provide the country with "specialists, technicians, and experts in various fields, and equip citizens with knowledge and advanced research methodologies". Being the sole provider of higher education degrees in the country for a long time, the university had been tasked with the responsibility of providing the cadres of a highly ambitious country, that has proved to be a key player in the region and worldwide in recent days. Those achievements were framed in the Qatar National Vision 2030 document, and Qatar University's reform efforts could not have come soon enough to contribute to this effort's four pillars: Human, Social, Economic, and Environmental development. Following its launching in 2008, QU's first strategic plan 2010–2013 (Preceded by an academic plan 2006–2009) was given a raison-d'etre, and clear guidance on the key performance indicators needed at the national level. The concept of "excellence" appeared in the foreword written by then President Dr. Al-Misnad, who stated that changes were "brought about by the dynamic environment in which today's universities must operate, and by opportunities to enhance our contribution to Qatar and its vision of 2030, as well as all its stakeholders". The next cycle of the strategic plan (2013–2016) moved to clearly position QU vis-à-vis QNV 2030, stating that: "As the largest and most comprehensive university in Qatar, QU contributes actively to the realization of the Qatar 2030 National Vision, which envisages a change from the current carbon-based economy to a knowledge-based economy characterized by innovation, entrepreneurship, excellence in education, a world-class infrastructural backbone, the efficient delivery of public services, and transparent and accountable government".

This statement acknowledges the fact that QU is no longer alone on the market; nevertheless, that it remains the leading source of expertise to help the country's efforts in moving into a knowledge-based economy. The KPIs (key performance indicators) have started to show collaborations, initiatives, and activities that aim to support the country's transition towards a knowledge community. To state but a few, to modify the facilities so as to accommodate students of varying abilities and competences, provide academic support services, invest in personal and career advice support systems, invest in a new library and IT facilities, prepare competent graduates by involving undergraduate students in research, and boost research activities across all academic units.

It is worth noting that, alongside these efforts, QU expanded its programmes offered to include 50 Bachelors and 47 Masters,tailoring them to meet the needs of Qatari society. The university now has eleven colleges expanding mainly in the areas of health and clinical medicine. That expansion is mainly the result of the "Transformation" effort undertaken by President Al-Derham's administration since 2018, stating in its strategy documents the fact that QU is now "a power house of intellectual and academic achievement", whose goal is to deliver the best possible education to its students, to energise them to use their talents, and to inspire them to achieve their potential. The seven goals of this "Transformation" strategy aim at creating graduates who are job creators rather than job seekers, a set of skills that is right in line with the country's efforts to be pioneers in many endeavours in the region.

The strategy included a set of new initiatives that would shape the transition to a knowledge-based economy even further, mainly in areas of entrepreneurship, innovation, digital education, and e-learning, in recognition of its importance and its impact in this era of technical and technological development. The purpose of these initiatives was to move the educational process from traditional methods to learning based on understanding, research, innovation, thinking, and experimentation. An effort that proved to be vital in the early days of the COVID-19 Pandemic, where the transition to online teaching went smoothly, but more importantly, the expertise provided by QU's health cluster to the pertinent ministry ensured that proper advice was given as to what was then a threat to humanity. This test of the readiness level QU had reached, despite its young age as an academic institution, proved to be a successful one, thanks to its personnel, but also revealed the extent of its international network of partners, who were willing to collaborate and share their experiences. The transformation of the university from a traditional comprehensive institution into one that wants to create job creators even led it to establish the Qatar University Holding Company, with the aim of embracing the creative ideas of faculty members, students, and researchers, within the framework of supporting Qatar's orientation towards a knowledge-based economy, as then President Al-Derham stated.

On the role of the university in Qatari society, and the expectations of the government of Qatar, HH Emir Tamim wrote that "humans are the most important aspect of building a nation and a country's biggest investment. Qatar has invested in its youth and with them it will develop further with the aid of science". The Emir expressed appreciation for the role of youth and the university in the development of the country, thereby acknowledging the contributions of QU. In fact, with more than 40,000 alumni, of whom nearly 150 occupy high-level government positions, and leading the country's main corporations and institutions, QU has played a major role in supporting the transition to a knowledge-based economy.

17.8 Positioning in Qatar's Higher Education Hub

Here we discuss the transformation of Qatar University into a prominent higher education hub and highlight its strategic partnerships, collaborations with branch campuses, and other institutions that have contributed to its unique position within the higher education landscape.

While Qatar may not have as many universities as some other hubs of higher education, its focused efforts, strategic investments, and commitment to academic excellence have allowed it to establish itself as an emerging hub for higher education and research in the Middle East.

The State of Qatar has made significant investments in higher education, research, and academic infrastructure, establishing itself as an emerging higher education hub in the Middle East.

Key elements contributing to Qatar's status as a higher education hub include:

- Education City, an education complex hosting the branch campuses of numerous prestigious international universities with the main target of attracting students from the region and around the world.
- The Qatar Foundation, which oversees Education City, has prioritised research and innovation. Its state-of-the-art research centres like the Qatar Computing Research Institute, Qatar Environment and Energy Research Institute, and the Sidra Medical and Research Center foster cutting-edge research and development.
- Scholarships and Opportunities: Qatar offers numerous scholarships to its nationals to pursue higher education studies abroad, but also opportunities for international students, researchers, and academics to pursue higher education and research in the country. This includes the Qatar Scholarship Program and

other initiatives aimed at attracting talent to Qatar, especially the Qatar National Research Fund.

- Investment in Infrastructure: Qatar has invested heavily in educational infrastructure, providing state-of-the-art facilities and resources for students, researchers, and faculty.

Qatar University positions itself centrally in this ecosystem. In fact, while the university had started its reform before the establishment of Education City, it has benefited from the support and initiatives of the Qatar Foundation, which has played a significant role in the development and advancement of higher education in Qatar. This is clear in two main areas: joint research initiatives and academic exchanges, fostering a culture of innovation and knowledge sharing, and the Qatar Foundation's focus on community development that aligns with Qatar University's commitment to community engagement and outreach. Both institutions have worked together on projects that have a positive impact on the local community.

The area of International Collaboration, while considered to be a by-product of the latter two is worth highlighting. Indeed, through the Qatar Foundation's initiatives, Qatar University has the opportunity to collaborate with renowned international universities and institutions located within Education City, enhancing its academic and research capabilities. This has led the university to top the charts on the International Outlook indicators across many rankings. The Qatar National Research Fund, established in 2006, had a clearly defined vision: "To enable research and development excellence in Qatar in order to achieve a knowledge-based economy". This encouraged international collaboration and could be said to be the main reason why Qatar University topped the list of the most international universities for many years and across rankings. "The generous funds provided to Qatari and international researchers encourages international collaboration and, as a result, Qatar tops the charts in international collaboration", said Dr. Al-Derham, and this was confirmed by the number of international research collaborations soaring to a record high 450 research projects with 319 collaborators, in the years 2010–2015. The ever-increasing research efforts have resulted in 3,200 co-authored publications, from 1,093 collaborating institutes, with a 246% increase. These numbers, while slowing down in the next few years due to budgetary issues, still make Qatar University one of the best proponents of international collaboration. As for its number of international students, QU hosts nearly 30% of them, which makes it one of the most international public universities in the region, and one of the main attractions for students from the region, despite not having a mandate to attract foreign students like the branch campuses.

Through its articulation agreement, Qatar University likes to play the role of a Big Sister to the universities in town and provides opportunities for students to enrich their academic experiences through cross-registration and transfer. Agreements in place with branch campuses as well as the Community College of Qatar provide for circulation between institutions and ensure complementarity and healthy competition.

Table 17.2 Comparison of Qatar University and other institutions for scholarly output and citations in the years 2013–2022

Institution	Scholarly output	Scholarly output (Growth %)	Citations	Value of awards (USD)
Qatar University	19,654	311	419,274	6,301,833
Hamad bin Khalifa University	7,221	567.9	207,965	290,028
Texas A&M University at Qatar	6,124	15.7	146,666	0
Weill Cornell Medicine-Qatar	3,794	254.3	146,097	0
Sidra Medical and Research Center	2,207	1,218.5	61,792	0
Qatar Science & Technology Park	350	−42.5	7,450	0
Aspire Academy	323	−39.5	10,896	0
Georgetown University in Qatar	234	68.8	1,820	0
University College London, Qatar	129	−50	1,651	0
Virginia Commonwealth University in Qatar	85	250	648	0
Qatar Cardiovascular Research Center	72	−80	1,096	0

Table 17.2, retrieved from Elsevier's SciVal for the period 2013–2022, shows the importance of both the scholarly output and the number of citations QU receives vis-à-vis other institutions in town.

Furthermore, in terms of number of students, Qatar University hosts more than 90% of the higher education student body in Qatar, making it safely the major contributor and beneficiary of the educational hub Qatar is becoming, in the region and worldwide, both on the academic and research levels.

17.9 Findings and Conclusion

As a new dawn starts to unfold on the university with the appointment of a new President, Dr. Omar Al-Ansari at the reins, Qatar University is set to keep on playing a major role in shaping the future of education in Qatar. That is not only due to the levels of flexibility, resilience, and capacity to adapt that it has shown over its young but dynamic history, discussed earlier in this chapter, but also due to the nature of the new leadership. In fact, Dr. Al-Ansari has played an essential role in history, be it in the Reform era or the Transformation era, having served, among other duties, as Vice President for Student Affairs under Dr. Al-Misnad's administration, and Vice President for Academic Affairs under Dr. Al-Derham's. With a professional career spanning over two decades, and a product of the university itself, and its scholarship programmes, he has shown an unwavering commitment to excellence, a diligent work ethic, and dedication to advancing science and knowledge. As such, the direction the university will take, and this is yet to be determined in its new strategy document, will be anchored in its rich past which helped it overcome, despite the efforts required that others need years for such a reform, a transformation, and despite the disruption of operations due to a pandemic, a blockade and the organisation of the most popular sports event around the globe.

Of course, that success would not have been possible without the excellent support provided by the State of Qatar primarily and its local partners, mainly QF, as well as the immense network of international collaborators (more than 400 institutions) it has been able to forge. Qatar University's role in shaping collaborations with the branch campuses is also expected to make the Education City project more sustainable and useful, as it continues to set standards of excellence; this is despite many doubts cast around the project, its needs, and whether it will stand the test of time. The last of these challenges was during the 2017 blockade, when Western news outlets predicted that the branch campuses would flee the country. On the contrary, the types of relationships these campuses were able to forge in Qatar at large and Qatar University in particular made it unthinkable for them to leave, especially as the country was facing one of its most critical national crises. Qatar University will continue to play this role, as the complementarity with the branch campuses is due to reach its maturity level, with the campuses becoming an enduring part of the Qatari educational ecosystem.

This ecosystem that is becoming a regional hub for education is growing more and more in symbiosis with Qatar University, which has always been at its centre and will continue to thrive, thanks to its own efforts and excellence, but also to its synergy with the existing (and soon to exist) higher education institutions in the country.

On the graduate profile level, Qatar University is set to continue its support for the efforts of the country and its need for cadres to take over and expand the existing initiatives, as Qatar plays a major role on the international scene in conflict resolution and humanitarian aid. In research, the country's commitment to the United Nations Sustainable Development Goals (SDGs) has been captured at all levels by the university. Alongside the previously stated research priorities that govern its grant awards system, QU has put in place an annual reporting system that tracks the alignment of these grants with the SDGs, helping in that sense with the implementation of the SDGs, not only at QU but also in the Qatari society as a whole.

Finally, as the State of Qatar keeps on playing a leading role in many areas worldwide, the correlation with Qatar University's role in shaping the future of education will keep on growing. The university tries to keep up with these expectations internally, through a flexible strategy allowing for agility in the process, a must in a dynamic environment like that of Qatar, and externally, through its network of trusted partners, equally involved in reaching the goals of QNV2030 and beyond.

Bibliography

Amiri Diwan. HH The Emir visits Qatar University.

Baker, D. (2024). *Making Qatar University (1973–2018) an informal history*.Qatar University Press.

Emergent Materials (2022). Qatar University.

Gulf Times (2012) 'A historic moment' for Qatar University. https://www.gulf-times.com/story/33438/a-historic-moment-for-qatar-university. www.qu.edu.qa

Lari, N., Awadalla, A., Al-Ansari, N., & Elmaghraby, E. (2022). Determinants of female labor force participation: Implications for policy in Qatar. *Cogent Social Sciences, 8*, 1.

Moini, J. S., Bikson, T. K., Neu, C. R., & DeSisto, L. (2009). *The reform of Qatar University*. Rand Corporation.

The Peninsula. Qatar University Completes Strategic Plan 2009–2013 (2013). https://thepeninsulaqatar.com/article/10/12/2013/qatar-university-completes-strategic-plan-2009-2013

Qatar National Vision 2030 https://www.gco.gov.qa/en/about-qatar/national-vision2030/ https://www.diwan.gov.qa/briefing-room/news/activities-and-occasions/2016/october/19/hh-the-amir-visits-qatar-university?sc_lang=en

QU SDG Report https://www.qu.edu.qa/static_file/qu/offices%20and%20departments/International%20Affairs%20Office/Documents/QU%20SDG%20Report%2023.pdf

QU Strategic plan 2010–2013 chrome-extension://efaidnbmnnnibpcajpcglclefindmkaj/https://www.qu.edu.qa/static_file/qu/About/documents/qu-strategic-plan-2010-2013-en.pdf

Times Higher Education. (2016) THE announces the most international universities in the world. (https://www.timeshighereducation.com/news/announces-most-international-universities-world-2016)

Part III
Sustainability

Chapter 18
The Evolution and Future of Education for Sustainable Development in Qatar

S. Duygu Sever, Hira Amin, and Evren Tok

Abstract This chapter examines the multifaceted landscape of Education for Sustainable Development (ESD) with a specific focus on the context in Qatar. The analysis offers an exploration of ESD with regard to its historical evolution, current status, and future prospects. Against the backdrop of Qatar's ambitious national development agenda, as reflected in Qatar National Vision 2030 (QNV 2030), this chapter scrutinises the pivotal role of education in the pursuit of sustainable development goals and the nation's broader development strategies. The chapter presents a comprehensive overview of the current state of ESD within Qatar's education system and the wider society, locating the discussion within the broader context of the country's socio-economic evolution and sustainable development. The study concludes by focusing on the future prospects for ESD in Qatar, offering insightful recommendations for its continued development and integration into the national educational policy.

Keywords Education for sustainable development (ESD) · Education policy · Qatar National Vision · Sustainable development goals

18.1 Introduction

The achievement of sustainable development entails a holistic transformation at the systemic, societal, and individual levels. Education emerges as a key factor to ensure that this transformation translates from policy papers into concrete action. To this

S. D. Sever (✉)
Arts & Sciences Department, Carnegie Mellon University in Qatar, Doha, Qatar
e-mail: duygusever@cmu.edu; sduygusever@gmail.com

H. Amin · E. Tok
College of Public Policy, Hamad Bin Khalifa University, Doha, Qatar
e-mail: hiamin@hbku.edu.qa

E. Tok
e-mail: etok@hbku.edu.qa

E. Zaidan et al. (eds.), *The Future of Education Policy in the State of Qatar*,
Gulf Studies 21, https://doi.org/10.1007/978-981-97-9667-0_18

end, Education for Sustainable Development (ESD) has been increasingly institutionalised and integrated into teaching and learning experiences. The implementation of ESD currently takes on a multifaceted character, entailing a blend of top-down, official initiatives, global paradigms, and extracurricular endeavours, within both formal and informal educational arenas. In examining ESD, the specific context of Qatar offers a nuanced perspective, not only with regard to how ESD is evolving in the country, but also to underscore the pivotal role of education as a transformative tool in guiding the nation towards advanced socio-political development, while achieving the requirements of Sustainable Development Goals (SDGs).

This chapter aims to unpack the intricate interplay between education and sustainable development, investigating how these dimensions intersect with the Qatari vision of the future. The chapter starts with a brief history of ESD in the global context, by explaining its roots, evolution, and its diffusion into the formal education systems. Next, the analysis focuses on the GCC to overview the uptake of ESD at the regional level.

For the context of Qatar, the chapter casts a panoramic view of extant ESD efforts. The analysis includes a top-down and bottom-up review of ESD in Qatar, explaining how ESD relates to the country's 2030 vision and overall development strategies. The section highlights how educational initiatives can be instrumental in cultivating environmental consciousness, fostering responsible consumption, and instilling a deep-seated commitment to environmental stewardship among Qatar's youth.

Furthermore, the chapter examines the dynamic educational landscape, encompassing formal and informal avenues for ESD, ranging from the UNESCO Associated Schools Network (ASPnet) initiatives to the Eco-Schools Program and grassroots efforts by local organisations. It underscores the pivotal role of institutions such as the Qatar Foundation as a critical nexus, bridging formal and informal educational practices, with an agenda to shape sustainability policy and promote behavioural change within the community. In looking towards the future, the chapter concludes by focusing on key lessons and policy recommendations to address Qatar's unique realities, challenges, and societal needs.

18.2 From Environmental Education to Education for Sustainable Development: A Brief History

Educating for a sustainable world has a long history. It began with environmental education (EE) in the mid to late 1960s (Sterling, 2004). One of the earliest definitions of environmental education was that it meant raising awareness and knowledge among citizens concerning the biophysical environment, and that humans are an integral part of a wider interdependent ecological system (Stapp, 1969). The focus on the physical environment led to EE being largely slotted into science classes, framing the issue under positivist epistemology and ontology (Gough, 2013). The first official international recommendation to establish an educational programme of EE around the world was the United Nations Conference on the Human Environment in Stockholm in 1972 (UNEP, 1972). Other international declarations and initiatives followed, gradually unpacking and unravelling the complex social, economic, and political forces behind the issue. One example is the 1975 Belgrade Charter, which framed EE within rising global inequalities, including class, gender, and racial discrimination, and challenged the hitherto orthodox doctrine of unbridled economic growth (Belgrade-Charter, 1975). These concepts were further refined at the first major international intergovernmental conference on environmental education in Tbilisi, Georgia in 1977. The declaration of this conference produced one of the most influential founding documents that outlined the role, objectives and characteristics of EE, particularly underscoring the complexity, interdependency, and multidisciplinary nature of the issue (Hume & Barry, 2015; Tbilisi, 2012).

A key milestone was the shift to sustainable development introduced by the Brundtland Report in 1987. This publication outlined the discussions of the World Commission on Environment and Development (WCED)—a body established by the UN to investigate the connections between environmental protection and development. The shift towards ESD was consolidated in the 1992 United Nations Conference on Environment and Development in Rio. Building on the Tbilisi Principles, Agenda 21 emphasised the critical urgency in "promoting sustainable development and improving the capacity of the people to address environment and development issues" (UNDCC, 1992, ch. 36, para. 3). This shift began the coalescence of EE and development studies that had separately evolved from charities, aid organisations, and NGOs, stressing the importance of collective empowerment, democratisation, and social action (Sterling, 2004). Again, this shift was followed by a slew of international conferences, country declarations, and initiatives, including research projects, networks, and publications (Hume & Barry, 2015). Filho (2011) further notes that from 1998 to 2002 there was a marked shift from the perception that countries ought to be the key instigators to individuals and institutions also engaging and taking on responsibility.

Yet these activities were not yielding the desired results. At the World Summit on Sustainable Development in Johannesburg in September 2002 (also called Rio + 10), it was noted that these pledges, initiatives, and commitments had not yet materialised into anything substantial (Filho, 2011). One of the major recommendations of the summit was a global push for ESD that led to the UN Decade of Education for Sustainable Development (DESD) from 2005 to 2015. Every member country of the UN was encouraged to pursue ESD in policies, practices, and initiatives, with guidance on national implementation strategies (UNESCO, 2007). A network of regional centres of expertise was also established including in the Arab world, as discussed in further detail below (Mochizuki & Fadeeva, 2008). Furthermore, the Rio + 10 Summit marked the integration of education for sustainable development with the education for all movement (2000–2015)—namely the UN's Millennium Development Goal 2 which aimed for basic skills and education for every individual (UNESCO, 2018). In short, concepts, practices, and ideas of education, sustainability, and development, as well as understandings of human happiness and well-being were all undergoing substantial changes, and interconnections and interdependencies were increasingly being realised.

The launch of the UN's 2030 Agenda for Sustainable Development, outlining 17 goals (SDGs) replacing the eight Millennium Development Goals, provided a fresh boost for ESD. The fourth SDG was quality education, which aims to "Ensure inclusive and equitable quality education and promote lifelong learning opportunities for all" (UN, 2015). Target 4.7, specifically, discusses ESD:

> By 2030, ensure that all learners acquire the knowledge and skills needed to promote sustainable development, including, among others, through Education for Sustainable Development and sustainable lifestyles, human rights, gender equality, promotion of a culture of peace and nonviolence, global citizenship and appreciation of cultural diversity and of culture's contribution to sustainable development.

Thus, ESD became an integral part of quality education as a whole. As DESD had now ended, the Global Action Programme (GAP) on ESD (2015–2019) was launched at the UNESCO World Conference on ESD in Aichi-Nagoya in 2014 (UNESCO, 2014a). Its aim was to "generate and scale up action in all levels and areas of education and learning to accelerate progress towards sustainable development" (UNESCO, 2014d). It is important to note that different sectors, such as the formal, informal, and non-formal educational spaces were stressed for a more holistic and impactful experience.

The GAP had four main implementation strategies (One-Planet, 2018), namely, building new momentum (with an open call for stakeholders to make voluntary commitments and pledges), harnessing partnerships (through networking with local, regional, and international partners and relevant entities), fostering a global community of practice (regularly meeting stakeholders and partners for the exchange of ideas, practices, and trends), and showcasing good practice (giving visibility to innovative and effective cases and practices).

The GAP focused on five priority areas (UNESCO, 2014c):

1. Advancing policy by organising dialogues between policymakers, civil society members, and educationalists to integrate ESD policies or revise existing policies.
2. Transforming learning and training environments especially promoting whole institutional approaches fostering sustainability at all levels and settings of the educational space, such as green campus facilities and engaging with community sustainable development stakeholders. This would ensure we go by the adage "learn what we live and live what we learn".
3. Building capacities of educators by empowering them with the knowledge, skills, and attitudes for sustainability.
4. Empowering and mobilising youth by giving them the role as agents of change possibly through leadership workshops and recognising them as key actors.
5. Accelerating sustainable solutions at the local level incorporating and engaging different sectors such as businesses and civil society organisations who can make local transformative impact.

Following the GAP (2015–2019), a new framework dubbed as "ESD for 2030" was adopted by the 206th UNESCO Executive Board and the 40th UNESCO General

Conference, and acknowledged by the 74th UN General Assembly. According to the UNESCO roadmap, ESD for 2030 continues and builds upon the GAP priority areas and implementation strategies outlined above (UNESCO, 2020). It has three additional key features. The first is reinforcing the role of education in achieving the SDGs by raising awareness of the SDGs, promoting their critical and contextualised understanding, as well as mobilising action. The second is to focus on major transformation through action-orientated approaches and deeper structural and societal changes that are the root causes of issues such as inequality. The third is the emphasis on the role and importance of strong leadership from the member states to initiate and implement the five priority areas laid out above (UNESCO, 2020, pp. 16–20).

The global ESD agenda is not without its critics. Some focus on the inherent contradictions within the concept of sustainable development. They argue that it is precisely the model of economic development, unlimited growth, and industrialisation that is the underlying catalyst for environmental destruction, the rapid decline of biodiversity, and climate change (Kahn, 2010; Kopnina, 2019a, 2019b; Kopnina and Gjerris, 2015, 2016; Washington, 2015, 2018). As the argument follows, it is oxymoronic to believe in limitless economic growth and in the possibility of delivering generational justice, that is, sustaining the same level and quality of resources for generations to come. This, they argue, translates into ESD that uncritically focuses on social and economic priorities, with a weak emphasis on eco-philosophical principles (Bonnett, 2015; Kahn, 2010; Kopnina, 2020). In this framework sustainable development is anthropocentric and economy-centric, reducing the environment to an object of exploitation, manipulation, and possession (Molina-Motos, 2019). Some have even called this "cognitive imperialism", which is the process where Western models of education foreground a certain epistemology and ways of being (Black, 2017). In doing so, it renders other, mainly indigenous knowledge, as mere mythology, folklore, or a relic of the past, as part of people's cultural heritage, in contrast to "real" knowledge that is applicable and relevant today (Black, 2017).

Apart from conceptual limitations, there are also numerous practical and educational policy challenges. These are illustrated in the numerous in-depth case studies which were conducted after DESD. One common thread was the announcement of grand national decrees and programmes for ESD in European countries, yet the degree of centralisation of the educational system led to various results and sets of challenges. Centralised systems, such as that of France, are more "demand-orientated" and mandated from "above" leading to lower stakeholder engagement, particularly in the design phase, thereby reducing its efficacy and the enthusiasm for implementation (Ricard & Dussaux, 2014). Additionally, there is considerable agreement among ESD scholars and practitioners that successful ESD is deeply rooted in local practices, community engagement, and contextualised issues. This requires local autonomy to deal with complex local dynamics. Top-down centralised efforts that aim to mainstream ESD inadvertently flatten and reduce local autonomy and dynamics. On the other hand, decentralised systems that grant greater autonomy to schools have different issues. In these cases, schools can choose the extent of their integration of national policies and standards, which leaves ESD as a "nice to have" (de Wolf and de Hamer, 2014). Without adequate teacher training—a problem highlighted in both systems—it is left down to the individual teacher and their expertise and experience in integrating ESD. To illustrate, the decentralised system in the UK (relative to France), led to mixed results, as the reality on the ground was mediated by schools' socio-economic situation and individual actors (Martin et al., 2014). Overall, the lack of robust teaching training programmes, vertical feedback loops, and evaluation structures is stated as key issues that stifled the execution of ambitious national plans and strategies, in both centralised and decentralised systems (Jucker and Mathar, 2014).

However, even if national strategic visions are followed by robust plans for structural integration into schools and educational institutions, there are still challenges. For example, Germany established a cross-party politically mandated national DESD committee and action plan, that was integrated into all German states alongside developing training and tools, such as an extensive national website for ESD, as well as nearly 2,000 certified individual localised ESD projects. Despite these impressive actions, Mathar (2014, p. 7) argues that "there is still a lack of understanding and acceptance of ESD amongst politicians and the general public, there is insufficient structural and institutional anchoring of ESD in schools". This connects with the debate discussed above about the dissonance between sustainable development and current meta frameworks of industrialisation and economic development. Without a transformative shift in the understandings of society, sustainability, and progress, ESD will remain marginal and relegated to changing simple behaviours (such as switching off lights) in the hearts and minds of people, as opposed to the revolutionary change of worldview that is required for tangible change. Vare and Scott (2007) refer to this as ESD 1 and ESD 2, where the former is promoting and facilitating a change of behaviour (such as insulating lofts). The latter is "building a capacity to think critically about and beyond what experts say and to test sustainable development ideas". ESD2 in other words is an open-ended, collaborative, and reflective process that recognises and embraces the complexity and unpredictability of the world. While these are complementary, Vare and Scott argue that educational policies are disproportionally skewed towards ESD 1 at the expense of ESD 2, resulting in little transformative change and action (Vare & Scott, 2007).

Austria is another exemplary case study. Despite its strong national strategy on ESD and various national programmes, they have faced similar hurdles to many other countries (Rauch & Pfaffenwimmer, 2014). Among them was the struggle for teachers to integrate ESD in already packed national curriculum. Cross-collaborative courses and projects that engage with outside stakeholders and multiple modules, courses, and disciplines require enormous time and effort from the teacher's side, as well as significant chunks of time taken from the core curriculum. Moreover, there is considerable pressure on teachers to focus on core subjects such as numeracy, literacy, and ICT. These practicalities on the ground led to a loss in the creativity required for transformative ESD, despite national government support (Rauch & Pfaffenwimmer, 2014).

In short, despite growing awareness and acceptance of the global agendas for ESD, implementation on the ground is stifled due to numerous and complex conceptual and practical hurdles.

18.3 Regional Uptake of ESD

Similar to most developing countries, those of the GCC region were initially reluctant to join agreements and pledges related to environmental sustainability, especially as their economies were based on high carbon-based growth and consumption models. However, there has been a shift from scepticism to positive engagement (Al-Saidi et al., 2019). The following is a brief outline of the key milestones for ESD in the wider MENA region.

The Arab world began to slowly incorporate EE in the 1980s, and ESD underwent a significant boost during the UN DESD (AFED, 2019). In September 2005, Bahrain hosted a regional meeting filled with educational and training specialists to kickstart discussions on an ESD strategy for this part of the world (UNESCO, 2014b). In April 2006, the Arab Conference on Education and Sustainable Development was held in Beirut, attended by regional ministers and experts (UNESCO, 2007). The following year, in May 2007, a regional workshop enhancing the ESD pedagogical skills for teachers was organised by the UNESCO regional office for Education in Beirut, in cooperation with the DESD secretariat and UNESCO offices in Cairo, Doha, and Rabat. The workshop was held in Egypt, where discussions on the regional ESD strategy also continued (UNESCO, 2007). These events culminated in the Regional Guiding Framework of Education for Sustainable Development in the Arab Region, published in Beirut in June 2008. The Guiding Framework defined ESD for the Arab region as encompassing "the acquisition and practice of knowledge, values and skills that ensure balance between the economic, social and environmental aspects of development and the observance of both individuals and society development and progress in life" (Beirut, 2008). Specifically, they identified the following regional challenges to focus on:

- Poverty, employment, and the brain drain
- Agriculture and water scarcity
- Desertification and the rational use of scarce water resources
- Health awareness
- Political security and demographic stability
- Environmental awareness
- Unsustainable consumption patterns
- Problems related to an increase in air pollution (water, air, and soil) (OECD, 2019).

Another ground-breaking effort in the region came with the launch of the World Innovation Summit for Education (WISE) in 2009. WISE was established by the Qatar Foundation, with the leadership of Her Highness Sheikha Moza bint Nasser. It was created as a multi-sectoral, international platform to serve as a reference point for innovation, partnerships, and new methodologies in education. The 2015 WISE Summit titled "Investing for Impact: Quality Education for Sustainable and Inclusive Growth" took place in Doha. WISE, 2015 was the very first international conference on education which was organised after the ratification of the UN Sustainable Development Goals (SDGs) in September 2015. Accordingly, the UN Sustainable Development Goals were one of the key pillars, along with "education and the economy" and "fostering innovation in education systems" (WISE, 2015).

The summit hosted hundreds of decision-makers, practitioners, and social figures creating a significant venue for discussing reform and innovation in education for Qatar, the MENA Region, and beyond. WISE, 2015 marked a benchmark in ESD for the region, as emerging priorities with regard to the achievement of the SDGs were articulated through debates on practical action items for the way forward (WISE, 2015).

18.4 ESD in Qatar: A Top-Down Review

As the focus on ESD has been increasing on both global and regional education agendas, Qatar has been no exception to the countries which showcase significant efforts to align their development strategies with sustainability and with quality education and to lead the way in this endeavour. Qatar embraces education's pivotal significance in empowering individuals, fostering civic engagement, nurturing human development, and establishing sustainable societies, guided by ethical and moral frameworks within the education sphere. Qatar's emphasis on ESD has been evident, not only in key national documents, but also in the country's engagements with global networks to enhance ESD.

18.4.1 Key National Documents

Qatar National Vision 2030 (QNV 2030) stands as the fundamental strategic blueprint that underpins Qatar's aspirations for a multifaceted and forward-looking economic development, rooted in diversification, innovation, and sustainability. The vision rests upon four key pillars of development, namely, human, social, economic, and environmental development, integrated with each other to orchestrate a holistic trajectory towards sustainable growth and elevated standards of living (Qatar General Secretariat for Development Planning, 2008).

Within the specific context of sustainable development, QNV 2030 recognises the importance of nurturing the ethical and moral underpinnings that underscore the conservation and protection of the environment and ecological diversity. Sustainable development emerges as a foundational element of Qatar's forthcoming trajectory. QNV 2030 also places significant emphasis on the key role of education in propelling social progress. Therefore, the national vision endorses the indispensability of ESD as part of the curriculum and training programmes that can respond to Qatar's present and future needs, by cultivating an environmentally conscious citizenship (Qatar General Secretariat for Development Planning, 2008, p. 20).

Expanding upon the framework set forth by QNV 2030, Qatar National Strategy Documents provide major themes, strategies, and objectives, intended to diversify Qatar's economic landscape and facilitate the nation's socio-economic advancement. This pursuit is intrinsically reliant upon comprehensive reforms, inclusive engagement with stakeholders, and localisation of international benchmarks showcasing good practices (Qatar General Secretariat for Development Planning, 2011, p. 139). Qatar's development plans entail simultaneous progress in institutional, financial, and human capabilities, in alignment with the pursuit of sustainable practices across all pillars of QNV 2030 (Planning & Statistics Authority, 2018, pp. 11, 302).

Qatar's dedication to the principles and ideals of sustainable development and ESD is also evident in the National Development Plans covering 2011–2016 and 2017–2022, which contain explicit and implicit references to education's role in nurturing sustainability. Although the precise label of "education for sustainable development" is not expressly stated, the strategy inherently encapsulates the motivations and targets underpinning this concept, manifesting a resonance with the spirit of ESD. A recurrent and major point raised under the theme of education in these major national documents concentrates on the need to equip Qatari citizens with a vision to see the future of the country beyond its hydrocarbon resources and to prepare the human capital for a shift towards a diversified and sustainable economy (Qatar General Secretariat for Development Planning, 2011). The economic diversification envisioned for the future of Qatar requires increasing the quality of education and ensuring its relevance to Qatar's current and emerging needs (Qatar General Secretariat for Development Planning, 2011, p. 139).

Parallel to quality education, the theme of environmental awareness is a prominent aspect repeatedly addressed as part of environmental sustainability. This creates a natural connection with education, as ESD emerges as a tool to enable enhanced environmental stewardship and responsible practices. To illustrate, Qatar National Development Strategy 2018–2022 clearly advocates a fundamental paradigm shift in the human relationship with the environment, which can be triggered by environmental awareness. To this end, it calls for active and determined participation of all actors and sectors within society. In this regard, Qatar's education strategy predominantly addresses the objective of equipping Qatar's youth not only with an enriched comprehension of the nation's values, culture, heritage, and global citizenship, but also with environmental awareness (Planning & Statistics Authority, 2018).

The combination of education with sustainable development as per Qatar's official strategies entails multifaceted policy avenues such as encompassing the increase in environmental consciousness, the advocacy for sustainable production of goods, or the enhancement of environmentally responsible consumption behaviours. At the level of formal education planning, these policy targets connect with action items specifically aiming to increase the enrichment of environmental studies within educational establishments, the integration of sustainability principles within curricula, and the fostering of connections between local and international educational initiatives (Qatar General Secretariat for Development Planning, 2011, pp. 245–252).

There has been also an increasing acknowledgement among the Qatari authorities of the role of initiatives, projects, and efforts, encompassing non-formal educational endeavours, external to conventional educational institutions. This includes the promotion of community-based environmental conservation initiatives, and the launch of environmental awareness campaigns designed to familiarise individuals with environmental consciousness, cultivate sensitivity to environmental imperatives, and disseminate knowledge on conservation methods and resource management (Planning & Statistics Authority, 2018 p. 318). The Qatari government acknowledges that this effort requires a comprehensive, long-term ESD plan, simultaneously reinforced through diverse channels such as curricula, school spaces, and informal initiatives (Planning & Statistics Authority, 2018, p. 305). The upcoming sections will elaborate more on these elements.

To translate the policy-level acknowledgment of the role of Education for Sustainable Development (ESD) in Qatar's development into actual implementation and educational practices, it is essential to crystallize the educational mandate within curricular frameworks. These frameworks should integrate Qatari national values, fostering a sense of belonging, accountability, citizenship, and a commitment to sustainable development among students. The execution of the national vision and development strategies in the realm of education is under the responsibility of the Ministry of Education and Higher Education (MOEHE). The MOEHE leadership is intrinsically oriented towards ensuring high-quality education and the cultivation of the required skills, knowledge, and capabilities across all educational levels, at K-12 public and private schools. Integral to MOEHE's mandate is a pronounced emphasis on cultivating "twenty-first century skills, values, and attitudes", which conceptually aligns with the international and regional efforts for ESD.

The General Framework for National Education, along with its main objectives of (a) successful lifelong learning, (b) personal, moral, and social growth, and (c) active and responsible citizenship; showcases a significant commitment to sustainable development. The need for students to understand the balance between economic development and environmental protection, the importance of a living ecosystem, their responsibility towards the next generations, and their active contribution towards achieving sustainable development goals are identified as important outcomes of an effective education strategy for Qatar. The document also acknowledges the critical role of teachers in achieving these goals and ensuring the youth's commitment to sustainability.

18.4.2 Organisations and Networks

Many international organisations and initiatives championing ESD have been established in Qatar. In the context of global networks for formal educational architecture, the UNESCO Associated Schools Network emerges as a central pillar for fostering ESD. Functioning as a conduit to link educational institutions worldwide, this network aligns with the overarching objective of achieving SDG target 4.7 on Global Citizenship Education (GCED) and ESD. Operating under the tenet that ESD empowers students to effectively confront contemporary global challenges and to contribute to sustainable and resilient societies, this network endeavours to transform educational environments. It aspires to integrate local and global sustainability imperatives into curricula, generate interactive pedagogies that foster participatory teaching and learning, encourage critical thinking, and cultivate shifts in attitudes. This initiative empowers associated schools with innovative educational resources, novel teaching methodologies, capacity-building initiatives, and platforms for exchanging experiences, insights, best practices, and knowledge with educational institutions, individuals, communities, policymakers, and society at large. Furthermore, the conceptual framework envisions the school campus as a dynamic learning environment, intrinsically intertwined with ecologically conscious activities aligning with community needs, values, and sustainability requisites. The overarching objective is to nurture agents of transformative change, capable of informed decision-making for sustainable development. As a global collaborative effort, over 11,500 member schools in 182 countries actively engage in fostering international understanding, intercultural dialogue, peace, sustainable development, and quality education (UNESCO, 2022).

Qatar, as an esteemed member of UNESCO since 1972, joined the UNESCO Associated Schools Network in 1983 (Qatar National Commission for Education, Culture and Science, 2022). As of 2022, 82 Qatari schools are affiliated with the network (UNESCO, 2022). The National Commission in Qatar articulates the roles of UNESCO schools in Qatar as covering environmental protection and conservation, the promotion of cross-cultural learning, advocacy for nonviolent conflict resolution, and the cultivation of solidarity, integration, and support for the community. The specific example of the Green School Program targets primary, preparatory, and secondary students within UNESCO Associated Schools, including objectives such as promoting eco-friendly behaviour, transforming school premises into green spaces, nurturing students' sense of environmental responsibility and societal affiliation, and increasing the school's role in sustainable development and environmental consciousness (Qatar National Commission for Education, Culture and Science, 2022).

Another globally connected initiative which contributes to ESD within Qatar is the Eco-Schools Program, an international accreditation for ESD. Established in Denmark in 1994, this programme envisions a world where ESD fosters positive transformation and empowers individuals. Active across 94 countries with over 55,000 registered schools, 19.8 million students, and 1.4 million educators, Eco-Schools resonates with the SDGs, urging students to champion action for tangible sustainability outcomes. This student-driven initiative urges participants to identify real-world challenges, draft action plans, and lead projects that advance sustainable development. An important component of this programme is the innovative utilisation of school premises as dynamic arenas for practical sustainability interventions. Eco-Schools Program in Qatar was introduced in 2018 by Qatar Green

Building Council and is operated today by Earthna. Although Eco-Schools provides an impetus for ESD, currently Qatar is home to 42 private, and 20 public registered Eco-Schools, which represents a relatively small share of the overall school system, which includes more than 332 private schools serving approximately 211,000 students and more than 207 public schools, serving more than 124,600 Qatari and non-Qatari students (MOEHE, 2023). The transfer of good practices and such ESD initiatives to the entire school system is a pressing issue for Qatar.

On the other hand, beyond formal educational endeavours which are either centralised by MOEHE or international education networks, a constellation of local initiatives, organisations, and informal actors within Qatar plays crucial roles in fostering ESD. In addition to K-12 schools, Qatar hosts multiple stakeholders and initiatives with ad-hoc efforts to underscore multifaceted dimensions of sustainable development, raise awareness, and cultivate a culture of responsible consumption. Qatar's ESD ecosystem is characterised by a diverse array of initiatives, projects, and non-institutional actors, which are involved in activities and partnerships. The stakeholder mapping of the ESD ecosystem in Qatar reveals more than 400 actors, campaigns, and initiatives, both online and on-site, which serve teaching and learning on sustainability, as summarised by the infographic (Fig. 18.1):

As the infographic illustrates, the ESD ecosystem in Qatar is dominated by K-12 school initiatives, but also consists of a vibrant community of ESD-related initiatives

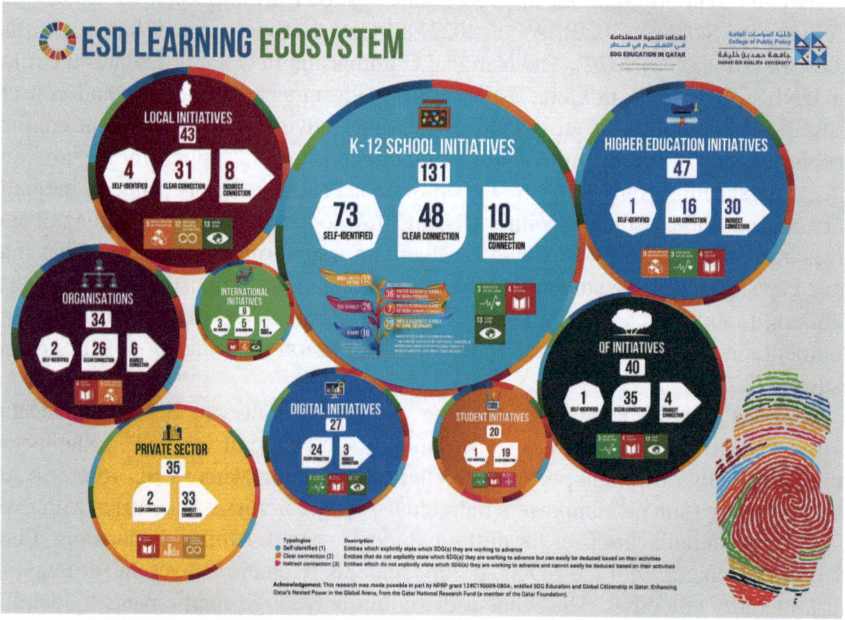

Fig. 18.1 ESD ecosystem in Qatar

or actors which have direct and/or indirect connections with the sustainable development goals. Creating synergies among these stakeholder groups and ensuring a mainstream progress for ESD for all the students in the country remains a challenge.

18.5 ESD in Qatar: A Bottom-Up Review

Few studies have explored bottom-up approaches to ESD in Qatar, in other words as it is actually practised in everyday classrooms or higher education settings. Relative to the top-down review, the findings are granular and not as coherent. The authors have conducted a systematic literature review of ESD in the region and found that the challenges within Qatar by and large chime with the rest of the GCC, and indeed echo the hurdles many other countries face, as discussed above (Amin et al., 2023). Relevant to this discussion, two key challenges have emerged from this review. The first is the lack of institutionalisation and alignment between master plans and bottom-up practicalities, and the second is the lack of effective pedagogies.

Institutionalisation and Alignment

One major finding underpinning most studies on the ESD in Qatar and the region as a whole is that most K-12 and higher education students, teachers, and university faculty have an—albeit varied and scattered—understanding of sustainability and its significance. Despite this, it is not being translated into concrete action; there seems to be a gap between knowledge and action. Some studies have argued that this chasm between knowing the importance of sustainability and tangible action is derived from the lack of institutionalisation in educational institutes. This leads to the faculty not perceiving it as a priority or as part of their role (Alsharif et al., 2020; Alshuwaikhat et al., 2016).

This spills over into student perception. When students do not explicitly see their institute taking bold steps towards sustainability, this contributes to their lack of motivation and action. In one study, 83% of university students at one institute felt that their university was not partaking enough in sustainability activities and teaching, which they cited as one of their reasons for their inactivity—if the university itself is not championing and prioritising the cause, why should they? (Almughairy, 2016; Radwan & Khalil, 2021).

This again has a spill-over effect whereby industries struggle to promote sustainable practices. For example, one study on architecture and engineering firms in Kuwait found little appetite for green design practices due to a lack of public education and awareness, a lack of governmental incentives and policies, and a lack of educational courses related to sustainability in general and green design specifically (Mahmeed, 2011).

Overall, there seems to be a misalignment on multiple levels. The impressive top-down global agreements and national visions are not reflected in bottom-up reviews and experiences on the ground. This includes inconsistencies between national visions and practices in schools (Al-Kuwari et al. 2021a; 2021b; Al-Abri, 2020; Al-Abri et al., 2022; Al-Maamari, 2016); between teacher training and development and national educational policy documents (Al-Maamari, 2014); between the concepts and frameworks of citizenship and sustainability in national policy documents and teaching curricula (Selaibeekh, 2017); between policy documents and educational implementation plans overall (Essa, 2021); and between national visions and learning outcomes and competencies in school subjects and university courses (Al-Kuwari, 2021; Al-Kuwari et al. 2021a). The quote below summarises this issue. Although this is taken from a specific study based on governmental schools, the description is reflective of the entire educational system:

> "...analysing the current assessment practices... reveals a clear gap between the assessment and goals in Qatar.... Although the MOEHE recognizes the importance of twenty-first-century skills and SDGs to develop the abilities of students, neither learning activities nor assessment methods are currently supporting these intended outcomes and, further, neither are aligned with the SDGs. Moreover, these educational practices need qualified teachers with sufficient knowledge in pedagogical approaches" (Al-Kuwari, 2021).

Effective Pedagogy

A central argument that underpins almost all studies on ESD is the need for innovative and engaging pedagogical techniques and resources that can instill in students a critical sustainability consciousness and understanding of its complexity and future possibilities. One major barrier studies have found is the varying degrees of teachers' understanding of sustainability-related issues, which directly impacts their level of confidence and subsequent integration in the classroom and thus their subjects' engagement (Alghamdi & El-Hassan, 2019; Alkhayyal et al., 2019; Alqahtani et al., 2017).

Furthermore, as has been mentioned above, the overwhelming majority of scholars state that successful ESD is a transdisciplinary endeavour involving local community stakeholders, businesses, and initiatives. This is where students at all levels can witness the complexity and tangible impact. This would also ensure that the educational sector is more aligned with local needs, environmental trends, and industries' stance. Studies have highlighted that this is severely lacking in the GCC context (Almughairy, 2016; Alsaati et al. 2020; Mohamed et al., 2022). Where connections have been made, there have been various challenges. One study in Oman exploring the use of university-industry partnerships to enhance innovative thinking among higher education students found that industry was not cooperative (Nikolaou, 2022). They lacked the ability to train, educate, or mentor students, and did not have well-thought-out internships that would enhance their experiences and contribute to their career development. Alternatively, a stronger partnership could have led to professors being conversant with the cutting-edge of the industry, that would have translated into better and more relevant content for their university courses (Nikolaou, 2022).

Even if the strength of these partnerships is resolved, how to fit in these cross-cutting initiatives and projects in an already tightly packed and designed curriculum is another serious challenge that requires re-organising every subject's aims, objectives, and outcomes (Ashraf & Alanezi, 2020). This is further challenging for facilitators dealing with students from multiple backgrounds and with varying abilities, as well as fitting the course into an already fully designed syllabus.

Developing critical thinking skills is another core aspect of ESD. Many scholars have advocated for multi-course project-based learning (MPL), or inquiry-based learning (IQL) related to sustainability issues. This enhances transdisciplinary, critical thinking, and collaborative decision-making skills—all of this essential to shape a sustainable future. These studies found that although some students gave positive feedback and demanded more similar projects, others underwent various challenges. Not all students had the same ability, capacity, and motivation, resulting in internal discord within the groups. Other students found the task too difficult without clear guiding instructions and steps, which goes against the principles of MPL which encourage self-learning and minimal monitoring (Alghamdi & El-Hassan, 2020; Khandakar et al., 2020).

18.6 Recommendations for Implementing ESD in Qatar

In alignment with its national vision, Qatar endeavours to establish an advanced and globally esteemed educational framework, designed to equip its citizens and residents with the essential competencies and knowledge required to navigate the challenging dynamics of the contemporary world. This endeavour embraces the cultivation of analytical and critical thinking, a robust civic consciousness and sense of belonging, and the fostering of creative thinking and innovation. In parallel, acknowledgement of local cultural heritage and societal values belonging to Qatari identity is consciously advocated (Qatar General Secretariat for Development Planning, 2008, 17). As the overview of key strategic documents has showcased, ESD constitutes a significant part of the country's effort for quality education.

These key points and the current landscape of ESD in Qatar signal certain policy recommendations that can enhance ESD, and contribute to the effectiveness of existing efforts in Qatar for future progress. At the policy level, it is crucially important to combine education and sustainable development strategies, in order to translate them into tangible action plans. This also necessitates the design of consistent monitoring and evaluation to realise the ESD goals as outlined in national documents. At the practice level, following a holistic approach for curriculum design at K-12 Schools is important, so as to integrate ESD into the entire system. This should also be in parallel with the efforts to use schools as "spaces" for ESD, by making sure that the students are engaged with all aspects of sustainable development.

Regarding institutionalisation, policies, and leadership, it is imperative that the overarching sustainability objectives and goals derived from national documents be firmly embedded at both school and university levels. The Ministry of Education and Higher Education (MOEHE) should obligate every institution, both governmental and private, to formulate comprehensive strategic plans that resonate with the national vision on sustainability. The inclusion of teacher training, essential resources, and an engagement with the sustainability ecosystem must be integral to each institution's education policy. To develop these strategic plans, continuous consultation with educators and school leaders from diverse backgrounds is crucial to guarantee optimal involvement and feasibility. Furthermore, it is advisable to set up departments dedicated to sustainability, connecting varied stakeholders, to supervise the intricate alignment of national documents, institutional aspirations, teacher training, and educational outcomes. The role of these departments is pivotal in streamlining the integration of sustainability-focused curriculum components and projects. For a seamless implementation, a robust system of evaluation, oversight, and feedback is paramount.

It goes without saying that the role the teachers are playing in ESD is critical, especially when the policies, strategies, and curriculum boil down to actual classroom experience and learning activities. Accordingly, teachers' readiness to teach within the scope of ESD must be addressed through training and professional development programmes across all educational institutions, including both public and private K-12 schools, as well as universities. In doing so, incentives to ensure collaboration and exchanges between peers, local institutions, and families must be in place to create exciting hands-on pedagogical experiences for ESD. With respect to teacher and faculty training and resources, a substantial commitment to enhancing K-12 teacher training and providing innovative localised tools is essential. This should cater to both centrally governed schools, necessitating a more structured approach, and private institutions, which may require a more adaptable framework. The design of training and resources should be a collaborative effort, considering the input of teachers from diverse backgrounds and institutions. A collaborative platform or "majlis" is recommended for educators to share insights, deliberate on challenges, and brainstorm solutions. At the tertiary level, university faculty should periodically undergo courses updating them on the latest sustainability trends relevant to their disciplines. Moreover, specialised training aimed at nurturing advanced critical thinking through engaging pedagogical techniques is essential for this academic tier.

Focusing on the sustainability ecosystem, a resilient and interlinked network connecting educational establishments at all tiers with local and regional industrial and organisational entities is crucial. This network would facilitate the seamless exchange of knowledge and offer holistic, real-world experiences to both educators and students. Faculty members from universities should actively participate in these networks to enrich their research and pedagogical approaches. Considering Qatar's monumental strides in sustainable initiatives, such as hosting an accessible FIFA World Cup, and launching the environmentally progressive Metro system, integrating students into these ventures can immensely amplify their academic journey, readying them for future challenges, and augmenting their professional prospects. Within academic premises, visible endorsement and execution of sustainable practices are vital. To bolster these affiliations, it is advisable that the government, in collaboration with the MOEHE, should either mandate, urgently advocate, or provide incentives for these partnerships.

Last but not least, each country faces unique challenges and issues in the realm of sustainable development. While ESD is initiated and promoted by international networks, and it aims to ensure the achievement of global efforts for sustainable development, the alignment of ESD strategies, materials, and techniques with distinctive contexts, requirements, and capabilities inherent to specific societies, is key in advancing the relevance and efficiency of future ESD practices in Qatar. Localisation of ESD to create the understandings and frameworks that relate to Qatar's historical culture, traditions, and sustainability challenges is one aspect, while the other is to ensure that all local communities are included in the ESD practices in both formal and informal settings.

18.7 Conclusion

This chapter has focused on education for sustainable development in Qatar. The chapter started with an extensive overview of the evolution of ESD in global and regional agendas. Next, the chapter showcased the connections between ESD and Qatar's national vision and development strategies. The analysis further investigated the organisations, initiatives, and the ESD ecosystem in Qatar. After that, the analysis was followed by a bottom-up review of ESD in Qatar, with regard to institutionalisation, alignment, and pedagogy for ESD. The chapter concluded with recommendations for implementing ESD in Qatar.

Acknowledgements This research was made possible in part by NPRP grant #12C-0804-190009, entitled SDG Education and Global Citizenship in Qatar: Enhancing Qatar's Nested Power in the Global Arena, from the Qatar National Research Fund (a member of the Qatar Foundation).

Bibliography

AFED. (2019). *Environmental education for sustainable development in Arab Countries.* Beirut, Lebanon: Arab Forum for Environment and Development.

Al-Abri, K. M., Ambusaidi, A. K., & Alhadi, B. R. (2022). Promoting global citizenship education (GCED) in the sultanate of Oman: An analysis of national policies. *Sustainability, 14*(12), 7140. https://doi.org/10.3390/su14127140

Al-Abri, M. (2020). *The role of UNESCO in sustaining cultural diversity in the sultanate of Oman, 1970–2020.* Bangor University.

Al-Kuwari, M. M., Al-Fagih, L. & Koç, M. (2021a). Asking the right questions for sustainable development goals: performance assessment approaches for the Qatar education system. *Sustainability 13* (7): 3883.

Al-Kuwari, M. M. (2021). *Education for Sustainability (EFS): Investigations on Performance Assessment Issues, Needs and Recommendations for the Qatar Education System (QES).* Hamad Bin Khalifa University (Qatar).

Al-Kuwari, M. M., Du X., & Koç, M. (2021b). Performance Assessment in Education for Sustainable Development: a Case Study of the Qatar Education System. *Prospects 52*: 513–527. [Crossref], [Google Scholar]

Al-Maamari, S. (2014). Education for developing a global omani citizen: current practices and challenges. *Journal of Education and Training Studies, 2*(3), 108–117.

Al-Maamari, S. (2016). Education for connecting Omani students with other cultures in the World: The role of social studies. *International Review of Education, 62*, 439–457. https://doi.org/10.1007/s11159-016-9577-2

Al-Mansoori, R. S., & Koc, M. (2019). Toward knowledge-based economy: Innovation and transformational leadership in public universities in Texas and Qatar. *Sustainability, 11*(23), 6721.

Al-Nuaimi, S. R., & Al-Ghamdi, S. R. (2022). Assessment of knowledge, attitude and practice towards sustainability aspects among higher education students in Qatar. *Sustainability, 14*(20), 13149.

Al-Saidi, M., Zaidan E., & Hammad S. (2019). Participation modes and diplomacy of Gulf Cooperation Council (GCC) countries towards the global sustainability agenda. *Development in Practice 29* (5): 545–558. https://doi.org/10.1080/09614524.2019.1597017.

Alghamdi, A., & El-Hassan, W. S. (2019). Raising Saudi students' (Energy) sustainability awareness through ESL—Teachers' Thoughts. *Journal of Teacher Education for Sustainability, 21*(1), 137–154. https://doi.org/10.2478/jtes-2019-0011

Alghamdi, A., Hamdan, K., & El-Hassan, W. S. (2020). Interdisciplinary inquiry-based teaching and learning of sustainability in Saudi Arabia. *Journal of Teacher Education for Sustainability, 22*(2), 121–139.

Alghamdi, A., & El-Hassan, W. S. (2020). Interdisciplinary inquiry-based teaching and learning of sustainability in Saudi Arabia. *Journal of Teacher Education for Sustainability, 22*(2), 121–139. https://doi.org/10.2478/jtes-2020-0020

Alkhayyal, B., Labib, W., Alsulaiman, T., & Abdelhadi, A. (2019). Analyzing sustainability awareness among higher education faculty members: A case study in Saudi Arabia. *Sustainability, 11*(23), 6837. https://doi.org/10.3390/su11236837

Almughairy, A. (2016). The gap of teaching sustainability in planning and urban design programs in Saudi Arabia. *The International Journal of Design in Society* 10 (1): 1. https://doi.org/10.18848/2325-1328/CGP/v10i01/1-11.

Alqahtani, A., Alqahtani, F., & Alqurashi, M. (2017). The extent of comprehension and knowledge with respect to digital citizenship among middle eastern and US students at UNC. *Journal of Education and Practice, 8*(9), 96–102.

Alsaati, T., El-Nakla, S., & El-Nakla, D. (2020). Level of sustainability awareness among university students in the Eastern Province of Saudi Arabia. *Sustainability, 12*(8), 3159. https://doi.org/10.3390/su12083159

Alsharif, M. A., Peters, M. D., & Dixon, T. D. (2020). Designing and implementing effective campus sustainability in Saudi Arabian Universities: An assessment of drivers and barriers in a rational choice theoretical context. *Sustainability, 12*(12), 5096. https://doi.org/10.3390/su12125096

Alshuwaikhat, H. M., Adenle, Y. A., & Saghir, B. (2016). Sustainability assessment of higher education institutions in Saudi Arabia. *Sustainability, 8*(8), 750. https://doi.org/10.3390/su8080750

Amin, H., Zaman, A., & Tok, E. (2023). Education for sustainable development and global citizenship education in the GCC: A systematic literature review. *Globalisation, Societies and Education,*. https://doi.org/10.1080/14767724.2023.2265846

Ashraf, M. W., & Alanezi, F. (2020). Incorporation of sustainability concepts into the engineering core program by adopting a micro curriculum approach: A case study in Saudi Arabia. *Sustainability, 12*(7), 2901. https://doi.org/10.3390/su12072901

Beirut, U. N. E. S. C. O. O. (2008). *Regional guiding framework of education for sustainable development in the Arab Region*. UNESCO.

Belgrade-Charter. (1975). *International Workshop on Environmental Education – Belgrade Charter*. UNESCO.

Black, C. (2017). Land-based pedagogies and the culture of schooling. In H. Kopnina & E. Shoreman-Ouimet (Eds), *Routledge handbook of environmental anthropology* (pp. 452–466). Routledge Earthscan.

Bonnett, M. (2015). Sustainability, the metaphysics of mastery, and transcendent nature. In H. Kopnina & E. Shoreman-Ouimet (Eds), *Sustainability: Key issues* (pp. 25–39). Routledge Earthscan.

de Wolf, M. and de Hamer, A. (2014). Education for Sustainable Development in the Netherlands. In *Schooling for Sustainable Development in Europe,* edited by R. Jucker and R. Mathar. Springer.

Essa, S. (2021). *Education for Sustainable Development Integration Plan in Vision 2030 in Saudi Arabia: A Discourse Analysis of Media and Government Documents.* McGill University (Canada).

Filho, W. L. (2011). "Chapter 1: Applied Sustainable Development: A Way Forward in Promoting Sustainable Development in Higher Education Institutions." In *World Trends in Education for Sustainable Development* edited by W. L. Filho. Peter Lang.

Gough, A. (2013). "Historical, contextual and theoretical orientations that have shaped environmental education research." In *International Handbook of Research on Environmental Education,* edited by R. B. Stevenson, M. Brody, J. Dillon & A. E.J. Wals. Routledge.

Hume, T., & Barry J. (2015). Environmental Education and Education for Sustainable Development. *International Encyclopedia of the Social & Behavioral Sciences 7.*

Jucker, R. & Reiner M. (2014). Introduction: From a single project to a systemic approach to sustainability—an overview of developments in Europe. In *Schooling for Sustainable Development in Europe,* edited by Jucker R. & Mathar R. Springer.

Kahn, R. (2010). *Critical pedagogy, ecoliteracy, & planetary crisis: The ecopedagogy movement.* Peter Lang.

Khandakar, A., Chowdhury, M. E. H., Khalid, M. S., & Zorba, N. (2022). Case study of multi-course project-based learning and online assessment in electrical engineering courses during COVID-19 pandemic. *Sustainability, 14*(9), 5056.

Khandakar, A., Chowdhury, M. E. H., Gonzales, S. P., A., Touati, F., Al, E. N., & Ayari, M. A. (2020). Case study to analyze the impact of multi-course project-based learning approach on education for sustainable development. *Sustainability, 12*(2), 480.

Kopnina, H. (2019a). Ecocentric education: Student reflections on anthropocentrism-ecocentrism continuum and justice. *Journal of Education for Sustainable Development, 13*(1), 5–23. https://doi.org/10.1177/0973408219840567

Kopnina, H. (2019b). Green-washing or best case practice? Using a circular economy and Cradle to Cradle case studies in educational practice. *Journal of Cleaner Production, 219*, 613–623. https://doi.org/10.1016/j.jclepro.2019.02.005

Kopnina, H. (2020). Education for the future? Critical evaluation of education for sustainable development goals. *The Journal of Environmental Education, 51*(4), 280–291. https://doi.org/10.1080/00958964.2019.1710444

Kopnina, H., & Cherniak, B. (2016). Neoliberalism and justice in education for sustainable development: A call for inclusive pluralism. *Environmental Education Research, 22*(6), 827–841. https://doi.org/10.1080/13504622.2016.1149550

Kopnina, H., & Gjerris, M. (2015). Are some animals more equal than others? Animal Rights and deep ecology in environmental education. *Canadian Journal of Environmental Education, 20*, 108–123.

Mahmeed, T. (2011). *Opportunities and challenges for green design practice in the State of Kuwait: An exploratory study of design firm perceptions.* University of Florida.

Martin, S., Dillon, J., Higgins, P., Strachan, G. & Vare, P. (2014). Reflections on ESD in UK Schools. In *Schooling for Sustainable Development in Europe,* edited by Jucker R. & Mathar R.. Springer.

Mathar, R. (2014). Project variety and established structures: development and actual practice of ESD in Germany. In *Schooling for Sustainable Development in Europe,* edited by Jucker R. & Mathar R. Springer.

Mochizuki, Y., & Fadeeva, Z. (2008). Regional centres of expertise on education for sustainable development: An overview. *International Journal of Sustainability in Higher Education, 9*(4), 369–381.

MOEHE. (2023). Schools. https://www.edu.gov.qa/en/Pages/pubschoolsdefault.aspx?ItemID=48

Mohamed, B. H., Disli M., Al-Sada M. bin S., & Koç, M. (2022). "Investigation on Human Development Needs, Challenges, and Drivers for Transition to Sustainable Development: the Case of Qatar." *Sustainability 14* (6): 3705.

Molina-Motos, D. (2019). Ecophilosophical principles for an ecocentric environmental education. *Education Sciences, 9*(1), 37. https://doi.org/10.3390/educsci9010037

Najam, A. (2005). Developing countries and global environmental governance: From contestation to participation to engagement. *International Environmental Agreements: Politics, Law and Economics, 5*, 303–321.

Nikolaou, P. (2022). Sustainable development: A developmental evaluation of logistics higher education in the Sultanate of Oman Based on two Innovation Approaches (the Triple Helix of Innovation and Innovation Competencies). *Higher Education Evaluation and Development, 16*(2), 102–120. https://doi.org/10.1108/HEED-07-2021-0056

OECD. (2019). *The United Nations Decade of Education for Sustainable Development (2005–2014): Developing Strategies for Education for Sustainable Development*. OECD.

One-Planet. (2018). *Global Action Programme (GAP) on Education for Sustainable Development (ESD).*

Planning and Statistics Authority. (2018). Qatar National Development Strategy 2018–2022.

Qatar General Secretariat for Development Planning. (2008). Qatar National Vision 2030.

Qatar General Secretariat for Development Planning. (2011). Qatar National Development Strategy 2011–2016: Towards Qatar National Vision 2030.

Qatar National Commission for Education, Culture and Science. (2022). Associated Schools https://qnc.edu.gov.qa/En/AssociatedSchools/Pages/default.aspx.

Radwan, A. F., & Khalil, E. M. A. S. (2021). Knowledge, attitude and practice toward sustainability among university students in UAE. *International Journal of Sustainability in Higher Education, 22*(5), 964–981. https://doi.org/10.1108/IJSHE-06-2020-0229

Rauch, F. & Pfaffenwimmer, G. (2014). Education for sustainable development in austria: networking for innovation. In *Schooling for Sustainable Development in Europe,* edited by Jucker, R. & Mathar, R. Switzerland: Springer.

Ricard, M. & Dussaux, M. (2014). "Education for Sustainable Development in France" in *Schooling for Sustainable Development in Europe,* edited by Jucker R. & Mathar, R. Switzerland: Springer.

Selaibeekh, L. (2017). *Citizenship Education in Bahrain: an Investigation of the Perceptions and Understandings of Policymakers, Teachers and Pupils.* University of Surrey (United Kingdom).

Sitka-Sage, M. D., Kopnina, H., Blenkinsop, S., & Piersol, L. (2017). Rewilding education in troubling times; or, getting back to the wrong post-nature. *Visions for Sustainability, 8*, 1–19.

Stapp, W. (1969). The concept of environmental education. *Journal of Environmental Education, 1*(1), 33–35.

Sterling, S. (2004). An analysis of the development of sustainability education internationally: evolution, interpretation and potential transformation. In *The Sustainability Curriculum: the Challenge for Higher Education,* edited by Blewitt, J. & Cullingford C., 43–62. Earthscan.

Tbilisi. (2012). Tbilisi+35 a history of education for sustainable development. Ministry of Environment Protection of Georgia.

UN. (2015). *2030 Agenda for Sustainable Development and its 17 Sustainable Development Goals.* United Nations.

UNDCC. (1992). Agenda 21, Chapter 36. New York: United Nations Documents Cooperation Circles.

UNEP. (1972). Declaration of the United Nations Conference on the Human Environment. Nairobi, Kenya: United Nations Environment Programme.

UNESCO. (2007). The UN Decade of Education for Sustainable Development (DESD 2005–2014): The First Two Years. Paris: UNESCO.

UNESCO. (2014a). *Aichi-Nagoya declaration on education for sustainable development.* UNESCO.

UNESCO. (2014b). Shaping the Future we want: UN Decade of Education for Sustainable Development (2005–2014) Paris: UNESCO.

UNESCO. (2014c). Teaching and Learning: Achieving Quality for All. EFA Global Monitoring Report 2013–2014. Paris: UNESCO.

UNESCO. (2014d). *UNESCO Roadmap for Implementing the Global Action Programme on Education for Sustainable Development*. UNESCO.
UNESCO. (2018). Issues and trends in Education for Sustainable Development Paris: UNESCO.
UNESCO. (2020). Education for Sustainable Development: a roadmap #ESDfor2030. Paris: UNESCO.
UNESCO. (2022). UNESCO Associated Schools Network. https://aspnet.unesco.org/en-us/Pages/Education%20for%20Sustainable%20Development.aspx.
Vare, P., & Scott, W. (2007). Learning for a Change. *Journal of Education for Sustainable Development., 1*(2), 191–198.
Washington, H. (2015). Is 'sustainability' the same as 'sustainable development'? In H. Kopnina & E. Shoreman-Ouimet (Eds.), *Sustainability: Key issues* (pp. 359–376). Routledge.
Washington, H. (2018). *Education for Wonder. Education Sciences, 8*(3), 125–139. https://doi.org/10.3390/educsci8030125
WISE. 2015. The 2015 World Innovation Summit for Education. https://www.wise-qatar.org/app/uploads/2019/06/english_2015_wise_summit_curtain_raiser.pdf.

Chapter 19
Higher Education and Sustainability in the GCC Countries, the Case of Qatar

Igor Jacky Dimitri Michaleczek and Abdellatif Sellami

Abstract This chapter brings into question the role that institutions of higher education in the Gulf Cooperation Council (GCC) states play in promoting sustainable development (SD). The chapter discusses the challenges and issues associated with higher education and sustainable development in the region, arguing that any realistic approach to sustainability will have to provide contextualised solutions that reflect the local realities of the GCC countries, and at the same time take into account the dynamics surrounding these challenges. Underlying the present chapter is the argument that, across institutions of higher education, the promotion of sustainability has so far been limited to activities, interventions, and general strategy, but no specific focused policies have been put into place, and there are further major problems. Currently, there is no proper evaluation of the impact of higher education institutions on sustainable development, not only in Qatar and the GCC, but across the world as well.

Keywords Higher education · Sustainable development · Gulf cooperation council · Qatar

19.1 Introduction

With the increasing pressures to rethink progress and development, in the light of the world's shrinking resources, and the recognition of the importance of achieving sustainable growth and prosperity, countries around the globe have embarked upon a wide range of policy interventions to enhance economic, human, and social development. At the core of these endeavours lie the efforts made by (higher) education institutions, which have been identified as a key agent that serves to enact and effect change in the domains established as crucial for sustainable development (SD) (Shields, 2019). As Al-Misnad (in Sillitoe, 2014, p. XV) suggests, educational institutions are "primary units of socialization, the place where values are shaped and

I. J. D. Michaleczek (✉) · A. Sellami
Educational Research Center, College of Education, Qatar University, Doha, Qatar
e-mail: imichaleczek@qu.edu.qa

E. Zaidan et al. (eds.), *The Future of Education Policy in the State of Qatar*, Gulf Studies 21, https://doi.org/10.1007/978-981-97-9667-0_19

visions of the future are nurtured [they] have a natural role and even a moral obligation to lead the intellectual and values debate about sustainability" (p. xv).

In the face of heightened concerns raised about the myriad and complex social, economic, and environmental plights crippling the world, the demand for urgent action has now become more pronounced than it has ever been. Herein lies the importance of HEIs—and universities in particular—in providing the knowledge, skill sets, and values required for sustainable ways of living. Notwithstanding the scope, scale, and sophistication of the interconnectedness between universities as drivers of the knowledge economy and SD, there are stark reminders of the central place HEIs occupy in furthering a nation's human, economic, and social development. Arguably, whereas universities are increasingly recognised as a vehicle of progress and development and a catalyst for economic diversification, their influence as a driver of SD is still not fully acknowledged (Zaidan et al., 2019).

While international research has examined ways in which HEIs can contribute to SD goals in various parts of the globe (Berchin et al., 2021; Findler et al., 2018), limited work has been done on the role of HEIs in SD in the Arab world. Among some of the work done on sustainability and policies, Al-Hababi and Olawuyi (Al-Hababi, Chap. 2, and Olawuyi and Athmal, Chap. 3, in Cochrane and Habab 2023) both acknowledge the implementation in Qatar of various strategies, laws, and institutions to address the environmental challenges and to promote balanced social, economic, and environmental development. They also highlight the need for ongoing monitoring and assessment policies that will be able to provide reliable and accurate statistics which would allow the development of meaningful and relevant policies for SD. Much less is known about how and what SD policies are operational in Qatar and the larger Gulf Cooperation Council (GCC) and, by extension, the larger Arab region.

Amid the dominant official discourse condoning the proliferation of multiple education reforms, and the flow of countless international branch campuses (IBCs) within the GCC states, ongoing efforts by GCC leaders have not yet succeeded in yielding a blueprint for sustainability. Indeed, the meagre presence of GCC states, much like other Arab countries, among other green campuses in various world regions speaks volumes about their achievement regarding national capacity building and economic development.

This study problematises the role that current university models based in the GCC can play in promoting sustainability. The study makes a case against emulating "best practices" in implementing the commonly held vision of education for SD in a context such as that of the Arabian Gulf states. In so doing, it questions the utility of borrowing foreign sustainability models in mitigating the consequences of many of the region's escalating social problems and environmental crises. Undoubtedly, any realistic approach to sustainability will have to tackle the dynamics surrounding these challenges, and to be conceived to adapt to emerging and sudden changes. The need to lessen the effects of and/or stabilise erratic human activity within the social and natural environments cannot be overstated.

Following the preceding introduction, the next section there is a background section to set the scene for the remainder of the study, looking at the history and recent past, offering a contextual profile of Qatar and a description of its higher education system. Then there follows a presentation of sustainability, policy understanding within the context of this chapter, the specific challenges for SD in Qatar and Qatar HE, and finally some proposals for future policies and actions in supporting HE and SD.

19.2 Qatar: Country Profile and Its Higher Education

This chapter does not intend to offer a complete or detailed description of the SDGs in Qatar, as may have already been addressed in other chapters of this edited book. Instead, it highlights pertinent aspects that are relevant to understanding the current situation of Higher Education Institutions (HEIs) and SD in the country. In so doing, the aim is to explore potential policies, measures, or actions that could support HEIs in implementing SD Goals (SDGs).

Owing to its economic policies of investment and diversification, Qatar has over the years become a key player in the liquefied natural gas (LNG) market, as well as in other sectors of the economy in the region, Europe and the United States. Under Sheikh Hamad's rule, fundamental changes and strategic orientations were introduced in various educational, cultural, socio-economic, political, and media-related fields, among others. One of his major strategic choices for the development of Qatar is a transition to a knowledge society and the development of HEIs.

One of the main socio-economic aspects relevant to this chapter is Qatar's unique demographic developments. After the discovery of oil, this small peninsula started to attract migrants from India and the wider Arab world (Ahmed, 2018); while the population was less than 50,000 in 1960, today it is closer to 3 million (World Bank 2022). The development of gas and oil exploitation, winning the bid to host the 2022 World Cup in Qatar in 2010, and the attribution of the Asian Games earlier, enhanced the already existing developments in population growth from 1,637,443 in 2010 to 2,773,598 in March 2022 (World Bank 2022).

This increase in population can partially explain the challenges Qatar is facing in organising society, and in providing adequate services in vital public sectors, such as education, health, national security, etc. At present, Qatar's demographic composition features an overwhelming majority of foreigners, representing a large proportion of the working-age male population, resulting in a starkly unbalanced gendered population. This imbalance has led to an ethnic and cultural melting pot with challenges and opportunities for Qatari society.

19.2.1 Higher Education in Qatar

Before the establishment of the first Ministry of Education in Qatar in 1951, education was provided primarily at home or in mosques by traditionally trained teachers (Mutawwa'). The first boys' primary school was established in 1952 and the girls' in 1954 (Ahmed, 2018, 2019). The first boys' secondary school was founded in 1961 and the girls' in 1965. By 2019, Qatar had a total of 687 schools, from primary to secondary, with 22,988 teachers providing instruction to 283,000 children (PSA, 2022). This progression is even more striking when considering the diversity of school curricula and languages of instruction on offer. In fact, in 2019, government schools, which represented 30% of all schools in Qatar, offered the national curriculum in Arabic. By contrast, private and international schools, which enrol around 70% of students, are supposed to offer a more diverse education with higher quality (Amin & Cochrane, 2023) with various curricula from a range of countries, including American, British, French, Bangladeshi, and Finnish curricula, among others. This diversity is representative of the social fabric of the country, where the majority of the resident population consists of expatriates. Arguably, these migrants have specific educational needs that government schools are not able to offer, particularly in terms of curriculum and language.

Since the beginning of the twenty-first century, Qatar's education system has undergone a series of reforms to ensure its sustainability and relevance to a changing society and a competitive world. In the 1960s and 1970s, a local curriculum was developed to replace curricula imported from other Arab countries, including Egypt. The RAND landmark Education for a New Era study, conducted in 2002, aimed at addressing low achievement and systemic teaching and learning issues. Following the RAND report, a new organisational structure for the management of education, titled the Supreme Council of Education, was adopted. Key aspects of the proposed school reform included student-centred learning, critical thinking, and improvement in the status of teachers. The decision made back then aimed to develop a more efficient education system, following external advice and adopting imported practices from Western countries. The education reforms were criticised for neglecting local culture and knowledge. Radical rectifications and adjustments were made to further reforms and orientations that promote the culture, language, and history of Qatar.

It is important to note that due to its relatively recent organisation, the educational system in Qatar is still trying to find relevant solutions to recurring issues, such as lower attainment, the development of identities in a diverse society, and the appropriation of foreign models and practices. There is a need to understand how these K-12 reforms affect higher education in the country, especially regarding areas related to language, identity, and attitudes towards teaching and learning. It is also important to note that the education system has been the centre of successive reforms that may make it more challenging to adopt radical innovative practices such as education for SD.

The path of higher education provision in Qatar differs from that of K-12 schooling. The first Higher Education Institution (HEI) in the country started through the establishment of Qatar University's College of Education in 1977 with a small cohort of 57 males and 93 females. Interestingly, Qatar University remained the only HEI in the country until 2001. Today, it is still the main HEI in the country, with 10 colleges, offering courses in subjects ranging from Health Sciences, Engineering, Arts and Sciences, Education, and Sharia & Islamic Studies to Medicine and Law.

Qatar University hosts the majority of higher education students in the country, with more than 23,000 registered at the institution.

Alongside Qatar University, a breakthrough came with the initiation of the Qatar Foundation in August 1995 by Her Highness Sheikha Moza bint Nasser. The main objectives of the organisation are to support innovative and open society, by preparing "new leaders of Qatar, developing sustainable human capacity, and social and economic prosperity for a knowledge-based society" (Ahmed, 2019, p. 52). The Qatar Foundation established Education City, which houses the Qatar National Library, an innovative hospital (SIDRA), the Convention Centre, a research and innovation centre, and different sports and religious facilities.

Education City is also a campus that hosts eight universities, including seven IBCs from prestigious Western universities that maintain Western standards and offer qualifications locally, with the aim of building local capacities. In so doing, Qatar has joined other GCC countries that started hosting more than a third of the world's IBCs (Miller-Idriss & Hanauer, 2011). Between 1998 and 2017, 17 HIEs opened in Qatar with the majority being branches of European or American institutions, and main campuses, including the Qatar Foundation IBCs. Alongside these structures, local institutions such as the National Defence University for the armed services, the College of Islamic Studies at Hamad Bin Khalifa University (2007), and the Community College of Qatar (2010) were launched, followed by the Doha Institute (2015). This trend has endured in recent years, with two new universities opening since 2020: Lusail University in 2020, and the University of Doha for Science and Technology (2022), a new professionally orientated institution.

In the 2017/2018 academic year, 26,000 students registered in public HEI, and 8,000 in private institutions (PSA, 2021). Of these, 20,880 were enrolled at Qatar University (QU 2023). It is interesting to note that, back then, there were only three government and 17 private HEIs, exposing the size and impact of government HEIs compared to the private ones. Even if Qatar University remains the main HEI, the diversification and multiplication of HEIs have provoked an evolution in the higher education landscape in the country. It also questions the sustainability, economic, and social impact of the oldest Higher Education Institution. Qatar's potentially perceived dichotomy between the attraction of the IBCs and persistence to open local HEIs can, in fact, be seen as a strategic desire to attract and gain knowledge from renowned international institutions, while developing local capacity, taking into account local or regional expertise.

While GCC countries have enjoyed unprecedented economic growth and prosperity levels over the past four decades, their economic successes coincide with different and complex issues. Chief among these is the ongoing demographic transformations, increasing pressures on finite local natural resources, and urbanisation challenges. Awareness of the need to diversify their economies and move their countries away from heavy reliance on the gas and oil industries towards sustainable knowledge-based societies (Khalid 2016) confirmed Qatar's decision-makers conviction that the key to achieving sustainability is quality education. Integral to the country's strategic SD policy agendas over the past three decades is the vital importance attached to education institutions, especially HEIs, as engines of development. The sustainability of HEI relates to two main different business models, with some of these institutions being funded by the Qatar Foundation, some being strictly private organisations, and some public structures. This diversity presents opportunities and challenges for their sustainability.

Despite the significant resources invested in HEIs to promote SD and diversify the economy, the meagre results gathered so far point to a different reality. Indeed, ongoing debates accompanying the transition to a knowledge-based society have not always yielded an accurate portrayal of the prominent challenges that are predicted to persist in the future. The official discourse has not always shown a desire or readiness to delve into the complex dynamics that govern what are viewed as sensitive issues deeply embedded in the local culture. The present research offers a critical scrutiny of how Qatar-based universities and other HEIs have failed to present—and represent—a local voice that reflects the unique needs and perspectives of the communities that they seek to serve. Universities are required to engage in an honest and genuine dialogue with the local community, including students, families, educators, stakeholders, and other community members.

19.3 A Definition of Sustainability

In the face of myriad social, economic, and environmental challenges, SD has gained prominence on the world scene as a pressing twenty-first-century issue (Shields, 2019). With growing evidence testifying to the relentless depletion of the world's resources in different parts of the globe, the topic of sustainability has garnered heightened attention and continues to gather momentum (Robertson, 2017; 2014). At present, it is widely acknowledged that educational institutions are spearheading efforts to help societies adapt to profound socio-economic and cultural changes. More than ever before, institutions of education are called upon to promote Education for SD (García-González et al., 2020).

The literature on sustainability is vast, spanning various spheres of modern society, such as the economy, health, education, culture, the environment, and so forth. The definitions of 'Sustainability' and 'Sustainable Development' can be as diverse as the people who define the two terms (Redclift, 2002: 275). SD encompasses a multiplicity of facets related to the interaction between individuals and society, as well as between their knowledge, practises, attitudes, and beliefs about others and the environment. Following the Brundtland Report of 1983, and 36 years after the subsequent report Our Common Future, the concept of SD has evolved to promote social and economic transformation at the forefront. However, the original definition of sustainability as imperative to meet "the needs of the present without compromising the ability of future generations to meet their own needs" (United Nations, 1987, p. 54) remains relevant and highlights the overall responsibilities associated with participating in SD.

The definition and appreciation of SD now encompass economic growth, social progress, and environmental protection (Kirkfeldt & Frazão Santos, 2021). This renders SD more receptive to engaging social and individual well-being, along with the protection of the environment. Three main dimensions must be maintained to ensure a balanced and sustainable society: environmental (human activity respecting natural boundaries), social (social capital preserved through community participation), and economic (maintaining the capital) (Ukko et al., 2018).

In 1999, through the adoption of the Millennium Development Goals, the focus of SD was on social (education, poverty, and health) and environmental dimensions. These two goals were separated at the UN Conventions on Climate Change, Biodiversity, and Desertification (Islam and Iverson 2018). Islam and Iverson note that the separation between social and environmental goals was not necessarily efficient and even increased disparities and injustices. Limited achievements led to the Rio + 20 summit and, three years later, the adoption of the 2030 Agenda was announced with seventeen SDGs and 169 targets (United Nations 2015).

Two important aspects should be stressed concerning SD. Firstly, SD is a *preservation* process, preservation of resources, ecological and human heritage, and economic capacities. Secondly, it is also the validation of just, fair, and positive sustainable practices for social organisation and access to resources, as well as the knowledge necessary for individuals to be active in this process. This can provoke tension and may have deep implications in the definition of policies for the delivery and achievement of goals.

The 2015 UN declaration introduced the concept of transformative change, in order to end poverty, reduce inequality, promote a fair, peaceful, and inclusive society, tackle climate change, and protect the environment (Purcell & Haddock-Fraser, 2023, p. 4; UN 2015). Transformative change implies the need to address the status quo and social structures, to bring about radical transformation. It calls for partnership and collaboration, as well as for socio-technical innovation that transforms technologies, practices, and behaviours (Schot & Steinmueller, 2018). Transformative change requires what Purcell and Haddock-Fraser (2023) call double-loop learning, entailing a critical analysis of the mismatch between the current situation and the expected goals, through examination and adjustment of the governing variables (Purcell & Haddock-Fraser, 2023, p. 5). The concept of transformative change is particularly relevant in exploring the operations and practices of HEIs, and the concept of double-loop thinking is significant in looking at policies governing and orienting HEIs.

Education in general and HEI in particular have always been important components of the SDGs, as called for in SDG4: "by 2030, ensure equal access for all women and men to affordable and quality technical, vocational, and tertiary education, including university" (UNESCO, 2017 p. 1). The SDGs that address poverty (SDG1), gender equality (SDG5), governance and economic growth (SDG8), climate change (SDG13), and peace, justice, and strong institutions (SDG16), among others, integrate HEI as a key component of their goals. The 2030 Education Framework for Action calls for international agreements and for the regulatory framework to align teaching and learning with SD.

The UN asserts that universities must play a responsible role in the implementation of the SDGs, and contribute to the necessary socio-economic and behavioural changes. This implies an assumption about the role of the university. Traditionally, universities were seen as seeking the establishment of objective truth through teaching and research, and today's HEI can also serve the same purposes (Sin et al., 2017). Recently, universities have had to adapt to the massification of student enrolment, tending to adjust to job market needs by offering degrees and qualifications in demand, and attempting to develop skills that would increase students' employability (Yap et al., 2014).

In addition to their traditional role of knowledge creation, training, and educating future leaders of societies (Baker-Shelley et al., 2017), HEIs are now responsible for contributing to ensuring the future of our societies, economies, and the environment, through the implementation of SDG. Mass enrolment in the HEI has led to its commercialisation and the redefinition of its role in promoting social mobility, increasing knowledge, and ensuring continuous economic growth and competitiveness (Purcell et al., 2023). With students becoming units of output in terms of university activities, ready and willing to participate in economic activity, they are treated as 'human capital' (Marks & Steuer, 2008).

Finally, the university's mission has had to be expanded, to include interaction with communities, and, more recently, they have become responsible for incorporating sustainability into their processes, systems, and strategies (Zaidan, 2019). The role of the university in SD has been acknowledged at different levels as a catalyst for macro- and mesoeconomic development, social cohesion, and innovation. HEIs can play an important role in the realisation of SDGs, as they have long been powerful drivers of global, national, and local innovation, economic development, and social well-being (Australia/Pacific SDSN, 2017). University teaching and learning activities, governance, research, and cultural contribution, are key in influencing and affecting societies (Ottersen, 2019). As centres of knowledge initiation, and research, HEIs are expected to guide and support societies in tackling environmental and societal challenges (Stephens et al., 2008).

As the locus of critical thinking, knowledge creation, innovation and training for the new generation, HEIs possess the necessary ingredients to facilitate the transformative change required for the successful implementation of the SDGs. To be able to do so, HEIs need not only to support and facilitate changes in society, but also to change within their own organisation so as to become more sustainable (Stephens et al., 2008). This implies the ability to think critically in developing curricula, changing practices, and analysing their environment. This will allow HEIs not only to be part of changes, but also to lead and sustain SD.

19.4 Sustainability in the Context of Qatar and the Broader GCC

The 17 SD Goals (SDGs) recognised by the United Nations General Assembly in 2015 were identified as critical aims to be achieved by 2030, in order to "secure a sustainable, peaceful, prosperous and equitable life on earth for everyone now and in the future" (United Nations, 2017, p. 6). The recognition of the SDGs as key targets established to ensure the world's mounting woes are addressed has triggered compelling demands for SD initiatives to preserve the planet's natural resources, protect the ecosystem, and enhance people's quality of life and well-being for the present and future generations. To attain these goals, governments, the private sector, and civil society are called upon to engage in various forms of SD endeavours.

The transition to a sustainable knowledge-based society postulates a profound transformation of the way people think and behave. Demonstrably, an ambitious initiative that has been distinguished as fundamental to effecting change on the path of SD is related to education. The onus rests on institutions of education to contribute to the development of sustainability. According to a United Nations report titled Education for sustainable development goals: Learning objectives, education is entrusted with the "responsibility to be in gear with twenty-first century challenges and aspirations and foster the right types of values and skills that will lead to sustainable and inclusive growth, and peaceful living together" (United Nations, 2017, p. 7).

Aware of the centrality of education in the process of SD and social progress, and the key role it plays in enabling societies to gain a competitive edge in the world today, educational institutions are taking the lead in efforts that strive to respond to the economic, environmental, and social needs of modern globalised societies. Higher Education Institutions (HEIs), in particular, are taking centre stage in leading global efforts dedicated to responding to growing concerns, essentially via teaching and research channels, along with community service activities. More specifically, universities occupy a unique place as key actors with great potential to guide SD policies and influence current practices at both individual and community levels.

In the GCC context, the need for sustainability is even more compelling compared to other world regions. There is ample evidence to suggest that the GCC states, much like other Arab countries, constitute an area that is particularly vulnerable to rapid climatic (Al-Maamary et al., 2017) and demographic changes (Zaidan et al., 2019). Exacerbating the problem is the region's low capacity to adapt to change (Al-Hababi, 2023). To address these challenges, GCC leaders adopted strategic measures to ascertain the protection of the environment and growth of their economies, without jeopardising the welfare and well-being of their populace. At the heart of these efforts is the realisation that educational institutions, specifically HEIs, hold the key to solving the region's existing and emerging environmental and socio-economic ailments.

It is true that successive GCC governments have embarked on implementing long-term social and economic strategic plans (i.e., Bahrain's Vision 2030, Oman's Vision 2020, Qatar's National Vision 2030, Saudi Arabia's Vision 2025, and the

UAE's Vision 2021). It is also true, however, that reality on the ground tells a different story. Several researchers (Hassen, 2022; Syed et al., 2023) point to a chasm between aspirations cherished by the leadership in GCC countries, and their potential frustrations with what has been accomplished. Political and public discourse and rhetoric within the realm of education are stark reminders of the discrepancy between visions, policy statements, and real-life practices. This study looks at some key barriers hampering ongoing attempts to transform GCC nations into sustainable knowledge-based societies.

19.5 Policy, Definition, and Analysis Framework

19.5.1 Policy Definition

Declarations and reports issued by the United Nations and UNESCO provide both a theoretical and practical framework guiding decision-makers and societies in understanding and developing SD. They have been the basis for the design of numerous policies around the world. Policy is a complex concept that cannot be understood solely from a positivist perspective as simple governmental directive. This is even more the case in the field of education, where policy implies comprehension of social, economic, and political contexts that can influence its formulation and implementation (Ozga, 2000). Bates and Eldredge (1980) define policy as a statement providing guidance to decision-makers in charge of operating the system. In other words, policy refers to what the government expresses and does about identified challenges (Theodoulou & Cahn, 1995:201).

This implies that policies tend to impact the way we act upon certain issues, according to the perception we have of both these issues and the potentially relevant way to address them. Policies should then be culturally and contextually relevant.

Through its different bodies and organised conferences, the UN has developed a corpus of texts and guidance for the past 50 years, starting with the Stockholm Human Environment (UNCHE) Conference in 1972. All of these texts, and more particularly following Chapter 36 of Agenda 21, elaborated during the Rio Conference in 1992, have strived to highlight and guide education and Higher Education's contribution to SD. A clearer institutional framework emerged at the Johannesburg, 2002 Conference, with the creation of the concept entitled World Decade. Ten years later, at the 2012 Rio + 20 Conference, additional guidance was provided, with insistence on the importance of education.

These authoritative texts and discourses influence the action and conceptualisation at a macro-level. However, they are general guidelines that need to be revised and adapted to the local context. The expansion of SD implementation has continued around the world, as mentioned in Dale's description of the five mechanisms of globalisation: harmonisation (through treaties), dissemination (through international organisation activities), standardisation (through charters), installation of interdependence (common curriculum and resources), and imposition (through international bodies) (Dale, cited in Fimyar, 2014, p. 13). SD may follow some of these patterns, so that designing adapted and appropriated policies at the local level may be the opportunity to appropriate and adapt external guidance to the local context.

19.5.2 Policy Analysis

The main purpose of this chapter is to explore the current challenges and difficulties in HEI support for SD and identify potential action that could facilitate an adapted implementation and appropriation of SD relevant to the local context of Qatar. Policy analysis is multidisciplinary and, as such, can draw on diverse fields; as this chapter is limited in scope, we will adopt a delimited approach with defined concepts. In doing so, we agree with Dunn (2015) that it is better to be pragmatic, by agreeing on expected desired outcomes and/or courses of action (Dunn, 2015).

Knowledge about policy analysis is based on five areas. These five areas guide and organise inquiry for policy relevance and inform research and choice.

Specific methodological approaches are connected with each of these knowledge areas:

- o *Problem structuring* supports understanding of policy problems, allowing for identification, organisation, and appreciation of the problems to be solved;
- o *Forecasting* allows for the identification of expected outcomes and allows for the evaluation of policy impact;
- o *Prescription* informs about the preferred policies identified to address a problem, while detailing the potential cost and income generated by the expected policies;
- o *Monitoring* is a framework allowing for the organisation and analysis of the observed outcomes, and the forecasting of potential additional expected outcomes;
- o *Evaluation analyses* the values and outcome, in relation to the expected performance of a policy (Dunn, 2015).

These areas are interdependent, and using one often implies using the others. However, the present chapter examines two main areas of policy analysis, namely problem structuring and prescription. The chapter focuses on identifying and indexing the main problems and challenges associated with SD in Qatar, in relation to the existing policy choices. Then, by looking at various policies and strategies selected in various contexts, potential solutions and prescriptions will be identified.

19.5.3 Sustainable Development in Qatar: Identified Problems and Challenges

Qatar's geographical location and socio-economic situation entail several challenges that must be addressed to engage with SD. In the face of these challenges, Qatar benefits from several assets that allow the government and its institutions to explore potential solutions. These challenges need to be acknowledged and structured, in order to identify potential avenues to inform decision-makers and policies for HEI. Indeed, HEIs have a role to play in supporting innovation and social change. The challenges may be organised into two categories: those inherent in the socio-economic and geographical situation of Qatar, and those emerging from the strategic and political choices made to steer the development of the country.

Qatar has the most unsustainable ecological footprint in the world (World Population 2017) and has been for most of the last ten years one of the countries with the highest per capita impact on the planet. Qatar's inhabitants need an average of 14.72 hectares to produce resources for their consumption and daily life. With 99.2% of Qatar's population living in urban areas, this pressure is increased by one of the highest urbanisation rates in the region. The urban population has seen rapid growth, passing from 622,000 in 2010 to 2,676,000 in 2022 (World Bank in Macro Trends, 2023). This rapid urbanisation increases pressure on the fragile local ecosystem, currently one of the weakest in the world, with difficulties over water supply, limitations in waste management, and rapid urban expansion. Even if careful planning and conceptualisation efforts have been put into urban centre development (Sillitoe, 2014), there remain management complexities, along with challenges over waste and water supply, economic growth, cultural diversity, and identity.

Another main challenge for SD in Qatar is its high dependence on hydrocarbons and its associated need for economic diversification. This has led to the development of alternative sectors such as industrialisation and tourism. Both of these sectors involve increased pressure on the already fragile ecosystem. Tourism increases pressures on the ecological environment through intensification in the need for water and improved infrastructure, as well as a surge in waste production. Industrialisation also increases pressure on water consumption with a high energy footprint and environmental footprint. Given that Qatar depends at a rate of more than 90% on imports for its internal consumption, agriculture, and the food agro-industry is also to be encouraged (Hassen et al., 2020).

Although this might address the dependency of Qatar on food imports, food production adds another strain on the environment and, more particularly, on the depletion of water reserves. Economic diversification also involves promoting privatisation and entrepreneurship, leading to increased demand for a foreign workforce possessing higher skills. Challenges to diversification and privatisation include bureaucratic, legal, and organisational issues. These add to highly subsidised sectors of the economy and monopoly situations (Ezmet et al., 2019). All of these factors make the development of the private sector more challenging.

To ensure the sustainability of Qatar's society and economy, Qatarisation of the workforce and management of the migrant influx remain a priority, because a reliance on migrant workers implies an ongoing dependency. However, this has proven to be a challenge, especially given that Qataris represented only 5% of the workforce population in 2019, for example, compared to 6% in 2011. While 79% of the total workforce is in the private sector, 81% of Qatari workers work in the public sector (PSA, 2020). Concerns about workforce qualification and population structure are another major challenge for SD implementation in Qatar. On the one hand, 47% of Qatari citizens in the workforce are university graduates, while on the other hand, 64% of the migrant workforce did not attend secondary school (PSA, 2019). These figures present challenges to guaranteeing a qualified workforce, but above all render Qatarisation nearly impossible. This is even more difficult when working for government institutions or the military promises a higher income than the private sector. Early retirement is made easier for the Qatari population.

Like other GCC countries, Qatar has often been described as a rentier state, a system that implies the redistribution of benefits from oil and gas to citizens, through an extended social welfare apparatus (Sillitoe, 2014). The rentier system is one of the main components of the social structure, and the relationship between the state and the population, and is a fundamental aspect of socio-economic systems (Hilal, 2020; Sillitoe, 2014). The mechanism results in a high dependency of the population on state and government support. It also results in increased consumption of highly subsidised water and energy, a high rate of waste production, and disengagement from work and education among the population. Rentierism is recognised as a challenge for the implementation of the SDG, as it negatively affects behaviours, mentalities, and practices, while ensuring a higher material quality of life. The ability to address these structural and contextual socioeconomic challenges in the context of Qatar will constitute a real milestone in the achievement of the SDGs. Overcoming their impacts is a prerequisite for successfully supporting a sustainable society, economy, and environment.

Alongside these structural and contextual challenges, Qatar is facing other challenges linked to the strategic choices made by the country's leadership, so as to transition from a hydrocarbon to a knowledge-based economy (Al-Thani et al., 2021, Mohamed et al., 2022). The 2007 report Turning Qatar into a Competitive-Based Economy: The Qatar Knowledge Economy Assessment, supported by the Qatar Foundation, the Qatar Planning Council, and the World Bank, influenced most policy and planning documents, in including the knowledge economy as a constitutive part of Qatar's future (Weber, 2014). The main orientation of public policies and regulation is based on Qatar National Vision 2030 (QNV 2030), published in July 2008, as the main document guiding upcoming policies and plans, and it integrates education and the knowledge economy in its core objectives. These objectives are organised around five main pillars articulating the development of human, social, economic, and environmental dimensions. Developing a knowledge economy and society demands addressing Qatar's main challenge, in order for it to become sustainable and end its dependence on hydrocarbons.

The first main challenge in developing a knowledge economy relates to the development of education and a skilled national workforce. Myriad other challenges continue to cripple the field of education, including a limited interest in HE and the low number of Qatari graduates in HEIs in key subjects needed for adaptation to a knowledge society, with a steady decline in students' interest in studying STEM subjects, and questionable critical thinking skills development. Other challenges comprise a lack of collaboration between HEIs within the country, despite their limited number, and the high proportion of migrant children and international students in HEIs, who are highly likely not to remain in the country in the long term. This list is not exhaustive and contributes to the difficulty of ensuring economic diversification. They add to repeated problems in the education system, manifesting themselves in consistently lower performance in international examinations, such as PISA and TIMSS, compared to other OCDE countries. These low attainments are challenging, as this also exposes teaching and learning issues.

The existence of these difficulties must not prevent the recognition of the progress made in the education sector or the development of institutional support for the advancement of research within a diverse dynamic HE sector in the country. Continuous efforts to improve education provision through comprehensive institutional reforms, massive investment in infrastructure improvement, and the number of education providers from primary to higher education must also be acknowledged (Fadlelmula and Koc, 2016). The inception of research funding, along with other initiatives with the Qatar National Research Fund, the establishment of research and technological hubs, and an increase in education spending in the national budget are prominent examples of this determination and strategic orientations assumed by the country's leadership.

19.5.4 Structuring Challenges and Problems

Identification of the problematic and challenging issues is the first step towards defining potential policies and their expected outcomes. As Dunn stresses, "policy problems are unrealized needs, values or opportunities for improvement" (Dunn, 2015, p. 69). Problems are perceived as opportunities for action and improvement, rather than as negative effects of deterministic reality. The challenges and problems related to successfully reaching the SDG and transitioning into a knowledge society have been the object of much research (Esmat et al., 2019; Mohamed et al., 2022; Sillitoe, 2022). In all of these studies, the problems and challenges have some commonalities and proposals to address them. However, rather than just listing these challenges and problems, in the following section, we will attempt to proceed to problem structuring, organising the problems, and hierarchising them, as a pre-step to problem-solving and the identification of solutions.

There is a need to understand an evident interdependency between the problems, for problems are part of an overall system. Therefore, it is logical that addressing a

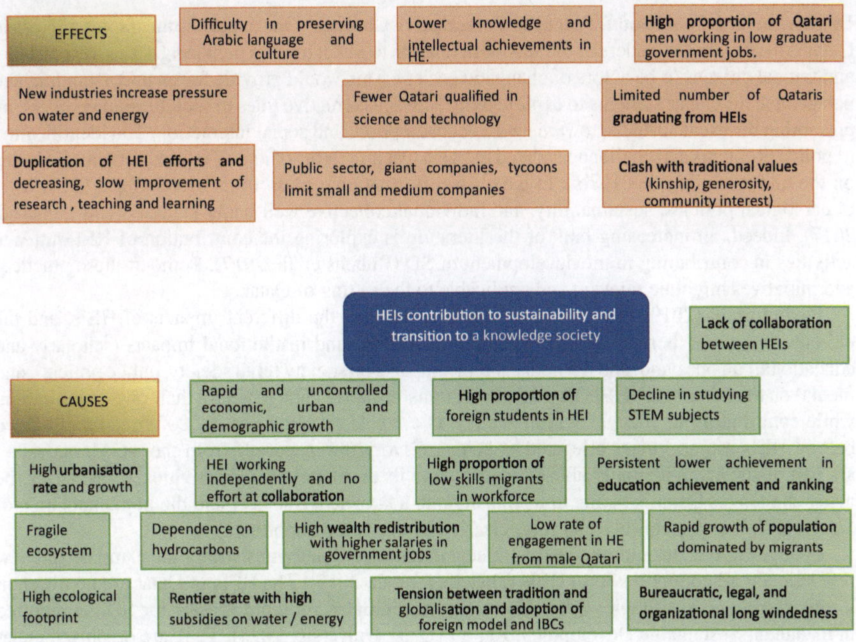

Fig. 19.1 Policy problems tree for developing HEIs contributing to SD Goals (based on Veselý, 2008)

problem using a reductionist and isolationist approach, i.e., taking problems independently of each other, will not be efficient (Russell in Dunn, 2015, pp. 72–73). The group that defines also construes problematic policies, because policy problems are socially constructed, resulting from a group's interpretation of social reality and expected reality. Thus, the definition and solution to policy problems can change and become irrelevant as the definition of the problem is subject to change, which means that policy problems are unstable.

We will use Veselý's approach, in order to gain a heuristic understanding of policy problems (2008), using a problem tree identifying the respective causes and effects of identified problem(s). This aims at identifying and structuring the problem to explore potential solutions. In Fig. 19.1, the challenges and problems identified in the literature are organised in terms of cause and effect. The problem tree is a thinking instrument and offers the option of determining causes and effects, and rather than focusing on the effect of the situation to define a potential solution, it directs attention to the causes of the problems. Under this approach, three main groups* of problems appear: those related to administration and governance, cultural and social, economic and workforce, teaching and learning, and finally, the environment. This allows the grouping of the problems without a necessary causal relationship between them, even if it can be set down in relation to plausible effects.

Discussion and Conclusion. Potential Prescriptive Choices One of the assumed roles of HEIs in Qatar is to prepare the leaders of tomorrow. In relation to SD, it implies preparing active, responsible, and critical citizens to be actors of change in society. This would provide for an understanding and perspective for Qatari students to explore more active and creative roles in society, among others, by prompting different attitudes towards work, consumption, and social interaction. The identification of policy problems and challenges related to SD in Qatar in terms of its society and economy is based on the understanding that HEIs can have a transformational impact in the real world at the level of ecological policies, sustainability, and individual/collective well-being (Thomas and Ormerod, 2017). Indeed, an increasing bulk of the literature is exploring the contribution of HEI outreach activities in contributing to the development of SD (Rubens et al., 2017). Some of these practices and initiatives might be relevant and applicable to the setting of Qatar.

Findler et al. (2019) insist on the distinction between the different impacts of HEIs, and the policies that guide them in SD; the traditional academic and institutional impacts (scholarly and educational) on one hand, and the nonacademic impacts on society (civil society, public policies, and media) on the other hand. HEIs have to impart transformative change within their own organisation, while contributing to changes within society as a whole (Stephens et al., 2008). As highlighted before, HEIs are no longer relegated to being an ivory tower, cut off from the world, but rather are part of society, potentially able to contribute to its development and activities. Considering the many challenges Qatar is facing in its transition to a knowledge society and the implementation of SDGs, HEIs' responsibilities and roles clearly become more important.

Changes to the curriculum, designing new modules, SD activities within the campus, and new research, are all common within HEIs (Disterheft et al., 2015). The HEIs in Qatar are no different in this respect. For example, Qatar University has initiated multiple actions for SD, as reported in its annual Sustainable Development Report (Qatar University 2023). They are organised under four categories, Research, Outreach, Students and Learning, Operation and Collaboration, for each of the 17 SDGs. While the university engages in long-term activities, such as finding financial support for students, many of the activities are one-off events or include activities already existing in the university that may not necessarily have been initiated while having SD in mind (Shawe et al., 2019). This may raise concerns about the potential dispersal of activities without an overall detailed strategic or policy framework. Such a framework would enhance the activities' impact by aiming at a common SD goal. It would also guide actors' interventions, by raising awareness of the importance of SD in achieving its goals. Other HEIs are also conducting SD activities and interventions. There may be dispersal, duplication, or even competition related to these activities. The definition of a common nationwide framework and reference policy for HEIs with a healthy leeway for each institution to act upon its own priorities and strategies could help in coordinating efforts and enhancing potential impact.

Another important aspect is the need for collaboration between the HEIs in the country. Education City IBCs collaborate and coordinate on specific programmes (Ahmed, 2019), and there is also the Education City Sustainability Network, involving the Qatar Foundation Institution and IBCs in SD. Yet there seems to be limited collaboration with other HEIs. Additionally, several networks for SD exist in Qatar, at the national level, with Sustainable Qatar or Qatar Sustainability Network (QSN), as well as at the regional level, with the Gulf Organisation for Research & Development. However, these networks are either not specifically focused on HE nationally or focusing at national level but not within HE.

Similarly, HEIs are individual members of international initiatives such as Nature Positive Universities. Let us highlight two aspects which can be beneficial in this regard. First, each institution remains isolated and takes action on its own, without institutional or formal collaboration, even if collaboration can happen at an individual level. Secondly, there is again risk of duplication or even competition in undertaking activities for SD, a risk already highlighted above. Regional and national HE networks exist in Europe or North America. The efficiency and benefits of international networks and collaboration for SD have been observed (Bizerril et al., 2018). HEI Networks can both be influenced by or influence policies. Indeed, policies can facilitate their emergence and efficiency, and they can contribute to policy elaboration and implementation. Institutional policies, from governments or HEIs, can support the development of an HEI network focused on SD (Fahim et al., 2021).

HEIs could also play a capital role in influencing attitudes towards work, career choices, and orientation in the direction of STEM-related academic subjects and associated jobs. HEIs aim at preparing responsible individuals, equipping students with lifelong learning abilities, and "empowering" them to become future leaders (Sever & Tok, 2023, p. 332), which are factors of change in society (Stephens et al., 2008). Universities develop critical and responsible citizens (Larraín et al., 2021), providing them with a comprehensive understanding of environmental issues, their causes, and potential solutions (Boca & Saraçli, 2019). This will be cardinal in providing students with the ability to reflect on their professional choices and take responsible sustainable action in society. This is a complex process that the policies at the MOEHE and HEI levels can help to develop.

While Education City IBCs have provided efforts in recent years to transform teaching and learning practices in HEIs (Chouchane et al., 2011), they have traditionally been teacher-centred, with the focus primarily on lectures and the transmission of knowledge from teacher to students (Sabah & Du, 2018). Efforts have been made to move towards student-centred learning, which emphasises active engagement, problem-solving, critical thinking, and collaboration among students (idem). There remains to be done to ensure that critical thinking skills are promoted within HEI practices. Critical thinking skills are key in ensuring that HEIs graduates and the rest of the society become actors in SD.

Finally, the evaluation of HEIs' impact on SD is challenging, as it is not possible to attribute results to the action itself without discarding other elements (international input, social change, economic evolution, and so on) (Findler et al., 2018). In Qatar, this is even more challenging as a result of diversity in the actions and approaches of the HEIs. Research into the impact of HEI outreach activities has gaps, among others, because of the difficulties in evaluating the impact of action within society. Informing policy and decision-makers, both at institutional and governmental levels would benefit from efforts from the academics in Qatar to evaluate, think, and explore HEI action over SDGs.

Bibliography

Ahmed, F. B. J. (2019). Foreign universities in Qatar: a critical review of policy and sustainability issues. *Challenges of the Knowledge Society*, 1158–1168.

Ahmed, F. B. J. (2018). Challenges of the knowledge society: Exploring the case of Qatar. *Global Economic Observer, 6*(1), 39–54.

Al-Hababi, R. (2023). The evolvement of Qatar's environmental sustainability policy: The strategies, regulations, and institutions. In: Cochrane, L., Al-Hababi, R. (eds) *Sustainable Qatar*. Gulf Studies, vol 9. Springer. https://doi.org/10.1007/978-981-19-7398-7_2

Al-Maamary, H. M., Kazem, H. A., & Chaichan, M. T. (2017). Climate change: The game changer in the Gulf Cooperation Council Region. *Renewable and Sustainable Energy Reviews, 76*, 555–576.

Al-Thani, W. A., Ari, I., & Koç, M. (2021). Education as a critical factor of sustainability: Case study in Qatar from the teachers' development perspective. *Sustainability, 13*, 11525. https://doi.org/10.3390/su132011525

Amerio, M. (2020). Doha metro project: 3TI lesson learned. *Transportation Research Procedia, 45*, 866–873.

Amin, H., & Cochrane, L. (2023). The development of the education system in Qatar: Assessing the intended and unintended impacts of privatization policy shifts. *British Journal of Middle Eastern Studies*. https://doi.org/10.1080/13530194.2023.2198688

Australia/Pacific SDSN. (2017) Getting started with the SDGs in universities: a guide for universities, higher education institutions, and the academic sector; Australia/Pacific SDSN: Melbourne, Australia.

Baker-Shelley, A., van Zeijl-Rozema, A., & Martens, P. (2017). A conceptual synthesis of organisational transformation: How to diagnose, and navigate, pathways for sustainability at universities? *Journal of Cleaner Production., 145*(1), 262–276. https://doi.org/10.1016/j.clepro.2017.01.026

Bates, D. L. & Eldredge, D. L. (1980). *Strategy and policy: Analysis, formulation, and implementation.* WC Brown.

Berchin, I. I., de Aguiar Dutra, A. R., & Guerra, J. B. S. O. D. A. (2021). How do higher education institutions promote SD? A Literature Review. *Sustainable Development, 29*(6), 1204–1222.

Bizerril, M., Joao Rosa, M., Carvalho, T., Pedrosa, J., (2018). Sustainability in higher education: A review of contributions from Portuguese speaking countries. *Journal of Cleaner Production, 171.* 10/1016/j.jclepro.2017.10.048.

Boca, G. D., & Saraçli, S. (2019). Environmental education and student's perception, for sustainability. *Sustainability, 11*(6), 1553. https://doi.org/10.3390/su11061553

Chouchane, L., Mamtani, R., Al-Thani, M., Al-Thani, A. A. M., Ameduri, M., & Sheikh, J. I. (2011). Medical education and research environment in Qatar: a new epoch for translational research in the Middle East. *Journal of Translational Medicine, 9* (1). https://doi.org/10.1186/1479-5876-9-16

Cochrane, L., & Al-Hababi, R. (Eds.). (2023). *Sustainable Qatar: Social, political and environmental perspectives.* Springer.

Disterheft, A., Caeiro, S., Azeiteiro, U.M., Leal Filho, W., (2015). Sustainable universities—a study of critical success factors for participatory approaches. *Journal of Cleaner Production 106*, 11–21. https://doi.org/10.1016/j.jclepro.2014.01.030

Dunn, W. N. (2015). *Public policy analysis: An integrated approach.* Routledge.

Fadlelmula, F., & Koc, M. (2016). *Overall review of education system in Qatar*. Lambert Academic Publishing.

Fahim, A., Tan, Q., Naz, B., Ain, Q. U., & Bazai, S. U. (2021). Sustainable higher education reform quality assessment using SWOT analysis with integration of AHP and entropy models: A case study of Morocco. *Sustainability, 13*(8), 4312. https://doi.org/10.3390/su13084312

Fimyar O. (2014). What is policy? In search of frameworks and definitions for non-Western contexts, *Educate Vol. 14*, No. 3, p. 6–21, http://www.educatejournal.org 6 Critical Review

Findler, F., Schönherr, N., Lozano, R., & Stacherl, B. (2018). Assessing the impacts of higher education institutions on sustainable development—an analysis of tools and indicators. *Sustainability, 11*(1), 59.

García-González, E., Jiménez-Fontana, R., & Azcárate, P. (2020). Education for sustainability and the sustainable development goals: Pre-service teachers' perceptions and knowledge. *Sustainability*, 12(18), 7741. https://doi.org/10.3390/su12187741

GCC-STAT. (2018). "Economic Performance and Outlook for the Gulf Cooperation Council (GCC) 2018." Accessed May 15, 2019. https://gccstat.org/images/gccstat/docman/publications/Economic_2018.pdf.

Hassen, T. B., Bilali, H. E., & Damalas, C. A. (2020). Impact of Covid-19 on food behavior and consumption in Qatar. *Sustainability, 12*(17), 6973. https://doi.org/10.3390/su12176973

Hassen, B. T., El Bilali, H. (2022). Sustainable development goals in the Middle East and North Africa (MENA) region: Policy and governance. In I. R. Abubakar, I. da Silva, R. Pretorius, K. Tarabieh (Eds.), *SDGs in Africa and the Middle East region. Implementing the UN sustainable development goals—regional perspectives.* Springer, Cham. https://doi.org/10.1007/978-3-031-17465-0_20

Hilal, N. (2020). Tourism in the Gulf Cooperation Council countries as a priority for economic prospects and diversification. *Journal of Tourism and Hospitality, 9*(451), 2167–2269.

Khalid, F. (2016). Transforming strategic outlook on sustainability: *The case of higher education institutions in the Gulf Cooperation Council (GCC) countries.* Also sponsored by Hosted and sponsored by, 45.

Kirkfeldt, T. S., & Frazão Santos, C. (2021). A review of sustainability concepts in marine spatial planning and the potential to supporting the UN sustainable development goal 14. *Frontiers in Marine Science, 8*, 713980. https://doi.org/10.3389/fmars.2021.713980

Larraín, A., Fortes, G., & Rojas, M. (2021). Deliberative teaching as an emergent field: the challenge of articulating diverse research agendas to promote educational experiences for citizenship. *Frontiers in Psychology, 12*. https://doi.org/10.3389/fpsyg.2021.660825

Macro Trends, (2023), Qatar Urban population 1960–2023, access 27/08/2023, Qatar Urban Population 1960–2023 | MacroTrends

Marks, N., and Steuer, N. (2008). 'University Challenge: towards a wellbeing Approach to Quality in Higher Education'. New Economics Foundation. https://neweconomics.org/2008/06/university-challenge

Meza, A., Koç, M., & Al-Sada, M. S. (2022). Perspectives and strategies for LNG expansion in Qatar: A SWOT analysis. *Resources Policy, 76*, 102633.

Miller-Idriss, C., & Hanauer, E. (2011). Transnational higher education: Offshore campuses in the Middle East. *Comparative Education, 47*(2), 181–207.

Mohamed, B.H., Disli, M., Al-Sada, M.b.S. & Koç, M. (2022). Investigation on human development needs, challenges, and drivers for transition to sustainable development: the case of Qatar. *Sustainability, 14*, 3705. https://doi.org/10.3390/su14063705

Obaidli, K. (2009). Implementation of the K-12 education reform in Qatar's Schools; RAND report prepared for the Supreme Education Council. RAND.

Olawuyi, D.S., Athwal, E.I. (2023). Law and governance innovations on sustainability in Qatar: Current approaches and future directions. In: Cochrane, L., Al-Hababi, R. (eds) *Sustainable Qatar*. Gulf Studies, vol 9. Springer. https://doi.org/10.1007/978-981-19-7398-7_3

Ottersen, O.P. (2019). Modernising higher education in the agenda 2030 era. In *Proceedings of the Rethinking Higher Education Inspired by the Sustainable Development Goals*, Solna, Sweden.

Ozga, J. (2000). *Policy research in educational settings: Contested terrain*. Open University Press.

Planning and Statistics Authority, Labor Force Sample Survey: Annual Report 2020 Doha—Qatar

Planning and Statistic Authority. (2021). *Education in Qatar, statistical profile*. Planning and Statistic Authority (PSA).

Planning and Statistic Authority. (2022). *Education in Qatar, statistical profile*. Planning and Statistic Authority (PSA)..

Purcell, W. M., & Haddock-Fraser, J. (Eds.). (2023). *The Bloomsbury Handbook of Sustainability in Higher Education: an Agenda for Transformational Change*. Bloomsbury Publishing.

Qatar University Sustainable Development Goals, Report 2022/2023, consulted July 2023, http://www.qu.edu.qa/static_file/qu/offices%20and%20departments/International%20Affairs%20Office/Documents/QU%20SDG%20Report%2023.pdf

Qatar University. (2023). *Book of trends. Spring 2022*. Department of Institutional Research and Analytics, Office of Strategy and Development, Qatar University.

QNV. (2008). *Qatar national vision 2030*. General Secretariat for Development Planning.

Redclift, M. (2002). Discourses of sustainable development. In V. Deasi & R. B. Potter (Eds.), *The Companion to Development Studies* (pp. 275–278). Arnold.

Robertson, M. (2017). A crash course in the science of sustainability. In D. Lerch (Ed.), *The community resilience reader*. Island Press, Washington, DC. https://doi.org/10.5822/978-1-61091-861-9_9

Rubens, A., Spigarelli, F., Cavicchi, A., & Rinaldi, C. (2017). Universities' third mission and the entrepreneurial university and the challenges they bring to higher education institutions. *Journal of Enterprising Communities: People and Places in the Global Economy, 11*(3), 354–372.

Sabah, S., & Du, X. (2018). University faculty's perceptions and practices of student-centered learning in Qatar. *Journal of Applied Research in Higher Education, 10*(4), 514–533. https://doi.org/10.1108/jarhe-11-2017-0144

Schot, J., & Steinmueller, W. E. (2018). Three Frames for Innovation Policy: R&D, Systems of Innovation and Transformative Change. *Research Policy, 47*(9), 1554–1567.

Sever, S.D., & Tok, M.E. (2023). Education for Sustainable Development in Qatar. In: Cochrane, L., Al-Hababi, R. (eds) *Sustainable Qatar*. Gulf Studies, vol. 9. Springer. https://doi.org/10.1007/978-981-19-7398-7_17

Shawe, R., Horan, W., Moles, R., & O'Regan, B. (2019). Mapping of sustainability policies and initiatives in higher education institutes. *Environmental Science & Policy, 99*, 80–88.

Shiel, C., Leal Filho, W., & do Paço, A., & Brandli, L.,. (2016). Evaluating the engagement of universities in capacity building for sustainable development in local communities. *Journal of Evaluation and Program Planning, 54*, 123–134. https://doi.org/10.1016/j.evalprogplan.2015.07.006

Shields, R. (2019). The sustainability of international higher education: Student mobility and global climate change. *Journal of Cleaner Production, 217*, 594–602. https://doi.org/10.1016/j.jclepro.2019.01.291

Sillitoe, P. (Ed.). (2022). *Sustainable development: an appraisal from the Gulf region.* Berghahn Books.

Sillitoe, P. (Ed.). (2014). *Sustainable development: an appraisal from the Gulf region.* Berghahn Books.

Sin, C., Tavares, O., & Amaral, A. (2017). Accepting employability as a purpose of higher education? Academics' perceptions and practices. *Studies in Higher Education, 1–12.* https://doi.org/10.1080/03075079.2017.1402174

Stephens, J. C., Hernandez, M. E., Roman, M., Graham, A. C., & Scholz, R. W. (2008). Higher education as a change agent for sustainability in different cultures and contexts. *International Journal of Sustainability in Higher Education, 9*(3), 317–338. https://doi.org/10.1108/14676370810885916

Syed, R. T., Singh, D., Agrawal, R., & Spicer, D. P. (2023). Entrepreneurship development in universities across Gulf Cooperation Council countries: A systematic review of the research and way forward. *Journal of Enterprising Communities: People and Places in the Global Economy, 17*(5), 1045–1062. https://doi.org/10.1108/jec-03-2022-0045

Theodoulou, S. Z., & Cahn, M. A. (1995). *Public policy: The Essential Readings.* Pearson

Thomas, R., & Ormerod, N. (2017). The (almost) imperceptible impact of tourism research on policy and practice. *Tourism Management, 62*, 379–389.

Ukko, J., Saunila, M., Rantala, T., & Havukainen, J. (2018). Sustainable development: implications and definition for open sustainability. *Sustainable Development.*

UNESCO. (2017) UNESCO Moving Forward the 2030 Agenda for Sustainable Development; UNESCO: Paris, France; Available online: https://tcg.uis.unesco.org/wp-content/uploads/sites/4/2019/08/sdg4-global-indicators-4.3.pdf

United Nations. (1987). Report of the world commission on environment and development: our common future. United Nations General Assembly.

United Nations. (2015). Resolution adopted by the General Assembly on 25 September 2015. Transforming our world: the 2030 Agenda for Sustainable Development (A/RES/70/1 Archived 28 November 2020 at the Wayback Machine.

United Nations. (2017). Resolution adopted by the General Assembly on 6 July 2017. Work of the Statistical Commission pertaining to the 2030 Agenda for Sustainable Development (A/RES/71/313).

Veselý, A. (2008). Problem tree: A problem structuring heuristic. *Central European Journal of Public Policy, 2*(02), 60–81.

Weber, A. S., (2014) "Education, development and sustainability in Qatar: a case study of economic and knowledge transformation in the Arabian Gulf" In *Education for a Knowledge Society in Arabian Gulf Countries.* Published online: 10 Oct 2014; 59–82.

World Bank. (2022). *Qatar population.* Access at Population, total—Qatar | Data

Yap, C., Ryan, S., & Yong, J. (2014). Challenges facing professional accounting education in a commercialised education sector. *Accounting Education, 23*(6), 562–581.

Zaidan, E. (2019). Cultural-based challenges of the westernized approach to development in newly developed societies. *Development in Practice, 29* (5), 670–681. Routledge.

Zaidan, E., Al-Saidi, M., & Hammad, S. H. (2019). Sustainable development in the Arab world—is the Gulf Cooperation Council (GCC) region fit for the challenge? *Development in Practice, 29*(5), 670–681. https://doi.org/10.1080/09614524.2019.1628922

Part IV
Career Pathways & Life Skills

Chapter 20
Entrepreneurial Education in the GCC: Review and Prospective Directions

Rana Sobh, Heba Younis, and Mohamed Keshta

Abstract In the Gulf Cooperation Council (GCC) countries, entrepreneurship education has emerged as a pivotal catalyst for driving economic diversification and promoting sustainable development. This chapter serves to synthesise the current body of literature on entrepreneurial education and tailor optimal models for integration into higher education institutions (HEIs) across the GCC. Through a comprehensive review of the current state of entrepreneurial education in the GCC, we aim to spotlight exemplary practices within public universities in the region. The ultimate objective is to put forth a holistic model for the design and implementation of effective entrepreneurial education within the GCC, and facilitate the transformation of higher education institutions into dynamic hubs of entrepreneurial learning, thus contributing to the region's economic diversification and sustainable development goals.

20.1 Introduction

Entrepreneurship plays a pivotal role in steering economic growth and fostering innovation within societies (Meyer & De Jongh, 2018; Tao et al., 2022). In the context of economies reliant on oil and gas reserves, such as countries of the Gulf Cooperation Council (Qatar, Bahrain, Kuwait, Oman, Saudi Arabia, and the United Arab Emirates), the emergence of entrepreneurship assumes an indispensable role as a catalyst for economic diversification, and acts as the driving force behind the transition to a knowledge-based economy (Alruabsihi et al., 2019; Tadros, 2015).

R. Sobh (✉)
College of Business and Economics, Qatar University, Doha, Qatar
e-mail: r.sobh@qu.edu.qa

H. Younis
Associate Professor, Entrepreneurship and Strategy, SP Jain London School of Management, London, UK

M. Keshta
Core Curriculum Program, Qatar University, Doha, Qatar

E. Zaidan et al. (eds.), *The Future of Education Policy in the State of Qatar*, Gulf Studies 21, https://doi.org/10.1007/978-981-97-9667-0_20

Recognising the compelling need to establish conducive ecosystems for entrepreneurship, this imperative has been notably ingrained within the national visions and strategic frameworks of the majority of GCC countries (Tok, 2018). These nations have strategically created frameworks for innovation in their national development strategies and policies, prioritising human capital and lifelong learning, entrepreneurship, and technology development (Lee & Mirza, 2021).

The need to propel GCC countries' economies towards sustained growth and resilience accentuates the pivotal role of education in fostering entrepreneurial capacity and building entrepreneurial ecosystems (Hammed & Irfan, 2019). Specifically speaking, entrepreneurship education in the region nurtures innovation, technological proficiency, and entrepreneurial skills among students, enabling them to create new ventures or secure job opportunities for others (e.g. Ahmed et al., 2020; Hammed & Irfan, 2019).

In response to this need, educational institutions across the GCC region have been progressively embracing the integration of entrepreneurship programmes within both tertiary and K-12 education. In the realm of higher education in particular, our focus in this chapter, entrepreneurship education has witnessed significant growth. Universities have allocated considerable resources to the establishment of comprehensive programmes, innovative courses, and dynamic incubation initiatives, all geared towards fostering the cultivation of entrepreneurial skills and mindset.

Nonetheless, these endeavours exhibit a fractured landscape, characterised by disparate frameworks, skills gaps, and a weak alignment between curriculum and job market needs, a multitude of learning objectives, varied curriculum design methodologies, and divergent pedagogical approaches. These challenges are exacerbated by the convergence of rooted traditionalism within the education system among public universities in the region, a lack of enthusiasm among students for STEM subjects, and a leadership model resistant to change (Saji & Nair, 2018). This combination hampers the adaptability and responsiveness of the educational framework to the evolving needs of the modern world, particularly in the face of technological disruptions (Lee & Mirza, 2021).

To address these challenges, the GCC universities are to be encouraged to adapt their education systems and pedagogical approaches to meet market needs and integrate entrepreneurial skills into their curricula. This requirement underscores the need for a more cohesive and holistic approach to curriculum design, to ensure consistency and effectiveness across diverse educational institutions within the region.

In this chapter, we attempt to consolidate the existing literature, and adapt the existing frameworks, so as to propose a holistic model for optimal implementation in higher education institutions within the GCC context. Additionally, we highlight the regional best practices that address the pivotal obstacles hindering the efficacy of entrepreneurial education. The primary goal is to provide a holistic model for producing entrepreneurial and highly skilled graduates who can make significant contributions to the region's economic progress and societal well-being.

We begin by examining the meaning and significance of entrepreneurship education, in order to fully understand its potential as a transformative force in the region. Drawing from established theories, frameworks, and empirical evidence, we identify key components and best practices crucial for effective entrepreneurial education. We then assess the current state of entrepreneurship education in the GCC, gaining insights into its unique challenges and opportunities. This assessment informs the development of our holistic model, aimed at addressing these challenges and capitalising on opportunities to enhance entrepreneurship capacity within the GCC. Lastly, we look ahead, discussing potential future directions for research and development in entrepreneurship education in the GCC.

20.2 The Essence and Significance of Entrepreneurial Education

In 1985, Peter Drucker, a prominent figure in management philosophy, predicted the pivotal shift towards an Entrepreneurial Society as a cornerstone of the twenty-first century (Joiner, 2019). The recognition of education's fundamental role in fostering entrepreneurial attitudes and behaviours has been widespread, leading emerging and developed economies to increasingly invest in nurturing a new generation of capable and ambitious entrepreneurs (Fayolle et al., 2014; Fellnhofer, 2019; Gielnik et al. 2021; Zamfir et al. 2013).

The concept of entrepreneurship has received many definitions in the literature, highlighting the broad and interchangeable nature of the term across various contexts and situations (Matlay, 2005; OECD, 2012). Despite the abundance of definitions and interpretations of the concept of entrepreneurship, a prevailing consensus crystallises around its core: the generation of value through innovation (Pittaway & Cope, 2007). This consensus underscores the pivotal role of learning and value creation, as indispensable components of entrepreneurship (Bruyat et al., 2001).

While there is significant interest in entrepreneurship education, a consensus regarding its essence remains elusive (Bridge, 2017; Wong & Chan, 2022). Consequently, diverse terminologies are employed to convey the endeavour of fostering entrepreneurship capacity among learners. Among these terms, enterprise education, entrepreneurship education, and entrepreneurial education stand out as the most prevalent (Lackéus, 2015). To be specific, the term 'enterprise education', adopted by the Independent Quality body for Higher Education in the United Kingdom, embraces a wide perspective, and defines it as the process of nurturing students in a way that equips them with an enhanced ability to conceive ideas and the necessary qualities, characteristics, and skills to turn those ideas into reality; this extends beyond mere knowledge acquisition. It encompasses a broad spectrum of emotional, intellectual, social, cultural, and practical behaviours, attributes, and proficiencies that are relevant to all students. These form the foundational elements that can not only boost employability prospects, but can also be further developed through 'entrepreneurship education' (QAA, 2018). 'Entrepreneurship education', on the other hand, is a narrower and more business-centric perspective, as it mainly aims to teach students about the intricacies of establishing ventures and self-employment, and focuses on training learners in opportunity recognition, commercialising a prototype or an idea, and securing tangible and intangible resources in the face of risk to start a new venture (Mahieu, 2006; The Quality Assurance Agency for Higher Education, 2012).

In this chapter, we adopt the term 'entrepreneurial education', encapsulating both enterprise and entrepreneurship education as proposed by Erkkilä (2000), to harmonise the terminology. Hence, hereinafter, 'entrepreneurial education' refers to a comprehensive approach equipping students with the requisite knowledge, skills, and mindset to foster innovation, capitalise on opportunities, and create new ventures and/or value in any context (Schramm, 2012).

While the frequently highlighted advantages of entrepreneurial education encompass economic growth and the generation of employment opportunities (Mwasalwiba, 2010; Wong et al., 2005), it is important to note that entrepreneurial education has also been associated with increased student motivation, enhanced depth of learning, a fostering of creativity, the nurturing of resilience, and refining problem-solving skills (Mwasalwiba, 2010). Research also highlights its potential to elevate the perceived relevance of subjects among learners, amplifying their motivation and engagement, while mitigating issues of student disinterest and dropout (Deuchar, 2007; Mahieu, 2006; Moberg, 2014; Surlemont, 2007). Besides the positive impact of entrepreneurial education on students' academic performance, possessing entrepreneurial competencies is being viewed as an imperative to prepare graduates for success in an increasingly globalised, intricate, and uncertain world, characterised by perpetual career transitions and lifelong learning needs (Jones & Iredale, 2010; Sobh, 2022).

Furthermore, entrepreneurial education transcends the aim of mere business pursuit; it moulds change agents and equips graduates to confront pressing global challenges such as climate change, environmental degradation, and inequality (Sobh, 2022). This aligns with UNESCO's recent appeal for renewed solidarity and cooperation, so as to fortify education as a collective and public good, giving rise to a burgeoning interest in social entrepreneurship (Rae, 2010; Tracey & Phillips, 2007).

The impact of entrepreneurial education extends beyond fostering entrepreneurship in individuals; it also harnesses *intra*preneurship within organisations and institutions. Employees ingrained with an entrepreneurial mindset possess an enhanced comprehension of their work environment, enabling more effective problem-solving, opportunity seizing, and substantive contributions to institutional advancement (Lackéus, 2020; Teece, 2012).

To conclude, the benefits of entrepreneurial education go beyond mere startups, innovative ventures, job opportunities, and economic expansion; they extend across various facets of modern life, creating value within economic, social, and cultural spheres (Lackéus, 2015).

20.3 Approaches to Entrepreneurial Education in the Literature

Despite substantial scholarly attention directed towards entrepreneurship education, there is still no clear agreement about what constitutes entrepreneurship competencies and skills. This lack of clarity has contributed to an ongoing ambiguity about the most effective approaches for entrepreneurial education, strategies for curriculum design, instruction pedagogies, and assessment methods (Bridge, 2017; Lackéus, 2015; Mwasalwiba, 2010; Wong & Chan, 2022).

Overall, entrepreneurial learning objectives have been grouped into two main dimensions: entrepreneurial intention and entrepreneurial competencies (e.g. Wong & Chan, 2022). Entrepreneurial intention, a key facet, often refers to students harbouring aspirations to start their own businesses in the future (Yar et al., 2008) or having the intention to embrace an entrepreneurial role (Matlay et al., 2015). At the heart of entrepreneurial education lies the nurturing of entrepreneurial competencies crucial for the effective perusal of entrepreneurial endeavours (Lackéus, 2018). As highlighted by (Mitchelmore & Rowley, 2010), pivotal entrepreneurial competencies encompass idea generation, environmental scanning, and the capacity to discern and envisage untapped opportunities. Expanding upon this perspective, a more comprehensive framework for entrepreneurial competencies includes skills, characteristics, attitudes, and values, that contribute to the development of an individual's personal attributes (Spencer & Spencer, 1993). This comprehensive view finds resonance in The European Framework for Entrepreneurship Competencies, which articulates competencies and associated learning outcomes across three core domains spanning knowledge, skills, and attitudes—encompassing ideas and opportunities, resources, and entrepreneurship in action (European Commission. Directorate-General for Employment, 2018). Prevalent curriculum approaches to deliver these competencies have been commonly classified under three main categories (Heinonen & Hytti, 2010; Johnson, 1988; O'Connor, 2013):

A theoretical approach ('Teaching about Entrepreneurship'): This approach aims to offer a comprehensive understanding of the entrepreneurial concept. It is commonly found in higher education institutions (Heinonen & Hytti, 2010; O'Connor, 2013).

An occupational approach ('Teaching for Entrepreneurship'): Geared towards equipping potential entrepreneurs with the necessary knowledge and skills for successful ventures, this approach is centred on practical skills and occupation-specific expertise.

An experiential approach ('Teaching through Entrepreneurship'): This method adopts a hands-on, experiential approach. It actively involves students in genuine entrepreneurial learning encounters, guiding them through the practical intricacies of the entrepreneurial process (Kyrö, 2005).

While the first two approaches, 'Teaching about' and 'Teaching for', are most applicable to subsets of higher-level students, the embedded 'Teaching through' approach emphasises the delivery of universal entrepreneurial learning outcomes.

These outcomes encompass critical proficiencies such as problem-solving, creative thinking, and leadership skills. Notably, this approach is capable of integration across disciplinary curricula, spanning diverse subjects and educational levels (Handscombe et al., 2008; Smith et al., 2006).

Within Higher Educational Institutions, these approaches are manifested through a variety of methods detailed below:

20.3.1 Integrating Entrepreneurship in General Education

The integration of entrepreneurship into general education programmes has gained traction, particularly in the United States. This emerging trend acknowledges the value of incorporating entrepreneurial thinking into general education, to enhance the practicality and relevance of students' learning experience (Godwyn, 2009). Furthermore, characteristics inherent to entrepreneurship, such as creativity, confidence, critical-thinking, opportunity recognition, leadership, and effective communication, stand as essential elements for attaining both professional excellence and responsible citizenship—two fundamental objectives of general education (Sobh, 2022). Advocates of this integration also contend that general education offers an ideal platform for incorporating entrepreneurial education, as it provides an avenue for students across diverse disciplines to nurture an entrepreneurial mindset. However, such integration must extend beyond a business-centric viewpoint, and accommodate various specialisations (Godwyn, 2009). The inaugural entrepreneurship course granted general education status was introduced by Rowan University in the United States in 2004 (D'Intino et al., 2010). Similar initiatives have been undertaken in the GCC region, like Qatar University and United Arab Emirates University, as elucidated in Sect. 18.4.

20.3.2 Entrepreneurship in the Disciplines

An alternative approach to entrepreneurial education involves incorporating entrepreneurship courses within specific disciplines, or the integration of entrepreneurial concepts into existing courses across the curriculum (Ahsan et al., 2018; Mehlhorn et al., 2015). For instance, business colleges, the traditional providers of business education, may deliver tailored courses or academic programmes centred on entrepreneurship, with the aim of developing the knowledge and skills necessary for thriving within a dynamic business landscape (Kirby, 2004). Similarly, this approach can be tailored to other disciplines, such as engineering, health, and humanities (Caseiro & Alberto, 2013). Engineering students, for example, could examine the transformation of innovative engineering concepts into viable business ventures, while students in humanities might explore the utilisation of their creative aptitude in entrepreneurial pursuits (Mäkimurto-Koivumaa & Belt, 2016). Within the healthcare realm, students could be prompted to identify unaddressed needs or disparities within the healthcare system, and devise inventive solutions (Suryavanshi et al., 2019). In the realm of creative industries such as the media and art, students would learn how to leverage their artistic talents and creative vision to conceptualise, launch, and sustain successful ventures that align with market demands and industry trends, in the context of artistic creation (Rae, 2004).

20.3.3 Fostering Entrepreneurial Skills through Teaching Approaches

Fostering entrepreneurial competencies can be powerfully nurtured through active teaching approaches and pedagogical strategies that depart from passive one-way teaching methods (Jones, 2007). In essence, the design and delivery of entrepreneurial curricula are rooted in creating an immersive learning experience akin to an entrepreneur's actual journey (Jones, 2007). The focus of learning will be a shift from passive learning to problem-solving, creativity, and active engagement (The Quality Assurance Agency for Higher Education, 2012). A dynamic approach, rooted in real-world application, interdisciplinary collaboration, and experiential components, encourages a cycle of iterative experimentation, collaboration with external stakeholders, and a focus on innovation facilitates the cultivation of key traits such as risk-taking, innovation, and leadership (Rae, 2007). The use of interactive learning activities such as games, case study discussions, simulations, role-play, and interactive online platforms, enhances critical-thinking and problem-solving skills, contributing to the development of a comprehensive entrepreneurial ecosystem (Chan and Luke 2013).

20.3.4 An Ecosystem Approach

An important piece of the puzzle in entrepreneurial education is the concept of an entrepreneurial ecosystem (Meshram & Rawani, 2019; Stam & Ven, 2021). The study of the entrepreneurial ecosystem emerged in the 1980s and 1990s (Stam & Ven, 2021) to shift the focus from the limited emphasis on the individual towards the wider boundaries of countries and regions, in order to investigate a set of factors that encourage or hinder the development of enterprise ventures (Nicotra et al., 2018). Creating a supportive environment that integrates programmes, policies, and institutions, is important for entrepreneurs' success, and is vital to the delivery and teaching of entrepreneurship in higher education settings (Audretsch & Belitski, 2017; Meshram & Rawani, 2019).

Up until now, there is no unified definition of the entrepreneurial ecosystem among researchers of entrepreneurial education (Meshram & Rawani, 2019). Higher education settings build on the Global Entrepreneurship Monitor (GEM) framework (Ali et al., 2019). Based on GEM's framework, the main pillars of the entrepreneurial ecosystem are 'availability of finance, supportive government policies and regulations, government programs for entrepreneurs, entrepreneurial education and training, supportive culture and social norms, and access to physical infrastructure'. Universities hence design their ecosystem to include entrepreneurs in residence, links with governmental bodies, networks with alumni and startups, business incubators, and funding schemes, along with the traditional timetabled lectures (Stam, 2015).

To conclude, the aforementioned approaches to entrepreneurial education underscore the multifaceted nature of entrepreneurial education and curriculum design. Integrating entrepreneurship into general education, in specific disciplines, and diverse courses, along with leveraging teaching methodologies, extracurricular activities, and a supportive ecosystem, are approaches that have been proposed to process a future generation of entrepreneurs, and prepare graduates for success in a rapidly evolving global landscape. Nevertheless, it is important to acknowledge the existing research gaps in terms of assessing their respective contributions to fostering new ventures and facilitating their growth (Ahmed et al., 2020; Fayolle & Gailly, 2015).

20.4 The Status of Entrepreneurship Education in the GCC

Despite the growing interest in fostering entrepreneurship and promoting entrepreneurial education, there is a dearth of research conducted in developing countries at large and notably in the Arab world, into the optimal design and delivery of entrepreneurial curricula (Al-Mamary et al., 2020; Sharahiley, 2020; Thani Al Dhaheri, 2020). There is a widespread recognition that the advancement of curricula in many Arab countries, including the GCC, is influenced by the modernised educational reforms prevalent in Western countries, such as in Europe and the United States, despite the cultural distinctions between the Western and Arab contexts (Dagher & Boujaoude, 2011).

Entrepreneurial curriculum design in GCC countries, and the frameworks for delivering entrepreneurial education, predominantly align with Ajzen's (1991) Theory of Planned Behavior (TPB), which elucidates and anticipates the factors influencing individuals' inclination to initiate new ventures (Barbosa et al., 2007; Eid et al., 2019). Consequently, entrepreneurial education delivery is designed with a focus on elements such as entrepreneurial self-efficacy, seen as a precursor to the establishment of entrepreneurial intentions among budding entrepreneurs (Barbosa et al., 2007).

In this section, we provide an overview of the current landscape of entrepreneurial education delivery, investigating exemplary and innovative approaches to entrepreneurial education, particularly within the context of public universities in GCC countries. Data collection is constrained by the availability of information on official university websites, related resources, and relevant research publications.

20.4.1 The Kingdom of Saudi Arabia

Saudi Arabia formulated a government policy aiming at cultivating entrepreneurship among Saudi youth, a directive underscored in Saudi Vision 2030 and the Ninth Development Plan of Saudi Arabia. This would elucidate the noticeable congruity on the findings of the 2019 GEM report and Saudi Arabia's Total Entrepreneurial Activity (TEA) rate; both indicating that approximately 33% of the Saudi population has signalled their intention to embark on entrepreneurial ventures within the upcoming three years (Balawi, 2021; Elnadi & Gheith, 2021).

The Saudi government allocated substantial investments to universities, to cultivate entrepreneurial mindsets among students, and motivate their active engagement in entrepreneurship (Ahmed 2019; Alharbi et al., 2018; Elnadi & Gheith, 2021). Consequently, several initiatives have been launched to support entrepreneurship education and innovation in Saudi schools and public universities (Almahdi, 2019; Almawishir & Messen, 2021).

The majority of Saudi public universities provide entrepreneurship courses exclusively to business students, and these courses are typically offered as electives rather than as mandatory subjects (Faghih & Zali, 2018). According to the Global University Entrepreneurial Spirit Students' Survey (GUESSS), Saudi National report 2021, the percentage of students who attended either an elective or a mandatory course on entrepreneurship is 32% and 12% respectively, and only 5% of the sample studied were studying specific programmes on entrepreneurship (Saudi & Report, 2021). Based on the report's findings, it was recommended that Saudi higher education institutions integrate entrepreneurship courses into the curriculum of all their academic disciplines. This suggestion is rooted in the empirical evidence highlighting the positive impact of entrepreneurial education on students' intentions towards entrepreneurship.

To support emerging entrepreneurs, many Saudi universities have embraced an ecosystem approach to entrepreneurial education, with the aim of fostering an environment conducive to techno-entrepreneurship (Hameed & Irfan, 2019). Among these institutions, King Abdullah University of Science and Technology (KAUST) stands out as a prominent centre for innovation and research in the Middle East, known for its state-of-the-art facilities. The Innovation and Economic Development Department at KAUST has launched an entrepreneurship programme with the overarching goal of cultivating a robust entrepreneurial culture, and fostering the creation of knowledge-based enterprises in the region. This programme includes initiatives such as a venture school, startup accelerators, and an innovation fund, all aimed at providing diverse opportunities for entrepreneurship and learning. It offers comprehensive training and facilitates access to funding for aspiring entrepreneurs (Faghih & Zali, 2018).

Furthermore, KAUST's Entrepreneurship Center provides elective courses in entrepreneurship and product innovation for its students. In 2021, the university expanded its reach, by offering a Massive Open Online Course (MOOC) titled 'Entrepreneurship Adventures' on the global educational platform edX.org, making it the first Arabic language entrepreneurship course featured on the platform. Additionally, KAUST has distinguished itself with its Research and Technology Park (KRTP), covering over 2.7 million square metres. This park offers versatile research and development space for corporate partners, small and medium-sized enterprises (SMEs), and startups (Faghih & Zali, 2018).

Regarding partnerships and collaborations in support of entrepreneurship education, the KAUST Entrepreneurship Center has forged significant collaborations with leading organisations, aimed at enhancing innovation training in Saudi Arabia. Since 2015, the centre has established partnerships with over 35 organisations (e.g. Microsoft, Stanford Center for Professional Development, Global Entrepreneurship Network, NEOM, Berkley Sutardja Center for Entrepreneurship and Technology, Saudi Venture Capital Company, Venture Capital and Private Equity Association, among others) spanning various industry sectors, with the objective of providing rigorous training to aspiring entrepreneurs participating in regional and international startup contests. These collaborative efforts encompass the delivery of bootcamps, acceleration programmes, pitch competitions, corporate innovation initiatives, and various other training activities.

An exemplary collaboration involves KAUST Ignite, which teamed up with the Saudi Smart Grid Conference 2022, to assemble a diverse cohort of students from universities across Saudi Arabia. These students share a common goal: addressing pressing societal issues and challenges.

Another noteworthy example is King Abdulaziz University (KAU), which has developed an entrepreneurship programme designed and delivered by experts from Babson Global, a top-ranked institution in the field of entrepreneurship (Almahdi, 2019).

Furthermore, the Riyadh Valley Company (RVC), a prominent initiative associated with King Saud University (KSU), aims to facilitate partnerships between educational and research institutions and the investment community. This initiative involves investing in joint projects that enhance the experiential and practical learning of both university students and professors. The RVC introduced the KSU Passport, a Virtual Entrepreneurship Platform, developed in collaboration with GSVLabs. This platform provides access to a global entrepreneurship community, Saudi and international resources, and mentors, all contributing to the development of a thriving entrepreneurship ecosystem within the kingdom.

20.4.2 United Arab Emirates

The UAE National Strategy for Higher Education 2030 made it crucial to nurture entrepreneurially oriented future generations, who are able to lead the nation's development and economic growth (Saji & Nair, 2018). This has resulted in a continuous impetus for higher educators to incorporate entrepreneurship into the curricula of all disciplines across all UAE universities (Eid et al., 2019; Thani Al Dhaheri, 2020).

The UAE universities mainly offer entrepreneurship courses within business programmes; however, responding to a directive from the UAE Ministry of Education, all federal and private universities and colleges should offer introductory entrepreneurship courses, as part of their general education programmes (Thani Al Dhaheri, 2020). Many other UAE universities offer a specialised mandatory course on entrepreneurship and innovation (Saji & Nair, 2018). Since 2016, the development of entrepreneurship curricula in the UAE has been aligned with programmes designed in conjunction with Stanford University. The Stanford University entrepreneurship curriculum rests on three components: (a) innovation, (b) entrepreneurship, and (c) growth and leadership. The main theme of the programmes is to rely on experiential learning and group dynamics as key factors to inculcate successful innovation in business education settings. The initiative started at Khalifa University in Abu Dhabi, and cascaded to other public universities, and to schools of business and nursing as exemplars (Thani Al Dhaheri, 2020).

There are three public universities in UAE, namely, the United Arab Emirates University, Zayed University, and Higher Colleges of Technology (HCT), which all together provide education to more than 50% of the UAE's national body of undergraduate students (Thani Al Dhaheri, 2020). The United Arab Emirates University College of Business and Economics offers an 18-credit hour minor in Entrepreneurship. The minor serves as an introduction to the various functional areas of business, and is open to all students from various university colleges. Students in this minor should complete two three-credit hour General Education courses: GEIE 222 Fundamentals of Innovation and Entrepreneurship and (2) GEIT112 Fourth Industrial Revolution, then they have to complete a 12-credit course, ENTR415 Developing an Entrepreneurial Venture, where they learn how to start up an innovative business. The university also encompasses the Department of Innovation, Technology and Entrepreneurship, with a mission of promoting entrepreneurship and encouraging technology-driven innovation through various initiatives, including incubators, accelerators, the Mobile FabLab, and outreach programmes, among others.

The Higher College of Technology (HCT) offers the Innovation and Entrepreneurship Management Bachelor's Program. What distinguishes HCT is the unique 'Education-to Employment' Strategy. One of the guiding pillars of this strategy is 'Graduate Companies'. This pillar is centred on bringing together business leaders and policy makers, to support entrepreneurs and to cultivate student innovation and entrepreneurship through incubation and startup funding. It is worth noting that entrepreneurship and innovation (Graduate Companies) is infused in their academic programmes' learning outcomes. Below are examples from the Pharmacy and Social Work Programs' learning outcomes:

PLO 6: 'Demonstrate entrepreneurship and innovation skills to contribute and manage practice challenges within local and global contexts respecting socio-cultural and relationships (Graduating Companies)'.

PLO 7: 'Apply entrepreneurship and innovation skills to identify, develop and create innovative opportunities and solutions that will strategically navigate the complex demands stemming from pharmacy and the healthcare overall. (Graduating Companies)'.

HCT also offers many other courses on entrepreneurship and innovation, including the BUS 2403 Innovation and Entrepreneurship course, which applies the Stanford-informed approach to learning entrepreneurship and innovation, using project-based learning, and the ENT 3033 Social Entrepreneurship course, which involves students in a learning journey to address pressing social, environmental and business issues within the UAE. According to HCT Open Data Sources, 92,319 students (comprising 56,200 males and 36,119 females) have successfully completed their studies at HCT for the period 1990 to 2022.

20.4.3 Bahrain

In the decade since the launch of Vision 2030, Bahrain has invested in a set of institutional frameworks that aim to support a robust entrepreneurship ecosystem (Hashimi et al., 2021). The education programmes in Bahraini public universities have been designed to equip learners with the following key entrepreneurship competencies: autonomy, teamwork, adaptability, management of resources, social networks, risk-taking, individual and social responsibility, interaction, and social skills (Hashimi et al., 2021). It is worth noting that entrepreneurship programmes in the Bahrain higher education system encourage cross-disciplinary entrepreneurship education, based on the premise that policymakers with higher education context believe collaboration across disciplines may lead to more successful outcomes (Hashimi et al., 2021).

The University of Bahrain (UoB), as the national flagship university, has increased its collaborations and partnerships with international organisations, in line with its strategic endeavours to drive entrepreneurship forward. Not only has it become an Amazon Web Services academy, to cultivate talented individuals, who have played a pivotal role in the launch of a new AWS data centre on the university campus, but it also partnered with Cisco, Microsoft, Huawei, and Dell and integrated teaching and learning innovation through deeper partnerships with these organisations (Lee & Mirza, 2021). To nurture and spur a culture of innovation and entrepreneurship, UoB has also incorporated entrepreneurship education into its curriculum, requiring all its students to complete a mandatory module on entrepreneurship (Lee & Mirza, 2021). It distinguished itself by establishing a coding academy for all students, in partnership with the Google Developers Group, aiming at fostering interdisciplinary learning as part of its focus on STEAM education (Lee & Mirza, 2021). The University of Bahrain also encompasses a Business Incubation Center, which provides an ideal entrepreneurial environment that supports students and graduates to develop their innovative business ideas, so as to create and grow business. The entrepreneurial journey goes through three phases: the awareness and discovery phase, pre-incubation, and the actual incubation phase. The centre provides various services, including mentorship and consultation, commercialisation assistance, specialised training, and bootcamps (University of Bahrain Business Incubation Center n.d.).

20.4.4 Oman

Recognising the inevitability of natural resource depletion, His Majesty Sultan Qaboos bin Said emphasised the importance of diversifying Oman's economy during his 2013 address. He specifically highlighted the significance of the SME sector, and directed the establishment of the Al Raffd Fund, which comprises three financial programmes: the Sanad Program, Mawarid Alrizq Program, and Rural Woman Program. These programmes aim to empower Omani youth, to initiate their own SME projects. Additionally, Oman has witnessed numerous entrepreneurial initiatives, including the SME Fund, Riyada Mandates, Sharakah, Grofin, Saud Bahwan Group, and Zubair SEC, among others (Khan et al., 2021).

In response to these directives, the Ministry of Higher Education mandated the inclusion of an entrepreneurship course titled 'Entrepreneurship: Creativity and Innovation' across all universities and colleges in the country (Khan et al., 2021). Sultan Qaboos University (SQU), Oman's premier higher education institution, plays a pivotal role in fostering entrepreneurship, by offering entrepreneurship education. For instance, SQU hosts 14 research centres, including the Innovation & Technology Transfer Centre, established to support young Omanis in pursuing innovation and entrepreneurship endeavours.

A recent research study investigated the impact of the entrepreneurship ecosystem at SQU on preparing future entrepreneurs. The findings revealed that only 22% of SQU students had engaged in an entrepreneurship course, 21% had attended entrepreneurship workshops, and 17% had participated in conferences and symposiums related to entrepreneurship. The study suggested that while the SQU ecosystem moderately contributed to entrepreneurship development among students, there is room for improvement. It proposed introducing dedicated mandatory entrepreneurship courses, and fostering campus-wide entrepreneurship awareness (Al Rabaani & Al Dayri, 2022).

20.4.5 Kuwait

Kuwaiti Vision 2035 constitutes a beacon that guides the Kuwaiti Leadership's notable endeavours and initiatives to stimulate entrepreneurship and facilitate the growth of small and medium-size enterprises. Consequently, Kuwait has recently witnessed a paradigm shift in the government policies towards entrepreneurship and SMEs (Asheim, 2015). This is exemplified in the establishment of the National Fund for the Care and Development of Small and Medium Enterprises (National Fund) in 2013 (Langworthy & Warnecke, 2020). In the same vein, Kuwait has been striving to attract investments exceeding 400 KD million in various sectors, including education and technology, over the past few years (Al Hussaini, 2019). Moreover, the Government of Kuwait and the World Bank have cultivated an enduring partnership through an Advisory Services Program. One of this programme's accomplishments entails providing an integrated SME capacity development programme, encompassing the training for over 1,000 Kuwaitis, and organising eight events to promote entrepreneurship, drawing the participation of over 500 entrepreneurs (World Bank, 2021).

Kuwait University (KU) is the flagship and the leading public higher education institution in Kuwait. KU's Global Center launched the Startup Kuwait Project in 2015. This project is funded and supported by private higher education institutions, to enable a compelling entrepreneurial ecosystem that motivates students across Kuwait to formulate creative and innovative business ideas, helping the academic institutions prepare the next generation of entrepreneurs to initiate their entrepreneurial projects, while facilitating the transformation of faculty and industrial research into commercial business ventures (SAAED, 2007). Startup Kuwait stands as a pioneer and unprecedented endeavour within Kuwait, bringing together experts from around the globe, entrepreneurs, accelerators, and investors to support young and fledgling Kuwaiti startups and entrepreneurs. The Office of Student Training and Alumni (OSTA) at Kuwait University strives to equip the KU students with the entrepreneurial skills required to engage in entrepreneurial activities leading to venture creation. Furthermore, the OSTA engages these students with a diverse community of accomplished Kuwaiti entrepreneurs, who share their success stories and thus serve as role models to motivate the students to create their own ventures. A study conducted by Ahmed (2019) on Kuwait University's College of Business students, indicated that KU offers some forms of entrepreneurship education, including entrepreneurship courses, workshops, seminars, and other events pertinent to entrepreneurship.

According to the Global Entrepreneurship Monitor (GEM) (2020) report, Kuwait scored a total early-stage Entrepreneurial Activity (TEA) rate of 19.2%, which was the highest of all peer GEM Middle East economies. The report also indicated that a significant percentage of aspiring Kuwaiti entrepreneurs do end up becoming early-stage entrepreneurs, which constitutes a significant amount of aspiring entrepreneurial potential (Ghura et al., 2020).

20.4.6 Qatar

Qatar emerges as a frontrunner among GCC countries with the most substantial GDP growth rate and the world's fastest growing economy (Saberi & Hamdan, 2019). The National Development Strategy 2011–2016 (NDS), stemming from Qatar's National Vision 2030, outlines specific goals to stimulate SMEs and entrepreneurial ventures, in the aim of diversifying its economy, while transitioning to a knowledge-based economy (Greene et al., 2016). Towards that end, the government has created some entities, including 'Qatar Development Bank (QDB), Enterprise Qatar, Silatech, Social Development Center, INJAZ Qatar, Center for Entrepreneurship (Qatar University), Qatar Business Incubation Center (QBIC), Digital Incubation Center

(DIC), Qatar Science and Technology Park (QSTP), and Qatar Foundation (QF)' (Greene et al., 2016; Hassen, 2020; Lee & Mirza, 2021).

To support the government quest for fostering entrepreneurial competencies among youth and aspiring entrepreneurs, a number of prominent higher education institutions, including Qatar University, Hamad Bin Khalifa University, and the Qatar Foundation and its seven branches of international universities have been striving to play a role in nurturing the growth of the entrepreneurial ecosystem in Qatar (Villegas-Mateos & Abdellatif, 2021).

A robust institutional commitment to entrepreneurship has been evident in Qatar University (QU), the national flagship university in Qatar. QU's 2018–2023 Transformational Strategy identifies innovation and entrepreneurship as one of the two core enabling strategies that cuts across all sectors. The other enabling strategy is digital transformation. As a result, the word 'entrepreneurial' has been identified as one of the five graduate attributes in QU's Qualifications Framework, in addition to 'competent', 'lifelong learners', 'well-rounded', and 'ethically and socially responsible'. To guide the implementation of this strategy in teaching and learning, QU developed a QU Entrepreneurship Competency Framework, including essential competencies and skills for cultivating entrepreneurial attributes, and chartered a roadmap for implementation. This framework has been translated into a curriculum integration strategy and a guide for implementing student-centred and entrepreneurial teaching approaches, across all courses throughout the students' academic journey.

In addition to those academic programmes offered by the College of Business and Economics to business and non-business students, QU has integrated entrepreneurial thinking into the core curriculum programme as part of a programme overhaul, wherein entrepreneurial thinking was introduced as a new Program Learning Outcome (PLO). A course about social entrepreneurship 'Civic Engagement, Leadership, and Innovation—UNIV 200' has been tailored to general education students and incorporated into the social and behavioural sciences curriculum since 2022. This project-based course is intended to foster interdisciplinary collaboration, prompting students to devise innovative solutions to address local or global challenges. Through the process, participants acquire insights into civic engagement mechanisms, hone leadership competencies, unleash creativity, and gain a good understanding of the entrepreneurial process. Interdisciplinary teams from business, engineering, health, the humanities and arts bring diverse perspectives and skills to these entrepreneurial projects.

To foster a comprehensive supporting ecosystem and enhance research into innovation and entrepreneurship, the university has established the Center of Entrepreneurship in 2013, lately renamed as the Center of Entrepreneurship and Organizational Excellence. QU has also recently established the Center for Innovation, Entrepreneurship, and Economic Development to accelerate faculty and students' spin-off ventures.

As a result, based on the Global University Entrepreneurial Spirit Students' Survey (GUESSS) 2021 National Report, students at Qatar University exhibit a relatively more positive attitude and greater self-efficacy to establish entrepreneurial ventures than that of their peers worldwide. Upon graduation, the prominent career choices for the students in Qatar are to become a business founder, or to work for a large business venture at the first instance. These preferences are found to be above the Middle East and North Africa (MENA) and the world average (Villegas-Mateos & Abdellatif, 2021).

To conclude this section, it is worth noting that education in the GCC seems to be increasingly emphasising the development of educational approaches geared towards building up entrepreneurial capacity. Nevertheless, these approaches remain fragmented, due to the absence of a common GCC arena for agreed upon definitions and approaches to entrepreneurship education. One potential solution to address this fragmented landscape entails embracing a forward-thinking framework that can accommodate the range of GCC contexts, and which allows for the ongoing refinement of definitions, pedagogical approaches, and outcomes, as students advance in their journey (Lackéus, 2013).

Such a reference framework should encompass GCC countries' commonly shared, comprehensive definitions of entrepreneurship, and entrepreneurship education, and identify the skills and competencies needed to shape the GCC nationals' entrepreneurial mindsets, enabling them to create financial and social values. Given these states' common economic and cultural challenges to entrepreneurship, this framework should contribute to the region's economic development and integration. Building on this framework, the establishment of regional centres for entrepreneurship and innovation is inevitable, to avoid duplication of effort, and enhance regional integration.

There is also an urgent need for research that focuses on benchmarking the entrepreneurial competences within the GCC, as opposed to the established global frameworks (e.g. the European Entrepreneurship Competence Framework, EU 2018). Additionally, this research should investigate the underlying attributes of an entrepreneurial ecosystem, to bring about a holistic model for an entrepreneurial university in the GCC.

20.5 Adopting a Holistic Model for Entrepreneurial Curriculum Integration

As highlighted in previous sections, effective entrepreneurial education in the region is hampered by fragmented strategies in curriculum design, alongside ambiguous definitions and learning objectives. A potential solution lies in embracing a progressive framework that accommodates diverse contexts and allows for the evolution of definitions and outcomes as students progress (Fuller, 2015; Nybye & Rasmussen, 2013). In this context, Lackéus (2013) introduces a European framework, featuring four progressive action-oriented pedagogical approaches that increase in complexity according to the educational level. The model is intended to guide educators in selecting suitable activities based on the context, available resources, and student readiness.

Building on Lackéus's foundational model from 2013, we present a holistic model, designed to effectively cultivate and advance students' entrepreneurship competencies along a continuum (See Fig. 20.1). The first stage in the continuum starts at the general education curriculum level, and should primarily aim to develop students' entrepreneurial thinking, that is, focus on the attitudes and ideas part of the entrepreneurship competencies, as indicated in the Entrepreneurship Competency Framework (European Commission. Directorate-General for Employment, 2018). However, such integration requires a coherent assessment rubric that can be used in a multidisciplinary context. In this rubric, the primary focus should be on developing students' entrepreneurial thinking, encompassing their capacity to identify innovative solutions to real-world challenges in various contexts. This should also encompass the demonstration of initiative, agency, and social responsibility. In pursuit of this goal, one potential avenue is the integration of a dedicated entrepreneurship course within the general education curriculum. To accommodate various specialisations and in order to have a broader outlook, an emphasis on social entrepreneurship within such a course holds promise. In this context, economic value and for-profit enterprise development represent just two facets of the potential outcomes (Tracey & Phillips, 2007). Such an approach nurtures entrepreneurial proficiency, prepares socially conscious agents, and injects pragmatism and significance into student learning experiences (Youniss et al., 2002).

As the foundational traits and skills take root and student engagement flourishes, a conventional business plan-centric education resulting in venture creation can be adopted. Indeed, the second phase involves integrating entrepreneurship concepts within specific academic disciplines. The emphasis now turns to cultivating the resources and execution aspects of the entrepreneurship competencies framework. Students will be immersed in entrepreneurial principles, and the process of venture creation across a diverse range of subjects, thereby highlighting the significance of entrepreneurship within their distinct areas of study. For instance, multidisciplinary courses offered in collaboration with business colleges, focused modules, and project-based assignments. Universities in the GCC region also have the opportunity to develop entrepreneurship-focused interdisciplinary programmes or programmes catering to non-business students, such as minors, or those offering graduate certificates. These programmes should be designed to ensure that graduates from diverse academic backgrounds are aptly prepared to seize opportunities and initiate innovative ventures within their respective domains.

As individuals progress through their educational journey, the last stage in the continuum would be to facilitate the development of ventures, startups, and commercial outputs, establishing infrastructures, programmes, and collaboration with the government, industry, and funding organisations. Complementary extracurricular activities, such as hackathons and competitions, can provide a fertile ground for nurturing students' entrepreneurial ideas. This pillar propels and shapes students' entrepreneurial aspirations, and encourages them to start and run their own enterprises (Lackéus, 2020). An example is QU Holding at Qatar University, which has been established to adopt and develop faculty and students' entrepreneurial ideas into new ventures and to sell the concept of research to industry and the public.

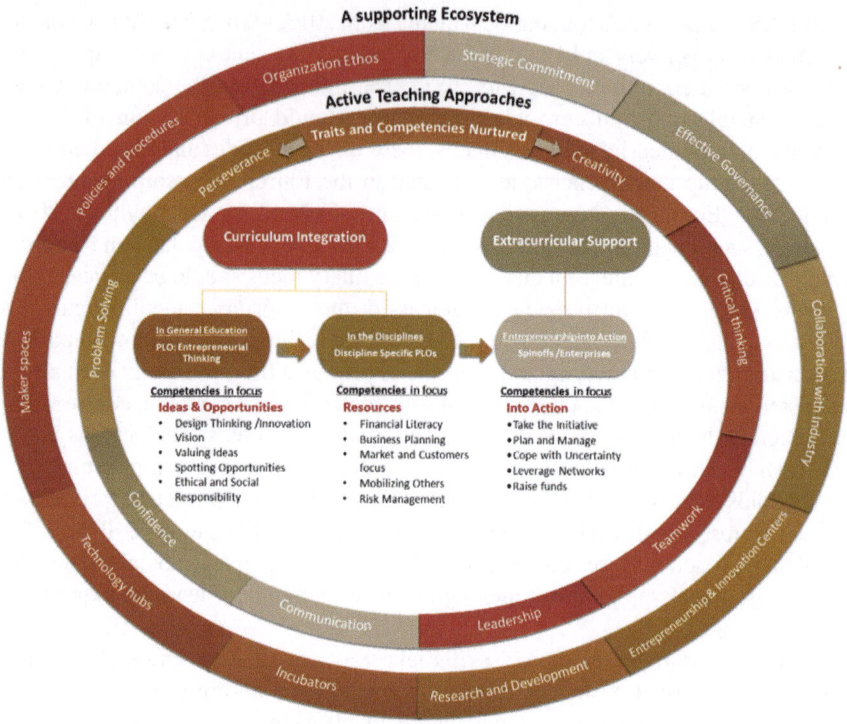

Fig. 20.1 A Holistic Model for Entrepreneurial Education

Consistently implementing active teaching methods across all courses throughout students' academic journeys, while also nurturing educators who embody qualities like openness, flexibility, and creativity, holds paramount importance. Such an approach is pivotal in cultivating and reinforcing the foundational entrepreneurial skills and competencies that span the entirety of the educational journey, such as creativity, problem-solving, critical-thinking, leadership, communication, and perseverance.

It is also essential to underscore the significance of fostering a comprehensive supporting ecosystem, and cultivating a firm university ethos. An entrepreneurial outlook within the university has to be championed at the highest level of the institution, and should be driven by the institution's mission and strategic goals. This requires a robust institutional commitment, manifest in clearly defined strategies, initiatives, and policies that mandate a coherent integration of entrepreneurial values throughout both academic and administrative domains (Shah & Nair, 2014).

More specifically, in addition to a cohesive curriculum, to ensure the sustainability and scalability of entrepreneurial education, there should be a strong institutional commitment towards fostering innovation. Taking Qatar University as an example, its 2018–2023 transformational strategy identifies innovation and entrepreneurship as one of the two core enabling strategies that cut across all sectors. The other enabling strategy is digital transformation.

To put strategic initiatives into action, and offer direction for those engaged in executing the change, it is vital to guide effective governance processes, develop conceptual frameworks, and to chart out clear roadmaps for implementation (Shah & Nair, 2014). However, to anchor the spirit of entrepreneurship within the institution, it must become ingrained within the fabric of the organisational culture, and permeate all sectors and facets of the university's operations (Kezar & Eckel, 2002; Kotter, 1995).

The strategic focus on entrepreneurial education demands substantial financial support, including investments in research and development, along with the establishment of accelerators, maker spaces, and technology hubs, the establishment of mentorship programmes, and provision of grants for experimental projects (Lackéus & Middleton, 2015). In this context, encouraging students to embrace STEM subjects, as well as integrating them within other disciplines, is essential for developing the skills and knowledge base necessary for entrepreneurial success in emerging industries and technology-driven sectors. Addressing this challenge involves not only promoting the value and opportunities within STEM fields, but also adapting educational approaches, to make these subjects more engaging and relevant to the entrepreneurial landscape. All these mechanisms serve as the lifeblood that nurtures the spirit of entrepreneurship within the institution.

A web of support is equally instrumental, and includes an array of elements, ranging from mentorship and networking opportunities to access to funding, research resources, and collaborative platforms. Collaboration with the governmental and non-governmental agencies, industry, and funding organisations is crucial, and allows for the transfer of the knowhow and resources to the university community, thereby strengthening the development of resources and execution aspects of entrepreneurial competencies (Jameson & O'Donnell, 2015).

In conclusion, this holistic model nestled within a comprehensive supportive ecosystem is instrumental to facilitate the transformation of individuals into not only entrepreneurs but also *intra*-preneurs and impactful social agents.

20.6 Concluding Reflections and Future Directions

Entrepreneurial education plays a crucial role in the economic development of the GCC countries. However, entrepreneurial education, like any educational reform, often faces challenges in achieving enduring shifts and the desired impact (Fullan, 2011; Kezar, 2011). To elevate the impact of entrepreneurial education within the GCC, a profound paradigm shift underscored by strategic, operational, and cultural adjustments becomes indispensable. The all-encompassing strategy establishes the significance of strategic congruence, when designing the individual, organisational, and systemic dimensions within the entrepreneurial university framework (Audretsch & Belitski, 2022).

In essence, the institution must not only promote but also embody an entrepreneurial ethos at its very heart. Nevertheless, due to their distinct institutional characteristics, scale, scope, and size, many universities in the region struggle to embody the entrepreneurial dynamism they seek to promote. The intrinsic need for systematic and structured operations in many public universities within the GCC may hinder the flexibility and agility essential for entrepreneurial pursuits. Consequently, several universities fall short of meeting the entrepreneurial aspirations they strive to cultivate among their students. The university governance structures, policies, and procedures should support the integration of innovation into the educational landscape, and enable agile adaptations (Gibb et al., 2009). To overcome resistance to change and enhance agility and adaptation, the GCC universities should strive to recruit and empower dynamic leaders from non-traditional backgrounds who can adapt to technological disruptions, and promote an ethos of innovation. This alignment with entrepreneurial values would create a holistic educational environment, and make entrepreneurship resonate through every corner of the university.

Innovative teaching or administrative practices should be encouraged and rewarded. This is of paramount importance, given that in GCC countries, cultural attitudes towards failure pose a significant challenge to entrepreneurial education, as the stigma associated with failure may hinder individuals from embracing the trial-and-error nature essential for entrepreneurship. Fostering an environment that encourages experimentation and embraces the notion of trial and error, is crucial for both faculty and students. To establish a culture where failure is seen as a valuable learning opportunity rather than a setback, universities can organise events dedicated to celebrating failure. During these gatherings, faculty and students can openly share their experiences and lessons learned from unsuccessful initiatives and ventures, promoting a mindset that values resilience and persistence. This paradigm shift entails the establishment of budgetary frameworks and incentive systems that not only recognise but also reward faculty and academic departments for their pioneering curricular endeavours. Faculty should think like entrepreneurs and pursue curricular innovations. Cultivating champions among the faculty is pivotal for fostering an entrepreneurial culture, and extending faculty endorsement beyond isolated pockets of excellence.

To this end, the higher education sector in the GCC should institutionalise the training and development of entrepreneurial educators qualified to deliver entrepreneurial education, and to assess its learning outcomes. This training should be designed and integrated with our proposed holistic framework for entrepreneurial education. Planning, implementing, and evaluating professional accreditation and recognition of the need for entrepreneurial educator training should enable higher institutions in the GCC to practically recognise and reward academics for curricular innovations. The recognition could be pathed and mapped to different tiers, such as fellowship, senior fellowship, etc. A good example of such training is the Enterprise Educators UK programme that aims to deliver excellence in enterprise education, by connecting academics to advance entrepreneurial education curriculum design and delivery (Enterprise Educators UK 2023). Academics who are certified as enterprise educators are trained in delivery processes, as well as the creation and evaluation of enterprise education materials and resources. The design of training sessions and the different tiers of enterprise education training programmes are informed by the needs of the entrepreneurial education community stakeholders, such as senior academics, students, assessors, curriculum committee, and funding bodies.

Initiating accreditation for universities and schools that champion and foster entrepreneurial education is crucial in this context. Take, for instance, accreditation bodies like the Small Business Charter in the United Kingdom, which celebrates world-class business schools demonstrating excellence and having impact in supporting small businesses, student entrepreneurship, and local economies. Accrediting schools and universities as 'centres of excellence in enterprise education' involves encouraging innovation, risk-taking, and the pursuit of opportunities in shaping their mission, vision, and systems.

If higher education institutions in the GCC aim to replicate such institutionalisation at the country and regional levels, a consistent and standardised approach can be applied. For instance, universities, especially business schools, could implement strategies like organising formal idea hackathons each semester, establishing incubators and pre-incubators to foster innovation and idea generation, launching think-tank initiatives, and integrating applied and formal entrepreneurial education in curricula across different academic programmes.

Certifying business schools as centres of excellence for entrepreneurial education can leverage existing tech ecosystems. In the unique context of the GCC countries, this presents an opportunity to integrate technological advancements into their curricula. A strategic focus on digital entrepreneurship, e-commerce, and emerging technologies, holds the potential to significantly enhance students' abilities to create innovative, efficient, and competitive businesses, thereby improving their employability prospects. Moreover, a deliberate shift beyond disciplinary confines, coupled with the establishment of interdisciplinary entrepreneurship programmes that blend business modules with those rooted in science and technology, as well as creative disciplines such as art and media, is likely to foster fertile ground for nurturing creativity and innovation. This approach also serves as a dynamic platform for the exchange of knowledge, ultimately creating an ecosystem conducive to the growth and development of high-tech enterprises and creative industries (Sibilla & Kurul, 2020).

Simply put, when cohesive curriculum integration models are combined with a robust supporting ecosystem, they create a holistic framework that facilitates the entrepreneurial journey at every stage. Within this dynamic and comprehensive setting, the prospects for the youth in GCC countries to transcend conventional boundaries and emerge as entrepreneurs, are not only conceivable but also inevitable. Only then can the universities stand out as pivotal catalysts for both national and regional socio-economic advancement in the region.

Bibliography

Ahmed, S. M. (2019a). Artificial Intelligence in Saudi Arabia: Leveraging Entrepreneurship in the Arab Markets. In *Proceedings—2019 Amity International Conference on Artificial Intelligence, AICAI 2019*, 394–398. https://doi.org/10.1109/AICAI.2019.8701348

Ahmed, T., Chandran, V. G. R., Klobas, J. E., Liñán, F., & Kokkalis, P. (2020). Entrepreneurship education programmes: how learning, inspiration and resources affect intentions for new venture creation in a developing economy. *International Journal of Management Education, 18* (1). https://doi.org/10.1016/j.ijme.2019.100327

Ahsan, M., Zheng, C., DeNoble, A., & Musteen, M. (2018). From student to entrepreneur: How mentorships and affect influence student venture launch. *Journal of Small Business Management, 56*(1), 76–102. https://doi.org/10.1111/jsbm.12362

Al Hussaini, A. N. (2019). Exploring the relationship between entrepreneurship development and financial management expertise among the students: a study from Kuwait. *Academy of Entrepreneurship Journal, 25* (Special Issue 2), 1–17.

Al Rabaani, A., & Al Dayri, H. (2022). Influence of entrepreneurship ecosystem at Sultan Qaboos University on preparing entrepreneurs. *International Journal of Education, Culture and Society, 7* (1), 63. https://doi.org/10.11648/j.ijecs.20220701.19

Alharbi, J., Almahdi, H., & Mosbah, A. (2018). The Impact of Entrepreneurship Education Programmes (EEPs) on the Entrepreneurial Attitudes among Higher Education Students in Saudi Arabia. *International Journal of Management, Economics and Social Sciences, 7* (3). https://doi.org/10.32327/ijmess.7.3.2018.16

Ali, I., Ali, M., & Badghish, S. (2019). Symmetric and asymmetric modeling of entrepreneurial ecosystem in developing entrepreneurial intentions among female university students in Saudi Arabia. *International Journal of Gender and Entrepreneurship, 11*(4), 435–458. https://doi.org/10.1108/IJGE-02-2019-0039

Almahdi, H. K. (2019). Promotion and participation of Saudi universities towards the development of entrepreneurial leadership—an empirical study in Saudi Arabian context. *Journal of Entrepreneurship Education, 22*(6), 1–15.

Al-Mamary, Y. H. S., Abdulrab, M., Alwaheeb, M. A., & Alshammari, N. G. M. (2020). Factors impacting entrepreneurial intentions among university students in Saudi Arabia: Testing an integrated model of TPB and EO. *Education and Training, 62*(7–8), 779–803. https://doi.org/10.1108/ET-04-2020-0096

Almawishir, N., & Messen, K. (2021). Impact of entrepreneurial education on potential entrepreneurs: The case of Jouf University's female students. *International Journal of Management (IJM), 12* (4), 673–686. https://doi.org/10.34218/IJM.12.4.2021.057

Alruabsihi, D., Lyons, R., Alarifi, G., & Largey, A. (2019). *The Entrepreneurial Career Intentions and Behavior of KSA University Students 2019.* 1–30. http://www.guesssurvey.org/resources/nat_2018/GUESSS_Report_2018_SaudiArabia.pdf

Asheim, B. (2015). An Innovation driven Economic Diversification Strategy for Kuwait. *Kuwait Foundation for the Advancement of Sciences, October*, 99.

Audretsch, D. B., & Belitski, M. (2017). Entrepreneurial ecosystems in cities: Establishing the framework conditions. *Journal of Technology Transfer, 42*(5), 1030–1051. https://doi.org/10.1007/s10961-016-9473-8

Audretsch, D. B., & Belitski, M. (2022). A strategic alignment framework for the entrepreneurial university. *Industry and Innovation, 29*(2), 285–309. https://doi.org/10.1080/13662716.2021.1941799

Balawi, A. (2021). Entrepreneurship ecosystem in the United Arab Emirates: an empirical comparison with Qatar and Saudi Arabia. *International Entrepreneurship Review, 7* (2), 55–66. https://doi.org/10.15678/ier.2021.0702.05

Barbosa, S. D., Gerhardt, M. W., & Kickul, J. R. (2007). The role of cognitive style and risk preference on entrepreneurial self-efficacy and entrepreneurial intentions. *Journal of Leadership & Organizational Studies, 13*(4), 86–104. https://doi.org/10.1177/10717919070130041001

Bridge, S. (2017). Is "entrepreneurship" the problem in entrepreneurship education? *Education and Training, 59*(7–8), 740–750. https://doi.org/10.1108/ET-02-2016-0037

Bruyat, C., Julien, P.-A., & Chair, B. (2001). Defining the Field of Research in Entrepreneurship. In *Journal of Business Venturing* (Vol. 16).

Caseiro, N., & Alberto, D. (2013). *Teaching entrepreneurship at non-business schools: a reflection*. In *Conference on Enabling Teachers for Entrepreneurship Education* (pp. 30–37). Guarda Polytechnic Institute.

D'Intino, R. S., Ross, L. W., Byrd, K. A., & Weaver, K. M. (2010). Entrepreneurship Is a general education course! The why, how and transferability of the concept. *Journal of Small Business and Entrepreneurship, 23*(sup1), 669–682. https://doi.org/10.1080/08276331.2010.10593508

Dagher, Z. R., & Boujaoude, S. (2011). Science education in Arab states: Bright future or status quo? *Studies in Science Education, 47*(1), 73–101. https://doi.org/10.1080/03057267.2011.549622

Deuchar, R. (2007). *Citizenship, enterprise and learning: harmonising competing educational agendas*. Trentham Books Ltd.

Eid, R., Badewi, A., Selim, H., & El-Gohary, H. (2019). Integrating and extending competing intention models to understand the entrepreneurial intention of senior university students. *Education and Training, 61* (2). https://doi.org/10.1108/ET-02-2018-0030

Elnadi, M., & Gheith, M. H. (2021). Entrepreneurial ecosystem, entrepreneurial self-efficacy, and entrepreneurial intention in higher education: Evidence from Saudi Arabia. *International Journal of Management Education, 19*(1), 100458. https://doi.org/10.1016/j.ijme.2021.100458

Erkkilä, K. (2000). *Entrepreneurial education: Mapping the debates in the United States, the United Kingdom and Finland*. Taylor & Francis.

European Commission. Directorate-General for Employment, S. A. and Inclusion. (2018). *EntreComp: the European Entrepreneurship Competence Framework*.

Faghih, N., & Zali, M. R. (2018). Entrepreneurship Education and Research in the Middle East and North Africa (MENA). In *Entrepreneurship Education and Research in the Middle East and North Africa (MENA)* (Issue April). https://doi.org/10.1007/978-3-319-90394-1

Fayolle, A., & Gailly, B. (2015). The impact of entrepreneurship education on entrepreneurial attitudes and intention: Hysteresis and persistence. *Journal of Small Business Management, 53*(1), 75–93. https://doi.org/10.1111/jsbm.12065

Fayolle, A., Liñán, F., & Moriano, J. A. (2014). Beyond entrepreneurial intentions: Values and motivations in entrepreneurship. *International Entrepreneurship and Management Journal, 10*(4), 679–689. https://doi.org/10.1007/s11365-014-0306-7

Fellnhofer, K. (2019). Toward a taxonomy of entrepreneurship education research literature: a bibliometric mapping and visualization. In *Educational Research Review* (Vol. 27, pp. 28–55). Elsevier Ltd. https://doi.org/10.1016/j.edurev.2018.10.002

Fullan, M. (2011). *Choosing the wrong drivers for whole system reform*. Centre for Strategic Education.

Fuller, A. (2015). Vocational Education. In *International Encyclopedia of the Social & Behavioral Sciences: Second Edition* (pp. 232–238). Elsevier Inc. https://doi.org/10.1016/B978-0-08-097086-8.92091-9

Ghura, H., Harraf, A., & Coduras, A. (2020). Global Entrepreneurship Monitor Kuwait Report. http://es.opinometre.com

Gibb, A., Haskins, G., & Robertson, I. (2009). *Leading the entrepreneurial university: meeting the entrepreneurial development needs of higher education institutions*. www.ncge.org.uk

Gielnik, M. M., Cardon, M. S., & Frese, M. (2021). *The Psychology of Entrepreneurship: New Perspectives*. Routledge.

Godwyn, M. (2009). Can the liberal arts and entrepreneurship work together? *Academe, 95*(1), 36–38.

Greene, P. G., Brush, C. G., Eisenman, E. J., Neck, H., & Perkins, S. (2016). *Entrepreneurship Education: a Global Consideration from Practice to Policy around the World with contributions from: The Finnish Lifelong Learning Foundation*. 10–11. www.XueTangX.com

Hameed, I., & Irfan, Z. (2019). Entrepreneurship education: A review of challenges, characteristics and opportunities. *Entrepreneurship Education, 2*(3–4), 135–148. https://doi.org/10.1007/s41959-019-00018-z

Handscombe, R. D., Rodriguez-Falcon, E., & Patterson, E. A. (2008). Embedding enterprise in science and engineering departments. *Education and Training, 50*(7), 615–625. https://doi.org/10.1108/00400910810909063

Hashimi, S. Al, Zaki, Y., Al Muwali, A., & Mahdi, N. (2021). The role of nurturing technopreneurship education and building university students' entrepreneurial mindsets and skill sets in fostering digital innovation and augmenting the tech start-up ecosystem in Bahrain. *International Journal of Learning, Teaching and Educational Research, 20* (6), 152–173. https://doi.org/10.26803/IJLTER.20.6.8

Hassen, T. Ben. (2020). *The entrepreneurship ecosystem in the ICT sector in Qatar: local advantages and constraints. 27* (2), 177–195. https://doi.org/10.1108/JSBED-04-2019-0119

Heinonen, J., & Hytti, U. (2010). Back to basics: The role of teaching in developing the entrepreneurial university. *The International Journal of Entrepreneurship and Innovation, 11*(4), 283–292. https://doi.org/10.5367/ijei.2010.0006

Jameson, J., & O'Donnell, P. (2015). *Stream 1: Enterprise and Engagement Higher Education in Transformation Conference.* https://arrow.tudublin.ie/st1

Johnson, C. (1988). Enterprise education and training. *British Journal of Education & Work, 2* (1). https://doi.org/10.1080/0269000880020106

Joiner, B. (2019). Leadership agility for organizational agility. *Journal of Creating Value, 5*(2), 139–149. https://doi.org/10.1177/2394964319868321

Jones, B., & Iredale, N. (2010). Enterprise education as pedagogy. *Education + Training, 52* (1), 7–19. https://doi.org/10.1108/00400911011017654

Jones, M. V. (2007). Should high-technology SMEs expect to internationalize by passing through a sequence of development stages that affect choice of export market and entry mode? *In Small and Medium-Sized Enterprises and the Global Economy.* https://doi.org/10.4337/9781847204431.00021

Kezar, A., & Eckel, P. (2002). Examining the institutional transformation process: The importance of sensemaking, interrelated strategies, and balance. In *Research in Higher Education* (Vol. 43, Issue 3).

Kezar, A. (2011). What is the best way to achieve broader reach of improved practices in higher education? In *Innovative Higher Education* (Vol. 36, Issue 4, pp. 235–247). https://doi.org/10.1007/s10755-011-9174-z

Khan, G. M., Al Abri, S., & Rahman, S. A. (2021). Entrepreneurship in Oman. In *Entrepreneurship in the Gulf Cooperation Council Region: Evolution and Future Perspectives* (Issue 2015). https://doi.org/10.1142/9781786348081_0004

Kirby, D. A. (2004). Entrepreneurship education: can business schools meet the challenge? *Education + Training, 46*, 510–519. https://doi.org/10.1108/00400910410569632

Kotter, J. (1995). Leading change: why transformation efforts fail. *Harvard Business Review, March–April.*

Kyrö, P. (2005). Entrepreneurial learning in a cross-cultural context challenges previous learning paradigms? In *The dynamics of learning entrepreneurship in a cross-cultural university context* (Issue 2005).

Lackéus, M. (2013). *Developing entrepreneurial competencies—an action-based approach and classification in entrepreneurial education.*

Lackéus, M. (2015). *Entrepreneurship in Education—what, why, when, how. Background paper for the OECD.*

Lackéus, M. (2018). *Classifying entrepreneurial education into five basic types through four questions grounded in definitions of entrepreneurship.* In Paper präsentiert im Rahmen der 3E2018 Konferenz (Vol. 16, No. 18.5).

Lackéus, M. (2020). Comparing the impact of three different experiential approaches to entrepreneurship in education. *International Journal of Entrepreneurial Behaviour and Research, 26*(5), 937–971. https://doi.org/10.1108/IJEBR-04-2018-0236

Lackéus, M., & Middleton, K. W. (2015). Venture creation programs: Bridging entrepreneurship education and technology transfer. *Education and Training, 57*(1), 48–73. https://doi.org/10.1108/ET-02-2013-0013

Langworthy, M., & Warnecke, T. (2020). Capabilities and (missed) opportunity for women's entrepreneurship in Kuwait. *Journal of Economic Issues, 54*(2), 404–412. https://doi.org/10.1080/00213624.2020.1752105

Lee, S. S., & Mirza, C. (2021). Pursuing innovation as a strategy for improving quality of higher education challenges and opportunities facing GCC countries. *Higher Education in the Gulf: Quality Drivers, June 2021*, 16–31. https://doi.org/10.4324/9781003174172-2

Mahieu, R. (2006). *Agents of change and policies of scale: a policy study of entrepreneurship and enterprise in education.* Umeå Universitet.

Mäkimurto-Koivumaa, S., & Belt, P. (2016). About, for, in or through entrepreneurship in engineering education. *European Journal of Engineering Education, 41*(5), 512–529. https://doi.org/10.1080/03043797.2015.1095163

Matlay, H. (2005). Researching entrepreneurship and education: Part 1: What is entrepreneurship and does it matter? In *Education and Training* (Vol. 47, Issues 8–9, pp. 665–677). https://doi.org/10.1108/00400910510633198

Matlay, H. N. Mohamad, H. E. Lim, N. Y., & Soon, J.J. (2015). "Estimating the Effect of Entrepreneur Education on Graduates' Intention to Be Entrepreneurs." *Education + Training 57* (8/9): 874–890. https://doi.org/10.1108/ET-03-2014-0030. https://www.tandfonline.com/doi/pdf/https://doi.org/10.1080/02602938.2021.2021583?casa_token=-GGV4i63TqkAAAAA:lvQ7tYi0AQx6zDcdBULZNPGrO5E0U7oqUo2uXA2Zu1z1sDKaTrJJpHkAmofqLHg_KumxjljXI7Zc5g

Mehlhorn, J. E., Bonney, L., Fraser, N., & Miles, M. P. (2015). Benchmarking entrepreneurship education in U.S., Australian, and New Zealand University Agriculture Programs. *Journal of Developmental Entrepreneurship, 20* (3). https://doi.org/10.1142/S108494671550017X

Meshram, S. A., & Rawani, A. M. (2019). Understanding entrepreneurial ecosystem. *International Journal of Social Ecology and Sustainable Development, 10* (3). https://doi.org/10.4018/IJSESD.2019070107

Meyer, N., & De Jongh, J. (2018). The importance of entrepreneurship as a contributing factor to economic growth and development: the case of selected European Countries. In *Journal of Economics and Behavioral Studies* (Vol. 10, Issue 4).

Mitchelmore, S., & Rowley, J. (2010). Entrepreneurial competencies: a literature review and development agenda. In *International Journal of Entrepreneurial Behaviour & Research* (Vol. 16, Issue 2, pp. 92–111). https://doi.org/10.1108/13552551011026995

Moberg, K. (2014). *Assessing the Impact of Entrepreneurship Education: From ABC to PhD.* Copenhagen Business School (CBS), Frederiksberg. https://hdl.handle.net/10398/8965

Mwasalwiba, E. S. (2010). Entrepreneurship education: A review of its objectives, teaching methods, and impact indicators. *Education and Training, 52*(1), 20–47. https://doi.org/10.1108/00400911011017663

Nicotra, M., Romano, M., Del Giudice, M., & Schillaci, C. E. (2018). The causal relation between entrepreneurial ecosystem and productive entrepreneurship: A measurement framework. *Journal of Technology Transfer, 43*(3), 640–673. https://doi.org/10.1007/s10961-017-9628-2

Nybye, N., & Rasmussen, A. (2013). Entrepreneurship education: progression model. *Odense: Young Enterprise Denmark. Lokaliseret På Dansk Den, 25* (3), 50–67.

O'Connor, A. (2013). A conceptual framework for entrepreneurship education policy: Meeting government and economic purposes. *Journal of Business Venturing, 28*(4), 546–563. https://doi.org/10.1016/j.jbusvent.2012.07.003

OECD. (2012). Entrepreneurship at a Glance 2012. *OECD*. https://doi.org/10.1787/entrepreneur_aag-2012-en

Pittaway, L., & Cope, J. (2007). *Entrepreneurship Education: a Systematic Review of the Evidence*. http://www.ncge.org.uk/research.php

Quality Assurance Agency for Higher Education. (2012). Enterprise and entrepreneurship education: guidance for UK higher education providers. *Handbook, September*.

Rae, D. (2004). Entrepreneurial learning: A practical model from the creative industries. *Education + Training, 46*, 492–500. https://doi.org/10.1108/00400910410569614

Rae, D. (2007). Connecting enterprise and graduate employability: Challenges to the higher education culture and curriculum? *Education and Training, 49*(8–9), 605–619. https://doi.org/10.1108/00400910710834049

Rae, D. (2010). Universities and enterprise education: Responding to the challenges of the new era. *Journal of Small Business and Enterprise Development, 17*(4), 591–606. https://doi.org/10.1108/14626001011088741

Saaed, A. (2007). Applied econometrics and international development. *Applied Econometrics and International Development, 7*, 2001–2014.

Saberi, M., & Hamdan, A. (2019). The moderating role of governmental support in the relationship between entrepreneurship and economic growth: A study on the GCC countries. *Journal of Entrepreneurship in Emerging Economies, 11*(2), 200–216. https://doi.org/10.1108/JEEE-10-2017-0072

Saji, B. S., & Nair, A. R. (2018). Effectiveness of innovation and entrepreneurship education in UAE higher education. *Academy of Strategic Management Journal, 17*(4), 1–13.

Saudi, G., & Report, N. (2021). *Student Entrepreneurship in Saudi Arabia GUESSS Saudi National Report 2021*.

Schramm, C. (2012). Expanding the entrepreneur class. *Harvard Business Review*. https://hbr.org/2012/07/expanding-the-entrepreneur-class

Shah, M., & Nair, C. S. (2014). Turning the ship around: Rethinking strategy development and implementation in universities. *Quality Assurance in Education, 22*(2), 145–157. https://doi.org/10.1108/QAE-03-2012-0016

Sharahiley, S. M. (2020). Examining entrepreneurial intention of the Saudi Arabia's University students: analyzing alternative integrated research model of TPB and EEM. *Global Journal of Flexible Systems Management, 21*(1), 67–84. https://doi.org/10.1007/s40171-019-00231-8

Sibilla, M., & Kurul, E. (2020). Transdisciplinarity in energy retrofit. A conceptual framework. In *Journal of Cleaner Production* (Vol. 250). Elsevier Ltd. https://doi.org/10.1016/j.jclepro.2019.119461

Smith, A. J., Collins, L. A., & Hannon, P. D. (2006). Embedding new entrepreneurship programmes in UK higher education institutions: challenges and considerations. *Education + Training, 48* (8–9), 555–567. https://doi.org/10.1108/00400910610710001

Sobh, R. (2022). Entrepreneurial education: preparing our graduates for the challenges of life in the 21st century. In *Paper presented at the Annual Conference of the American Association of Colleges and Universities, Washington*.

Spencer, L. M., & Spencer, S. M. (1993). *Competence at work: models for superior performance*. John Wiley & Sons.

Stam, E. (2015). Entrepreneurial ecosystems and regional policy: a sympathetic critique. *European Planning Studies, 23*(9), 1759–1769. https://doi.org/10.1080/09654313.2015.1061484

Stam, E., & van de Ven, A. (2021). Entrepreneurial ecosystem elements. *Small Business Economics, 56*(2), 809–832. https://doi.org/10.1007/s11187-019-00270-6

Surlemont, B. (2007). Promoting enterprising: a strategic move to get schools' cooperation in the promotion of entrepreneurship. In A. Fayolle (Ed.), *Handbook of Research in Entrepreneurship Education: Contextual Perspectives* (Vol. 2, pp. 255–265). Edward Elgar Publishing Limited.

Suryavanshi, T., Lambert, S., & Chan, T. (2019). P126: Entrepreneurship in healthcare and health education: A scoping review. *CJEM, 21*(S1), S109–S110. https://doi.org/10.1017/cem.2019.317

Tadros, M. E. (2015). The Arab Gulf States and the knowledge economy: Challenges and opportunities. www.agsiw.org.

Tao, Z., Zhi, X., & Shangkun, L. (2022). Digital economy entrepreneurship and high quality economic development: empirical evidence from Urban China. *Frontiers of Economics in China, 17*(3), 393–426.

Teece, D. J. (2012). Dynamic capabilities: Routines versus entrepreneurial action. In *Journal of Management Studies* (Vol. 49, Issue 8, pp. 1395–1401). https://doi.org/10.1111/j.1467-6486.2012.01080.x

Thani Al Dhaheri, R. (2020). *Exploring the Implementation of Entrepreneurship Education in the UAE's Higher Education Institutions: Perspectives of Faculty Members, Academic Leaders and Policymakers. September.* https://bspace.buid.ac.ae/handle/1234/1802

Tok, M. E. (2018). *Can GCC States Achieve Sustainable Economic Diversification and Development by Driving Entrepreneurship Efforts?* http://gulfnews.com/business/economy/gcc-economies-expected-to-post-modest-recovery-in-2017-

Tracey, P., & Phillips, N. W. (2007). *The Distinctive Challenge of Educating Social Entrepreneurs: a Post-Script and Rejoinder to the Special Issue on Entrepreneurship Education.* https://www.researchgate.net/publication/228315021

Villegas-Mateos, A., & Abdellatif, M. M. (2021). *The Entrepreneurial Career Intentions of University Students in Qatar*. Guesss 2021 National Report. .

Wong, H. Y. H., & Chan, C. K. Y. (2022). A systematic review on the learning outcomes in entrepreneurship education within higher education settings. *Assessment and Evaluation in Higher Education, 47*(8), 1213–1230. https://doi.org/10.1080/02602938.2021.2021583

Wong, P. K., Ho, Y. P., & Autio, E. (2005). Entrepreneurship, innovation and economic growth: Evidence from GEM data. *Small Business Economics, 24*(3), 335–350. https://doi.org/10.1007/s11187-005-2000-1

World Bank. (2021). World Bank Country Engagement Framework (2021–2025)—State of Kuwait. *Country Engagement Framework for the State of Kuwait.*

Yar, D. H., Wennberg, W., & Berglund, H. (2008). Creativity in entrepreneurship education. *Journal of Small Business and Enterprise Development, 15*(2), 304–320. https://doi.org/10.1108/14626000810871691

Youniss, J., Bales, S., Christmas-Best, V., Diversi, M., McLaughlin, M., & Silbereisen, R. (2002). Youth civic engagement in the twenty-first century. In *Journal of Research on Adolescence* (Vol. 12, Isue 1, pp. 121–148). https://doi.org/10.1111/1532-7795.00027

Zamfir, A.-M., Lungu, E.-O., & Mocanu, C. (2013). Entrepreneurship among higher education graduates in 13 European countries. In *Theoretical and Applied Economics: Vol. XX* (Issue 11).

Chapter 21
Navigating Entrepreneurship Ecosystems in the UAE, Qatar, and the UK

Said Elbanna, Heba Younis, Mohamoud Soufi, and Loreta Armstrong

Abstract In the Gulf Cooperation Council (GCC) nations, there is a growing emphasis on entrepreneurship education as a core component of their economic strategies. Despite its recognised significance, there is an evident research void about its practical implementation within the GCC region. Our chapter bridges this gap with a detailed analysis of entrepreneurial ecosystems in the UAE and Qatar. By juxtaposing these GCC nations with the UK, we aim to provide insights that can steer policymakers in honing entrepreneurship. Our in-depth assessment touches upon the strengths, weaknesses, government interventions, funding models, challenges, and upcoming needs of these countries. A key revelation from our study is the influence cultural norms hold on entrepreneurship. Both GCC countries are refining their entrepreneurial ecosystems, emphasising higher education and experiential learning. Taking cues from the UK and other relevant countries can enhance enterprise education. Stakeholder support is crucial for a robust entrepreneurial foundation.

S. Elbanna (✉) · L. Armstrong
Strategy & Development Office & College of Business & Economics, Qatar University, P.O.BOX 2713, Doha, Qatar
e-mail: selbanna@qu.edu.qa

L. Armstrong
e-mail: ll1805333@qu.edu.qa

H. Younis
Associate Professor, Entrepreneurship and Strategy, SP Jain London School of Management, London, UK
e-mail: younish3@lsbu.ac.uk

M. Soufi
Center for Entrepreneurship & Organizational Excellence, College of Business and Economics, Qatar University, P.O.BOX 2713, Doha, Qatar
e-mail: mohamoud.soufi@qu.edu.qa

E. Zaidan et al. (eds.), *The Future of Education Policy in the State of Qatar*, Gulf Studies 21, https://doi.org/10.1007/978-981-97-9667-0_21

21.1 Introduction

Recognising the transformative potential of entrepreneurship education, particularly in developing economies, has spurred global policymakers to reevaluate and restructure their economic strategies, so as to formulate a comprehensive national economic framework, with entrepreneurship education as its driving force. This is particularly pertinent to the Gulf Cooperation Council (GCC) nations, where entrepreneurship education has emerged as a strategic tool for economic diversification (Abubakar et al., 2022). However, there has been limited research conducted into education in the GCC region (Younis & Elbanna, 2023), especially concerning entrepreneurship. This highlights the significance of this chapter in exploring this crucial domain in today's global context, to contemplate how to further this field in alignment with the transformative plans of the GCC countries.

Entrepreneurial success hinges on a deep exploration of potential opportunities, and a readiness to learn from inevitable failures (Mauji et al., 2022). Success is rarely immediate; instead, failures often outnumber victories. While successful startups like Airbnb, Facebook, Twitter, or Uber are praised, many others fade into obscurity. Given these odds, traditional employment often provides greater financial security to graduates than venturing into the unpredictable world of startups. Nonetheless, the importance of entrepreneurship cannot be overstated, especially for emerging nations facing job scarcity and rising unemployment rates. Merely recognising the GCC's significant economic potential is the initial step. The effective harnessing of this potential to create value and employment opportunities is key to establishing a robust and fully industrialised region.

While the significance of entrepreneurship education is universally acknowledged, there is a noticeable research void when it comes to its examination within the GCC countries, as noted by Elbanna et al. (2020). To truly grasp its intricacies, it is imperative to study models from diverse countries, distilling knowledge from their achievements and setbacks. This chapter endeavours to bridge this research void. It starts by considering the UK's entrepreneurship landscape, and then focuses on the unique practices evident in the UAE and Qatar. A comparative study of the entrepreneurial paths of these nations provides rich insights, proving invaluable for policymakers in the GCC region who might be eager to cultivate a robust entrepreneurial ecosystem. For further discussions on the implications, reference should be made to the subsequent chapter.

21.2 Entrepreneurship Education in the UK

Paradoxically, in the UK, despite a significant expansion of the higher education system over the past three decades, the graduate job market has experienced a decline. This can be attributed to large public organisations' restructuring and privatisation strategies (Calo et al., 2022). Furthermore, despite the increasing demand for self-employment, official data indicate that less than 1% of university graduates were involved in entrepreneurial activities from 1995 to 2005. These alarmingly low statistics prompted the UK government to adopt several initiatives to foster entrepreneurship or enterprise education.

Higher Education Institutions (HEIs) in the UK opted to adopt an entrepreneurial ecosystem approach, to enable entrepreneurial activity and encourage business startups. Consequently, massive literature on the entrepreneurial ecosystem has emerged in recent years, such as Stam (2015) and Audretsch and Belitski (2017). To acquaint the reader with the ecosystem perspective, we adopt the definition of Mason and Brown (2014), which is widely adopted in this strand of the literature. An entrepreneurial ecosystem includes "a set of interconnected entrepreneurial actors (both potential and existing), entrepreneurial organizations (e.g. firms, venture capitalists, business angels, banks), institutions (universities, public sector agencies, financial bodies), and entrepreneurial processes (e.g. the business birth rate, numbers of high-growth firms, levels of 'blockbuster entrepreneurship', number of serial entrepreneurs, degree of sell-out mentality within firms and levels of entrepreneurial ambition) which formally and informally coalesce to connect, mediate and govern the performance within the local entrepreneurial environment" (p. 5). At the heart of this definition is the HEI as the key player in the system.

Enterprise education and HEIs have therefore evolved towards the view that "… entrepreneurship, or certain facets of it, can be taught … and business educators and professionals have evolved beyond the myth that entrepreneurs are born, not made" (Kuratko, 2003). Consequently, enterprise education in the UK aims to acquaint business graduates with knowledge and skills through formal and informal means, by them attending formal and vocational training courses, to ensure prospective entrepreneurs are acquainted with the wide portfolio of skills and inter-disciplinary knowledge needed to manage startup processes.

To promote the growth in number of entrepreneurial graduates, and highlight the importance of entrepreneurial education in enhancing their employability, the UK has implemented several initiatives. In the following sub-sections, we will discuss the key practices and initiatives established, to enhance enterprise education offerings within the UK's HEIs, with a particular focus on business schools. It is worth noting that enterprise modules are also integrated into the curriculum of other types of schools, including those offering engineering, nursing, hospitality management, and applied arts.

21.2.1 UK Professional Enterprise Accreditation Bodies

The HEIs collaborate with professional accreditation bodies to recognise and celebrate business schools that adopt best practices in their entrepreneurship education offerings. Two major accreditation bodies predominantly create value in this line of work: the Small Business Charter and the Institute of Enterprise and Entrepreneurs.

Small Business Charter (SBC): The Small Business Charter was developed by Lord Young, The Prime Minister's Enterprise Advisor in 2010, and the Chartered Association of Business Schools. The Small Business Charter award is a mark of excellence for business schools, which recognises their expertise in supporting small businesses, student entrepreneurship, and engagement with SMEs in the local economy. In addition to providing business schools with nationally recognisable accreditation, the SBC facilitates knowledge-sharing networks for business schools. To attain the SBC award, the applicant business school undergoes a rigorous assessment, to determine the effectiveness of their business support, entrepreneurship education, and engagement with their local economy. As of 2022, 62 business schools across the UK hold the Small Business Charter, and have directly supported more than 18,750 small businesses (Francis, 2022).

Applicant business schools undergo assessment and evaluation based on a comprehensive set of criteria, consisting of at least 16 dimensions, such as:

1. Presence of a dedicated enterprise unit: The school should have a specialised unit capable of addressing the specific needs of small businesses.
2. Active engagement of small businesses: The school should demonstrate the active involvement of small businesses in university activities, such as alumni support, mentoring, peer support, awareness raising, and serving as role models for university staff and students.
3. Provision for supporting small business growth: The school should provide wider support mechanisms aimed at meeting the growth needs of small businesses. Evidence of impact, such as organising small business events and workshops, should also be demonstrated.
4. General management leadership programmes: The school should offer leadership programmes and training courses in areas relevant to small businesses, including tax, international finance, and trade.
5. Strategic focus on small businesses and entrepreneurs: The school should have a strategic emphasis on small businesses and entrepreneurship, reflected in its programmes, curriculum, and initiatives.
6. Evidence of small business research outputs: The school should showcase research outputs that contribute to small businesses, including the co-production of knowledge that supports their development.

These dimensions collectively form the basis for evaluating applicant business schools, and determining their suitability and commitment to supporting small businesses and entrepreneurship. The assessment period is typically 3–5 years. There are now 57 business schools that hold the SBC Award in the UK and Ireland (SBC, 2023).

In recent years, to demonstrate commitment to entrepreneurship education, Enterprise Educators UK (EEUK) has emerged as a valued partner of SBC. EEUK is a national membership entity for entrepreneurship educators. EEUK's primary purpose is to train and support educators to develop and deliver entrepreneurship education within the curriculum and through extra-curricular activities. It is worth mentioning that trained directors of EEUK make up at least one-third of any SBC assessment panel, which also comprises a senior scholar and a member of the SME community, to ensure academic and practice views are well embedded in the SBC charter award process.

The Institute of Enterprise and Entrepreneurs (IoEE):

IoEE works with its members to verify that their enterprise programmes offer the most up-to-date approaches and contents. Once the applicant university/school is approved as an IoEE centre of excellence, graduates recognise that the taught programme and awarded qualifications are benchmarked and assessed against fundamental key criteria. These criteria demonstrate the higher education institution's (HEI) commitment to entrepreneurship in the UK, conveying this commitment to potential students and funders alike. This recognition adds value and credibility to the entrepreneurship education offered by the HEI, further solidifying its position as a trusted institution in the field.

Applicant higher education institutions (HEIs) in the UK are assessed based on 12 criteria organised into four dimensions. These dimensions are as follows:

1. Anticipation and understanding of the needs of entrepreneurship learners.
2. Delivery of promised enterprise education offerings.
3. Commitment to being a learning organisation focused on entrepreneurship education.
4. Establishment of an adequate and prevailing entrepreneurship culture.

The 12 criteria, spanning the four dimensions, serve as the basis for assessing applicant HEIs in the UK, ensuring their ability to deliver quality enterprise education and cultivate a strong enterprise culture within their institutions.

21.2.2 Entrepreneurship Education in Secondary Schools' Curricula

This body of work within the literature advocates the view that education before university often stimulates and encourages interest in entrepreneurial activities. The current streams address two major forefronts: (1) the role schools play in preparing students to engage with enterprise education at the university level and (2) the outreach and support activities universities assume, to support the development of students' entrepreneurial activities during their secondary education. It is worth mentioning that a major critique of this stream of work is that all related activities are classified as extra-curricular activities, and are not included in the formal curriculum. The "Enterprise for All" campaign was launched in 2007 and relaunched in 2014. Advocates of this campaign highlight the importance of establishing effective connections between schools, universities, and real businesses, to bring education offerings to life. They argue that such links are crucial for promoting the interests of the UK economy. The need to encourage entrepreneurship activity is framed as a response to circumstances, to develop graduate's employability skills and mindset, and to empower students with a "can-do attitude" (Young, 2014).

At school level, students develop skills, attitudes, and knowledge through extra-curricular entrepreneurial activities with three major aspects: entrepreneurship capability, financial capability, and economic and business understanding.

(1) entrepreneurship capability is the ability to handle uncertainty and respond positively to create and implement new ideas and ways of doing things, and make reasonable risk/reward assessments,
(2) financial capability is the ability to manage one's finances and to become questioning and informed consumers of financial services, and
(3) economic and business understanding is the ability to understand the business context and make informed choices between alternative uses of scarce resources (Gillie, 2012).

These three aspects are used by the Office for Standards in Education (Ofsted, 2015), the English schools' inspectorate, and the Department for Education. The main emphasis of the provisions prior to the university at the school level is its explicit focus on the personal development of young students, to ensure schools prepare pupils for the next stage of their educational and professional development, whether it be employment, self-employment, undergraduate or postgraduate education, or vocational training (Ofsted, 2015).

An integral aspect of this stream work is the initiation and advent of The Careers and Enterprise Company, incorporated in February 2015. The establishment of this company demonstrated the UK government's commitment to put enterprise education back on the schools' agenda in the UK. The Careers and Enterprise Company established the national Enterprise Advisers Network (EAN) to promote the agenda in schools. The Enterprise Advisers Network is a network of volunteers from the world of work and business, who work with schools to support the development of careers and enterprises, and undertake a "baseline audit", to identify gaps in practice and help set strategic priorities (Tait, 2017). There are 18 activities that Enterprise Advisers Network works on with schools, which relate directly to fostering and encouraging entrepreneurial intentions of pupils, such as Enterprise Competitions (described as "Longer term business competitions involving employers where groups of pupils develop and run a small enterprise"), and Entrepreneurship Activities (described as "Short term entrepreneurship activities delivered in schools that can involve simulation of business challenges") (Tait, 2017). Despite these pockets of initiatives, there remain debates currently about the necessity for an intervention designed to deliver an inclusive form of entrepreneurship education as part of the curriculum offerings at the secondary school level.

21.2.3 University Entrepreneurship Degrees in the UK

The in-depth analysis of several entrepreneurship courses offered in business schools across the UK identifies the fact that many degrees adopt a blend of theoretical and conceptual contents that is augmented and enhanced by extensive use of practical, real-life, and live case studies in the classrooms. The offerings are also supplemented with applied learning techniques such as simulation packages, which constitute part of the assessment briefs. Entrepreneurship undergraduate and postgraduate degrees that focus on enterprise development, idea generation, and innovation, tend to deviate from relying solely on exams as a form of assessment. Non-traditional forms of assessment include the use of practice frameworks such as Business Model Canvas, simulation games, and evaluation of enterprise startup success over a semester. They facilitate students' applied learning, to ensure the intended learning outcomes of the modules are met. According to Honig (2004), teaching the practicalities and monitoring the production of a "business plan" is the most popular curriculum format used by business schools in the UK in their enterprise offering. The main caveat which underpins entrepreneurship course design is the focus on developing an "enterprise entity that is economically feasible and profitable" (Peterman & Kennedy, 2003).

HEIs in the UK that offer entrepreneurship degrees demonstrate their solid commitment to entrepreneurship students, by having incubation facilities at the school's physical premises, and implementing initiatives such as Peer Support Mentoring Programme, to students with startup ideas. Furthermore, business schools host entrepreneurs in residence, and allow them permanent offices; in return they tap into their expertise to offer frequent guest lecturing to students during each semester, and utilise the entrepreneurs in residence as a "live case study" in the classroom, for teaching and assessment purposes. Business schools also offer extra-curricular activities in various inter-disciplinary aspects, such as leadership masterclasses on a range of subjects, including resilient leadership, responsible leadership, and emotional leadership, to enhance a student's entrepreneurial intentions and skills. It is worth noting that most of the business schools that offer specialised entrepreneurship degrees at the undergraduate level have opted to enable their students to gain valuable industry experience in years 3 and 4, through placements with SMEs.

Entrepreneurship courses in the UK are designed so that learners can apply the transferrable skills they acquire throughout their learning journey, to create a new venture, and to review its commercial and technical feasibility through idea hackathon hubs, guest lecturing and applied learning workshops. Once an idea is confirmed or formulated, learners are required to develop a viable business plan, and model tapping into the various resources available through a supportive ecosystem. Courses offer a balance between theory and practice, and modules aim to achieve learning outcomes such as developing a student's proactivity, problem-solving skills, creativity, resilience, and sense of self-initiation and development.

21.2.4 Government Support—Help to Grow Scheme

In the 2021 Budget, Rishi Sunak, the Chancellor of the Exchequer at the time, announced the introduction of a new scheme called Help to Grow, aimed at assisting small businesses in enhancing their productivity. The fund does not provide direct funding to SMEs, rather, Help to Grow provides access to management training and technology advice, through top-tier business schools in the UK. The plan enables entrepreneurs to access some of the UK's top business schools, while encouraging more companies to adopt technology, in the hope of opening up opportunities to SMEs that larger businesses can normally afford. Business schools also offer one-to-one support from a mentor to applicant SMEs. Scholars and practitioners command the initiative, since it facilitates a reflective exercise by academics to design and prepare content for delivery to participant businesses.

The 12-week management training programme provides subsidised training and consolidated knowledge, to enhance small businesses' capability in financial management, digital marketing, accounting, tax filing, and venture capital funding. The Help to Grow scheme is widely regarded as a flagship initiative, with the aim of enhancing the productivity and growth of small businesses in the UK. An important aspect of this scheme is the establishment of closer network links between small businesses and HEIs in the UK, through direct mentoring provided by business schools. The programme is 90 per cent subsidised by the government, and payments are directed to the business schools upon the participant's successful completion of the 12-week programme. The in-depth curriculum through the 12-week course supports entrepreneurs to build their capabilities in areas such as leadership, innovation, digital adoption, employee engagement, marketing, responsible business, and financial management. Each participant is mandated by the end of the programme to submit a redesigned business growth plan, based on their 12-week learning journey.

The programme is delivered by the UK's leading schools that have Small Business Charter (SBC) accreditation. This requirement provides an assurance that the delivery schools have a proven track record in delivering programmes to SMEs, nascent and serial entrepreneurs. We list here a few aspects of the Help to Grow scheme:

The programme is delivered through a combination of virtual webinar sessions, peer-networking calls, and face-to-face case study workshops. Participants also receive ten hours of 1:1 support from a business mentor to ensure they apply what they have learnt.

By the end of the course, participants will have developed a Growth Action Plan that is customised to support them in leading and growing their business in the future.

Following completion of the programme, businesses gain access to expert advice through the alumni network.

The structure of the programme is outlined below (Gov.uk.2023).[1]

Part 1—Strategies for Growth and Innovation

Module 1—Strategy and Innovation.

Module 2—Digital Adoption.

Module 3—Internationalisation & Winning New Markets.

Part 2—Engaging with Customers

Module 4—Vision, Mission, and Values.

[1] https://www.gov.uk/business-finance-support/help-to-grow-management-uk. Accessed 6 June 2023.

Module 5—Developing a Marketing Strategy.
Module 6—Building a Brand.
Part 3—Building a Sustainable and Agile Business
Module 7—Organisational Design.
Module 8—Employee Engagement and Leading Change.
Module 9—High Performance Workplace.
Part 4—Operations and Financial Strategies
Module 10—Efficient Operations.
Module 11—Finance and Financial Management.
Module 12—Implementing Growth Plans.

21.3 Entrepreneurship Education in the UAE

In November 2015, the UAE Ministry of Higher Education and Scientific Research enacted a compulsory "Innovation and Entrepreneurship" curriculum for all higher education institutions within the UAE. This directive applied to all students attending public and private universities within the nation, irrespective of their chosen fields of study, and required them to enrol in an "Innovation and Entrepreneurship" course commencing in January 2016. The curriculum, a collaborative effort between the UAE Prime Minister's Office, the Ministry of Higher Education and Scientific Research, and Stanford University, sought to foster an innovative and entrepreneurial culture, to supplement the government's endeavours in establishing a knowledge-based economy (United Arab Emirates Ministry of Education, 2015). This initiative involved Stanford University developing an Innovation and Entrepreneurship curriculum tailored to the UAE, facilitating faculty training and an online guide for educators, and aiding the growth of the UAE Innovation and Entrepreneurship ecosystem.

Subsequently, a course on Innovation and Entrepreneurship, informed by Stanford's methodology, was developed with comprehensive syllabus and learning materials, including lecture notes, key concepts, reading resources, activities, instructional videos, and additional instructor resources, all made accessible online. Additionally, a selected group of thirty faculty members underwent rigorous training on the course contents, pedagogical approach, design thinking, crafting effective learning experiences, and classroom management. The training included a five-day trip to Silicon Valley, offering faculty members firsthand exposure to the Stanford and Silicon Valley ecosystems, and the chance to observe Stanford courses and workshops focusing on high-impact projects and commonly encountered obstacles in teaching entrepreneurship and innovation. Following the initial two years after the programme's introduction, new faculty cohorts underwent analogous training, programme "ambassadors" were chosen from the inaugural and second cohorts, and an annual symposium on innovation and entrepreneurship education was organised to display programme accomplishments, exchange knowledge and experiences on student learning, and discuss trends in education and potential strategies to bolster the entrepreneurial ecosystem.

21.3.1 The Contents and Learning Objectives

The Innovation and Entrepreneurship course is structured into three distinct modules, to be covered over a span of a 15-week semester. The three modules are as follows:

1. "Design Thinking": Primarily centred on understanding the creative mindset to apply the design thinking methodology for innovation.
2. "Entrepreneurship": This module provides a comprehensive overview of the creation of startup organisations, covering idea generation and analysis, vision and strategy formulation, customer development using lean startup methods, go-to-market strategies, financing startups, team management, and legal issues associated with startups, as well as effective pitching and presentation skills for entrepreneurs.
3. "Growth and Leadership": Concentrates on the challenges faced while leading and fostering innovation in an organisational context.

Under the heading of "knowledge", the course clarifies the fundamental concepts of "innovation" and "entrepreneurship", elucidates the distinctions between these two terms and their respective processes, debates the origins of innovative ideas, and differentiates between an idea and a viable business opportunity. Additionally, the course introduces students to the design thinking approach to innovation and the lean startup methods for initiating a business venture. As for mindset and skills, the course seeks to bolster students' confidence in their creative thinking capacities, highlight the significance of innovation and entrepreneurship in their careers and the economy and society, employ design thinking skills and methodologies, and foster teamwork and self-reflection. It also aims to leverage frameworks for evaluating a business idea, and apply lean startup skills and techniques such as developing a business model, validating assumptions, customer discovery, devising go-to-market strategies, and pitching business concepts.

21.3.2 Delivery and Assessment

The course adopts a blended teaching approach that includes lectures, interactive discussions, in-class activities, open-ended Q&A sessions, and guest speaker presentations. Each session plan in the course syllabus includes a list of reading materials, recommended videos, and sets of reflective study questions to consider before the class. These questions guide students' pre-class preparation, and lay the groundwork for class discussions.

While the course does make use of traditional teaching methods such as lectures, group discussions, and guest speakers—techniques generally classified as passive/traditional and perceived to have less impact on students' entrepreneurial attributes (Bennett, 2006; Mwasalwiba, 2010)—it also adopts an active learning approach. Active learning models necessitate students taking on a higher level of responsibility for their education (Curtis et al., 2021; Michel et al., 2009). Upon closely examining the teaching methods and learning activities in light of this dichotomy between active and passive approaches, we observe the following:

Students are expected to complete assigned readings, watch videos, and prepare responses to study questions prior to class. This implies a shift of responsibility onto the students for their learning, with instructors serving primarily as facilitators.

Students are required to form groups, work collaboratively on tasks and projects, and engage in small group discussions, activities generally classified as active learning.

Therefore, although the course ostensibly incorporates traditional teaching methods, how these methods are applied suggests a significant utilisation of active teaching techniques.

Regarding assessment methods, the course uses the following:

1. Class Participation: Accounting for 20% of the student's final grade, this component is based on the student's attendance, punctuality, and active class contributions. Active contributions are gauged by the student's individual, or group written responses to study questions, presentations, and participation in in-class interactive activities.
2. Team-Based Projects: This component contributes 40% to the student's grade, and involves two practical team-based projects, aligned with the lessons taught in the modules.
3. Final Exam: This constitutes 40% of the student's grade and requires the student to craft a personal business plan that includes a long-term vision, identification of external opportunities, an assessment of personal strengths, a five-year personal strategy, and reflections on the role entrepreneurship and innovation could play in their life and career paths.

Given the course's use of active teaching methods, we surmise that the course's primary objective is to cultivate an entrepreneurial mindset and skills within students.

21.3.3 University-Based Science and Innovation Parks

The UAE University (UAEU), recognised as the UAE's Flagship University, proudly announced the inauguration of The UAEU Science and Innovation Park in November 2016. According to the International Association of Science Parks (IASP), a science park is defined as:

"an organization managed by specialized professionals, whose main aim is to augment the wealth of its community by encouraging a culture of innovation and enhancing the competitiveness of its associated businesses and knowledge-based institutions" (International Association of Science Parks and Areas of Innovation, 2023).

The scholarly literature emphasises science parks' vital role in supporting small to medium-sized firms, particularly technology-based ones. Science parks offer invaluable services such as business incubation, access to research facilities, funding, networking, mentorship, and professional support opportunities (Lecluyse et al., 2019; Theeranattapong et al., 2021).

Consequently, the launch of the UAEU Science and Innovation Park represents a significant step forward, marking the establishment of the country's first university-based science park. As an integral part of the UAE national innovation and entrepreneurship agenda outlined by UAE Vision 2021 and the National Innovation Strategy, the Science and Innovation Park's objective is to facilitate the nation's transition towards a knowledge-based economy. This objective is achieved by promoting scientific research, encouraging the commercialisation of innovative products and services, fostering an entrepreneurial and innovative culture within the university, and assisting in the growth and establishment of businesses across various sectors (UAEU 2023).

21.3.4 Science Park Approach to Entrepreneurship Education

Although the UAEU Science and Innovation Park provides services commonly associated with a typical science park—such as startup incubation, seed funding, and an array of physical and administrative support resources—it also functions as a hub for entrepreneurship education within the university. This latter role has become a prominent aspect of the science park's operations in recent years. In this capacity, the Science and Innovation Park serves as a business incubation programme that aids individuals and teams of façulty members and students, to leverage their ideas into high-growth startups. This is primarily achieved through two interconnected entrepreneurship programmes known as "Ideas to Prototype" (I2P) and "Prototypes to Market" (P2M).

21.3.5 Ideas to Prototype (I2P)

The principal objectives of the "Ideas to Prototype" (I2P) programme is to foster an entrepreneurial mindset within the university's ecosystem, and ensure a consistent influx of applicants for the "Prototypes to Market" (P2M) phase. As such, this programme is dedicated to inspiring potential entrepreneurs to embark on entrepreneurial journeys, fostering their innovative ideas, aiding them in acquiring project management skills, and providing them with the necessary tools and techniques to create and test their prototypes. The programme is structured as a series of workshops and boot camps facilitated by accomplished entrepreneurs and leaders in innovation. Furthermore, participants are offered opportunities to compete in national and international entrepreneurship competitions.

Students, faculty, and community members can join the Ideas to Prototype programme, through direct application or by participating in competitions. One such competition is the annual "Challenge for Innovation" event. In this competition, participants with business ideas vie for a spot on the 12-month incubation programme, where a panel of entrepreneurial judges selects only the top ten venture ideas. Finalists who do not make it into the incubator are eligible for other pre-incubation programmes designed to assist these participants in refining their ideas and developing market-awareness acumen.

21.3.6 Prototypes to Market (P2M)

Conversely, the "Prototypes to Market" (P2M) programme aims to aid incubated projects to finalise their business models and launch enterprises. In this phase, participants or ventures are provided with an educational programme that includes inspirational talks, workshops on various topics related to venture creation, an immersion week in Silicon Valley, and access to a global advisory panel comprising expert entrepreneurs, professors, and business professionals. Typically, each team is assigned three members from the global advisory panel, to provide mentorship and coaching, evaluate the startup's progress, and facilitate conditions conducive to its success. Moreover, participants in Prototypes to Market are given access to a resource pool that includes incubation space, internet services, web hosting, legal assistance, seed capital, and networking opportunities with potential strategic partners, customers, or investors.

In essence, the UAE Science and Innovation Park's primary objective is to foster the creation of high-growth startups in areas of priority, as designated by the UAE National Innovation Strategy and in response to global challenges, such as healthcare, space exploration, and enhancing life quality through education, inclusive economic growth, and sustainable job creation. The programme is an immersive, hands-on entrepreneurship experience, whereby participants develop an entrepreneurial mindset and skills while actively building a startup.

21.3.7 Entrepreneurship Education in Qatar

Qatar is aiming to transform into a knowledge-based economy, guided by Qatar National Vision 2030, launched in 2008 (GCO 2023). Developing entrepreneurship is viewed as critical for this transition (Kaya & Tsai, 2016), and the government has initiated numerous programmes to promote it, making innovation and entrepreneurship the primary goals of the economy. Entrepreneurship can significantly contribute to economic development by creating employment, driving innovation, fostering competition, empowering individuals, and transforming traditional economies into industrial ones. In Qatar, it can also help reduce the burden on the government, and decrease public sector employment (Elbanna et al., 2023; Elsharnouby et al., 2023).

Furthermore, Al-Housani et al. (2023) highlighted the growing interest in fostering entrepreneurship in Qatar. However, despite the government's efforts to promote innovation, Qatar was listed as one of the countries with a below-projected innovation output in the 2020 Global Innovation Index, indicating a limited number of innovation indicators, such as patents and publications, and with limited technology exports (Hassen, 2020). For instance, Qatar was ranked in 129th place for patent registration, 125th for trademark origins, and 88th for scientific publications. It is suggested that limited cooperation between R&D centres, universities, and businesses could be one of the reasons (Hassen, 2022). Similarly, the Second National Development Strategy (2018–2022) emphasises the need for cooperation and integration between sectors. The 2020 World Bank report ranked Qatar in 77th place out of 190 countries in terms of ease of doing business, considering procedures to start a business, policies, and laws. Therefore, transitioning towards a knowledge-based economy is challenging for the country, and necessitates consideration of multiple factors, such as entrepreneurship motivation and supporting institutions. International experiences show that entrepreneurship and innovation require the support of institutions, human capital, research, knowledge, and infrastructure.

This section provides an overview of Qatar's recent institutions and initiatives, and their role in developing the entrepreneurship ecosystem. Additionally, barriers to entrepreneurship are discussed.

21.3.8 Government Policies and Promotion of Entrepreneurship

Qatar has made significant attempts to introduce new policies regarding entrepreneurship and innovation. However, certain challenges lie within Qatar's economic structure, as its small local population presents an obstacle (Ennis, 2015). This demographic structure, combined with limited potential and incentives for innovation and entrepreneurship, poses challenges for economic diversification away from oil and gas revenues. Despite strong government intervention to develop the entrepreneurship ecosystem and diversify the economy, many entrepreneurs face barriers and difficulties, due to Qatar's characteristics as a rentier state, relying on hydrocarbon resources (Hassen, 2020). Since 2008, the government has heavily invested in infrastructure development, skills development, e-government, and the communications technology sector, positively impacting all sectors.

Numerous efforts have also been dedicated to developing the entrepreneurship ecosystem, through the establishment of various institutions and supporting organisations, including incubators and funding structures such as the Qatar Development Bank (QDB), Enterprise Qatar, Silatech, the Social Development Center, INJAZ Qatar, the Center for Entrepreneurship at Qatar University, Qatar Business Incubation Center (QBIC), the Digital Incubation Center (DIC), Qatar Science and Technology Park (QSTP), and the Qatar Foundation (QF) (Hassen, 2020). As a result of these initiatives, Qatar was ranked first among GCC countries, second in the MENA region, and 22nd globally in the Global Entrepreneurship Index in 2018 (GEDI, 2019). However, while Qatar has achieved a favourable ranking in this respect, there is no specific technological entrepreneurship strategy or policy, and technological entrepreneurs still face barriers and challenges.

The study by Pradhananga et al. (2017) describes Qatar's entrepreneurial regulation and institutional environment in relation to entrepreneurship. It highlights the fact that there are no legal obligations or incentives for actors in the Innovation-University-Government Partnership to participate, and that Qatar National Vision 2030 and the Qatar National Research Strategy do not include a regulatory framework. Nonetheless, universities, industries, and government sectors are aware of the importance of Qatar National Vision 2030 and are working towards its goals and objectives on various levels. Certain specific regulations, such as the Trademark Law enacted in 2002, and the Law for Protection of Trade Secrets enacted in 2005, support the development of knowledge competence. Efforts have also been made to encourage knowledge creation and transfer, including the enactment of the Law for Establishment of Free Zone for Qatar Science and Technology Park (QSTP) in 2005, the Patent Law in 2008, and the Law for Establishment of Qatar Development of SMEs in 2011 (QSTP 2023). However, the effective utilisation of these laws has been limited, due to cultural contexts related to business and entrepreneurship. There are several measures to promote entrepreneurship and support small and medium-sized enterprises (SMEs) in the country, including revised regulations in 2015 to encourage SME participation in public procurement and the MANATEQ project, initiated in 2011 to develop Special Economic Zones with tax exemptions for approved businesses.

Al Sharif et al. (2022) stresses the importance of government support for startups, particularly through consultancy and advisory services related to product compatibility and digital safety in transactions. The legal system in Qatar is described as a contributing factor to problematic processes and procedures faced by entrepreneurs (Al-Housani et al., 2023). It is noted that policies, laws, and regulations in Qatar and the GCC countries are often formulated without public consultation or input, resulting in limited awareness and understanding among the public. Additionally, the distribution of business licences by the government is often biased towards specific families or tribes, including the ruling family, as part of the rentier agreement in Qatar (Tok, 2020). This practice limits opportunities for other businesses to engage in domestic ventures outside of oil and gas-related industries. To raise awareness and interest in innovation and entrepreneurship, it is important to continue investment in education, particularly in vocational training and government-led initiatives in response to economic and political developments.

According to the Global Entrepreneurship Monitor and Qatar National Report in 2019, Qatar showed significant progress in terms of early-stage entrepreneurial activity, ranking 15th among the 50 participating countries (Qatar Development Bank, 2020). However, there was an increase in the discontinuation of businesses, ranking Qatar as 4th among the participating countries, indicating the lack of profitability as the primary reason for business exits. The sociocultural characteristics of Qatari society, such as a preference for public sector employment, and the perception of entrepreneurship as an additional source of income, can influence entrepreneurship (Hassen, 2020). The Global Entrepreneurship Monitor (GEM) report in 2017 indicates positive attitudes towards entrepreneurship in the MENA region, with governments making efforts to stimulate entrepreneurship in their respective countries. The United Arab Emirates and Qatar are regarded as having the most enabling environments and ecosystems for entrepreneurship (Aminova et al., 2020).

Policies, legal systems, and various financial, market, and interpersonal factors have hindered the promotion of entrepreneurship in developing countries like Qatar. The GEM 2020/2021 Global Report identified government regulations, lack of financial support, and inadequate entrepreneurship education as critical constraints for entrepreneurship in Qatar. Despite significant government funding and support in Qatar, the impact on the entrepreneurial system has been limited (Al-Housani et al., 2023). Insufficient planning, execution, and quality of initiatives may contribute to the under-realisation of entrepreneurs' success and fulfilled potential. Entrepreneurs also face significant obstacles due to unclear processes and procedures in Qatar. Al Sharif et al. (2022) emphasises the importance of collaboration with external parties, flexible technology markets, and environments that promote knowledge migration and competitiveness.

21.3.9 The Vital Role of Educational Entities in Fostering an Entrepreneurial Culture in Qatar

Investments in world-class research and education, such as the universities in Education City and Qatar University, play a key role in the aim of diversifying the economy and promoting entrepreneurship. However, the tangible results of these investments take time to be realised (Tok, 2020). Al-Housani et al. (2023) argue that the development of quality human capital through the educational structure can provide a competitive advantage and reduce reliance on a hydrocarbon-based economy. However, there is still a limited number of entrepreneurship-focused education opportunities in schools and universities in Qatar. Additionally, GCC countries, alongside the global community, face a pervasive challenge regarding the purpose of research, including studies on entrepreneurship. The critical objective should be to prioritise research that maximises societal benefits, rather than utilising it as a means to secondary ends, such as inflating university rankings, serving the broader interests of the publishing industry, or accommodating the personal ambitions of individual scholars (Elbanna & Child forthcoming).

There are several educational programmes designed to promote entrepreneurship. For example, Hamad bin Khalifa University, in collaboration with other institutions, is undertaking a three-year project to develop a local entrepreneurship platform, tailored to the country's needs, strengths, and traditions. The project aims to foster a culture of entrepreneurship, aligned with Qatar National Vision 2030, creating supportive environments and capacity-building tools (HBKU 2023). The Education City Innovative Entrepreneurship Program, in partnership with Qatar Development Bank, supports innovative entrepreneurs within the Qatar Foundation, and promotes the development of an entrepreneurship ecosystem.

The Ministry of Education and Higher Education should take a lead role in integrating entrepreneurial education into primary and secondary schools. Cooperation between institutions, particularly between educational institutions and businesses investing in R&D and fostering innovation, is crucial (Al-Housani et al., 2023; Hassen, 2022).

21.3.10 The Ministry of Education and Higher Education

The Ministry of Education and Higher Education has implemented several initiatives to enhance business thinking and an entrepreneurial mindset among students in public schools. One of these initiatives is the establishment of a Business Administration Department within the ministry, aimed at better preparing students for the business world, and providing them with relevant business knowledge. In line with this, the ministry recently launched a new initiative called "I am an Entrepreneur", which aims to promote and foster entrepreneurship among students in Qatar's public schools.

Furthermore, the Center for Entrepreneurship & Organizational Excellence at Qatar University has actively engaged in meetings with representatives from the Ministry's Business Administration Department, to explore new opportunities for collaboration and to work together so as to enhance entrepreneurship activities within the ministry. This collaborative effort highlights the integration and synergy between different entities in the country that are dedicated to promoting entrepreneurship, and fostering a culture of innovation. By leveraging their collective expertise and resources, they strive to create a more robust ecosystem for entrepreneurial development and support in Qatar.

21.3.11 Qatar University

Qatar University has played a crucial role in establishing the nation's entrepreneurial knowledge (Pradhananga et al., 2017). Its Entrepreneurship and Innovation strategy focuses on transforming the university, by fostering an entrepreneurial culture, building entrepreneurial competency, facilitating venture creation and commercial outcomes, and contributing to the national entrepreneurship and innovation ecosystem (Qatar University 2023). The strategy is guided by four key objectives, each supported by specific strategic initiatives.

The first objective is to cultivate an entrepreneurial mindset through targeted programmes at Qatar University, and the wider community in Qatar. The university aims to instil entrepreneurial thinking and values among its students, and provide them with the necessary skills and mindset to succeed as entrepreneurs. The second objective focuses on developing entrepreneurial competency within Qatar University itself and the country. The university provides training, workshops, and educational programmes, to enhance entrepreneurial skills and knowledge among its students, faculty, and staff. This objective is intending to develop a pool of talented and capable entrepreneurs. The third objective involves implementing programmes to support the growth of ventures, startups, and commercial outputs. Qatar University provides resources, mentorship, and support for aspiring entrepreneurs to start and scale their ventures. This includes incubation programmes, funding opportunities, and access to networks and partnerships. The fourth objective is to enhance Qatar's overall entrepreneurship and innovation ecosystem through collaboration with government, industry, and funding organisations. Qatar University works closely with various stakeholders, to create a conducive environment for entrepreneurship and innovation.

To achieve these objectives, the university established the Center for Entrepreneurship in 2013. Initially, the centre's mission was to assist aspiring entrepreneurs in implementing their ideas and supporting the sustainability of SMEs in Qatar. In 2022, the centre was repositioned and titled the Center for Entrepreneurship and Organizational Excellence, to enhance its role in serving the Qatari economy. The new vision of the centre is "to be an exemplary research center, regionally and internationally, by creating and applying new knowledge and engaging in value-added activities that lead to sustainable business and societal impact". The new direction of the centre indicates that while entrepreneurship remains a focal point, the centre has broadened its focus to include other disciplines within the College of Business and Economics, in order to have a greater impact. This broadening of scope might either dilute the centre's initial emphasis on entrepreneurship, or potentially enrich it, a determination we are currently unable to make.

In late 2021, the Office of Strategic Innovation, Entrepreneurship & Economic Development (SIEED) was established within the QU President's Office. Its purpose is to initiate, manage, and coordinate the development of enabling strategies and cross-sector initiatives, such as the entrepreneurship and innovation strategy and the digital transformation strategy. This is in line with QU's Transformation Strategy 2018–2022 and Qatar National Vision 2030. The SIEED Office supports the university-wide and cross-sector implementation of key drivers of socio-economic development in knowledge-based economies, including digital, innovative, and entrepreneurial initiatives[2]. There appears to be an overlap between the Center for Entrepreneurship in the College of Business and Economics and the entrepreneurship activities of the newly established SIEED Office. This is evident from the transfer of staff from the centre to the office. It raises the question of whether these two entities within the university, with similar missions associated with entrepreneurship activities, should be integrated or better distinguished, to ensure clarity and effectiveness in their respective roles.

21.3.12 The Qatar Foundation

The establishment of the Qatar Foundation's universities plays a central role in the country's objective to diversify its economy, and in supporting entrepreneurship and innovation in Qatar (Tok et al., 2021). It has fostered a culture of creativity and research excellence within Qatari society (Pradhananga et al., 2017). The Qatar Foundation is committed to transforming the economy into a knowledge-based model, and offers services to provide knowledge and technology support to startups. One of their programmes, the Al Khabeer Program, offers intellectual property (IP) and technical consultation to local technology-based startups. They also provide technology licences that present opportunities for entrepreneurs. Additionally, the Qatar Foundation possesses research and development (R&D) capabilities, supporting researchers, companies, and entrepreneurs in transforming technologies and discoveries into market-ready innovations, thus facilitating commercial success (Qatar Foundation, 2023).

The Qatar Foundation established Education City in 1997, housing the campuses of renowned US universities, to facilitate knowledge transfer and education (Ennis, 2015). Various agencies and centres have been established by the Qatar Foundation to foster commercialisation and the establishment of small and medium-sized enterprises (SMEs). These include Qatar Science and Technology Park (QSTP), Qatar Business Innovation Center (QBIC), Bedaya, Injaz, the Roudha Center, and Qatar Social Development Center (QSDC). Financial support and guidance for entrepreneurs are provided through Qatar Development Bank (QDB) and Qatar Financial Center (QFC), offering favourable packages and flexible payment plans, particularly for Qatari nationals (Ennis, 2015). QSTP, an organisation under the Qatar Foundation, plays a vital role in supporting the development of high-tech products and services, and commercialising market-ready technologies. They offer various programmes, including innovation programmes, tech incubation, and funding programmes, to support entrepreneurship and diversify the economy. QSTP has successfully incubated 22 tech-focused Qatari startups, and commercialised 31 products through its programmes (QSTP 2022). An example is Fatora.io by MaktApp

[2] Http://www.qu.edu.qa/offices/president/sieed as accessed on July 15, 2023.

QSTP-LLC, an online application that simplifies the payment process for SMEs (QSTP, 2022). Although the Qatar Foundation is considered non-governmental, it is connected to government agencies (Ennis, 2015).

In 2006, the Qatar Foundation established the Qatar National Research Fund (QNRF), serving as the country's primary research funding agency. QNRF's mission is to support and promote research and development in Qatar, and facilitate international scientific cooperation. It offers diverse funding programmes across various disciplines, aiming to foster original and competitive research in fields such as engineering, technology, life sciences, medicine, humanities, and social sciences (Hassen, 2022).

21.3.13 Incubators in Qatar and Their Role in Promoting Entrepreneurship

Qatar has witnessed the emergence of numerous incubators, dedicated to promoting entrepreneurship (Al Sharif et al., 2022). For example, Qatar Development Bank (QDB) is a government-owned entity established in 1997, with a mission to enhance the economic development cycle, by providing entrepreneurs and SMEs with a comprehensive range of financial and advisory products. QDB aims to develop and empower Qatari entrepreneurs and innovators to contribute to the diversification of the Qatari economy, through successful small and medium enterprises that can compete globally. They offer various programmes, packages, training courses, workshops, conferences, and competitions, to support entrepreneurs and SMEs (QDB 2023).

Another notable organisation is the Qatar Business Incubation Center (QBIC), a mixed-use incubation centre in Qatar founded by QDB and the social development centre, NAMA. QBIC's mission, vision, and strategy align with QDB's goal of contributing to the diversification of the economy, by ensuring the long-term success of startups. QBIC provides resources and programmes to support startups and SMEs, including incubation services, investment opportunities, courses to help entrepreneurs start or grow their businesses, and networking with partners (QBIC 2022).

INJAZ is another organisation that collaborates with various stakeholders, to educate the youth in succeeding in the global economy. As a member of Junior Achievement Worldwide and INJAZ Al-Arab, their vision is to inspire and prepare the Arab youth to become future business leaders and entrepreneurs who empower their community. Bedaya Center, founded by QDB and Silatech, serves as an advisory resource for supporting entrepreneurs in Qatar to launch their startups and pursue their careers. Bedaya's main objective is to help young people discover their path, and they offer sessions and workshops to guide entrepreneurs in starting and sustaining their startups (Al Sharif et al., 2022).

In addition, Qatar has several incubation centres including the QSTP Incubation Center, Qatar Business Incubation Center, Digital Incubation Center, and QBIC Tourism and Business Incubation at Qatar University. The QSTP Incubation Center focuses on technology-oriented startups, providing support and resources to accelerate their growth (QSTP 2022). The Qatar Business Incubation Center, a joint initiative of QDB and NAMA, offers incubation programmes to support entrepreneurs, startups, and existing SMEs. The Digital Incubation Center aims to advance innovation, and support digital startups in Qatar (Al Sharif et al., 2022). The Qatar Tourism Authority, in partnership with QDB, has initiated QBIC Tourism, an incubator dedicated to nurturing tourism startups. This programme empowers SMEs and budding entrepreneurs to refine and create offerings that enrich the tourist experience in Qatar, by equipping them with essential business tools, initial funding, and insights from QTA's leadership. Recently, Qatar University introduced a Business Incubation programme, stemming from the activities of the Center for Entrepreneurship (now renamed as the Center for Entrepreneurship and Organizational Excellence). The programme is still navigating its path, exploring various potential scenarios.

21.3.14 Accelerators and Funding Programmes

The investment ecosystem for startups in Qatar is still in its early stages. Seed funding is primarily obtained through incubators and accelerators, with support from institutions such as Qatar Development Bank (QDB), QBIC Entrepreneur Fund, and QSTP Tech Venture Fund. Private entities like Ooredoo and the Qatar Financial Centre (QFC) also contribute to the development of incubators and accelerators.

Qatar Development Bank (QDB) plays a significant role in venture capital financing through its USD100.3 million Venture Capital Fund. Additionally, individual players in the market, including Doha Tech Angels, 360 Nautica, Qatar Business Angel Network (QBAN), and Goodforce Labs, have formed partnerships with startups, to provide funding and support (Al Sharif et al., 2022). Qatar Development Bank (QDB) provides a variety of funding programmes to assist entrepreneurs and SMEs. The Ithmar Program provides seed funding for projects, covering up to 90% of the project cost, with a maximum investment of USD247.2 K. The Istithmar Program focuses on SME equity and provides growth capital to Qatari entrepreneurs. QDB also offers specialised programmes such as the Alfikra-Qatar National Business competition, the JAHIZ Program, the SME Equity Program, and TASDEER, which aim to support innovation and entrepreneurship in Qatar (Pradhananga et al., 2017).

Doha Tech Angels (DTA) is the first organisation in Qatar that invests in early-stage technology startups using private angel funds. They provide financial and non-financial support, including funding, guidance, mentoring, and overall assistance, to ensure the growth and development of startups. QFC (Qatar Financial Centre) provides non-financial services, to support local and international businesses, with a vision of establishing Doha as a leading global financial and commercial capital

(Qatar FinTech Hub, 2021). Qatar FinTech Hub collaborates with global partners, to offer opportunities to FinTech companies, and runs innovation programmes for them. The Digital Incubation Center, in collaboration with AstroLabs and MOTC, has incubated numerous startups, created jobs, and received significant investments. Qatar Sports Tech, introduced by Qatar Development Bank, is an accelerator for sports-tech companies, which provides mentorship, office space, seed funding, and access to a global sports network (Al Sharif et al., 2022).

As shown by the preceding discussion, diverse stakeholders contribute to supporting entrepreneurship through different mechanisms, such as education, training programmes, funding, incubation, and entrepreneurial guidance. However, it is crucial to analyse how the different initiatives carried out by the different actors in the entrepreneurial ecosystem interact. Understanding the inputs and outputs of these programmes can help structure them effectively, enhancing synergy and avoiding redundancy, thereby amplifying their overall impact. To demonstrate, Fig. 21.1 illustrates Qatar University's interaction with other actors within the entrepreneurship ecosystem. In this model, the university receives input from the Ministry of Education in the form of high school graduates. Then, through entrepreneurial education, training, and experiences provided to these graduates, the university transforms these inputs into tangible outcomes, such as startup projects, entrepreneurs, and graduates equipped with entrepreneurial skills. Subsequently, these outcomes become inputs that contribute to or are further served by other ecosystem actors, such as accelerators and incubators, government support and funding programmes, and private and public sector organisations. The figure also shows how other outcomes of the university, such as research and policy publications, can inform the policies and practices of policymaking institutions and legislators; how entrepreneurship awareness programmes spread the entrepreneurial spirit in the community; and how patents, copyrights, and inventions translate into university spin-offs, licensed or sold to private and public sector organisations.

This chapter suggests that conducting similar analysis at the level of each actor in the country's entrepreneurship ecosystem will not only enhance the quality and the impact of the ecosystem as a whole, but also aid in identifying and understanding the role of the various ecosystem actors.

21.4 Conclusion

This chapter considers the state of entrepreneurship across two distinct regions: two countries from the GCC and the UK, representing the Western context. These diverse perspectives underscore the importance of crafting and stewarding a comprehensive ecosystem that bolsters entrepreneurial growth. Within this ecosystem, myriad stakeholders—from governmental institutions, educational entities, both public and private financial institutions, to innovators and incubators—collaborate to coalesce and create value.

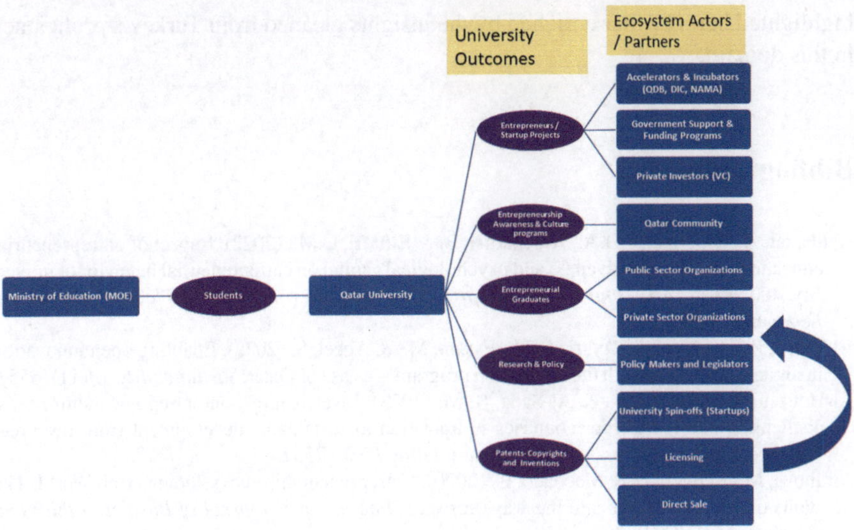

Fig. 21.1 Qatar University interaction with ecosystem partners

A review of these three nations accentuates the pivotal role of networking and cultural norms in shaping entrepreneurial mindsets and fortifying self-efficacy. By dissecting the entrepreneurial ecosystems in each of these countries, we have gleaned profound insights regarding their respective strengths, weaknesses, current governmental initiatives, funding mechanisms, educational endeavours, challenges, future requirements, and factors that either facilitate or obstruct entrepreneurial growth.

Our examination suggests that the GCC, particularly the UAE and Qatar, should continue refining their entrepreneurial ecosystems. Emphasis should be put on evaluating how extant support structures are adapting and identifying areas needing enhancement. One focal point should be higher education, where there is an opportunity to broaden experiential learning for entrepreneurship students. Taking cues from the UK's "Help to Grow" initiative, there is potential in connecting enterprise education directly to venture creation and seed financing.

While both the UAE and Qatar have made notable strides in fostering innovation and entrepreneurship, as part of their economic diversification strategies, the support from formal and informal institutions, along with other stakeholders, is crucial. A bolstered entrepreneurship ecosystem can cement its role as a significant pillar in the economy. It is pertinent to note that as we move into the era of artificial intelligence (AI), there will be significant implications for entrepreneurship education, as well as education at large (Elbanna & Armstrong, 2023). Educational leaders and policymakers in our region must carefully consider these evolving dynamics.

Building on the discussions in this chapter, the subsequent chapter will explore in greater depth the policy implications vital for fortifying Entrepreneurship Education in the GCC. Our analysis will draw from the experiences of the three countries highlighted here, further enriched by the insights gleaned from Turkey's proficiency in this domain.

Bibliography

Abubakar, A., Mourgan, F.H.A., Al Kharusi, B. & Elfitori, C.M., (2022). Impact of entrepreneurial education, trait competitiveness and psychological capital on entrepreneurial behavior of university students in GCC. *Journal of Enterprising Communities: People and Places in the Global Economy.*

Al Sharif, R., Pokharel, S., Ayari, A. M., Essam, M., & Aqeel, S. (2022). Enabling open innovation in digital startups through the incubation program—a case of Qatar. *Sustainability, 14*(11), 6557.

Al-Housani, I. M., Koç, M., & Al-Sada, S. M. (2023). Investigations on entrepreneurship needs, challenges, and models for countries in transition to sustainable development from resource-based economy — qatar as a case. *Sustainability, 15*(9), 7537.

Aminova, M., Mareef, S., & Machado, C. (2020). Entrepreneurship ecosystem in Arab World: The status quo, impediments and the ways forward. *International Journal of Business Ethics and Governance (IJBEG), 3*(3), 1–13.

Audretsch, D. B., & Belitski, M. (2017). Entrepreneurial ecosystems in cities: Establishing the framework conditions. *The Journal of Technology Transfer, 42*, 1030–1051.

Bennett, R. (2006). Business lecturers' perceptions of the nature of entrepreneurship. *International Journal of Entrepreneurial Behavior and Research, 12*(3), 165–188. https://doi.org/10.1108/13552550610667440

Calo, F., Montgomery, T., & Baglioni, S. (2022). "You have to work... but you can't!": contradictions of the active labor market policies for refugees and asylum seekers in the UK. *Journal of Social Policy*, 1–19.

Curtis, V., Moon, R., & Penaluna, A. (2021). Active entrepreneurship education and the impact on approaches to learning: Mixed methods evidence from a six-year study into one entrepreneurship educator's classroom. *Industry and Higher Education, 35*(4), 443–453. https://doi.org/10.1177/0950422220975319

International Association of Science Parks and Areas of Innovation. (2023). *Definitions.* https://www.iasp.ws/our-industry/definitions

Elbanna, S. & Armstrong, L. (2023). Exploring the integration of ChatGPT in education: adapting for the future. *Management & Sustainability: an Arab Review* (ahead-of-print).

Elbanna, S., Obeidat, S. M., Younis, H., & Elsharnouby, T. H. (2023). Development of Gulf Cooperation Council human resources: an evidence-based review of workforce nationalization. *Employee Relations: The International Journal* (ahead-of-print).

Elbanna & Child (forthcoming). From 'publish or perish' to 'publish for purpose'. *European Management Review.*

Elbanna, S., Abdelzaher, D. M., & Ramadan, N. (2020). Management research in the Arab World: What is now and what is next? *Journal of International Management, 26*(2), 1–21.

Elsharnouby, T. H., Younis, H., Elbanna, S., & Mauji, N. (2023). Voicing perceptions of local job seekers and employees toward workforce nationalization and employment. *Employee Relations: The International Journal, 45*(3), 615–636.

Ennis, C. A. (2015). Between trend and necessity: Top-down entrepreneurship promotion in Oman and Qatar. *Muslim World, 105*(1), 116–138.

Qatar University (2023). *Entrepreneurship and Innovation Strategy.* https://www.qu.edu.qa/offices/president/sieed/entrepreneurship-innovation.

Francis, M. (2022). Research and Innovation (R&I) and place-final report to UKRI. https://www.ukri.org/wp-content/uploads/2022/08/UKRI-050822-ResearchInnovationPlace-FinalReport.pdf

GEDI (2019). *Global Entrepreneurship Index report.* http://thegedi.org/global-entrepreneurship-and-development-index/.

GEM Global Entrepreneurship Monitor (2023). *Global Entrepreneurship Monitor*. https://www.gemconsortium.org/.

Gillie, C. (2012). Enterprise education in schools. Social Policy Section, London: House of Commons.

Government Communications Office (2023). *Qatar National Vision 2030.* https://www.gco.gov.qa/en/about-qatar/national-vision2030/.

Hamad Bin Khalifa University (2023). *Localizing Entrepreneurship Education in Qatar.* https://research.hbku.edu.qa/projects/12.

Hassen, B. T. (2022). A Transformative State in the Wake of COVID-19: what Is Needed to Enable Innovation, Entrepreneurship, and Education in Qatar? *Sustainability, 14* (7953).

Hassen, B. T. (2020). The entrepreneurship ecosystem in the ICT sector in Qatar: Local advantages and constraints. *Journal of Small Business and Enterprise Development, 27*(2), 177–195.

Honig, B. (2004). Entrepreneurship education: Toward a model of contingency-based business planning. *Academy of Management Learning & Education, 3*(3), 258–273.

Hamad Bin Khalifa University (2023). *Innovation at HBKU.* https://innovation.hbku.edu.qa/entrepreneurship-program.

Kaya, A., & Tsai, I. T. (2016). Inclusive economic institutions in the Gulf Cooperation Council States: Current status and theoretical implications. *Review of Middle East Economics and Finance, 12*(2), 139–173.

Kuratko, D.F. (2003). Entrepreneurship education: emerging trends and challenges for the 21st century. In *2003 Coleman Foundation White Paper, US Association of Small Business Education* (1–39).

Lecluyse, L., Knockaert, M., & Spithoven, A. (2019). The contribution of science parks: A literature review and future research agenda. *Journal of Technology Transfer, 44*(2), 559–595. https://doi.org/10.1007/S10961-018-09712-X

Mason, C., & Brown, R. (2014). Entrepreneurial ecosystems and growth-oriented entrepreneurship. *Final Report to OECD, Paris, 30*(1), 77–102.

Mauji, I. N., Elbanna, S., & Al Shamari, J. (2022). The COVID-19 pandemic and the case of Little Birds Kindergarten in Qatar: Strategic challenges and the role of leadership. *Emerald Emerging Markets Case Studies, 12*(4), 1–24.

Michel, N., Cater, J. J., & Varela, O. (2009). Active versus passive teaching styles: An empirical study of student learning outcomes. *Human Resource Development Quarterly, 20*(4), 397–418. https://doi.org/10.1002/HRDQ.20025

Mwasalwiba, E. S. (2010). Entrepreneurship education: A review of its objectives, teaching methods, and impact indicators. *Education and Training, 52*(1), 20–47. https://doi.org/10.1108/00400911011017663

Ofsted (2015). Ofsted Annual Report 2014/15: education and skills—GOV.UK (www.gov.uk)

Peterman, N. E., & Kennedy, J. (2003). Enterprise education: Influencing 'students perceptions of entrepreneurship. *Entrepreneurship Theory and Practice, 28*(2), 129–144.

Qatar Science & Technology Park (2022). *Incubation Center-Qatar Science and Technology Park.* https://qstp.org.qa/incubation-center/.

Pradhananga, R., Nawaz, W., Batur, I., & Koç, M. (2017, December). *Industry-University-Government Partnership (IUGP)—trends, drivers, and policy recommendations for Qatar* [Paper presentation]. The ISPIM innovation symposium 2017, Melbourne, Australia.

Prime Minister's Office. (2015). UAE National Innovation Strategy (pp. 5–13). Prime Minister's Office at the UAE Ministry of Cabinet Affairs. http://www.uaeinnovates.gov.ae/docs/default-source/pdfs/national-innovation-strategy-en.pdf?sfvrsn=2

Qatar Development Bank. (2020). *Global entrepreneurship monitor—Qatar national report.* https://www.qdb.qa/en/Documents/Qatar_GEM_2020_en.pdf.

Qatar FinTech Hub (2021). *From Qatar to the World: a report on the state of FinTech in Qatar.* https://fintech.qa/.

Qatar Business Incubation Center (2022). *Incubation at QBIC*. https://qbic.qa/en/programs.

Qatar Science & Technology Park (2023). *About Us*. https://qstp.org.qa/.

Qatar Development Bank (2023). *Mission and Vision*. https://www.qdb.qa/en/mission-and-vision.

Qatar Foundation (2023). *For entrepreneurs*. https://www.qf.org.qa/idkt/for-entrepreneurs.

SBC (2023). SBC Awards 2023. https://sbcevents.com/sbc-awards/.

Stam, E. (2015). Entrepreneurial ecosystems and regional policy: A sympathetic critique. *European Planning Studies, 23*(9), 1759–1769.

Tait, P. (2017). After the reboot: the state of computing education in UK schools and colleges. https://news.microsoft.com/wp-content/uploads/2017/11/computing-education-report.pdf

Theeranattapong, T., Pickernell, D., & Simms, C. (2021). Systematic literature review paper: The regional innovation system-university-science park nexus. *Journal of Technology Transfer, 46*(6), 2017–2050. https://doi.org/10.1007/S10961-020-09837-Y

Tok, E. (2020). The incentives and efforts for innovation and entrepreneurship in a resource-based economy: A survey on perspective of Qatari residents. *Sustainability, 12* (626).

Tok, E., Koç, M., & D'Alessandro, K. (2021). Entrepreneurship in a transformative and resource-rich state: the case of Qatar. *The Extractive Industries and Society, 8* (1).

United Arab Emirates Ministry of Education. (2015). *Higher Education organizes Innovation and Entrepreneurship Forum in collaboration with Stanford University and UAE University.* https://Www.Moe.Gov.Ae/.

United Arab Emirates University. (2023). *Science and Innovation Park.* https://www.uaeu.ac.ae/en/sip/about-us.shtml

Young, D.I. (2014). *Enterprise for all: the relevance of enterprise in education.* https://assets.publishing.service.gov.uk/government/uploads/system/uploads/attachment_data/file/338749/EnterpriseforAll-lowres-200614.pdf

Younis, H., & Elbanna, S. (2023). Teaching in times of crisis: The impact of the COVID-19 pandemic on higher education. *Innovations in Education and Teaching International, 60*(4), 451–462.

Chapter 22
Higher Education and Workforce Participation in Qatar: Unravelling the Disparity

Rula Momani

Abstract This chapter explores women's education versus workforce participation in Qatar. Despite advances in female education, the degree of women's career engagement remains low. Using extensive literature and the author's interactions with Qatari female students and employees, this study uncovers the reasons behind this. The author's experience at Zayed University highlights empowerment through education, but statistics reveal underemployment among Emirati women. Qatar shows a similar pattern: female students dominate, but faculty and local female employees are scarce. The chapter probes the perception of education as 'private' and work as 'public,' possibly explaining the presence of educational enthusiasm, but job market reluctance. It dissects societal expectations discouraging employment. This research emphasises educated women's untapped potential for economic and knowledge-based growth. By identifying the barriers, it offers insights into gender equality and workforce participation. This chapter enriches the discourse on women's empowerment and gender dynamics, in Qatar and globally.

22.1 Introduction

In the year 2000, the researcher was offered a position at Zayed University in the UAE, a newly established institution with an all-female school. The university's vision aimed to capitalise on the market to be found in the female population, comprising 50% of the workforce, for employment and economic growth. To prepare the students for remote learning and potential work-from-home scenarios, the institution implemented technology, and utilised laptops in every lesson plan. The objective was to offer female students the flexibility to choose between traditional career pursuits, or working from home and contributing to the economy. This strategic vision aligned with Zayed University's broader goal of fostering a culture of innovation, creativity, and research and development, to contribute to a knowledge-based economy.

R. Momani (✉)
Gulf Studies Program, Qatar University, Doha, Qatar
e-mail: rmomani@qu.edu.qa

E. Zaidan et al. (eds.), *The Future of Education Policy in the State of Qatar*, Gulf Studies 21, https://doi.org/10.1007/978-981-97-9667-0_22

However, despite an increase in the number of female college graduates, employment statistics for Emirati women remained low. Reports indicate that only 28% of female nationals were engaged in the workforce in the United Arab Emirates, despite their possessing higher educational qualifications compared to men (Paschyn, 2013). This raises the pertinent question of why qualified women are not participating in the workforce. Fast forward to the present, when the researcher currently works at Qatar University, where 73% of the student body is female. According to World Bank (2022) records, female employment in Qatar has reached an impressive 60.4%. On the surface, this looks like a remarkable achievement, but their statistics include all females working in Qatar, and this research is focused on Qatari nationals only. Even with the numbers currently available in 2023, and which will be presented below, some researchers are led to question if these women are contributing to the knowledge-based economy, or is it merely "window-dressing" to meet a 60% employment quota for nationalisation purposes (Elbanna & Fatima, 2023).

The 2023 statistics for economically active females (See Table 22.1 below), according to the Ministry of Development, Planning and Statistics, show that the Qatari female labour participation numbers have reached 42.7%. This is a slightly higher percentage than the 36% reported in 2015.

The same 2023 study by the ministry identifies the top reasons for unemployment among women: the highest at 42.8% being 'lack of suitable work' and the second highest at 29%, 'lack of experience.' Additionally, among the unemployed female population over 15 years of age, the categories identified for unemployment were 'homemakers' at 43%, 'students' at 43.6%, 'disabled' at 1.4%, 'retired' at 10.3% and 'other' at 1.8%.

Most studies on women's employment, including the most recent research papers from 2020 (Majbouri, 2020) and 2022 (Lari et al., 2022), reveal persistently low levels of female participation in the labour market. These low participation rates are attributed to enduring conservative cultural norms, the demands of household responsibilities, and the influence of religion and religiosity, which collectively act as barriers to women's engagement in the workforce. Furthermore, recent studies highlight that the challenges are particularly pronounced among married women. This is because the decision to work is often a collaborative process within the family (Gebel & Heyne, 2014), and it can be subject to veto by the husband, who typically serves as the primary breadwinner (Lari et al., 2022).

Nevertheless, progress has indeed been made, but the journey towards full gender equality in employment remains an ongoing one in the GCC countries. These nations aspire to a transition to knowledge-based economies, reducing their dependency on oil and gas. Achieving this objective necessitates fully harnessing the capabilities of their national workforces, particularly those of educated women. However, despite female nationals outnumbering their male counterparts in higher education institutions, their labour market contributions continue to lag behind, signalling an underutilisation of valuable human capital. It is worth mentioning that persistent legal and societal barriers hinder the complete economic integration of Gulf women (Ben Memoune & Kabbani, 2023).

Table 22.1 Figures for economically active and inactive Qataris, male and female, in 2023. *Source* Ministry of Development, Planning and Statistics (2013)

Nationality	Sex	Economically Active				Economically Active						Grand total
		Employed	Seeking work for first time	Unemployed with previous employment	Total	Home-maker	Student	Disabled	Retired	Other	Total	
Qatari	Males	71,661	121	62	71,834	0	22,221	1,771	11,360	1721	37,073	108,907
	Females	48,743	428	0	49,171	27,820	28,236	897	6668	1136	64,757	113,928
	Total	120,394	549	62	121,005	27,820	50,457	2668	18,028	2857	101,830	222,835

To reiterate, remarkable strides have been taken to increase female participation in the labour force in Qatar. Yet, a perplexing question remains: why do some Qatari women pursue higher education but choose not to seek employment? Qatar, with its relatively small native population compared to the influx of foreign workers, places a premium on the employment of nearly every Qatari national, to reduce reliance on expatriate labour and achieve the ambitious goal of 60% Qatarisation outlined in the 2030 national vision.

It is important to clarify that higher education serves a multitude of purposes beyond solely that of securing employment. Education has far-reaching benefits, including but not limited to, improving problem-solving skills, fostering life skills, enhancing flexibility, and promoting open-mindedness. Moreover, education's impact extends to various social outcomes, such as reduced mortality rates, improved fertility rates, better education for children, more equitable income distribution, and increased life expectancy at birth (World Bank 2004). Given these advantages, Qatar's pursuit of a small yet highly educated population becomes paramount in realising its goals of Qatarisation, economic stability, and a knowledge-based economy (Ince, 2010). It is essential to recognise the multifaceted role of education, and its profound impact on both individuals and society as a whole, even though this research primarily focuses on female employment trends in Qatar.

Drawing from a comprehensive literature review, and the researcher's extensive experience working with female university students and Qatari female employees over the last two decades, the chapter aims to offer insights into the underlying reasons behind this phenomenon of under-representation of females in the workforce. By exploring the domains of education in Qatar, with a specific focus on women's education, the labour market in Qatar, and women's position in the knowledge-based economy, the research endeavours to provide a nuanced understanding of the factors influencing women's workforce participation in the country. The chapter will culminate in an attempt to answer the central question, shedding light on why women receive education but refrain from joining the workforce in Qatar.

22.2 Research Question and Methodology

The general question the researcher wishes to answer through this study is: why are more women than men receiving higher education degrees (Ministry of Development, Planning and Statistics, 2013) and outperforming their male counterparts in k-12 standardised tests such as PISA, but they represent only just over a third of the workforce (34.7% UNDP 2014, 36% QNDS 2015, 42.7% Ministry of Development, Planning and Statistics, 2013). This is despite the Qatari government emphasising the importance of building human capital and a knowledge-based economy and calling specifically on women to be involved in it. The QNDS (2015) recognises this issue, and clearly articulates it, affirming, "As educational attainment increases among young Qatari women, their low labor force participation rates reflect an untapped reservoir of knowledge and skills." Moreover, it goes on by saying "The challenge is how to transform their educational gains into better economic empowerment" (p.55).

This chapter uses existing sources to explore women's participation in education and employment in Qatar. The adoption of the Qatarisation policy, which does not differentiate employees by gender, has opened many employment opportunities for Qatari women in the public and private sectors. Laws were enacted to give women retirement benefits and pensions, and a new Labor Law was enacted in 2004 (Law No. 14), which provides equality in working rights to men and women, including pay, training opportunities, and job advancement (Felder & Vuollo, 2008). These laws are meant to encourage women's economic participation, so that they reach the 60% Qatarisation goal in all sectors.

There is no dispute regarding the amount of female educational attainment and labour force participation; therefore, the question of why the majority of women are absent from the public sphere remains valid. Through a review of the literature, available survey results and the researcher's twenty years of experience teaching and working with female students and staff at higher education institutions in the GCC, this research will attempt to answer this question. This will help in arriving at a better understanding of the reason women are under-represented in the workforce, and the issues women grapple with when making the decision to work or not to work. From the review of the literature, there are a lot of mentions of the domestic or private sphere versus the public sphere. The literature explains that education is not considered the public sphere; it is not until women enter the workforce that they are considered active in the public sphere. Kemrava (2020) states that "education has increasingly been conceived as a 'private' realm, especially when gender segregation is maintained (p.145)," while the job market is considered 'public' because of the nature of the interactions with the opposite sex and the economic independence it provides. If this observation is correct, this might be one reason why women are not hesitant to enrol in higher education, but reluctant to take part in the workforce.

22.3 Education in Qatar

Formal education for women began in 1955, which followed the opening of the first boys school by only a few years; although in other Islamic countries, educating girls was opposed and considered as 'disruptive innovation' as Sir Charles Belgrave called it in 1928, the founding of the first girls school would have major repercussions. This comes from the "confusion in understanding and differentiating between what may be termed Islamic, emanating from the Qur'an and authentic sources and what may be viewed as a particular society's assumptions, originating from its sociohistorical experiences" (Abukari, 2014, p.5). In the early 1950s, there was a reluctance to send girls to school in Qatar, but those attitudes were superceded after a leading religious figure, Shaikh Mohammed Almanaa, issued a religious declaration (fatwa) clarifying the importance of education to all humans, male and female, as stated in the Qur'an (Al-Muftah, 2017; Kobaisi, 1979). The Shaikh addressed the ruler of Qatar in 1957 stating, "the education for girls was entirely consonant with the principles of Islam" (Kobaisi, 1979, p.36). As Abukari (2014) asserts, the Qur'an emphasises the importance of the acquisition of knowledge; for example, it states: "Are those equal, those who know and those who do not know? It is those who are endued with understanding that receive admonition" [Qur'an 39: 9]. This opened the doors for girls wide open, and it explains the number of female students receiving degrees since the mid-1950s in Qatar.

Before the establishment of higher education in Qatar, students had to study abroad, in Egypt and Lebanon, until the opening of the first national university in the country that later was known as Qatar University (Al-Misnad, 2012). Before having the option to study at Qatar University, the government began offering scholarships for women, as long as the family approved, and did not object to allowing their daughter to study abroad (Al-Misnad, 1984). The establishment of the national university in 1973 allowed the ministry to suspend the scholarship programme for women to study abroad, but it remained open to men. Currently, the scholarship programme is open for both genders; ironically, more women are currently receiving the Qatar University Scholarships than men (Qatar University Scholarship Office). The Director of Qatar University Scholarship office stated that the current percentages stand at 38% men and 62% female who are in the scholarship program, but she also mentioned that these numbers are constantly changing. This is a clear demonstration that not all policies in Qatar are top-down, because "women played a role in opening up spaces for their own education" (Al-Muftah, 2017, p. 22).

Women's university enrolment rate, educational attainment, and numbers of higher education graduates have been cited in a plethora of situations and literature, to show the advancements, gender equality, and modernity of the society in which they live. Now, for every male, two females are receiving a higher education degree. Perhaps the literature should focus on the "reverse gender gap" and ask where the men are in education and how education has failed them (Al-Muftah, 2017). Al-Misnad (2012) states that there are serious ramifications for the labour market and the society as a whole of the phenomenon she calls "the missing boys."

Al-Misnad (2012) speaks about the huge gender discrepancy at the universities, and that the ratio is so skewed. The trend of female domination at the university level has been going on since the inception of the university, so it is not new. Some have explained this by saying more men go abroad on scholarships, while the women stay at home. Dr. Buthaina Al Ansari, strategic expert and founder and chairperson of Tamkeen Training and Consulting Solutions, agrees that these statistics are misleading, because they do not include the numbers of young Qatari men studying abroad in the United States, United Kingdom, and Europe, but this, as mentioned above, is no longer true. In the summer of 2023, Qatar University enrolled 62% women on the scholarship program compared to 38% men (Qatar University Scholarship Office). This shows that women are also dominating the study abroad percentages. Furthermore, Dr. Amal Al-Malki, founding dean of the College of Humanities and Social Sciences at Hamad Bin Khalifa University (HBKU), disagrees that the statistics are skewed, and stresses that the students sent on scholarships abroad "are a fraction of the students that are studying in Qatar" (Abdallah, 2021). Abdallah (2021) states that, comparable to Qatar University, in the academic year 2020–2021, Qatar Foundation (QF) universities saw significantly more women enrolling than men, with nearly 75% of Qatari students being female. Specifically, 1,078 Qatari women were enrolled across all QF universities, compared to only 371 Qatari men. Additionally, 254 women graduated, while only 98 men did (Abdallah, 2021). The reasons behind this substantial gender gap in higher education involve intricate interactions between social, cultural, and economic factors.

The 'feminisation of Higher Education' denotes a global trend wherein higher education institutions experience a rising proportion of female students in comparison to their male counterparts. This phenomenon is not exclusive to the Gulf region or Qatar; rather, it has been observed in numerous countries worldwide, indicating that women are enrolling in and successfully completing higher education at higher rates than men. This trend has been recognised as a pivotal step towards empowering women and expanding their horizons. Dr. Al-Malki emphasises the significance of this development in her interview with Gulf News titled "Girls Need to See Early on in Their Lives That Everything is Open to Them, and Nothing Will Stop Them" (Ibrahim, 2020). She advocates for educators to play a transformative role in promoting social equality and broadening the possibilities and opportunities available to young minds.

Despite the fact that women are breaking traditional barriers every day, in addition to Shaikha Moza bint Nasser's encouragement and influence on women taking a bigger role in society, traditional values and societal norms continue to reinforce distinct gender roles, emphasising specific expectations for women and men. Women are largely depicted as homemakers, caregivers, and nurturers, whereas men are perceived as guardians and providers. This entrenched perspective reflects the prevailing image of gender roles in conservative Qatari society (Abdallah, 2021). Higher education attainment for women does not seem to contradict traditional norms but taking on the roles of "guardians and providers" does.

22.4 Contributing to a Knowledge-Based Economy

The discovery of oil has transformed the GCC economic environment in a very short period. This rapid economic transformation has catapulted the gulf from a simple agriculture and pearling society to a society aspiring to become a knowledge-based economy (Elessawy & Zaidan, 2014). Qatar National Vision 2030 emphasises the necessity to diversify the economy, and transform it into a knowledge-based economy (GSDP 2008). "The main strength of the knowledge-based economy in Qatar is the strong determination of the Qatari government to diversify the economy with top down intervention" (Hassen, 2020). Hassen (2020) also adds that the main weakness is the shortage of qualified workers.

An important aspect of economic development is utilising human capital efficiently; more emphasis is placed on knowledge (Abulibdeh et al., 2024), skills, and ideas in the modern economic environment than was the case only fifty years ago

(Zaidan et al., 2023). This is noticeable in the wage gaps between high school graduates and university graduates, or between high school graduates and high school dropouts (Psacharopoulos, 2006). The fact that more women are receiving a higher education than men means the success of a knowledge-based economy relies on their full participation. Men are choosing public sector jobs that have better pay and shorter hours, over private sector jobs that are more competitive and require higher qualifications. Private sector jobs are the foundation of a knowledge-based economy. This means special attention and incentives must be given to women, to take on the private sector challenge.

Table 22.2 shows that 87.3% of Qatari females work in public sector jobs while only 12.7% work in the private sector; the opposite percentages are true for non-citizens. Not surprisingly, the percentages are very similar to date. The reasons mentioned in the 2023 labour force survey for not being willing to work in the private sector are number of working days, hours of work, the low wages, and the need for the English language (Ministry of Development, Planning and Statistics, 2013).

The generous reserves in the GCC enable them to borrow and imitate the technologies, manpower, and inventions of developed countries without having experienced inventions and innovations of their own (Hvidt, 2015). Qatar National Vision 2030 is aiming to change that, and has focused one of its four pillars on human development. It is determined to provide world-class education, in order to equip and incentivise Qataris to take up prominent positions in the workforce. Women are taking on the education challenge, but they are under-represented in the labour force at a rate of 42.7% (Ministry of Development, Planning and Statistics, 2013). Their percentage increases with higher education attainment (Table 22.3).

There is no doubt that Qatar will spare no expense to bring into fruition a well-educated, capable, motivated workforce, but this potential can only be unlocked if the person is willing to do so (Hvidt, 2015). Subsequently, this is where the leadership's plans become complicated, when it comes to changing the mindset and increasing the female employment numbers in the private sector.

It needs a strong participatory role from leadership in government; in return, the government relies on the citizen to mobilise the vision. In the case of Qatar, it is clear

Table 22.2 Employed female citizens and non-citizens by work sector, 2016. *Source* GCC Statistical Center "Labor Statistics in GCC Countries 2016"

	Bahrain	Kuwait	Oman	Qatar	Saudi Arabia	UAE
Citizens						
Governmental sector	42.8%	92.0%	58.2%	87.3%	81.0%	75.0%
Private sector	57.2	8.0	41.8	12.7	19.0	25.0
Non-citizens						
Governmental sector	5.9%	16.9%	6.9%	12.2%	4.0%	4.0%
Private sector	94.1	83.1	93.1	87.8	96.0	96.0

Table 22.3 Qataris' labour force participation rate (%) by sex and educational level (age group 20–59 years)

Year	Primary school		Secondary school		Tertiary		All levels	
	Male	Females	Male	Females	Male	Females	Males	Females
2008	71	9	79	34	91	91	63	36
2010	70	10	81	39	90	72	66	36
2012	52	8	85	39	87	69	68	35
2014	60	7	87	40	92	61	69	35
2015	70	10	86	45	91	57	69	36

Source Retrieved from GSDP (2018)

that strong leadership is present in QNV 2030, in financing and building state-of-the-art facilities, institutions, and services, and in their sincere intention to develop human capital in all sectors, including healthcare and education. What is yet to be seen is if the national population is up to the challenge, and if they can be described as "knowledge workers" (Drucker, 2018).

Culture and religion exert a significant influence on women's labour market participation, particularly in contexts where more women than men are attaining higher education degrees, and outperforming their male counterparts. Given this logical scenario, one would expect women to play a pivotal role in driving the knowledge economy. However, instead of an increase in their importance, various factors are contributing to an actual decline in their workforce engagement, including the persistence of conservative groups advocating for the preservation of patriarchal systems. Such groups maintain the belief that a woman's rightful place is within the home, and if she does work, it should be in a gender-segregated environment.

In societies like Qatar, where women are primarily responsible for household and childcare duties, their opportunities for achievement and innovation may become limited. Numerous obstacles and constraints hinder female competitiveness in the private sector, where longer working hours and limited time off are prevalent. The interview with Dr. Amal Al-Malki highlights the challenges faced by Qatari females, with her personal experience of having been initially rejected for a job at the Qatar Foundation (QF) due to her nationality and gender. Subsequently, she had to demonstrate her capabilities through acquiring work experience in the United States, before being offered a position. Not all Qatari females possess the ambition, perseverance, and opportunities to pursue employment abroad, qualities which are often essential to compete for positions in the private sector. Consequently, the pool of qualified and motivated female candidates for such roles may remain restricted. These factors collectively contribute to the complex dynamics of women's workforce participation in Qatar, reflecting the intersection of cultural norms, religious beliefs, societal expectations, and economic considerations.

22.5 Results and Discussion

In attempting to answer the question why women are not joining the workforce despite the fact that they spend years receiving an education that prepares them for the labour market, the reasons have been categorised into three groups: personal choice, cultural barriers, and gender discrimination.

Personal choice

Throughout my six years of teaching women in Qatar and five years in Bahrain and the UAE, I engaged in numerous conversations with students regarding their aspirations and future plans. I took a particular interest in trying to understand their motivations for pursuing higher education when some did not envision joining the workforce after graduation. Curiosity led me to inquire about their reasons for obtaining a degree if not for career purposes. To my surprise, I encountered a range of responses that sparked intriguing discussions.

Several students revealed that acquiring an education served as a means of socialising, and provided them with a reason to venture beyond their homes. Others expressed the opinion that pursuing a degree was a way to fill their time until they found a suitable partner and entered into marriage. Some candidly admitted viewing their degree as a status symbol, akin to their material possessions, a way of elevating their standing in society. Initially, I questioned as to whether these remarks were mere jest or intended to provoke reactions, but their sincerity was never in doubt.

Conversely, another group of students sought education for personal enrichment, and the opportunity to contribute to the education of future generations. For them, higher education also represented a symbol of elevated status. On the other hand, there was a segment of the student population who demonstrated strong career aspirations, driven by a desire to make a meaningful impact and actively participate in the development of their country.

An interview with Todd Kent, Ph.D., the assistant dean for academic affairs at Texas A&M University at Qatar, provided further insights. He mentioned the fact that the majority of Qatari students who sought employment found opportunities readily available. However, he noted that a small percentage, typically female students, had different apparent intentions. These students seemed to want to pursue their education, obtain a degree, and perhaps get married, ultimately expressing no need or desire to engage in the workforce (Paschyn, 2013).

Cultural Barriers

There is a disconnect between the government's push towards increased female participation in the workforce and education, and the traditional societal expectations to prioritise the domestic sphere which remains solely the responsibility of the woman, wife, mother, and daughter. This is clearly stated in all sections of QNDS and QNV2030. In addition to taking part in the economic development of the country, "Women are central to the evolving Qatari family. Even as they maintain an adherence to valuable traditions, women are adapting to the impacts of modernization. Through their nurturing of language, codes of ethics, behavioral patterns, value systems and religious beliefs, women continue to play an indispensable role in upholding traditional familial and cultural values" (Ministry of Development Planning and Statistics, 2015, p.50). In contrast to the above, men are only expected to focus on their education, jobs, and careers. Cultural norms may place a strong emphasis on family and community responsibilities, leading women to prioritise these roles over career pursuits. As previously mentioned, conservative societal norms in Qatar often dictate specific roles for women as homemakers and caregivers. These traditional expectations might discourage some women from pursuing careers outside the home.

In an ethnographic study in women's 'majaalis' (discussion rooms), the researchers reported that many girls "expressed discontent with the unfairness of being driven by parents, schools and society towards higher education," especially when they later "realized that most men wanted a wife who did not go beyond secondary school" (Mitchell et al. 2015, p. 12). Higher education is perceived as decreasing their 'marriageability.' Reasons can vary: either because the men are less educated than the women, and do not want to marry a woman who is more educated than they are. Or the social stigma of leaving the house for an education, socialising and meeting people, being open-minded and perhaps more independent, not to mention being older than a girl who has just graduated from high school. We have come a long way from the 1950s, when teaching girls to write was considered dangerous because it allowed them to communicate more freely and even encouraged immoral behaviour (Golkowska, 2014). However, some still think today that going to college and working affects a woman's reputation. In an interview with female Qatari university students, many agreed that working in a mixed gender workplace negatively affects the woman's reputation and her marriage prospects (Mehran Kamrava, 2020):

> Listen when a man wants to propose to a woman who works in a hospital…as a receptionist for example. There is no man in the world that has not been in that hospital and does not know her name. So a man will not marry a woman everyone knows and has talked about (p. 256)

Implicitly and explicitly, women's education and employment are blamed for any societal problems, such as divorce, low fertility rates, and women remaining unmarried. This is even mentioned in QNV2030. Regarding fertility rate, which is the number of children a woman has in her lifetime, the number "fell from 4.5 in 2000 to just 3.6 in 2012. One factor in the fall in the total fertility rate is the rising proportion of women remaining permanently unmarried." QNV2030 goes on to explain why women remain unmarried by postulating that, "in most societies lesser-educated men tend to be reluctant to marry higher educated women and vice-versa." Women remain unmarried, because men are more likely to marry non-Qatari women than Qatari women, who have more to lose if they marry a non-Qatari man.

This is also explained by other factors such as smaller families to accommodate a new lifestyle where the woman works outside of the home, with delayed marriage and higher divorce rates. The contradiction between what society and the government expect of Qatari women is obvious, which is something every woman has to deal with when making these life decisions. In a survey item that stated, "There is social pressure for women to focus on family instead of Work" the majority of women (73%) agreed or strongly agreed. This is despite the fact that 97.7% agreed or strongly agreed with the statement "the state of Qatar encourages women to work" (Paschyn et al., 2015, p. 8). Women have to figure out how to balance both personal and professional responsibilities, or to decide which one they want more.

Gender Inequality

The leadership's vision expressed in QNV2030 defines gender equality clearly. It speaks of enhancing women's potential and empowering them "to participate fully in the political and economic spheres, especially in decision-making roles." It also affirms that "Progress towards gender equality before the law is fundamental for realizing human rights for all, creating and sustaining peaceful societies, and for socially inclusive and sustainable development" (GCO-QNV, p. 22). QNV2030 also admits that Qatar ranks in a high place internationally in human development, but low in terms of gender equality and women's empowerment. Nevertheless, gender bias and discrimination in the workplace could hinder women's advancement and opportunities for career growth. Moreover, despite educational preparation, there may be limited job opportunities available that align with women's qualifications and interests. Some industries might have a bias towards hiring men, or have work conditions that are not accommodating to women's traditional roles.

A Dubai Market survey in 2007 that covered 282 public and private institutions, was administered to investigate gender discrimination in the workplace. They confirmed that the percentage of female Emirati managers was significantly lower than that of male managers, they received lower wages than men, and they faced a stricter selection process than men (Mehran Kamrava, 2020). Employees are hesitant to hire or promote females, due to the unwritten segregation rule that might cause logistical issues, such as inadequate infrastructural settings. Concerns about females not being as mobile or freely available are problematic in certain jobs. Although this study was conducted in the UAE, most of the issues are the same in all GCC states. According to the survey, there are more female managers in the UAE than in Qatar, at 5.1% in the UAE, and fewer in Qatar at 3.6%.

The 'patriarchal gender contract' is implicit in GCC societies, in relation to male and female roles. It is explicit in the region's Islamic-based family laws, which place women and girls under the guardianship of fathers, husbands, or other male relatives (Moghadam, 2005). Surprisingly, women have pointed out that they need to present a 'no objection letter' to the human resources department of their new job, obtained from a male guardian. However, and at the same time, the state is advocating for 'women's empowerment.' Al-Muftah states that most of these empowerment policies have been written in a way not to challenge the local traditions. She goes on further to say that some accept these policies as symbolic in nature, and directed at an international audience more than at the local one (Al-Muftah, 2017).

It is important to recognise that societal changes take time, and breaking away from deeply rooted norms and expectations can be a gradual process. Addressing these challenges requires a comprehensive approach involving cultural shifts, supportive policies, and efforts to combat discrimination, thus enabling women to fully participate in the workforce, and contribute to the nation's economic growth and development. However, the increasing rates of women in higher education and the labour market indicates that Qatari men are becoming more accepting of women's engagement in wider society—outside of the home (Abdallah, 2021). Dr. Al-Malki commented, "The move from university into the job market is a big one and involves lots of challenges. One of the things that our education system fails to teach Qatari women is how to represent themselves in a patriarchal world and how to mitigate challenges and face gender discrimination." This may explain the low number of women in leadership roles and the fact that "the glass ceiling in certain fields does exist, and women are still struggling to break through gendered perceptions of leadership and society's lack of trust in their abilities" (Ibrahim, 2020).

22.6 Policy Implications

The labour force participation rates of Qatari women exhibited substantial growth from 2001 to 2008. However, in subsequent years, despite the attainment of higher educational levels and a multitude of employment opportunities, these participation rates have stabilised (Ben Memoune & Kabbani, 2023). It is worth noting that due to the relatively small proportion of Qatari nationals compared to the expatriate workforce in the country, even achieving 100% labour force participation among Qatari nationals would remain insufficient to meet the nation's labour needs. This underscores the critical importance placed on the education, training, and career advancement of women in Qatar.

To alleviate the reliance on foreign labour, and to harness the full potential of its educated female workforce, Qatar must strive for a more significant contribution from its women in the nation's development. As mentioned above, and taken from the Ministry of Planning, Development and Statistics, female labour force participation in Qatar stood recently at 42%, with the majority employed in the public sector. The GCC nations have all taken active steps in recent years to enhance women's participation in the labour force, aligning with broader economic and social development objectives. These endeavours have encompassed initiatives aimed at improving educational and training opportunities for women, as well as fostering more inclusive work environments conducive to female employment.

Despite growing aspirations among young females, there is a cultural backlash that does not support the employment of women (Williams et al., 2011). Even with the increased push for female employment by the government, the patriarchal culture in Arab countries that discourages women from coming into contact with men who are not their close kin has been a hurdle to overcome (Elbanna & Fatima, 2023). This can be reconciled in one of two ways by policy makers. The first is through public awareness campaigns that challenge the cultural and societal norms that prevent female labour force participation and forestall emphasis on the benefits of female involvement in developing the country. The alternative solution is by requiring the labour market to create more jobs that are deemed culturally fit for local women (Harry, 2007) such as facilitating the limits on the mixing of males and females in the workplace, creating an infrastructure that makes separating the males and females logistically possible, and providing work-from-home opportunities and flexible hours.

In Qatar, the public sector has historically served as the preferred employer for local Qataris, offering environments where gender interactions can be more limited and structured within formal group settings. However, the private sector, which aligns better with the higher education degrees and skills of the workforce, particularly women, poses unique challenges. It is in the private sector where the true potential of higher education can be realised, as opposed to the clerical service jobs typically offered by the public sector, which, although providing attractive benefits, may not fully utilise the qualifications and skills of educated women. To address this, the government can play a pivotal role by categorising specific jobs in the private sector as 'family-friendly,' thus emphasising a work-life balance for women (Hendy, 2016). Additionally, incentives can be extended to private sector companies that adhere to the criteria outlined by the government in offering such family-friendly positions.

In patriarchal societies like Qatar, women often grapple with the challenge of reconciling conflicting expectations. On one hand, government policies such as Qatar National Vision 2030 (QNV 2030) and Qatar National Development Strategy 2015 (QNDS 2015) encourage women to take on significant roles in various aspects of life, including active participation in the labour market and contributions to the knowledge-based economy. On the other hand, these expectations coexist with those cultural, societal, and family obligations that emphasise the importance of women in raising the next generation, with a strong grounding in Qatari moral and ethical values, traditions, and cultural heritage. An essential component of this expectation is that women should fulfil these roles without an over-reliance on nannies to care for their children.

Internationally, women have successfully balanced their professional and personal lives. Qatar can adopt a similar approach, provided that there is adequate support in the form of accessible nurseries, kindergartens, and elementary schools, that assist in, not only childcare, but in instilling the morals and values that are important to the nation. This approach will have a greater chance of success if Qatari females actively engage in the roles of nannies, teachers, and caregivers, to nurture, develop, and preserve Qatari identity and moral Islamic values within the younger generations. This perspective is in line with the dual roles of education, which extend beyond economic considerations.

Education in Qatar should serve two critical functions, to allow mothers to be successful in their professions. Firstly, it should impart valuable aspects of Qatari culture, fostering identity awareness among new generations, and emphasising socialisation as a fundamental element of the curriculum design (Bandiera et al., 2019; Perić Romić, 2020). Secondly, education must contribute to building a knowledge economy, equipping individuals with market-oriented skills, and driving economic diversification through education (Becker, 2002; Psacharopoulos & Woodhall, 1985; Weber, 2011). However, it is vital to strike a balance between these roles, to avoid reducing education solely to a means of acquiring skills for the workforce. Emphasising identity, culture, and moral values is equally crucial.

To achieve this balance, the Qatari curriculum should incorporate Arabic language, Islamic values, national identity, and Qatari heritage, all taught by Qataris who share the same cultural background and traditions. This approach will ensure continuity between home and school, promoting a holistic education that nurtures not only professional skills but also personal identity, culture, and moral values.

The proposed policy recommendations represent critical steps towards enhancing Qatari female labour force participation, and realising the nation's economic and social development goals. By classifying 'family-friendly' jobs in the private sector, and incentivising their provision, we can tap into the full potential of educated women and foster gender-inclusive workplaces. Moreover, supporting the availability of nurseries, kindergartens, and schools that promote national identity along with Arab and Islamic values, culture, and traditions, which are run by Qatari nationals, will empower Qatari women to balance their professional and family roles effectively. Recognising that education serves both to impart skills and to nurture cultural identity, moral values, and national heritage, underscores the need for a balanced approach in the curriculum. These policy measures are vital, not only for gender equality, but also for the broader economic growth and to foster the knowledge-based economy envisioned by Qatar's development strategies.

22.7 Conclusion

The trend of increased female education, coupled with lower female participation in the labour force, is not limited to Qatar or the Arab world; it is a global phenomenon. Dr. Al-Malki, in an interview during the Dean's Global Forum at Northwestern University in Qatar, discussed the fact that women worldwide face similar challenges, but the crucial factor is the chronological progression towards removing these obstacles. European women have encountered and overcome these hurdles earlier than their Gulf region counterparts. However, despite the varying timelines, all women ultimately confront comparable experiences.

An illustrative example from Cambridge, UK, highlights a reverse temporal disparity. During my summer visit to Cambridge, the tour guide pointed to Magdalene College, and mentioned the fact that the first cohort of female students was admitted in 1988. At that time, men characterised this as the 'death of education.' In contrast, Qatar admitted more women than men to its first college in 1971, and even prior to that, women were allowed to study abroad. Hence, Europe did not always have a chronological advantage in overcoming gender disparities in education and workforce participation. The reference to 'time' here underscores the importance of recognising historical variations in women's progress towards gender equality, emphasising that societal transformations occur at different paces in diverse regions across the globe.

Women in patriarchal societies, such as that of Qatar, encounter challenges in reconciling conflicting messages from the government and leadership on one side, and cultural, societal, and family obligations on the other. The navigation of ways to deal with these pressures varies among women, based on their individual circumstances and surroundings. Some advocate for prioritising family commitments, opting for shorter work hours, and dedicating more time to familial responsibilities, at the potential expense of their careers. Conversely, others believe in the possibility of successfully balancing both career and family responsibilities. The intricate balancing act between family and employment is universally difficult for most women worldwide, and within a patriarchal society, these difficulties are further intensified due to cultural expectations.

It is essential to differentiate between cultural pressures and religious teachings, as evidence shows that education and work are not inherently at odds with the Quranic verses or religious declarations (fatwa), except when male guardians exploit such interpretations to exert control over women. In Gulf Cooperation Council (GCC) countries, the significant oil wealth bestowed by the government on male guardians often renders the need for women to make a personal choice, unlike in other regions within the Middle East and North Africa (MENA) area, where women may have fewer options.

Al-Misnad (2012) emphasised the fact that the increasing educational attainment of women relative to men can have significant societal ramifications, as evidenced by educated women choosing to delay marriage or not marry at all. Men, in turn, may be less inclined to choose highly educated women as spouses, and might opt for non-Qatari partners, leading to an increase in 'unmarried women' statistics. Consequently, the number of Qatari women marrying foreign husbands has risen notably, becoming a relevant concern addressed in Qatar National Vision 2030. To fully utilise human capital in Qatar, and support the nation's Qatarisation project, there is a need to revise the Qatari National Law, which currently grants nationality based on the father's nationality rather than that of the mother.

Over the years, the cultural perception of education for women has evolved, and it has gained acceptance. However, acceptance of women in certain private sector jobs remains a bone of contention. Society has justified the acceptance of certain roles while rejecting others, categorising them into private versus public spheres. While Islam encourages women's participation in public roles, some private sector jobs are still deemed unsuitable for women. Women who are determined to pursue careers are increasingly stipulating such conditions in their marriage contracts, demanding the right to work after marriage. Consequently, some men accept this condition, while others choose to move on. However, there are instances where men demand that women quit or change jobs to more 'appropriate' roles before marriage, such as becoming a teacher in an all-girls school. This illustrates the contradictions between the government's efforts to empower women in the knowledge economy, and the enduring presence of the 'patriarchal gender contract,' and cultural expectations prevailing in society.

Bibliography

Abdallah, H. (2021). Qatari women "outnumber men" at local universities. Doha News | Qatar. https://dohanews.co/qatari-women-outnumber-men-at-local-universities/

Abukari, A. (2014). Education of women in Islam: A critical Islamic interpretation of the Quran. *Religious Education, 109*(1), 4–23. https://doi.org/10.1080/00344087.2014.868203

Abulibdeh, A., Zaidan, E., & Abulibdeh, R. (2024). Navigating the confluence of artificial intelligence and education for sustainable development in the era of industry 4.0: Challenges, opportunities, and ethical dimensions. *Journal of Cleaner Production, 437*, 140527.

Al-Misnad, S. A. (1984). The development of modern education in Bahrain, Kuwait and Qatar with special reference to the education of women and their position in modern Gulf society [Doctoral, Durham University]. http://etheses.dur.ac.uk/10485/

Al-Misnad, S. (2012). *The Dearth of Qatari Men in Higher Education: Reasons and Implications*. Middle East Institute. https://www.mei.edu/publications/dearth-qatari-men-higher-education-reasons-and-implications

Al-Muftah, E. (2017). Qatari women in a corporatized higher education setting: International reforms and their local bearings. *Journal for Critical Education Policy Studies*, 15.

Ben Memoune, N., & Kabbani, N. (2023). *Women of the Gulf Break Labor Market Barriers: a Journey in Progress*. Middle East Council on Global Affairs. https://mecouncil.org/publication/women-of-the-gulf-break-labor-market-barriers-a-journey-in-progress/

Drucker, P. F. (2018). *The effective executive*. Routledge.

Elbanna, S., & Fatima, T. (2023). Quantifying people in the GCC region: The uses, challenges, and efficacy of the quota system policy. *Human Resource Development International, 26*(3), 292–316. https://doi.org/10.1080/13678868.2022.2116260

Elessawy, F., & Zaidan, E. (2014). Living in the move: Impact of guest workers on population characteristics of the United Arab Emirates (UAE). *The Arab World Geographer, 17*(1), 2–23.

Felder, D., & Vuollo, M. (2008). Qatari women in the workforce. *SSRN Electronic Journal*. https://doi.org/10.2139/ssrn.1291424

GCO-QNV-English.pdf. (n.d.). Retrieved April 3, 2022, from https://www.gco.gov.qa/wp-content/uploads/2016/09/GCO-QNV-English.pdf

Gebel, M., & Heyne, S. (2014). *Transitions to adulthood in the Middle East and North Africa*. https://doi.org/10.1057/9781137355560

Golkowska, K. (2014). Arab Women in the gulf and the narrative of change: the case of Qatar. International Studies. *Interdisciplinary Political and Cultural Journal*, 16. https://doi.org/10.2478/ipcj-2014-0004

Harry, W. (2007). Employment creation and localization: The crucial human resource issues for the GCC. *The International Journal of Human Resource Management, 18*(1), 132–146. https://doi.org/10.1080/09585190601068508

Hassen, T. B. (2020). The state of the knowledge-based economy in the Arab world: cases of Qatar and Lebanon. *EuroMed Journal of Business*.

Hendy, R. (2016). Female labor force participation in the GCC. Doha International Family Institute

Hvidt, M. (2015). The state and the knowledge economy in the gulf: structural and motivational challenges. *The Muslim World, 105*(1), 24–45. https://doi.org/10.1111/muwo.12078

Ibrahim, M. (2020). Interview: "Girls need to see early on in their lives that everything is open to them, and nothing will stop them." Doha News | Qatar. https://dohanews.co/interview-girls-need-to-see-early-on-in-their-lives-that-everything-is-open-to-them-and-nothing-will-stop-them/

Ince, M. (2010). How the education affects female labor force? Empirical evidence from Turkey. *Procedia—Social and Behavioral Sciences, 2*(2), 634–639. https://doi.org/10.1016/j.sbspro.2010.03.076

Kamrava, M. (Ed.). (2020). Routledge handbook of Persian Gulf politics. Routledge. https://doi.org/10.4324/9780429201981

Kobaisi, A. J. (1979). The development of education in Qatar, 1950—1977 with an analysis of some educational problems. [Doctoral, Durham University]. http://etheses.dur.ac.uk/1856/

Lari, N., Awadalla, A., Al-Ansari, M., & Elmaghraby, E. (2022). Determinants of female labor force participation: Implications for policy in Qatar. *Cogent Social Sciences, 8*(1), 2130223. https://doi.org/10.1080/23311886.2022.2130223

Majbouri, M. (2020). Fertility and the puzzle of female employment in the Middle East and North Africa. *Economics of Transition and Institutional Change, 28*(2), 225–244. https://doi.org/10.1111/ecot.12243

Ministry of Development Planning and Statistics. (n.d.). Retrieved April 29, 2022, from https://www.psa.gov.qa/en/knowledge/Doc/HDR/Qatar_Fourth_National_HDR_Realising_QNV2030_The_Right_to_Development_2015_EN.pdf

Ministry of Development Planning and Statistics. (2023). https://www.psa.gov.qa/en/statistics/Statistical%20Releases/Social/LaborForce/2023/LF_Q1_2023_AE.pdf

Moghadam, V. M. (2005). Women's economic participation in the Middle East: What difference has the neoliberal policy turn made? *Journal of Middle East Women's Studies, 1*(1), 110–146.

Paschyn, C. (2013). Women in the Gulf: better educated but less employed. Al-Fanar Media. https://www.al-fanarmedia.org/2013/10/women-in-the-gulf-better-educated-but-less-employed/

Paschyn, C., Mir, S., Pike, K., Kane, T., & Mitchell, J. (2015). In majaalis al-hareem: The complex professional and personal choices of Qatari women. *DIFI Family Research and Proceedings, 2015*, 4. https://doi.org/10.5339/difi.2015.4

Psacharopoulos, G. (2006). The value of investment in education: Theory, evidence, and policy. *Journal of Education Finance*, 32 (2), 113–136.

Williams, J., Bhanugopan, R., & Fish, A. (2011). Localization of human resources in the State of Qatar: Emerging issues and research agenda. *Education, Business and Society: Contemporary Middle Eastern Issues, 4*(3), 193–206. https://doi.org/10.1108/17537981111159966

Zaidan, E., Belkhiria, E., & Wazen, C. (2023). Universities of the Future as Catalysts for Change: Using the Sustainable Development Goals to Reframe Sustainability-Qatar University as a Case Study. *The Sustainable University of the Future: Reimagining Higher Education and Research* (pp. 1–23). Springer International Publishing.

Chapter 23
Charting the Future of GCC Entrepreneurship: Insights for Education and Incubation Centers from the UK, Turkey, and the Region

Said Elbanna, Mohamoud Soufi, Mustafa Colak, and Heba Younis

Abstract This chapter underscores the critical role of entrepreneurship in driving economic growth and development in the Gulf Cooperation Council (GCC) countries. It emphasises two key mechanisms for promoting entrepreneurship: entrepreneurship education and incubation centres/business acceleration programmes. Specifically, the chapter focuses on the policy implications and orientations derived from an in-depth study conducted across diverse contexts, including the UK, Turkey, UAE, and Qatar, to inform and guide GCC policy-makers. By examining the barriers to entrepreneurship, designing effective entrepreneurship education programmes, and presenting case studies, this chapter provides valuable insights for crafting and implementing impactful policies. It highlights the need for tailored approaches to address the unique challenges faced by GCC nations and underscores the importance of context in entrepreneurship education. Ultimately, the chapter seeks to empower policy-makers. The goal is to strengthen entrepreneurship ecosystems in the GCC

S. Elbanna
Research Professor of Strategic Management & Director of Center for Entrepreneurship & Organizational Excellence, College of Business and Economics, Qatar University, P.O.BOX 2713, Doha, Qatar
e-mail: selbanna@qu.edu.qa

M. Soufi (✉)
Senior Learning and Development Specialist, Center for Entrepreneurship & Organizational Excellence, College of Business and Economics, Qatar University, P.O.BOX 2713, Doha, Qatar
e-mail: mohamoud.soufi@qu.edu.qa

M. Colak
Department of Business Administration, Faculty of Political Science, Social Sciences University of Ankara Ulus, Ankara, Turkey
e-mail: mustafa.colak@asbu.edu.tr

H. Younis
Associate Professor, Entrepreneurship and Strategy, SP Jain London School of Management, London, UK
e-mail: younish3@lsbu.ac.uk

E. Zaidan et al. (eds.), *The Future of Education Policy in the State of Qatar*, Gulf Studies 21, https://doi.org/10.1007/978-981-97-9667-0_23

region, driving sustainable economic growth, job creation, and prosperity through informed policy choices.

Keywords GCC entrepreneurship · Entrepreneurship education · Incubation centres · Policy implications · Economic growth

23.1 Introduction

The field of entrepreneurship has attracted substantial interest from scholars, policy-makers, and educators, owing to its newly recognised role in stimulating economic growth and development. Increasing levels of entrepreneurship lead to enhanced innovation, competitiveness, wealth creation, productivity, job creation, and the establishment of new industries (Kuratko & Morris, 2018; Kuratko, 2005). Such findings have spurred governments and policy-makers worldwide to prioritise and promote entrepreneurship, in order to stimulate economic activity, employment rates, and international competitiveness (Arshed et al., 2014; O'Connor, 2013). As a result, efforts to promote entrepreneurship have seen a significant surge worldwide. The Gulf Cooperation Council (GCC) countries are no exception to this global trend. The desire to diversify their economies beyond hydrocarbon exports and transition from resource-based to knowledge-based economic models has fueled interest in promoting entrepreneurship (Hamdan, 2019; Saberi & Hamdan, 2019).

The literature points to two particular mechanisms for promoting entrepreneurship, namely, entrepreneurship education programmes, and incubation centres and business acceleration programmes. In contrast to conventional beliefs, which argue that entrepreneurs are born, not made (see Otache, 2019), there is an increasing belief in the power of entrepreneurship education to nurture individuals into becoming entrepreneurs. Entrepreneurship education, defined as a collection of educational and training activities aimed at informing, training, and educating participants (e.g., students) interested in starting a business, plays a crucial role in bridging learning gaps, mitigating risks associated with potential failures, and fostering startup growth (Ramayah et al., 2012). Many startups fail, not due to a lack of customers, but primarily because of cash flow issues (see Bryant, 2022). Entrepreneurship education emphasises the critical distinction between revenue and cash, highlighting the latter's pivotal role in a startup's survival. Furthermore, entrepreneurship education serves as a crucial bridge, facilitating the transition from graduate employment to self-employment (Ferreira, 2020). Similarly, pre-incubation centres, which offer assistance (e.g., training, consultancy, office space, etc.) to nascent entrepreneurs, from the idea development stage to the startup stage (Kirby, 2004), incubation centres, which provide support from the startup stage to the expansion stage (Birden et al., 2020; Kepenek & Eser, 2018), and startup acceleration (or accelerator) programmes, which support technology startups in their initial stages of product development and commercialisation (Cubukcu & Gulsecen, 2018), can significantly contribute to entrepreneurial activities.

However, there has been limited research into entrepreneurship in the Arab world, including GCC countries (Elbanna et al., 2020; Younis et al., 2022), resulting in limited implications for policy-makers, entrepreneurs, governments, and other stakeholders. To address this critical gap, we conducted an in-depth study of the entrepreneurial trajectories of four countries—the UK, Turkey, UAE, and Qatar—while also considering other relevant experiences. Through this rigorous analysis, we aim to extract valuable insights that can profoundly influence policy-makers in their efforts to promote entrepreneurship within the GCC region. These valuable lessons learned from diverse contexts could serve as a foundation for formulating effective policies and strategies to foster economic diversification and growth through entrepreneurship education in the GCC countries.

Given this, the rest of this chapter is organised into three sections: in the first section, we address the barriers to entrepreneurship in the GCC region, by identifying the challenges that must be overcome. In the second section, we draw attention to the issue of designing entrepreneurship education programmes to promote entrepreneurship in the GCC region. In particular, we emphasise the significance of context in entrepreneurship education, and examine how various factors influence the success of such initiatives within the GCC countries. We also delve into designing an effective entrepreneurship curriculum and teaching methodologies, recognising their pivotal role in nurturing entrepreneurial mindsets and skills. Moving forward, we examine, particularly in Turkey, the case of incubation centres and acceleration programmes for promoting entrepreneurship. The findings and suggestions from the Turkish context could provide valuable insights for policy-makers in the GCC region. Finally, in the third section, we discuss specific implications for policy-makers in the GCC region, drawing from the experiences of the UK and Turkey, providing essential considerations and recommendations for developing and implementing effective entrepreneurship education policies. Our intention in this study is to fortify the growth and development of entrepreneurship ecosystems in the GCC region, so that they can empower aspiring entrepreneurs, and foster innovation and creativity. By optimising the impact of entrepreneurship education, incubation centres, and acceleration programmes on economic and social advancement, we envision a vibrant entrepreneurial landscape that drives sustainable growth and prosperity in the GCC region.

23.2 Barriers to Entrepreneurship

The GCC countries face challenges in their pursuit of a knowledge-based economy, due to limited in-house capacity for innovation and entrepreneurship, which hampers efforts towards sustainable economic diversification (Tok et al., 2021). The United Nations defines economic diversification as 'the process of shifting an economy away from a single income source toward multiple sources from a growing range of sectors and markets.' Despite efforts to diversify through high-value-added industries, the GCC countries still heavily rely on oil and gas extraction, which provides the main export revenue (Elbanna & Abdel-Maksoud, 2020). Volatile oil prices in international markets, a growing local population, and increasing domestic consumption of oil and gas all pose challenges. Additionally, the expected greater take-up of electric vehicles worldwide threatens GCC countries like Saudi Arabia and the UAE, by substantially reducing the demand for oil (Tok et al., 2021). Consequently, economic diversification has become a priority for GCC states (Hertog, 2013; Tok, 2020).

Several studies have identified various barriers to entrepreneurship in the GCC countries. Al-Mulla et al. (2022), for example, suggests that a lack of accessible and affordable funds, limited mentor support, and low female participation are common barriers to entrepreneurship in developing countries. In Qatar, the lack of financing options for startups is recognised as a major challenge (Kebaili et al., 2015), with entrepreneurs often resorting to self-funding due to difficulties in securing loans or grants (Al-Housani et al., 2023). The funding and support ecosystem for entrepreneurs in the GCC confronts constraints and limitations concerning incubation centres and banks, limited accessibility to information, absence of centralised procedures and guidelines, and limited connectivity between entrepreneurs and investors (Al-Mulla et al., 2022). Gender-related barriers for female entrepreneurs and investors have also been highlighted (Al-Housani et al., 2023; Al-Mulla et al., 2022; Elbanna et al., 2023).

Diversity and access to innovative sources of finance are found to impact the performance of social enterprises. These tend to prefer equity financing over debt, potentially due to limited access to debt. The availability and efficient utilisation of new sources of funds can positively influence businesses (Elgammal et al., 2023). Other factors impeding entrepreneurship in the GCC region (e.g., in Qatar and the UAE) include cumbersome procedures, inflexible and volatile laws and regulations, the absence of a centralised system, trust issues, limited access to pertinent information, the fear of potential failure (see Alnassai, 2023; Mehrez, 2019), and societal biases against female entrepreneurs (see Al Boinin, 2023; Tlaiss, 2015). There is also a lack of relevant entrepreneurial mentorship and guidance in certain GCC countries, emphasising the need for effective mentorship and guidance programmes, specifically designed for the local entrepreneurial context in each GCC country. Collaboration between different institutions is crucial for developing the entrepreneurial ecosystem (Al-Housani et al., 2023; Al-Qahtani,2022; Villegas-Mateos, 2021). These barriers highlight the need to prioritise the development of entrepreneurial education, provide continuous support, and foster a culture of entrepreneurship in the GCC countries. Additionally, stronger cooperation between institutions is necessary, to facilitate the growth and sustainability of the entrepreneurial ecosystem in the region.

23.3 Designing an Effective Entrepreneurship Curriculum and Programmes

23.3.1 The Role of Context in Entrepreneurship Education

The practice of entrepreneurship and the design of entrepreneurship education programmes and curricula are significantly affected by the context (Leitch et al., 2012; Walter & Dohse, 2012). Thomassen et al. (2020) suggest that the context of entrepreneurship education can be analysed on three levels. On the macro level, the focus is on broader social, cultural, and economic factors. These factors may include government policies, economic conditions, and the overall entrepreneurship ecosystem in the country or region. The meso-context level involves the organisational and institutional factors influencing entrepreneurship education, such as the resources available to students, the physical infrastructure of universities, and the adopted entrepreneurship curriculum. Finally, the micro-context level concerns individuals' or interpersonal factors, such as students' attitudes towards entrepreneurship, and the individual characteristics of students and instructors.

Understanding the contextual factors that influence entrepreneurship education can help educators and policy-makers design programmes and curricula that are relevant and effective for their populations. For instance, Walter and Dohse (2012) draw an association between the regional context and the mode of entrepreneurship education. Their study suggests that reflective methods of entrepreneurship education—those models that emphasise students' reflections on their own experiences, and analysis of different perspectives and theories related to entrepreneurship—are more effective in regions with a high degree of entrepreneurial activities; while active modes of entrepreneurship education—those that encourage hands-on and experiential learning—are effective regardless of the regional context.

Therefore, universities in the GCC region should consider customising their entrepreneurship education programmes to cater to their local region's specific needs and characteristics. For instance, gaining insight into cultural and economic conditions can aid educators in identifying areas of opportunity (such as high-growth technology startups or women's entrepreneurship), potential challenges in teaching entrepreneurship (e.g., local culture), and suitable contents to incorporate into the curriculum (such as how to manage a family business). Likewise, understanding students' traits and attitudes towards entrepreneurship can assist in making decisions about relevant individual learning outcomes (e.g., overcoming fear of failure) and suitable contents and teaching methods. This could be achieved by thoroughly examining contextual factors in existing entrepreneurship and entrepreneurship education research. Indeed, numerous research publications have addressed a wide range of contextual factors influencing entrepreneurship and entrepreneurship education in the GCC (see Abubakar et al., 2022; Alanzi et al., 2022; Alghamdi et al., 2021; Alkaabi et al., 2023; Faisal et al., 2017). These publications could serve as a starting point for mindful analysis of the context and its implications, and serve as resources to gather valuable insights that guide the design of effective programmes and curricula in the GCC. Additionally, we advocate for a more deliberate effort to expand scholarly inquiry and policy research that addresses the regional and country-specific context at various levels. This task should be undertaken by policy-makers, industry professionals, and academic scholars, to ensure a comprehensive understanding and effective action plan.

23.3.2 Curriculum Development and Components

There are different models of entrepreneurship education, such as including a mandatory single-course programme, a dedicated bachelor's degree in entrepreneurship, or an entrepreneurship track or minor within other courses in higher education institutions, or providing entrepreneurship centres—incubators or science parks—that support student entrepreneurs and others to act on their own business ideas or commercialise their research. These different models perhaps imply different perspectives on the nature and purpose of entrepreneurship, hence the approaches to programme and curriculum design and appropriate delivery modes, i.e., teaching methods. This outcome is not surprising, in light of the findings in the entrepreneurship education literature. For example, there is little agreement among entrepreneurship education programmes in universities worldwide on their learning objectives, contents, teaching methods, and assessment practices (Mwasalwiba, 2010; Sirelkhatim & Gangi, 2015). This lack of consensus is partially attributed to the multi-definitional perspective of entrepreneurship (O'Connor, 2013), which results in varied understandings of entrepreneurship education objectives, choice of target audience, design of learning outcomes and course contents, teaching and learning activities, and assessment practices. The simplified conceptual typology of 'about,' 'for,' and 'through' entrepreneurship is commonly used in academic literature to classify and account for the wide variety of entrepreneurship education programmes (Mwasalwiba, 2010; Pittaway & Edwards, 2012; Robinson et al., 2016; Sirelkhatim & Gangi, 2015). This typology classifies entrepreneurship education programmes according to the goals and objectives they pursue, and the contents they tend to emphasise. 'About' entrepreneurship programmes aim to educate and increase students' awareness *about* the entrepreneurship phenomenon, and encourage them to consider it a potential career choice; thus, the contents tends to focus on theories and themes about entrepreneurship. 'For' entrepreneurship courses aim to influence students' propensity and intentions to become entrepreneurs, and to prepare them *for* entrepreneurial careers (Sirelkhatim & Gangi, 2015); thus, the programmes' contents focus on developing students' entrepreneurial skills and abilities (Mwasalwiba, 2010). 'Through' entrepreneurship courses aim to help students acquire relevant skills, competencies, and business understanding while creating new ventures (Mwasalwiba, 2010); thus, the contents focus on various aspects of venture creation.

Considering the preceding discussion, realising entrepreneurship education's wide range of objectives and expected learning outcomes by adopting a standardised form, model, or curriculum proves unattainable (Jones & Matlay, 2011). Therefore, we advocate using the teaching model framework (Fayolle & Gailly, 2008) as a guide to developing robust and relevant curricula in the GCC, instead of a uniform curriculum template. This framework requires stakeholders to reflect on and answer questions on two levels: the ontological level and the educational level.

On the ontological level, stakeholders must reflect on what entrepreneurship is, as different constituents may have divergent views on what constitutes entrepreneurship. For instance, while a policy-maker may perceive entrepreneurship as a venture creation process, an educator may view it as the embodiment of specific attitudes and mindsets (see the next section for further illustration). Similarly, stakeholders may have different conceptions about what constitutes entrepreneurship education. Hence, answering the question 'what does education mean in the context of entrepreneurship?' is imperative. Finally, they should consider the educators' and participants' roles in entrepreneurship education. Educators may assume the roles of knowledge transmitters, facilitators, or coaches, among others. At the same time, participants may take roles such as passive recipients, active participants, or co-creators of knowledge.

On the other hand, questions on the educational level concern the design and structure of the curriculum or the educational programme. On this level, stakeholders must address the following questions:

Why? What are the goals and objectives of the curriculum?

What? What contents will be covered?

For whom? Who is the target audience for the programme/curriculum?

How? What teaching methods and pedagogies will be used?

For what results? How will the programme be assessed and evaluated?

Ensuring the internal consistency of the curriculum necessitates that the 'why' should define the contents, how it should be taught, and how the impact should be assessed.

23.3.3 Designing a Curriculum for Diverse Entrepreneurial Endeavours in the GCC

Designing an appropriate curriculum must account for the diverse conceptualisation and types of entrepreneurial endeavours of interest to the target audience. Many often associate 'entrepreneurship' solely with small and medium-sized enterprises (SMEs) and new business ventures. However, when entrepreneurship is viewed as a mindset, its definition becomes more inclusive, encompassing varieties with distinct characteristics and objectives. Here, we provide several examples of different types of entrepreneurship (Covin & Miles, 1999; Dees, 2001; Demircioglu & Chowdhury, 2021; Hisrich et al., 2017; Shane & Venkataraman, 2000).

Small business entrepreneurship: This type typically involves starting and operating a small-scale business. Small business entrepreneurs often focus on local markets, provide products or services to meet specific community needs, and strive for sustainable growth within their chosen market niche, which is extended to international markets in several cases.

Large company entrepreneurship: Large company entrepreneurship refers to the entrepreneurial activities within established corporations. These entrepreneurs aim to foster innovation, create new products or services, and explore new market opportunities, all while operating within the framework of a larger organisational structure.

Scalable startup entrepreneurship: Scalable startup entrepreneurs are driven by ambitious visions and aim to build high-growth ventures with the potential for rapid expansion. They often seek external funding, leverage technology and innovation, and prioritise scalability and market disruption.

Social entrepreneurship: Social entrepreneurs are dedicated to creating positive social impact and addressing societal challenges. They combine entrepreneurial principles with a mission to bring about positive change, focusing on social, environmental, or cultural issues, and pursuing innovative solutions to address them.

Innovative entrepreneurship: Innovative entrepreneurs emphasise creating and implementing novel ideas, products, or services. They push boundaries, challenge existing norms, and introduce disruptive innovations that can transform industries and markets.

Hustler entrepreneurship: Hustler entrepreneurs are characterised by their resourcefulness, tenacity, and ability to navigate challenges. They thrive in highly competitive environments, often leveraging their persuasive skills, networking abilities, and relentless drive to pursue opportunities and succeed.

Imitator entrepreneurship: Imitator entrepreneurs observe successful business models or ideas and replicate them in different markets or industries. While they may not introduce radical innovation, imitators can create value by adapting and applying proven concepts to new contexts.

Public sector entrepreneurship: Public sector entrepreneurship refers to the application of entrepreneurial principles and practices within the public sector, typically by government entities or public organisations. It involves the adoption of innovative and entrepreneurial approaches to address public problems, deliver public services, and drive social and economic development.

Researcher entrepreneurship: Researcher entrepreneurs bridge the gap between academia and the business world. They leverage their expertise in a particular field, to develop and commercialise innovations, turning scientific discoveries or research outcomes into viable businesses or products.

Managerial entrepreneurship: This refers to the application of entrepreneurial principles and behaviours within an established organisation. Managerial entrepreneurs demonstrate an entrepreneurial mindset and apply entrepreneurial skills to look at and work differently on their organisational problems, challenges, and opportunities for growth. They strive to create new realities in their organisations, instead of solely following the traditional achievements of conventional leaders.

The diversity in entrepreneurship types, and the varied stakeholders involved, indeed call for different designs for entrepreneurship curricula. Each entrepreneurship type necessitates specific knowledge, skills, and competencies, making it crucial to customise educational programmes to match these needs. Furthermore, entrepreneurship education programmes in the GCC region may have diverse objectives based on pedagogical and socioeconomic factors. In order to construct a curriculum that meets the particular needs of the intended audience, stakeholders should utilise tools such as the teaching model framework. This framework aids in structuring the curriculum, by posing essential questions throughout the design process. Contemplating these matters helps establish a shared understanding and definition of entrepreneurship that aligns with the programme's objectives. Additionally, this framework ensures consistency among the curricula's contents, teaching methods, and assessment and evaluation techniques. Such alignment enhances the effectiveness and relevance of the entrepreneurship education programme. By utilising suitable frameworks and engaging in reflective curriculum design, stakeholders can devise entrepreneurship programmes that cater to the unique prerequisites of different types of entrepreneurship and the diverse audiences in the GCC region. This method improves the educational experience, endows individuals with pertinent skills and knowledge, and magnifies the impact of entrepreneurship education on economic and social development.

23.3.4 Teaching Methods

Any initiative to introduce or revamp entrepreneurship education in the GCC must consider the extensive variety of available teaching methodologies, and adopt strategies that align with specific purposes and contexts (Hoppe et al., 2017; Jones & Lourenço, 2006). This could involve various teaching methods, from traditional lecture-based approaches to experiential and learning-by-doing methods. Entrepreneurship education literature emphasises the superiority of active experiential learning and real-world pedagogies, over traditional methods like lectures, workshops, case studies, and seminars (Fayolle, 2013; Maritz & Brown, 2013). While traditional approaches can effectively help present information, they may be less effective in influencing entrepreneurial attributes (Bennett, 2006). On the other hand, experiential teaching methods mirror the unpredictable nature of entrepreneurship, helping students learn from failure. Furthermore, collaborative activities and group projects can enhance opportunities and synergy, stimulate creativity, and teach students how to handle real-world problems and cope with uncertainty (Pittaway & Cope, 2007).

The selection of teaching methods is highly dependent on the type, goals, and context of the entrepreneurship education programme. For instance, programmes teaching 'about' entrepreneurship might lean towards traditional teaching methods, such as lectures and case studies. Programmes that are 'for' entrepreneurship, on the other hand, often advocate for experiential teaching methods, such as simulations and mentoring. Meanwhile, programmes that teach 'through' entrepreneurship focus on real-life entrepreneurship experiences, such as venture creation and internships (Mwasalwiba, 2010; Sirelkhatim & Gangi, 2015). Additionally, Bennett (2006) outlines two broad approaches to entrepreneurship teaching: the 'skills-training' approach, which focuses on teaching small business management skills, and the 'attributes development' approach, which emphasises developing entrepreneurial attributes such as creativity and risk-taking. Despite the growing body of research about entrepreneurship education in the GCC, exploration into preferred teaching methods and their impact remains limited. Current research predominantly focuses on specific cases of entrepreneurship education within universities. However, existing research, along with the broader literature on entrepreneurship education, provides initial evidence of a positive impact on students' self-efficacy, attitudes, and intentions towards entrepreneurship (Abubakar et al., 2022; Belwal et al., 2015; Elali & Al-Yacoub, 2016). Comprehensive research is still needed to explore various teaching methods and their effects within the context of entrepreneurship education in the GCC.

Pertinent to the above discussion, it is evident that adopting appropriate teaching methods is crucial to the success of any entrepreneurship education programme. As entrepreneurship educators are tasked with delivering a wide range of objectives, such as instilling in students the necessary entrepreneurial mindset and skills to achieve self-employment, commercialise intellectual property, enter into venture creation, etc., it is imperative that they employ appropriate pedagogical practices in teaching and assessment to realise the intended objectives. Consequently, faculty training and development, especially on educational theories and practices, assume paramount importance in ensuring the effective delivery of entrepreneurship education. Training and development not only equip faculty members with a comprehensive understanding of pedagogical theories and practices, but also enable them to determine the appropriate teaching approach for particular learning outcomes or learners, reflect on their practices, learn from others, design activities to motivate students, and acquire skills, as well as create a community of practitioners.

23.4 Incubation Centres and Acceleration Programmes to Promote Entrepreneurship

Establishing pre-incubation and incubation centres and acceleration programmes can play a significant role in entrepreneurship promotion. This is particularly true in developing countries, where entrepreneurship faces numerous challenges, such as limited human capital, high levels of macroeconomic volatility, and poorly functioning formal institutions. These obstacles are either nonexistent or less prevalent in developed countries (Akcomak 2009). Pre-incubation centres typically provide assistance to nascent entrepreneurs in the form of training, consultancy, and office space, to help them develop their business ideas and business plans, hence improving the likelihood of a successful startup. In contrast, incubation centres provide assistance to entrepreneurs, from the startup stage to the expansion stage (Birden et al., 2020).

In recent years, there has been an increase in incubation programmes, including those based at universities, government-supported initiatives, and independent science and technology parks in the GCC. The UAE Science and Innovation Park, and Qatar Science and Technology Park, discussed in the previous chapter, are recent examples of this trend. Furthermore, many universities run entrepreneurship centres that provide pre-incubation support activities. These programmes and initiatives promote entrepreneurship, thus contributing to the GCC national entrepreneurship and economic agenda. Nonetheless, despite the popularity of incubation and acceleration programmes as mechanisms to promote and support entrepreneurship, scholarly literature on their adoption and impact in developing countries is scant. There are very few available empirical studies that examine these programmes. Fortunately, the Turkish experience of establishing incubation and accelerator programmes can provide initial insights into the use and effectiveness of these programmes. According to Birden et al. (2020), Turkey has witnessed a rapid increase in incubation and accelerator programmes, which has triggered research in the area. Therefore, this section is devoted to reviewing the empirical literature on the impact of pre-incubation, accelerator, and incubation programmes, unveiling insights about the efficacy of these programmes as tools to promote entrepreneurship in the GCC from the Turkish context and other relevant global examples.

Pre-incubation and incubation centres can be established as business-based—i.e., private sector centres, university-based centres, university-private sector-based centres, or municipality-based centres, as is the case in Turkey. Such centres can provide valuable contributions to the entrepreneurial ecosystems in a country. Kepenek and Eser (2018) investigated the impact of university-based pre-incubation centres on entrepreneurial activities in Turkey, by interviewing 23 pre-incubation managers. The authors reported that the types of services offered by the pre-incubation centres—e.g., office space, secretarial services, writing support, hardware, internet, networking opportunities, education/training, mentorship, and international collaboration—were valuable to the incubatees. However, some problems were observed; for instance, incubatees' unwillingness to make changes to their projects or business models, a restrictive bureaucracy that resulted in slow decision-making at the pre-incubation centres' management level, and a lack of specialisation, as pre-incubation centres tended to offer their services to all types of sector, rather than being sector specific. Kepenek and Eser (2018) also identified the needs of incubatee firms. In particular, they reported that incubatees need support in finding potential customers (45%) and developing a proper business model (37%). Based on these findings, the study suggested various mechanisms to increase the effectiveness of university-based pre-incubation centres in Turkey (pp. 74–80) as follows:

> Empowering pre-incubation centres' management for them to become more autonomous, thus reducing the time for decision-making.
>
> Becoming more specialised (i.e., sector selective) in providing them with assistance.
>
> Enlisting skilled specialists and professional managers along with academics in pre-incubation centre management teams.

Establishing partnerships and strengthening links between universities that host pre-incubation centres and various entities, including regional and local organisations, SMEs, social enterprises, entrepreneurs, university innovation clusters, and international networks.

Supporting pre-incubation centres to establish and sustain their own financial funding through the government, universities, and public institutions.

Focusing on offering networking opportunities and market research to incubatees, to enhance their potential to find real customers, understand sector dynamics, and match their product with the right customer.

Engaging alumni of the host university with entrepreneurship or business coaching experience in the training and mentoring activities of the pre-incubation centres.

Monitoring and evaluating the impact of these centres on entrepreneurship activities regularly, focusing on assessing changes in the incubatee firms' motivation and level of competence in skills gained through the training and mentorship activities.

Business accelerator programmes, which emerged in the mid-2000s, represent a new generation of support mechanisms provided by incubation centres. These programmes provide support in the initial stages of technology startups that have created products and started commercialising them, but need help gaining market share and growth (Cubukcu & Gulsecen, 2018). The support, which includes office use, infrastructure, seed financing, networking, mentoring, and training, is provided to qualifying startups for a short period—generally up to 9 months—to help them achieve progress and business growth through an intensive consultation process (Cubukcu & Gulsecen, 2018). Literature examining accelerator programmes in developing countries is scarce, as there are few available empirical studies on them. One of the few such is the study of Beyhan et al., (2022), in which 10 managers of accelerator programmes in Turkey were interviewed. The study found that accelerator managers need to develop networks with other agents in the entrepreneurship ecosystem, in order to take collective action to overcome the institutional voids in the emerging country context and, thus, gain legitimacy. In another study, Cubukcu and Gulsecen (2018) analysed six accelerator programmes in Turkey. The authors found that (i) coordinators of accelerator programmes are well informed about the difference between incubator programmes and accelerator programmes; (ii) they view themselves as partners of startup firms, not as entities competing with incubation centres or other acceleration programmes; (iii) most of the startups participating in accelerator programmes are in the prototype phase or are ready to launch their products; (iv) investment support, mentorship networks, and networking and cooperative opportunities for going global are the most common expectations of startups from accelerator programmes; (v) startups benefit more from accelerator programmes if they participate in mentor meetings more frequently, and if they meet with different mentors; (vi) relationships with government representatives and different stakeholders are important for obtaining support; and (vii) all six accelerator programmes studied experienced the same common problems, the most challenging ones being the difficulty in recruiting quality entrepreneurs into their programmes, entrepreneurs being inexperienced and unable to explain their startup ideas fully, finding financing for entrepreneurs, and signing a contract with entrepreneurs (Cubukcu & Gulsecen, 2018, pp. 15–29). Similarly, Cubukcu and Gulsecen (2019) found that entrepreneurs participating in accelerator programmes in Turkey are generally young, male, well-educated, have little professional work experience, and come from families without entrepreneurship experience. These findings suggest that managers of accelerator programmes should consider the demographic characteristics of the startup firms attending their accelerator programmes.

As for incubators, Birden et al. (2020) reported that the incubation centres in Turkey provide invaluable support mechanisms to entrepreneurs, such as physical resources, office space, support services, access to capital, process and legal support, and networking services. Yıldırım and Keskin (2021) further add that university-based incubation centres allow startups to benefit from universities' facilities, such as laboratories, workshop rooms, and academic support; hence, they can be more advantageous than other types of incubation centres. Akcomak and Taymaz (2007) assessed the effectiveness of technology incubators located in Technology Development Centers (TEKMERs) in Turkey, by comparing the performance of 48 firms incubated by TEKMERs with 41 firms established without this support. The study found that the incubated firms enjoyed stronger economic performance in terms of growth, sales revenues, and employment growth. Similarly, Şehitoğlu and Özdemir (2013) found that incubated firms in 11 different TEKMERs located in six different locations in Turkey outperformed non-incubated firms both in terms of average annual sales growth and employment growth. Other studies in the context of Turkey have also reported on the positive impact of incubation centres—i.e., that of TEKMERs—not only on incubated firms' performance, but also on job creation, growth, and regional development (e.g., Akcomak 2009; Cansiz 2008).

In a study of private-sector-based incubation centres, Yıldırım and Keskin (2021) found that the most challenging issues for incubatee firms were networking, finding customers, introducing products, testing products, and creating a team. These findings align with those reported in Kepenek and Eser's (2018) study of pre-incubation centres in Turkey. Accordingly, Yıldırım and Keskin (2021) suggested that incubation centres should provide more networking opportunities, training in marketing and market research, and access to laboratory and workshop facilities, where the incubatee firms could develop and test their prototype products. In addition, the authors suggested establishing incubation centres for potential entrepreneurs in more remote regions of Turkey, given that incubation centres are concentrated in the three largest cities in Istanbul, Ankara, and Izmir.

Unlike the seemingly positive impact of incubation programmes in Turkey, several studies on the performance of incubation centres in other countries have provided opposite results or mixed findings. For example, earlier empirical studies conducted in the UK compared the performance of science-park-based firms, as a particular type of university-based incubator, to that of off-park companies (Massey et al., 1992; Westhead & Storey, 1994; Vedovello, 1997, cited in Huibing & Nengli, 2005). The findings of those studies showed that the incubator's location—i.e., association with a higher education institution—did not influence the survival rate or profitability of the on-park firms. In addition, the relationship with science park companies and higher education institutions did not necessarily lead to greater innovation or more effective technology transfer between academia and startups. Furthermore, the influence of the on-park firms on the regional unemployment rate was small, despite their outperforming off-park firms in employment growth and sales, as most of these firms were SMEs.

Tamasy (2007) synthesised evidence from studies that evaluated the performance of technology-oriented incubators in Germany and the US. The studies cited showed that the potential of technology-oriented startups in Germany was too small to create sustainable success in promoting the incubation industry, as many of these startups were engaged in low-value service sectors. In the US, an evaluation of 116 science parks indicated that only a quarter were successful, another quarter needed to change their goals due to failure in realising the anticipated impact, and half were deemed failures.

Schwartz (2011) argues that the data collection in studies suggesting superior employment and sales growth performance of on-park firms compared with off-park ones, often suffers from a survivor bias, as they tend to exclude less successful firms. Subsequently, they overestimate the growth potential of incubated firms. The study's investigation of five incubator graduated firms indicated a limited contribution to regional employment generation from a long-term perspective. Only a few firms experienced strong growth, while most showed moderate growth or even weaker performance several years after graduation. These findings challenge the assumptions that incubators automatically lead to sustained and high firm growth beyond graduation.

Several studies have reiterated the idea that the success of incubators depends on various factors, including the quality of personnel, integration with broader policies for promoting entrepreneurship, and innovation enhancement. In specific contexts, investing in education policies to cultivate a receptive environment for creativity and entrepreneurship, or improving financial institutions to facilitate venture capital, may yield better results than solely investing in incubators. For instance, Aernoudt (2004) suggests that the low entrepreneurship level in Europe negatively impacts the supply of startup projects to incubators. Accordingly, alternative policy measures to bridge the entrepreneurship gap must be prioritised. Similarly, in his study of incubator firms in Brazil, China, India, and Turkey, Akçomak (2009) suggests that incubators in many developing countries face various weaknesses, including a focus on tangible services, a heavy reliance on governments for promotion and funding, the lack of qualified personnel, and insufficient problem-solving creativity. Therefore, the success of an incubator policy is intertwined with broader entrepreneurship promotion, innovation, and economic development policies.

To conclude, incubators, including university-based incubators, may not automatically ensure higher success rates for firms. While some studies have reported that incubation centres contribute to the success of incubatee firms, other studies have indicated that incubator location, specifically its association with an educational institution, does not necessarily influence the key success metrics of incubatees. Performance data might be skewed due to survivor bias, potentially overestimating incubated firms' growth. Therefore, the success of incubators hinges on various factors, including staff quality and alignment with broader entrepreneurial policies. Subsequently, policy decisions should consider a holistic approach, integrating incubator strategies with overarching economic and innovation agendas.

23.5 Policy Implications for Entrepreneurship Education Policies

The transition of GCC countries, such as Qatar and UAE, into knowledge-based economies, as detailed in their national visions, highlights the central role of entrepreneurship in the countries' economic diversification (Elbanna & Fatima, 2023). However, despite several programmes designed to stimulate innovation and entrepreneurship, there have been obstacles in augmenting innovation output, primarily due to a lack of collaboration between research and development centres, universities, and businesses. Furthermore, there is a noticeable absence of a comprehensive national entrepreneurship education policy that articulates a well-defined vision, and facilitates coordinated efforts to integrate entrepreneurship education at the primary, secondary, and tertiary levels. For guidance on how to embed entrepreneurial education into these levels of the education system, refer to Gibb's theory of entrepreneurial behaviour (2008), which has garnered high levels of acceptance across European and Anglo-Saxon contexts (Klapper & Farber, 2016). We envisage that such a policy should not be prescriptive; instead, it should offer a strategic framework, to guide education institutions in effectively implementing entrepreneurship education programmes, while ensuring alignment with other pertinent policies. This leads to the following key policy implications for Qatar and the UAE, and by extension, for other GCC countries:

Cross-sector integration: Policies should advocate and simplify inter-sector partnerships, to ensure knowledge exchange and resource pooling, thereby enhancing innovation and entrepreneurial endeavours.

Business accessibility: Considering Qatar's position on the ease of doing business index, policies should be revamped to ease business startups' processes and regulatory constraints. Streamlining bureaucracy and implementing more transparent rules can help to realise this.

Technological entrepreneurship policy: Despite numerous initiatives to boost entrepreneurship, a distinct policy or strategy for technological entrepreneurship is absent. Therefore, a comprehensive, well-defined roadmap for promoting and supporting technological entrepreneurship that considers potential hurdles and impediments is needed.

Legal and regulatory infrastructure: Clear legal requirements and incentives for essential stakeholders in the innovation-university-government partnership should be established. A regulatory framework encouraging and rewarding involvement in entrepreneurship-related activities should be developed. Public awareness of these regulations should also be heightened, to encourage more public participation and engagement.

Transparent practices: The distribution of business licences should be transparent and equitable. By eliminating favouritism towards specific families or tribes, a level playing field for all entrepreneurs can be secured. This will foster more business ventures in industries beyond oil and gas.

Entrepreneurship education at an early stage: By recognising the vital role of education in nurturing an entrepreneurial culture, policies should incorporate entrepreneurship education in the primary and secondary school curricula.

Recognition of the influence of macro-, meso-, and micro-contextual factors: GCC universities should tailor their entrepreneurship programmes to local conditions, leveraging insights into cultural, economic, and individual aspects, to optimise teaching methods and curriculum content, thereby boosting the effectiveness and relevance of entrepreneurship education.

Proper teaching methods: To bolster entrepreneurial learning, GCC universities should encourage the incorporation of a blend of teaching methods, emphasising experiential learning techniques that mirror the complex nature of entrepreneurship.

Adoption of a flexible curriculum framework: Given the vast array of entrepreneurial avenues, GCC entrepreneurship education programmes should consider adopting a flexible curriculum framework, such as the teaching model framework, to cater to diverse entrepreneurial endeavours and objectives.

Cultural barrier mitigation: Sociocultural aspects, such as a preference for public sector jobs, can impede entrepreneurial growth. Policy-makers should strive to alter these perceptions, and promote entrepreneurship as a legitimate career choice rather than as a supplementary income source.

Role clarification: Enhancing coordination between bodies currently involved in entrepreneurial activities, such as those at Qatar University, is crucial. It is essential to revisit the existence of two organisational units at Qatar University with similar missions, to establish a more cohesive entrepreneurship ecosystem. Clear policies are needed, to prevent role confusion among stakeholders. Policy-makers should define the roles of these entities, to avoid duplication and ensure efficient functioning. Figures 23.1 and 23.2 depict the crucial role of Qatar University, as an example of academic institutions, in supporting the entrepreneurship ecosystem in Qatar. Figure 23.1 illustrates how Qatar University can engage with the entrepreneurship ecosystem so as to enhance it. Meanwhile, Fig. 23.2 delineates the present and potential roles of the Center for Entrepreneurship & Organizational Excellence, located within the College of Business and Economics, in contributing to the entrepreneurship ecosystem both internally and externally.

Easy access to supportive entities: The existence of multiple incubators and funding programmes calls for policy measures that promote transparency and facilitate easy access to these initiatives. Ensuring that startups are well informed about and capable of utilising these opportunities is essential for fostering entrepreneurial success.

Addressing entrepreneurship challenges: Overcoming barriers to entrepreneurship, including limited financial access, low female participation, and the lack of

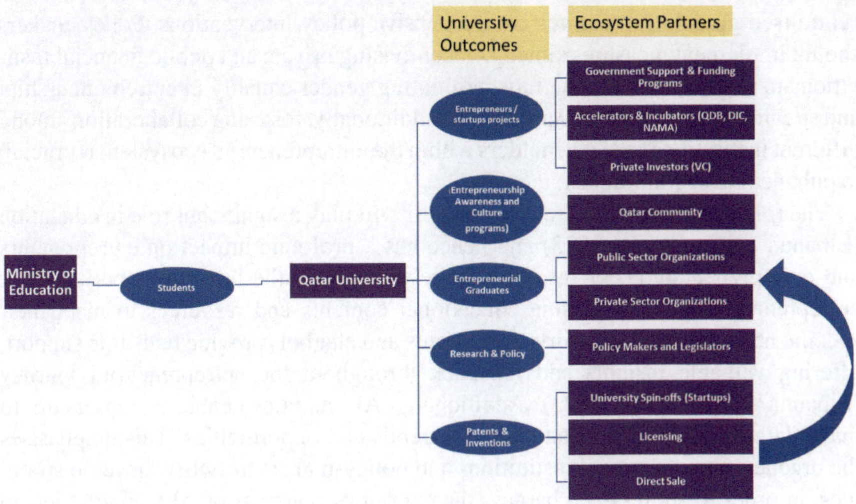

Fig. 23.1 The role of Qatar University in the entrepreneurship ecosystem in Qatar

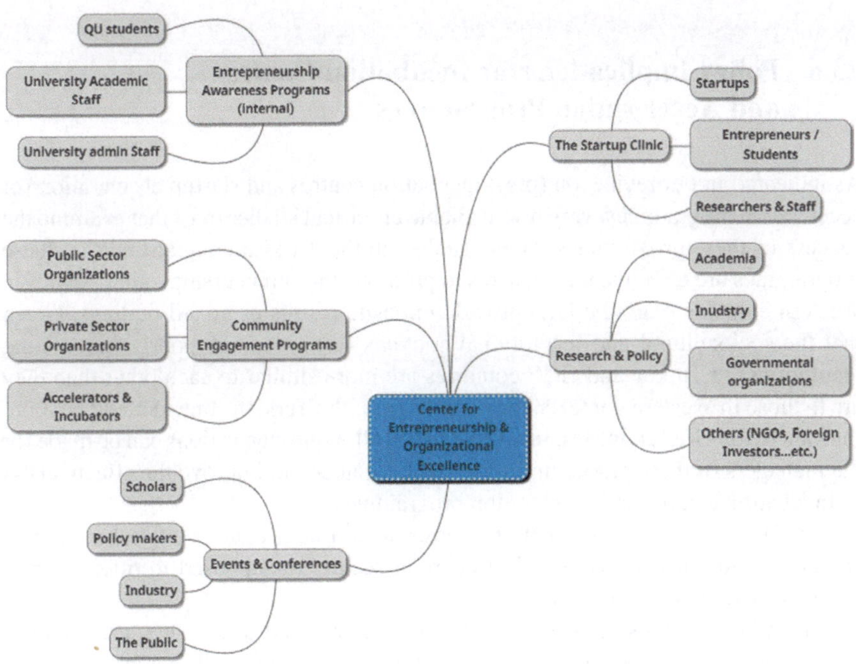

Fig. 23.2 The role of the center for entrepreneurship & organizational excellence in the entrepreneurship ecosystem in Qatar

centralised guidelines, requires comprehensive policy interventions. Policy-makers should implement measures aimed at incentivising private and public financial institutions to provide credit to startups, promoting gender equality in entrepreneurship, and streamlining bureaucratic processes. Additionally, fostering collaboration among different institutions and stakeholders within the entrepreneurial ecosystem is crucial, to enhance its overall impact.

The role of artificial intelligence (AI): AI will play a significant role in education (Elbanna & Armstrong, 2023) and hence has a profound impact on entrepreneurs and entrepreneurship. For instance, AI revolutionises the learning experience for entrepreneurs, by personalising educational contents and resources to meet their specific needs. AI-powered virtual assistants and chatbots provide real-time support, offering valuable insights and feedback throughout the entrepreneurial journey (Elbanna & Armstrong, 2023). Additionally, AI analytics enable entrepreneurs to make data-driven decisions and identify trends and opportunities. This emphasises the urgency for educational institutions and policy-makers to swiftly develop strategies, in order to address and harness the enormous potential of AI technologies, to adapt and leverage AI's capabilities.

23.6 Policy Implications for Incubation Centres and Acceleration Programmes

As indicated in our review on (pre-) incubation centres and startup acceleration (or accelerator) programmes, very few available empirical studies exist that examine the success of these programmes. While studies in the Turkish context find that these programmes are effective mechanisms to promote entrepreneurship, other studies in Western contexts (e.g., the UK) provide opposite results or mixed findings. Given that the sociocultural and institutional contexts (e.g., weak or poorly functioning institutions) in Turkey and GCC countries are more similar to each other than they are to those in Western contexts, the insights from the Turkish context could be more relevant to the GCC countries. In this vein, the following suggestions can be made for the managers of these programmes and policy-makers, to improve the effectiveness of incubation centres and acceleration programmes:

Specialization: (Pre-) Incubation centres should be sector specific; that is, they should provide their services only to certain sectors (as opposed to offering their services to all types of sectors).

Streamlined processes: (Pre-) Incubation centres and accelerator programmes should be free from bureaucratic procedures that slow down decision-making, in order for opportunities not to slip away.

Autonomy: Grant greater autonomy to the management teams of (pre-) incubation centres and acceleration programmes, to expedite decision-making.

Diverse funding sources: Funding from various entities (e.g., governments, universities, public institutions, etc.) should be obtained for (pre-) incubation centres and acceleration programmes, to establish and sustain their financial funding.

Collaborative Partnerships: (Pre-) Incubation centres and acceleration programmes should establish partnerships with various entities (e.g., with universities, regional and local organisations, SMEs, social enterprises, entrepreneurs, and international networks), to take collective action, in order to overcome the institutional voids in the country concerned, and to gain legitimacy in the eyes of the government, public institutions, and entrepreneurs.

Skill development: (Pre-) Incubation centres should focus particularly on offering networking opportunities and training in marketing and market research, so that their incubatees could enhance their ability to understand the dynamics of their industry, identify real customers, and match their product with the right customer. Acceleration programmes should pay particular attention to providing frequent mentor meetings with different mentors.

Ongoing evaluation: Implement regular monitoring and evaluation of (pre-) incubation centres and acceleration programmes, to assess their impact on entrepreneurial activities. These assessments should particularly focus on changes in the motivation and skills of incubated firms.

Enhance the entrepreneurship ecosystem: Determine the challenges and needs of present incubators and accelerators that might be underperforming or facing hurdles. Before introducing additional entities, especially within academic realms, it is essential to gauge the cumulative impact on the ecosystem. In this context, educational institutions, including universities, should conduct a comprehensive analysis of the prevailing entrepreneurship landscape. Their efforts should be directed towards addressing and filling existing gaps. This is better than accidentally making things more complicated, by adding redundant entities.

These recommendations aim to optimise the effectiveness of (pre-) incubation centres and acceleration programmes in fostering entrepreneurship and economic growth.

23.7 Conclusion

The preceding section presented key recommendations aimed at prevailing barriers to GCC entrepreneurship. These recommendations emphasise the dominant role which GCC governments can play in shaping an environment favourable to entrepreneurship. A significant number of these recommendations revolve around government policies and interventions, encompassing strategies to shape a resilient entrepreneurship ecosystem, tackle legal and regulatory challenges, enhance entrepreneurial skills through education, and boost collaboration among supportive entities. While these government-backed interventions are crucial, their introduction naturally leads to another pressing question: How often do these government-led policies meet their desired outcomes? It is important to realise that these initiatives, however well-intentioned, are not infallible. Many instances in the literature, such as those documented by Lerner (2013), underscore failed or abandoned government ventures. The potential pitfalls and challenges that might hinder the effectiveness of entrepreneurship policies need careful attention. Thus, this section seeks to spotlight these challenges, hoping to enhance policy-makers' perspectives on the complexities involved. By shedding light on these aspects, the aim is to shape policies and interventions that are not only well-conceived but also effectively executed, therefore driving sustainable economic growth.

At the outset, Lucas et al. (2018) argued that policy research tends to view the government as a single, coherent entity, possessing the perfect requisite knowledge to design and implement policies and interventions in the best interest of all stakeholders. However, this view is not supported by empirical evidence. Instead, an 'individualistic' view of government as a set of state actors with limited knowledge of the needs and incentives of the market participants is proposed. This view recognises that, in addition to having knowledge constraints, policy-makers may be susceptible to self-interest, potentially leading them to adopt measures that do not align with the target beneficiaries' best interests, perhaps due to political considerations. As such, government entrepreneurship policies, initiatives, and programmes may fail, due to limited knowledge of market participants' needs and incentives, a lack of skills, and self-serving behaviour by policy-makers and actors.

Similarly, Lerner (2013) notes that government policies and initiatives are prone to common conceptual and implementation pitfalls. According to this work, conceptual issues are those caused by policy-makers' lack of understanding of the realities of entrepreneurship. For instance, discontinuing entrepreneurial initiatives prematurely on the grounds that these initiatives fail to create immediate impact, is, according to the author, a reflection of policy-makers' lack of understanding that entrepreneurship initiatives may take many years to produce the intended results. Additional dimensions of the conceptual issues are evident in many documented practices within the framework of entrepreneurial policy and initiative design, for instance, the misallocation of funds and support towards industries, geographic regions, or types of startups that are less likely to attract private sector interest. Furthermore, counterproductive stipulations imposed on startups benefiting from public funding could hinder their growth. Another dimension pertinent to the scale of government programmes is the creation of programmes insufficient in scale to engender any impact or, conversely, of an excessive magnitude that consumes the lion's share of the existing funds or resources.

On the other hand, implementation issues are those problems that arise as programmes or initiatives enter operation. In this regard, Lerner (2013) perceives that government-led initiatives often lack a level of meticulousness when crafting incentives, compared with those practices adopted by private investors. Subsequently, there is less scrutiny of the participants in publicly supported entrepreneurship schemes, and less alignment of their behaviour with public sector objectives. In addition, the lack of appropriate programme evaluation mechanisms further impairs the implementation of those programmes. Finally, the author notes the importance of international investors' participation in the local entrepreneurial sector. However, public initiatives to promote entrepreneurship and build an entrepreneurial ecosystem frequently ignore this international dimension of entrepreneurship.

Likewise, Smallbone (2016) cites other issues that impact entrepreneurship policy design and implementation. Primarily, a lack of clarity regarding policies' objectives and the delineation of the roles and responsibilities assigned to the numerous entities involved in their implementation, can lead to a lack of transparency that erodes the confidence of target groups, ultimately undermining the efficacy of policies. Furthermore, the weak involvement of stakeholders, such as SMEs, in the policy design and implementation process, may hinder the ability of policies to encompass the diverse perspectives and needs of the actors they are intended to serve. Thirdly, insufficient awareness among the general public about public policy measures introduced by the government underscores the need for more effective communication and dissemination strategies. Lastly, empirical evidence is lacking for assessing public policy effectiveness.

Transitioning from these identified challenges, Kantis et al. (2020) pivot the focus towards the broader context in which these policies operate: the entrepreneurship ecosystem. Highlighting the importance of tailoring policies to specific ecosystems, they note that the characteristics of ecosystems, which are instrumental for the rise and growth of novel, dynamic firms, vary notably across countries and regions. This variation is a product of the intricate balance between three pivotal variables: entrepreneurial human capital, opportunities for nascent firms, and factors either promoting or obstructing new firm creation. The interrelation of these variables with numerous societal, economic, cultural, political, and regulatory elements is detailed in Table 23.1.

Table 23.1 Factors influencing the dimensions of entrepreneurial ecosystems

Dimensions (variables) of the ecosystem	Influencing factors
Level of entrepreneurial human capital	Family background The education system and work experience Social culture
Opportunity space for new firms	Income level of the population Industry and business structure Level of industry concentration or fragmentation Exit and entry barriers Strategies of large firms Proportion of large firms, SMEs, and micro-enterprises Presence of science, technology, and innovation platforms
Factors that can foster or inhibit the establishment of new firms	Role of government in promoting entrepreneurship Country's norms and regulations

Adapted from Kantis et al. (2020).

An empirical analysis of this study reveals that the entrepreneurship ecosystem in many emerging countries exhibits a degree of asymmetry in the development of the three dimensions essential for creating and growing new dynamic firms. Subsequently, a standardised or uniform approach to entrepreneurship policy design and intervention is not universally applicable. Instead, a systemic approach that directs efforts to the dimensions of the entrepreneurship ecosystem that are least developed, while persevering in areas of strength is warranted. For instance, those countries or regions with conditions that are favourable to the creation of new firms, such as having government support and market opportunities, but suffer from a shortage of motivated and skilled individuals, may focus on directing their social capital, education system, and overall culture towards entrepreneurship in the country or region. In contrast, those countries that suffer deficits in their entrepreneurial ecosystem due to structural barriers that inhibit entrepreneurial endeavours may focus on alleviating those deficits via policies and long-term strategies. In either case, intervention should be based on a thorough examination of the characteristics and state of the ecosystem, and must follow a systemic approach to policy design. In this regard, it is worth mentioning that many GCC countries, like the Kingdom of Saudi Arabia, the UAE, and Qatar, participate in the Global Entrepreneurship Monitor (GEM) survey-based research into entrepreneurship and entrepreneurship ecosystems. The GEM national reports provide data and insights into some of the ecosystem dimensions discussed above, such as the level of entrepreneurship activities in the country, societal attitude towards entrepreneurship, and the general entrepreneurship framework conditions (the ecosystem), which provide a basis to evaluate the comparative standings of the country in terms of supporting entrepreneurship and policy-making (Qatar Development Bank, 2021). Indeed, the UAE incorporates some of the GEM indicators in its entrepreneurship policy-making (Chabrak, personal communication May 24, 2018).

In summary, the studies mentioned above provide nuanced insights that enhance our comprehension of common issues that can constrain the efficacy of entrepreneurship policies and interventions. While the precise causes of the ineffectiveness or failure of specific policies or initiatives may vary due to a multitude of factors, an awareness of common pitfalls in policy design and intervention will aid policy-makers, to avoid practices that may inhibit the development of thriving entrepreneurship. In this spirit, we offer further policy design and implementation recommendations.

A systemic approach to policy design: Entrepreneurship policies in the GCC must account for the heterogeneity of the entrepreneurship ecosystem across their countries and regions. The type of intervention prioritised must be based on an understanding of the state of the entrepreneurship ecosystem and the degree of asymmetry in the development of its constituent dimensions. While reports such as those published by the GEM may provide some guidance, it is imperative that ecosystem conditions are thoroughly analysed by the government, academic institutions, and the private sector, to ensure they can provide strong foundations for entrepreneurship policy-making.

Wider stakeholder involvement: Creating shared platforms and venues for stakeholder involvement in policy design will enable a better understanding of the needs and challenges faced by the target groups, create a sense of shared purpose, promote transparency, and empower market participants to assume an active role in policy formulation. This consultative approach cultivates a more informed, inclusive, and robust policy design.

Experimentation in policy implementation: As discussed earlier, it is not uncommon to abandon major policies and initiatives even after committing significant resources, only to later discover that they were built upon incorrect assumptions. Therefore, implementing small-scale interventions, coupled with systematic feedback collection, is worth considering.

Monitoring and evaluation: Appropriate evaluation criteria and mechanisms must be in place, to ensure the success of entrepreneurship programmes, and gauge the extent to which they meet their objectives. In this process, rigorous evaluation of the policy's or programme's impact must be emphasised over low-level monitoring (Smallbone, 2016).

Encouraging academic research: Given the heightened efforts to promote entrepreneurship in the GCC, and the lack of sufficient context-specific research in various aspects of entrepreneurship (Elbanna et al., 2020), there is a need to encourage research in all areas of policy design, formulation, implementation, and effectiveness. Research centres like the Center for Entrepreneurship and Organizational Excellence at Qatar University can play a vital role in conducting theoretical and applied research.

Acknowledgements This paper was made possible by NPRP grant [# 14S-0317-210062] from the Qatar National Research Fund (a member of Qatar Foundation). The statements made herein are solely the responsibility of the authors.

Bibliography

Abubakar, A., Mourgan, F. H. A., Al Kharusi, B., & Elfitori, C. M. (2022). Impact of entrepreneurial education, trait competitiveness and psychological capital on entrepreneurial behavior of university students in GCC. *Journal of Enterprising Communities: People and Places in the Global Economy., 17*(5), 1022–1044. https://doi.org/10.1108/JEC-03-2022-0047/FULL/HTML

Aernoudt, R. (2004). Incubators: Tool for entrepreneurship? *Small Business Economics, 23*(2), 127–135. https://doi.org/10.1023/B:SBEJ.0000027665.54173.23

Akçomak, I. S., & Taymaz, E. (2007). Assessing the effectiveness of incubators: The case of Turkey. In V. V. Ramani & A. V. Bala Krishna (Eds.), *Business incubation: an introduction* (pp. 234–265). ICFAI University Press.

Akçomak, I. S. (2009). Incubators as tools for entrepreneurship promotion in developing countries. In N. Szirmai & E. Goedhuys (Eds.), *Entrepreneurship, Innovation and Economic Development* (pp. 228–264). Oxford University Press.

Al Boinin, H. (2023). Women's entrepreneurship in the GCC: A literature analysis from a socio-cultural perspective. *Journal of Enterprising Communities: People and Places in the Global Economy, 17*(5), 999–1021. https://doi.org/10.1108/JEC-03-2022-0048

Alanzi, S., Ratten, V., D'Souza, C., & Nanere, M. (2022). Culture, economic, and entrepreneurial environment in the gulf cooperation council (GCC) countries. In Ratten, V. (Ed.) *Strategic Entrepreneurial Ecosystems and Business Model Innovation*, (pp. 61–74). Emerald Publishing Limited. https://doi.org/10.1108/978-1-80382-137-520221005

Alghamdi, A. K. H., El-Hassan, W. S., Aldossari, A. T., & Alrasheed, H. S. (2021). Toward a Female Entrepreneurship Education Curriculum in Saudi Arabia. *Journal of Entrepreneurship Education, 24*(5), 1–22.

Al-Housani, I. M., Koç, M., & Al-Sada, S. M. (2023). Investigations on entrepreneurship needs, challenges, and models for countries in transition to sustainable development from resource-based economy—Qatar as a case. *Sustainability, 15*(9), 7537. https://doi.org/10.3390/su15097537

Alkaabi, K., Ramadani, V., & Zeqiri, J. (2023). Universities, entrepreneurial ecosystem, and family business performance: evidence from the United Arab Emirates. *Journal of the Knowledge Economy*. https://doi.org/10.1007/S13132-023-01384-9

Al-Mulla, A., Ari, I., & Koç, M. (2022). Sustainable financing for entrepreneurs: Case study in designing a crowdfunding platform tailored for Qatar. *Digital Business, 2*(2), 100032. https://doi.org/10.1016/j.digbus.2022.100032

Alnassai, J. M. I. A. (2023). Study on the barriers to entrepreneurship in the UAE. *Journal of Risk and Financial Management, 16*(3), 146. https://doi.org/10.3390/jrfm16030146

Al-Qahtani, M., Zguir, M. F., Al-Fagih, L., & Koç, M. (2022). Women entrepreneurship for sustainability: investigations on status, challenges, drivers, and potentials in Qatar. *Sustainability, 14*(7), 4091. https://doi.org/10.3390/su14074091

Arshed, N., Carter, S., & Mason, C. (2014). The ineffectiveness of entrepreneurship policy: Is policy formulation to blame? *Small Business Economics, 43*(3), 639–659. https://doi.org/10.1007/S11187-014-9554-8/METRICS

Belwal, R., Al Balushi, H., & Belwal, S. (2015). Students' perception of entrepreneurship and enterprise education in Oman. *Education and Training, 57*(8–9), 924–947. https://doi.org/10.1108/ET-12-2014-0149/FULL/HTML

Bennett, R. (2006). Business lecturers' perceptions of the nature of entrepreneurship. *International Journal of Entrepreneurial Behavior and Research, 12*(3), 165–188. https://doi.org/10.1108/13552550610667440

Beyhan, B., Akcomak, I. S., & Cetindamar, D. (2022). How do technology-based accelerators build their legitimacy as new organizations in an emerging entrepreneurship ecosystem? *Journal of Entrepreneurship in Emerging Economies,* (ahead-of-print). https://doi.org/10.1108/JEEE-06-2022-0161

Birden, M., Bastug, M., & Yazici, S. (2020). The impact of incubators on entrepreneurial process in Turkey: a guide for startups. *Journal of Business, Economics and Finance, 9* (2), 132–142. https://doi.org/10.17261/Pressacademia.2020.1219

Bryant, S. (2022). How many startups fail and why? *Investopedia*. https://www.investopedia.com/articles/personal-finance/040915/how-many-startups-fail-and-why.asp

Cansız, M. (2008). Türkiye'de KOBİ'ler ve KOSGEB, DPT Uzmanlık Tezi. DPT Müsteşarlığı, Ankara. https://www.sbb.gov.tr/wp-content/uploads/2022/08/Turkiye-de-KOBILER-ve-KOSGEB_Mehmet-Cansiz.pdf

Covin, J. G., & Miles, M. P. (1999). Corporate entrepreneurship and the pursuit of competitive advantage. *Entrepreneurship Theory and Practice, 23*(3), 47–63. https://doi.org/10.1177/104225879902300304

Cubukcu, C., & Gulsecen, S. (2018). An analysis of the business accelerator programs in Turkey. *International Journal of Computer, 29*(1), 13–32.

Cubukcu, C., & Gulsecen, S. (2019). Accelerator programs in Turkey: Who benefits the most from these programs? In T. Giuseppe, F. Corona, S. Bakhtiar, A. P. Marques, C. T. Silva, I. Cheremisova, D. Pazder, A. Ecirli, M. Aparicio, W. K. Mujani, K. Ovesni, E. Gulua, & T. Kopecka (Eds.), *New Studies and Research in Economics* (pp. 5–16). Revistia.

Dees, J. G. (2001). The meaning of social entrepreneurship. *Stanford Social Innovation Review, 1*(1), 30–39. https://doi.org/10.1108/14720700510604733

Demircioglu, M. A., & Chowdhury, F. (2021). Entrepreneurship in public organizations: The role of leadership behavior. *Small Business Economics, 57*(3), 1107–1123. https://doi.org/10.1007/s11187-020-00328-w

Elali, W., & Al-Yacoub, B. (2016). Factors affecting entrepreneurial intentions among Kuwaitis. *World Journal of Entrepreneurship, Management and Sustainable Development, 12*(1), 18–34. https://doi.org/10.1108/WJEMSD-07-2015-0029/FULL/HTML

Elbanna, S., & Armstrong, L. (2023). Exploring the integration of ChatGPT in education: adapting for the future. *Management & Sustainability: An Arab Review,* (ahead-of-print). https://doi.org/10.1108/MSAR-03-2023-0016

Elbanna, S., Obeidat, S. M., Younis, H., & Elsharnouby, T. H. (2023). Development of Gulf Cooperation Council human resources: an evidence-based review of workforce nationalization. *Employee Relations, 45*(5),1129–1160. https://doi.org/10.1108/ER-08-2022-0401

Elbanna, S., & Abdel-Maksoud, A. (2020). Organizational resources and performance: The case of an oil-rich country. *Public Performance & Management Review, 43*(3), 713–739. https://doi.org/10.1080/15309576.2019.1660187

Elbanna, S., Abdelzaher, D. M., & Ramadan, N. (2020). Management research in the Arab World: What is now and what is next? *Journal of International Management, 26*(2), 1–21. https://doi.org/10.1016/j.intman.2020.100734

Elbanna, S., & Fatima, T. (2023). Quantifying people in the GCC region: The uses, challenges, and efficacy of the quota system policy. *Human Resource Development International, 26*(3), 292–316. https://doi.org/10.1080/13678868.2022.2116260

Elgammal, M. M., Al Bakri, A. A., & Al Janahi, Y. A. (2023). The effect of new and traditional sources of financing on the performance of small and entrepreneurship businesses: The case of Qatar. *Cogent Economics & Finance, 11* (2). https://doi.org/10.1080/23322039.2023.2183667

Faisal, M. N., Jabeen, F., & Katsioloudes, M. I. (2017). Strategic interventions to improve women entrepreneurship in GCC countries: A relationship modeling approach. *Journal of Entrepreneurship in Emerging Economies, 9*(2), 161–180. https://doi.org/10.1108/JEEE-07-2016-0026/FULL/HTML

Fayolle, A. (2013). Personal views on the future of entrepreneurship education. *Entrepreneurship and Regional Development, 25*(7–8), 692–701. https://doi.org/10.1080/08985626.2013.821318

Fayolle, A., & Gailly, B. (2008). From craft to science: Teaching models and learning processes in entrepreneurship education. *Journal of European Industrial Training, 32*(7), 569–593. https://doi.org/10.1108/03090590810899838

Ferreira, C. C. (2020). Experiential learning theory and hybrid entrepreneurship: Factors influencing the transition to full-time entrepreneurship. *International Journal of Entrepreneurial Behavior & Research, 26*(8), 1845–1863.

GEM Global Entrepreneurship Monitor (2023). Global Entrepreneurship Monitor. https://www.gemconsortium.org/.

Gibb, A. (2008). Entrepreneurship and enterprise education in schools and colleges: Insights from UK practice. *International Journal of Entrepreneurship Education, 6*, 101–144.

Hamdan, A. M. (2019). Entrepreneurship and economic growth: An emirati perspective. *The Journal of Developing Areas, 53* (1), 65–79. https://muse.jhu.edu/article/702364/summary

Hertog, S. (2013). The private sector and reform in the Gulf Cooperation Council. LSE Research Online. http://eprints.lse.ac.uk/54398.

Hisrich, R. D., Peters, M. P., & Shepherd, D. A. (2017). *Entrepreneurship* (10th ed.). McGraw-Hill Education.

Hoppe, M., Westerberg, M., & Leffler, E. (2017). Educational approaches to entrepreneurship in higher education: A view from the Swedish horizon. *Education and Training, 59*(7–8), 751–767. https://doi.org/10.1108/ET-12-2016-0177

Huibing, X., & Nengli, S. (2005). Exploration of science parks. *Chinese Journal of Population Resources and Environment, 3*(1), 55–59. https://doi.org/10.1080/10042857.2005.10677406

Jones, O., & Lourenço, F. (2006). Developing entrepreneurship education: Comparing traditional and alternative teaching approaches. *International Journal of Entrepreneurship Education, 4*, 111–140. https://www.researchgate.net/publication/265820943

Jones, C., & Matlay, H. (2011). Understanding the heterogeneity of entrepreneurship education: Going beyond Gartner. *Education and Training, 53*(8), 692–703. https://doi.org/10.1108/00400911111185026/FULL/HTML

Kantis, H. D., Federico, J. S., & García, S. I. (2020). Entrepreneurship policy and systemic conditions: Evidence-based implications and recommendations for emerging countries. *Socio-Economic Planning Sciences, 72*, 1–11. https://doi.org/10.1016/J.SEPS.2020.100872

Kebaili, B., Al-Subyae, S. S., Al-Qahtani, F., & Belkhamza, Z. (2015). An exploratory study of entrepreneurship barriers: The case of Qatar. *World Journal of Entrepreneurship, Management and Sustainable Development, 11*(3), 210–219. https://doi.org/10.1108/wjemsd-03-2015-0014

Kepenek, E. B., & Eser, Z. (2018). Impact of pre-incubators on entrepreneurial activities in Turkey: Problems, successes, and policy recommendations. In D. Meissner, E. Erdil, & J. Chataway (Eds.), *Innovation and the entrepreneurial university* (pp. 57–82). Springer International Publishing.

Kirby, A. D. (2004). Creating Entrepreneurial Universities in the UK: Applying entrepreneurship theory to practice. *Journal of Technology Transfer, 31*(5), 599–603. https://doi.org/10.1007/s10961-006-9061-4

Klapper, R. G., & Farber, V. A. (2016). In Alain Gibb's footsteps: Evaluating alternative approaches to Sustainable Enterprise Education (SEE). *The International Journal of Management Education, 14*(3), 422–439. https://doi.org/10.1016/j.ijme.2016.09.001

Kuratko, D. F. (2005). The emergence of entrepreneurship education: Development, trends, and challenges. *Entrepreneurship Theory and Practice, 29*(5), 577–598. https://doi.org/10.1111/j.1540-6520.2005.00099.x

Kuratko, D. F., & Morris, M. H. (2018). Examining the future trajectory of entrepreneurship. *Journal of Small Business Management, 56*(1), 11–23. https://doi.org/10.1111/jsbm.12364

Leitch, C., Hazlett, S. A., & Pittaway, L. (2012). Entrepreneurship education and context. *Entrepreneurship and Regional Development, 24*(9–10), 733–740. https://doi.org/10.1080/08985626.2012.733613

Lerner, J. (2013). The Boulevard of broken dreams: Innovation policy and entrepreneurship. *Innovation Policy and the Economy, 13*(1), 61–82. https://doi.org/10.1086/668239

Lucas, D. S., Fuller, C. S., Piano, E. E., & Coyne, C. J. (2018). Visions of entrepreneurship policy. *Journal of Entrepreneurship and Public Policy, 7*(4), 336–356. https://doi.org/10.1108/JEPP-D-18-00034/FULL/HTML

Maritz, A., & Brown, C. R. (2013). Illuminating the black box of entrepreneurship education programs. *Education and Training, 55*(3), 234–252. https://doi.org/10.1108/00400911311309305/FULL/HTML

Massey, D., Quintas, P., & Wield, D. (1992). *High-Tech Fantacies: Science Parks in Society, Science and Space*. Routledge.

Mehrez, A. (2019). Investigating critical obstacles to entrepreneurship in emerging economies: A comparative study between males and females in Qatar. *Academy of Entrepreneurship Journal, 25*(1), 1–15.

Mwasalwiba, E. S. (2010). Entrepreneurship education: A review of its objectives, teaching methods, and impact indicators. *Education and Training, 52*(1), 20–47. https://doi.org/10.1108/00400911011017663

O'Connor, A. (2013). A conceptual framework for entrepreneurship education policy: Meeting government and economic purposes. *Journal of Business Venturing, 28*(4), 546–563. https://doi.org/10.1016/J.JBUSVENT.2012.07.003

Otache, I. (2019). Entrepreneurship education and undergraduate students' self- and paid-employment intentions: A conceptual framework. *Education & Training, 61*(1), 46–64. https://doi.org/10.1108/ET-10-2017-0148

Pittaway, L., & Cope, J. (2007). Entrepreneurship education: A systematic review of the evidence. *International Small Business Journal, 25*(5), 479–510. https://doi.org/10.1177/0266242607080656

Pittaway, L., & Edwards, C. (2012). Assessment: Examining practice in entrepreneurship education. *Education and Training, 54*(8), 778–800. https://doi.org/10.1108/00400911211274882/FULL/HTML

Qatar Development Bank. (2021). GEM Qatar National Report 2021. https://www.qdb.qa/en/Documents/Qatar_GEM_2021_en.pdf

Ramayah, T., Ahmad, N. H., & Fei, T. H. C. (2012). Entrepreneur education: Does prior experience matter? *Journal of Entrepreneurship Education, 15*, 65–82.

Robinson, S., Neergaard, H., Tanggaard, L., & Krueger, N. (2016). New horizons in entrepreneurship: From teacher-led to student-centered learning. *Education and Training, 58*(7–8), 661–683. https://doi.org/10.1108/ET-03-2016-0048/FULL/HTML

Saberi, M., & Hamdan, A. (2019). The moderating role of governmental support in the relationship between entrepreneurship and economic growth. *Journal of Entrepreneurship in Emerging Economies, 11*(2), 200–216. https://doi.org/10.1108/JEEE-10-2017-0072

Schwartz, M. (2011). Incubating an illusion? Long-term incubator firm performance after graduation. *Growth and Change, 42*(4), 491–516. https://doi.org/10.1111/j.1468-2257.2011.00565.x

Şehitoğlu, Y., & Özdemir, O. C. (2013). The impact of business incubation on firm performance during post graduation period—Turkey example. *British Journal of Arts and Social Sciences, 12*(1), 171–190.

Shane, S., & Venkataraman, S. (2000). The promise of entrepreneurship as a field of research. *Academy of Management Review, 25*(1), 217–226. https://doi.org/10.2307/259271

Sirelkhatim, F., & Gangi, Y. (2015). Entrepreneurship education: a systematic literature review of curricula contents and teaching methods. *Cogent Business and Management, 2* (1). https://doi.org/10.1080/23311975.2015.1052034

Smallbone, D. (2016). Entrepreneurship policy: Issues and challenges. *Small Enterprise Research, 23*(3), 201–218. https://doi.org/10.1080/13215906.2016.1270227

Tamasy, C. (2007). Rethinking technology-oriented business incubators: developing a robust policy instrument for entrepreneurship, innovation, and regional development? *Growth and Change, 38*(3), 460–473. https://doi.org/10.1111/j.1468-2257.2007.00379.x

Thomassen, M. L., Williams Middleton, K., Ramsgaard, M. B., Neergaard, H., & Warren, L. (2020). Conceptualizing context in entrepreneurship education: A literature review. *International Journal of Entrepreneurial Behavior and Research, 26*(5), 863–886. https://doi.org/10.1108/IJEBR-04-2018-0258/FULL/PDF

Tlaiss, H. A. (2015). Entrepreneurial motivations of women: Evidence from The United Arab Emirates. *International Small Business Journal, 33*(5), 562–581. https://doi.org/10.1177/0266242613496662

Tok, E., Koç, M., & D'Alessandro, K. (2021). Entrepreneurship in a transformative and resource-rich state: the case of Qatar. *The Extractive Industries and Society, 8* (1). https://doi.org/10.1016/j.exis.2020.04.002

Tok, E. (2020). The incentives and efforts for innovation and entrepreneurship in a resource-based economy: A survey on perspective of Qatari residents. *Sustainability, 12*(2), 626. https://doi.org/10.3390/su12020626

Vedovello, C. (1997). Science parks and university-industry interaction: Geographical proximity between the agents as a driving force. *Technovation, 17*(9), 491–502. https://doi.org/10.1016/S0166-4972(97)00027-8

Villegas-Mateos, A. (2021). Qatar's Entrepreneurial Ecosystem - 2021 Edition: Empowering the Transformation. *HEC Paris, Doha, Qatar.* https://doi.org/10.5281/zenodo.4719075

Walter, S. G., & Dohse, D. (2012). Why mode and regional context matter for entrepreneurship education. *Entrepreneurship and Regional Development, 24*(9–10), 807–835. https://doi.org/10.1080/08985626.2012.721009

Westhead, P., & Storey, D. J. (1994). *An Assessment of Firms Located on and off Science Parks in the United Kingdom: Executive Summary Report*. HMSO Publications Centre.

Yıldırım, F. & Keskin, G. (2021). Entrepreneurship ecosystem in Turkey: Accelerators & incubation centers. *Fenerbahçe Üniversitesi Sosyal Bilimler Dergisi, 1*(1), 19–49. https://dergipark.org.tr/en/pub/fbujoss/issue/66543/1041500

Younis, H., Dimitratos, P., & Elbanna, S. (2022). International entrepreneurial SMEs in the Muslim world: The role of religion in the GCC countries. *International Business Review, 31*(5), 102023. https://doi.org/10.1016/j.ibusrev.2022.102023

Chapter 24
Arabic Language Proficiency in Higher Education: A Study of Employers' Expectations and Programme Improvements in Qatar

Hany Fazza, Mohamed Mahgoub, and Abdul Rahman Mahmoud Chamseddine

Abstract This chapter presents the outcomes of the initial exploratory phase of a mixed-method research project that identified employer perceptions of Arabic language competencies of an International Branch Campus' (IBC) graduates. The aims were to compile a list of employer needs, inform Arabic programme planning in relevant IBCs in Qatar and the Gulf Cooperation Council (GCC) countries, and contribute to the literature. The chapter aligns with the book's objective of identifying the 'transformations' needed so that the outcomes of the higher education system in Qatar align well with the job market's needs. The study identified a need for graduates to have better Arabic language skills that meet employers' requirements, including fluency in written and spoken Modern Standard Arabic. It also observed a discrepancy in the qualifications of Arabic Heritage Language learners who graduated from an IBC in Qatar compared to Qatar's Second National Development Strategy objectives for 2018–2022 (Planning and Statistics Authority, 2018).

Keywords Employers' perceptions and needs · Language competencies and skills · Heritage learners · Employability · Modern Standard Arabic

24.1 Introduction

In line with its National Vision 2030 (General Secretariat for Development Planning, 2008), the State of Qatar issued Law No. 7 of 2019 by Amiri Decree to protect the Arabic language. The law made it mandatory to use Arabic in all "ministries…

H. Fazza · A. R. M. Chamseddine (✉)
Georgetown University in Qatar, Ar-Rayyan, Qatar
e-mail: amc276@georgetown.edu

M. Mahgoub
Pre-University Education, Qatar Foundation, Ar-Rayyan, Qatar

E. Zaidan et al. (eds.), *The Future of Education Policy in the State of Qatar*, Gulf Studies 21, https://doi.org/10.1007/978-981-97-9667-0_24

government agencies, and educational and public bodies and institutions, to preserve the Arabic Language and strengthen its role" (Hukoomi Qatar E-Government 2021).

Some research has been conducted in Qatar into language skills and students' perceptions (Mustafawi et al., 2022; Al-Buainain et al., 2011). Mustafawi et al. (2022) identified Qatar University student perceptions of the importance of Arabic and English in educational and social communication, and the importance of English in professional settings, while Al-Buainain et al. (2011) addressed how students at Qatar University perceived the essential requirements and practical applications of English in the workplace. Both studies addressed these issues only from the student's perspective and included English in their research into language skills, whereas our current research focuses specifically on Arabic language skills and competencies concerning employment, and examines them from the perspective of employers.

Qatar's Second National Development Strategy (QNDS) for 2018–2022 identified "human capital development under the existing economic environment" as one of the challenges facing the education and training sector. More specifically, "many graduates need to be equipped to meet the private sector's requirements of practices and values" (QNDS p. 180). The strategy also highlighted the role of higher education institutions in enabling society to maintain its culture and values, primarily through teaching the Arabic language at higher education institutions (QNDS p. 197).

In higher education, with its pivotal role in shaping the future workforce, understanding the ever-evolving needs of the job market is crucial. This chapter describes a small-scale research journey driven to attempt to bridge the gap between academia and industry, specifically focusing on the Arabic language skills and competencies of graduates from an International Branch Campus (IBC) in Qatar. Our primary objectives are twofold: first, to discover how employers perceive the Arabic language proficiency of such an IBC's graduates, and second, to elicit recommendations from these employers for improvements to the IBC's Arabic Programmes, to align more closely with their needs and expectations.

To identify employers' needs and perceptions regarding our target IBC graduates' Arabic language competencies and skills, our study asked the following research questions:

RQ1: How do employers perceive the Arabic language skills and competencies of the target IBC graduates? (RQ1-1: What they liked; and RQ1-2: What they did not like).

RQ2: What changes/improvements would these employers recommend to the target IBC's Arabic Programme so that its graduates can meet their needs and expectations?

By exploring the intricacies of employer perceptions and expectations, we aspire to contribute meaningfully to the discourse on Arabic language education in higher education settings, particularly where native speakers of Arabic are navigating the complexities of becoming 'Heritage Learners' of written and spoken Modern Standard Arabic (MSA). Furthermore, this chapter serves as a launchpad for further investigation.

In the following sections, we will review the relevant literature, describe our methodology, and present the rich and varied insights gleaned from employer perspectives, followed by the actionable recommendations and policy implications generated from this project.

24.2 Literature Review

Effective communication is crucial for teamwork and organisational growth. Language skills and cultural awareness have been shown to impact professional interactions. Research demonstrates a connection between English proficiency and employment opportunities, but little is yet known about employers' needs for Arabic language skills among graduates.

24.2.1 The Importance of Language Competence in the Workplace

In a globalised job market, a proficient level of English, widely recognised as the 'lingua franca' of international business, has gone beyond being merely a linguistic skill, to become a significant factor in determining employability (Joshi, 2013). Arkoudis et al. (2009) highlight the essential connection between English language competence and workplace readiness, particularly for international students. Effective English communication is critical for interactions with colleagues, clients, and customers, thereby enhancing job prospects. Graduates need strong language skills because employers prioritise candidates who can communicate effectively in English (Sarudin et al., 2008). The corporate landscape increasingly perceives English as a gatekeeper in recruitment processes (Khan and Chaudhury, 2012).

Nevertheless, a disconcerting gap frequently exists between the language skills imparted by educational institutions and the intricate demands of real-world communication in professional settings (Hart-Rawung & Li, 2008). This dissonance can severely impede one's prospects in the job market, a challenge acutely felt in fields like engineering, where technical expertise must be complemented by effective communication (Zaharim et al., 2010). The alignment between language education and workplace requirements is critical to employability. Patil and Karekatti (2012) recommend periodic evaluations of language courses, in order to ensure relevance to the current demands of the workplace. It is crucial to adopt teaching methodologies that simulate real-life workplace communication scenarios (Rajprasit et al., 2014).

24.2.2 *Global Perspectives on Language and Employability: An Integrated Review*

Mustafawi et al. (2022) studied how Qatar University students perceived the importance of Arabic and English in educational and communicative contexts, through a questionnaire that covered topics such as language preference, the role of Arabic in society and education, and the significance of English in professional settings. The study offered insights into the challenges and opportunities of implementing English-medium education in Multilingual University Settings (EMEMUS) in Qatar. The study also examined policy concerning EMEMUS in Qatar, analysing how various framework components intersected. This included the significance of English and Arabic in education and the workplace, the decision-making processes of academic disciplines regarding the primary medium of instruction, and the confusion resulting from top-down imposition of the policy, without adequate input or preparation from those on the ground.

A 2017 study conducted in Malaysia by Ting et al. aimed to investigate the correlation between university graduates' language proficiency, communication skills, and employability. They found that interpersonal communication skills and practical learning experiences facilitated effective communication, thus emphasising the importance of soft skills and attributes essential for employment and career growth. Cheong et al. (2016) investigated employers' perceptions regarding the employability of Malaysian graduates, exploring diverse theoretical frameworks concerning employability, such as trait factor theory, self-concept theory, developmental theory, opportunity structure theory, social learning theory, planned happenstance theory, and chaos theory, complemented by an examination of Malaysia's economic trajectory and educational challenges over the past decade. The study also chronicled government responses to private sector concerns. It revealed a deficit in essential skills and attributes among Malaysian graduates, including communication skills, critical thinking, and problem-solving acumen.

In 2023, Wu et al. adopted the Outcome-Based Education (OBE) theory as a foundational framework, to investigate the impact of verbal and non-verbal communication skills on the employability of engineering students. They introduced the USEM Model, which comprises four key components that help develop holistic employability attributes: understanding, skills practice, efficacy beliefs, and metacognition. The study concluded that communication skills are paramount in shaping engineering students' employment prospects, and articulated how the OBE theory and USEM framework bolstered communication skills within engineering education.

Aldashev et al. (2009) investigated the correlation between language proficiency and labour market outcomes among foreign workers in West Germany. The researchers explored the nexus between language proficiency and productivity, earnings, participation, and employment. They showed that "foreigners with better language command have a significantly higher participation rate and employment chances. Moreover, better language proficiency increases the probability of working in white-collar occupations" (Aldashev et al., 2009, p.3).

Cai (2013) proposed a new framework for institutionalism that offers a comprehensive understanding of how employers perceive the value of graduates in the workforce. The study emphasised the significant impact that international higher education providers have in moulding employers' perspectives and improving graduates' employment prospects, by offering up-to-date curricula, providing work-integrated learning opportunities, and collaborating with industry partners, to ensure their graduates possess the skills and knowledge required for the labour market.

In 2012, Singh and Choo (2012) studied the significance of English language proficiency for employers in the Malaysian manufacturing industry, and found that graduates with solid English skills had better communication skills, productivity, and competitiveness. Graduates needed to improve their English language skills to meet employer expectations. The study recommended intervention in language teaching at the primary, secondary, and tertiary levels, to improve the English language syllabus in line with globalisation and current workplace demands. Interestingly, and of relevance to our research project, the authors suggested that while English proficiency is essential, attention should also be paid to Malaysia's national language.

24.2.3 Work-Related Language Competencies Which Employers Seek

Numerous studies have emphasised the importance of language proficiency, specifically in English, for effective workplace communication and enhanced employment opportunities (Alih et al., 2018; Thomas et al., 2016; Ting et al., 2017). Proficient communication is a crucial asset in professional environments, as it correlates with problem-solving and interpersonal aptitudes. In global organisations, language proficiency can be a competitive advantage (Thomas et al., 2016; Senanayake, 2019; Al-Mahrooqi & Tuzlukova, 2014; Alih et al., 2018).

Al-Mahrooqi and Tuzlukova (2014) studied the communication abilities of higher education students in Oman. The study evaluated linguistic skills, including writing reports, emails, and letters; expressing opinions and ideas in writing; summarising other people's thoughts; paraphrasing and restating other people's opinions; and creating meeting agendas. Additionally, the researchers evaluated professional communication skills such as effective oral and written communication, accurately interpreting client needs, presenting alternatives to clients, and verbal negotiation abilities. The authors emphasised the importance of developing practical communication skills, to help students perform well in job interviews and communicate better in the job market (Al-Mahrooqi & Tuzlukova, 2014).

Alih et al. (2018) in Malaysia found that possessing excellent oral communication skills in English has become crucial for job seekers, particularly in the service sector. Furthermore, employers highly value individuals with a humanities, politics, and economics background. As a result, graduates possessing these qualities, and a broad spectrum of knowledge, are in high demand. Clokie and Fourie (2016) also studied the development of written and oral business communication skills, such as academic and business writing, referencing norms, proofreading and editing, research skills, text analysis, and speech types in the undergraduate curriculum, and their importance for employability. They addressed these language skills in emails, memos, letters, and reports (pp. 14–15). In England, Webb (2010) studied employers' expectations regarding the language skills of graduates from Leeds University.

Nour (2014) highlights the need for education and training policies in Gulf countries, to align with employer demands. Karoly (2010) argues that Gulf education systems require improvement to prepare students for modern jobs. Aldosary (2006) suggests a communicative approach to address graduate unemployment in Saudi Arabia, emphasising the importance of careful planning and consultation. Dakhli and El-Zohairy (2013) acknowledge that higher education across the GCC, including in privatised and foreign partnerships, still needs to meet expectations and address young people's aspirations. They nevertheless believe that such partnerships can significantly enhance both teaching and research.

Research indicates a notable disparity between the competencies and expertise of Omani graduates, and the requirements of employers in Oman. Belwal et al. (2017) found a mismatch between university teaching and employer expectations, with employers seeking more technical and computing skills and English proficiency—many recent graduates in Oman still struggle to find jobs, especially in the private sector. Also, Edwin and Sabura (2019) asserts that Oman's education system must adequately equip students for the job market. Some studies have examined strategies to enhance the employability of Omani graduates. Ali et al. (2017) suggests improving coordination between the government, the education system, and the private sector, to align study programmes with labour market needs. Al-Barwani et al. (2009) proposes that private universities pay greater attention to raising the quality of education, so as to better prepare students for employment, even if it results in lower profits. Al-Jahwari et al. (2020) recommend that the government create more jobs that match the skills of graduates, and that universities adjust their curricula to be more practical and career-oriented.

In conclusion, while much research has been done on graduate employability and the vital importance of language competence in the workplace, this review reveals a gap in the literature regarding employers' perceptions of the Arabic language skills and competencies of graduates in the GCC countries, like those graduating from IBCs, and employers' recommendations for improving Arabic language teaching programmes.

24.3 Methodology: Methods, Sample, and Ethics

This chapter is based on the initial exploratory phase of a two-phased research project that adopts a mixed-method exploratory, sequential approach (Fetters et al. 2013). Due to the scarcity of published data on this specific topic in this context, the research team adopted—at the initial phase, which this chapter is based on—an exploratory, constructivist, interpretivist approach. It is exploratory in that it did not start with any theories or hypotheses that it took to the participants; rather, it identified answers to the research questions from the participants, based on their experiences and perceptions. The constructivist nature of this approach is that it bases the findings on the meanings made by the study participants (Crotty, 1998). The interpretivist aspect relates to the multi-layered interpretation process that takes place when each of the participants interprets the researcher/interviewer's questions during the interview; the researchers interpret the meanings made by the participants during the data analysis process; and finally, the readers interpret the findings emerging from the study to see if they relate to their own experiences (Stake, 1995).

To collect data during the initial phase of our research project, we used semi-structured, individual, face-to-face interviews with our target IBC graduates, who were—at the time of the interview—in senior and midlevel leadership positions in government, semi-government, non-government, and private international organisations in Doha, Qatar. An interview protocol was developed by the research team and implemented by the lead author of this chapter. In addition to using the interview protocol questions, the researcher/interviewer asked probing questions as necessary, and in a way that helped the research team to get rich data from the participants. Interviews were conducted in English, to avoid any loss in meaning during the translation and back-translation process, that would have been necessary had interviews been conducted in Arabic (Al-Amer et al., 2016). Interviews were conducted in participants' offices, each lasting around an hour. Individual interviews were considered the most suitable data collection method in this context, especially since these participants had first-hand experience with the IBC Arabic Programme in Qatar; they were involved in recruitment panels in their respective organisations, and the topic could be perceived as sensitive (Merriam & Tisdell, 2016).

For the initial phase of our research project, the team adopted a purposive sampling approach: 11 participants were recruited, each meeting the criteria of being a graduate of the target IBC, bilingual and a native Arabic speaker, having participated in job interview/selection panels; and working in government, semi-government, non-government, private, or international institutions, located in Qatar. This sampling approach is the most suitable when little is known about the phenomenon under investigation, and the participants are the best placed to increase our knowledge and understanding of it (Merriam & Tisdell, 2016). While using this approach, the team applied the selection and exclusion criteria mentioned above, to control the researcher's bias (Robson & McCartan, 2016).

The team transcribed the interviews using Microsoft Word 365 and reviewed the transcripts for accuracy. For the analysis stage, we adopted a thematic analysis approach (Braun & Clarke, 2012). We familiarised ourselves with the interview narratives, identified the recurrent patterns in the data in an inductive way and coded them, then identified the overarching themes, and finally reviewed the coded narratives to ensure their alignment with the themes, through applying deductive analysis.

The research team received Institutional Review Board (IRB)/ethical application approval. Informed consent forms were taken from each participant before the interview. Participants' agreement was obtained before audio-recording each interview.

None of the participants refused to let the interview be audio-recorded, and they all answered all the interview and probing questions.

24.4 Findings and Discussion

This section presents the key findings that emerged from the accounts of the study participants. These findings are presented below, arranged by themes, under each research question. The findings are also discussed considering the literature above, which focused on something other than exactly this same topic: employer perceptions of university graduates' Arabic language competencies and skills.

24.4.1 RQ1: How Do Employers Perceive the Arabic Language Skills and Competencies of the Target IBC Graduates?

24.4.1.1 What They Liked

Writing Formal, Official Content

Participants (N = 5) highlighted that they liked graduates' skills in writing technical and official Arabic content, e.g., emails and reports, in various fields. For public media purposes, Participant 1 (P1) (P1) said, "*They can ...create content... [such as] statements, press releases, media messages, and tweets.*" This was part of their jobs assisting '*public figures*' in government positions. This echoes Al-Mahrooqi and Tuzlukova (2014), who underscored the need to train university students in writing reports, emails, and letters; however, the latter study focused on English rather than Arabic. Another participant (P8) commended the graduates' technical Arabic writing skills in social science research. Praising a female graduate, they said, "*She is capable of writing research proposals, analysing research data, and contributing to research reports.*" In the civil aviation sector, P10 praised graduates' skills in writing official correspondence to airline and civil aviation authorities across the Arab world in Arabic.

Editing Formal, Official Content

The participating employers (N = 3) also praised graduates' skills in editing formal Arabic content. An employer at a government department said, "*They edit Arabic content... [such as] an article, ... a tweet, [or an official] statement.*" P7 found the IBC graduates "*extremely capable*" of editing formal and official Arabic content. The same participant highlighted that his organisation relied on these graduates to "*proofread*" their Arabic materials. Editing written content was also emphasised as a skill favoured by employers by Clokie and Fourie (2016), who conducted their study in New Zealand.

Translation and Editing Translations

In addition to writing and editing written formal content in Modern Standard Arabic (MSA), most participants (N = 8) commended the IBC graduates on their competency in translating and editing English-Arabic and Arabic-English documents and content. One participant said graduates working with them "*translate from English to Arabic.*" Another (P6) claimed that they "*rely on them to check [other people's] translation*" and added, "*we think they are highly proficient.*" As for the type of content, participants said the IBC graduates translated and edited the translation of tweets and business correspondence. Webb (2010) highlights the importance of graduates' skills in translating and interpreting English into other languages; however, he did not include Arabic in his studies. A third employer considered the IBC graduates an "*asset*" to their institutions because of their skills in translation, especially from Arabic into English, so senior executives can read technical documents.

Summarising Formal Content

Working closely with senior staff and executives, the IBC graduates were applauded for their skills in summarising content originally written in Arabic. Some participants (N = 3) claimed that these graduates would be asked to summarise a law "*in half a page…or a paragraph*" in English (P4). In high government offices and research institutions, these graduates were praised for their skills in providing English summaries of lengthy Arabic interviews. Al-Mahrooqi and Tuzlukova (2014) also underscored the importance of building graduates' capacity to summarise oral discourse. Participants also commended these graduates' skills in reading and conceptualising technical documents, which enables them to provide appropriate summaries of technical documents, as Participant 3 emphasised.

Strategic Oral Skills

Some participating employers (N = 3) applauded the IBC graduates for their Arabic-speaking skills. In the business and government sectors, these employers appreciated these graduates' skills in giving oral presentations, making phone calls, conducting dialogues, and engaging in and leading negotiations. They were also assigned the tasks of leading interviews and moderating panels, which required them to write, ask questions, and provide summaries in MSA. Oral skills have been accentuated by many scholars, such as Al-Mahrooqi and Tuzlukova (2014), Leveson (2000), and Clokie and Fourie (2016).

24.4.2 What They Did not like

Inability to Write Formal Arabic Content

Almost half of the participating employers (N = 5) expressed dissatisfaction with what they considered the IBC graduates' inability to write formal, technical content in MSA. P1 said, "*If most of the [graduates] aren't able to formulate five sentences back-to-back in Arabic without making a mistake, it becomes a problem.*" Another participant (P2) specified the weaknesses in these graduates' writing, saying, "*Most of [them] cannot write in Arabic… I mean writing…technical reports or articles in newspapers.*" The same participant considered writing in MSA to be "*the main challenge to most [graduates].*" In addition to not being able to write reports and newspaper articles in Arabic, P8 explained that some of these graduates would be "*a little terrified*" if they were asked to write business letters in MSA; these graduates would write the letter in English and get it Google translated into Arabic and then review it. The same objection was expressed about their inability to write business emails and speeches.

Inability to Speak in MSA and Having to Code-Switch

Some participating employers (N = 4) voiced dissatisfaction with the IBC graduates' spoken MSA. P2 gave the example of "*delivering a class or a lecture*" in national institutions such as Qatar University, where most of the students needed the lecture to be in Arabic; the same participant claimed that these graduates "*have problems with speaking [MSA] in public… unlike [their] colleagues on the Main Campus.*" Employers found these graduates also lacking the ability to give "*official presentations*" in MSA and said they have to code-switch using their regional dialects, or English. One participant (P3) explained how this had caused them some embarrassment when they were attending a conference and started asking the panellists a question in MSA, but could not complete it and had to code-switch to English, saying, "*I couldn't… convey my message… fluently in Arabic.*" the IBC graduates were portrayed as being "*more comfortable to discuss [things] in English rather than in Arabic.*" It is very interesting that these graduates, who are considered 'AHL speakers' (Arabic Heritage Language speakers) while they live and study in their own country and society, code-switch to English, which is not their own country's first language, in addition to reverting to their regional dialect when they attempt to speak in MSA. Normally, speakers code-switch from a foreign language to their native language (Kroon et al., 2015).

RQ2: What Changes/Improvements Would These Employers Recommend to the IBC Arabic Programme so that Its Graduates Have the Skills and Competencies to Meet Their Needs and Expectations?

An integrated set of suggested improvements emerged from the participants' accounts. They covered course design and development, pedagogy, assessment, and certification.

- **Design and Develop Career-Oriented, Multi-Level, Interdisciplinary Arabic Courses**

Most participants (N = 6) recommended that the design and development of Arabic courses should closely align with the student's core requirements and career needs. They advocated basing the content of Arabic courses on professional business needs in economics, politics, government, communication, and culture. They believed—according to their interviews—that the programme content—during the interviews—focused "*basically on literature*" (P3), which they thought did not help prepare the IBC graduates for the job market. Another employer (P4) expressed the need for more knowledge of Arabic "*business and tax terms [economics].*" A third participant recommended that students be allowed to choose the field and skills in which they needed to build up their Arabic vocabulary and language competency.

P5 advocated designing Arabic courses to fit local job requirements, saying, "*make the courses more career-based...[specifically] in Doha.*" P9 recommended building Arabic courses around the same core requirements, to enable students to enhance their vocabulary and communication skills, such as debating and discussing "*international politics.*" Participants thought that graduates who wanted to be prepared for work in Qatari government offices needed to improve their communication skills and be proficient in business communication in Arabic. Linking course content and outcomes with job market needs in the Gulf countries has been emphasised by Nour (2014) and Karoly (2010).

Most participants (N = 9) also recommended that Arabic courses should reflect interdisciplinary integration. In other words, these participating employers thought that the IBC Arabic Programme needed to base its course design on core requirements and job needs; these courses should also present well-integrated materials that are not limited to specific subjects and fields. For example, if Middle Eastern studies are addressed in an Arabic course, they should be addressed from political, socio-economic, and cultural aspects. Participants also suggested training the IBC Arabic students in translation and editing translations. Our participants' recommendation of considering interdisciplinary integration on Arabic courses does not align with Wu's claim that this aspect does not affect employability (Wu et al., 2023).

Participants also suggested adding more advanced and post-advanced courses, to enable Arabic students to further develop their language skills. They recommended that the IBC Arabic Programme ought to require students on the MSA Non-native Learners and MSA AHL Learners tracks to take as many post-advanced courses as possible. They claimed that job readiness requires taking several of these courses. P1, for example, said, "I've *seen people who graduated...who have done their post-advanced, but they took one... course and did their proficiency [test]... but once they go to the real life, it's not enough.*" Hart-Rawung & Li, (2008) emphasise the fact that educational institutions need to enable students and graduates to build high levels of the language competencies needed for employment.

- **Further Improve Arabic Teaching to Nurture and Develop Skills Needed in the Job Market.**

How these courses are taught and assessed is closely linked to the types and themes of Arabic courses. All participants (N = 11) recommended that Arabic course instructors and professors should simulate real-life situations in their classes and courses, using presentations, debates, and discussions related to job market needs. This has been mentioned by Rajprasit et al. (2014). Such role plays and cooperative learning would help Arabic students become more fluent in spoken Arabic through practice. P3 said, "*[They] should all be capable...[and] fluent.*" They also recommended that Arabic Programme faculty apply pedagogical methods and techniques that help students build vocabulary bases and language skills—including syntax and language functions—in the area of employment they wish to specialise in. For example, P9 said, "*I think building up... vocabulary [base] is important, especially in politics and culture.*" The participating employers underscored the need for the IBC Arabic Programme graduates to be trained as much as possible in using relevant terminology to be used when engaged in a business meeting, presentation, or panel. This would require them to conceptualise key terms and concepts in Arabic, and practise and engage in debates and dialogues using these key concepts and terms.

In order to develop graduates' oral skills in MSA, P1 recommended that instructors need to train students more in preparing and giving job-related "*3/5/10-min presentations and ... speech[es].*" They recommended training them to "*organise a speech and cite Arabic references*" and deliver a [coherent] speech "*without reading off a piece of paper.*" Highlighting the significance of oral communication/messages in Arabic, P1 said, "*Speaking in Arabic is very, very important.*" Participants believed that the way to fluency is through practice, which requires Arabic instructors and professors to make their classes and courses more learner-centred. Alih et al. (2018) and Al-Mahrooqi and Tuzlukova (2014) highlight the importance of oral communication skills for graduates' employability; however, it should be noted that their studies looked at English, not Arabic.

Similarly, P5 recalled that the lack of a variety of student-led and engaging activities had made him feel "*bored*" in Arabic classes. They also recommended that students do more reading, to be ready and able to lead debates and discussions. For training IBC Arabic students in institutional archiving and storytelling, P2 recommended training them in writing stories documenting positive change and development. First, they would be trained in how to describe the situation before change and development took place; second, to explain what had changed; third, to focus on the hero/heroine of change and fourth, how the hero/heroine led the change; and finally, how the changes subsequently took root. Participants recommended that Arabic students ought to learn how to participate and conduct interviews, as part of their oral skills training.

Participants also emphasised the importance of practising writing emails in response to "*real [business] scenarios*" (P11). They recommended that advanced job-related skills, such as text analysis and writing, be developed throughout the programmes' courses. P5 highlighted the need to help AHL learners practise writing "*diplomatic emails and ... messages.*" Moreover, they suggested that these students be trained in summarising books and reports, writing critiques of books or reports, and writing technical reports. More specifically, they called for students to be trained in writing "*memos, emails, formal correspondence, and minutes of meetings*" (P10). Research has also highlighted the importance of business writing skills, such as Al-Mahrooqi and Tuzlukova 2014 and Clokie and Fourie (2016).

The participants also suggested that the target IBC Arabic Programme students need to be trained in conducting research in Arabic. This would require them to conduct a literature review, identify a problem, develop a research question(s), develop data collection tools, collect and analyse data, write up their research findings/results, discuss them concerning the relevant literature, and finally write a study report. This could be part of advanced or post-advanced level courses. P4 said, "[*This*] *requires a very good level of Arabic.*" The employers also recommended helping students to develop the same skill set in market research projects. Getting AHL learners trained in reading, analysing, and evaluating policies, and writing policies and policy briefs, is closely related to conducting and reporting research in Arabic. P3 highlighted these and other relevant skills, when they listed the skills and competencies an Arab employee would need to have to join their social research institution; they said they would focus on "*how to …write reports, how to write policy briefs…how to conceptualise policy recommendations… so that [these recommendations] reach the policymakers in the right format.*" The participants' call for training IBC Arabic students in research aligns with Dakhli and El-Zohairy's (2013) claim that branch campus partner universities in the GCC need to improve teaching and research skills. They recommended that students be trained to practise and master the conceptualisation of key terminologies, and the arts of communication and debating, summarising, reviewing, critiquing, writing academic research in Arabic, and writing relevant business documents.

- **Make the Arabic Proficiency Test Compulsory for All MSA Programme Students**

Being IBC graduates themselves, some participants recommended that all IBC Arabic students should take and pass the proficiency test. They explained that this would encourage them to take post-advanced Arabic courses in preparation, and ultimately help them develop advanced MSA skills. P1 said, "*A lot of heritage speakers don't [take the Arabic proficiency test]; it is a choice. Some of them do French.*" Another participant (P3) considered it a matter of identity and said, "*You're in an Arab society, an Arab country; you're going to have to work with Arab entities that deal with Arabic.*"

- **Identify and Apply for Internationally Recognised Arabic Certification**

Most participants (N = 7) recommended that the IBC Arabic Programme align and benchmark its Arabic courses, especially the proficiency test, to internationally recognised certification. Some of them mentioned English language tests such as the Test of English as a Foreign Language (TOEFL) and International English Language Testing System (IELTS), apparently unaware that no Arabic proficiency tests are as internationally recognised as these English tests. However, P4 said, "*There is no … certification like TOEFL or IELTS [in Arabic].*" These participants highlighted the need for valid certification of Arabic language skills and competencies that employers and institutions can rely on as indicative of candidates' knowledge and skills relating to MSA. Highlighting this, P8 said, "*[There needs to be] a certificate that [certifies that] you know how to write and speak fluently.*" They added that any such certification would need to set precise proficiency levels.

- **Advocate for a More Robust Arabic Programme During the K-12 Stage**

Almost half of the participants (N = 5) recommended that the IBC Arabic Programme—given its scholarly wealth of professors and faculty—should advocate for stronger Arabic language programmes to be developed in K-12, especially for international schools where English is used more than Arabic. P1 explained that it is challenging to "*nurture*" and develop students' MSA competencies in a four-year programme, when these competencies and skills have not been developed throughout an almost 18-year period. In other words, fixing what has not been developed over

18 years is difficult. P3 pinpointed international school graduates' main issue, saying, "*they think in English,*" not in Arabic. This can be attributed to extended exposure and practice in English compared to Arabic, in addition to using English as a medium of instruction and learning in most school subjects. The same view was expressed by P9, who said, "*Students coming from ...international schools...do not have as solid a foundation in the Arabic language [as their counterparts coming from national and government schools].*" P2, who had graduated from our target IBC but completed their K-12 in a government, public school system, said, "*My Arabic skills and competencies have gone [down] since I joined [IBC]...but the MSA foundation I had earlier could help me compete in the [job] market.*"

24.5 Recommendations and Policy Implications

This section presents the policy and programme planning recommendations and implications. Although these are based on the findings emerging from the exploratory initial phase of our research, they can be relatable and relevant to similar Arabic programmes as much as they are applicable to our target IBCs.

Employers seek candidates who can navigate diverse cultural and social landscapes, and graduates with a strong foundation in Arabic and social studies, who can excel in roles requiring international engagement and cross-cultural communication. This integration directly meets employers' needs for well-prepared graduates, equipping them with practical skills in line with the evolving demands of the job market. Based on the findings of the initial phase of our study, and to enhance the MSA competencies and skills of the IBC graduates, who have been perceived as having low MSA proficiency levels, as Mustafawi et al. (2022) claim, we would like to propose the following recommendations for improving the target IBC's Arabic Programme and other similar programmes.

24.5.1 Design and Develop a Career-Oriented Curriculum

Designing and developing a career-oriented curriculum for Arabic studies is essential, to prepare students for the demands of the job market. To achieve this, it is crucial to align course content with the professional business needs in economics, politics, government, communication, and culture. Taking courses in Arabic social studies can significantly contribute to graduates' job readiness, by providing them with a comprehensive understanding of societal and political intricacies. This equips them to tackle real-world issues and navigate professional challenges effectively. Although studying literature is culturally enriching, it may not directly contribute to the practical job skills sought in today's globalised workforce. Therefore, a curriculum should prioritise those subjects directly addressing the skills and knowledge needed in professional settings.

Additionally, offering interdisciplinary integration is key to addressing subjects like Middle Eastern studies comprehensively. Incorporating social studies content into the Arabic Programme reflects the type of interdisciplinary approach that employers highly value. This approach enables students to apply their Modern Standard Arabic (MSA) skills seamlessly across diverse social science disciplines, making them more versatile and adaptable in their careers. To further enhance language skills, the curriculum should introduce advanced and post-advanced courses. This ensures that Arabic students on both the Foreign Language and Heritage Language tracks take multiple post-advanced courses, thereby increasing their job readiness by providing them with advanced language proficiency skills.

24.5.2 Improve Teaching Methods to Enhance Education in the Arabic Language

Improving teaching methods is another vital aspect of enhancing Arabic language education. Instructors should be encouraged to simulate real-life situations in their teaching contexts through presentations, debates, and discussions. Implementing learner-centred approaches fosters active engagement and better retention of language skills. Building fluency in spoken MSA can be achieved through role plays and cooperative learning, effective in real-world communication scenarios. Furthermore, the curriculum should focus on building vocabulary and language skills relevant to specific job areas.

24.5.3 Enhance Oral and Written Communication

Effective oral and written communication is crucial in professional contexts. Graduates should be adept at navigating discussions on economic policies, historical events, political developments, and cultural intricacies within the Arabic-speaking world. Therefore, students should be trained in preparing and delivering job-related presentations and speeches in MSA. They should also practise organising and citing Arabic references in speeches, thus ensuring their communication skills are polished and professional. Encouraging more student-led and engaging activities in Arabic classes—using role play and experiential learning—can further enhance communication abilities. Additionally, developing business writing skills for real business scenarios, and teaching students to summarise books, write critiques, and produce technical reports are essential skills.

24.5.4 Build Students' Capacity for Research and Policy Analysis

It is crucial to provide a strong Arabic foundation in the social field of the Arab world. This foundation is instrumental in policy analysis, particularly in politics and economics, allowing graduates to understand and articulate policy implications within Arabic-speaking regions. Therefore, the curriculum should introduce courses in research methodology and data analysis in Arabic. Students should be trained to prepare literature reviews, write research reports, and conduct market research projects. Furthermore, fostering skills in policy analysis and policy brief development is essential, to prepare students for careers in research and analysis.

24.5.5 Align the Programme with Internationally Recognised Certification

Aligning with internationally recognised certification standards is essential for the credibility and global recognition of the Arabic programmes. This can be achieved by benchmarking Arabic courses and proficiency tests to internationally recognised standards, and developing a valid Arabic language certification with precise proficiency levels. The certification should indicate fluency in writing and speaking, providing employers with a clear understanding of students' MSA language competencies.

24.5.6 Advocate for Stronger Arabic Programmes in K-12

Lastly, advocating for stronger Arabic programmes in K-12 education is essential to nurture Arabic language proficiency from an early stage, especially in international schools. This can be accomplished by addressing the challenge of students primarily thinking in English, and enhancing their exposure to and practice of MSA. Collaboration with international schools, to improve the foundation of MSA language skills, can create an assembly line of proficient Arabic speakers, ensuring a continuous supply of well-prepared students for higher education in Arabic studies and related fields.

24.6 Limitations and Further Research

The findings presented in this chapter are not meant to be generalisable, as they are based on qualitative data produced by a limited number of participants. However, being based on the participants' lived experiences, these findings are trustworthy, and can be relatable to readers and researchers from similar contexts. As stated above, this chapter is based on the initial phase of our research project that adopts a mixed-method exploratory, sequential approach (Fetters et al., 2013) whereby we collected qualitative data from a selective sample of eleven participants—who represented employers who had graduated from the target IBC in Qatar. We will use the findings presented in this chapter to inform our next phase, wherein a comprehensive questionnaire will be designed along the lines of these findings. We plan to administer that questionnaire on a large scale, and get it completed by a representative sample of employers. As such, the final findings of our project can be generalisable. It will contribute to providing evidence-based and actionable policy recommendations for a variety of contexts.

24.7 Conclusion

Designing a career-oriented curriculum for Arabic studies is paramount in preparing students for the challenges of the modern job market. By aligning course content with professional needs in economics, politics, government, communication, and culture, we can equip graduates with the skills and knowledge necessary to excel in their careers. Emphasising interdisciplinary integration, offering advanced language courses, and improving teaching methods, further enhance students' job readiness. Effective communication, research, and policy analysis skills are essential in professional contexts, and the curriculum should focus on developing these competencies. Additionally, aligning Arabic programmes with international standards, and advocating for stronger Arabic education at the K-12 stage will ensure a continuous supply of well-prepared students for higher education and beyond. With these policy recommendations in place, we can truly empower students to navigate the diverse and dynamic Arabic-speaking world with confidence and competence. These recommendations address employers' concerns and needs for aligning the target IBC Arabic Programme with job market needs and expectations.

Bibliography

Shura Council Discusses Activation of Law No. 7 of 2019 on Protecting Arabic Language (2021) *Hukoomi Qatar E-Government*. Available at: https://hukoomi.gov.qa/en/news/shura-council-discusses-activation-of-law-no-7-of-2019-on-protecting-arabic-language. Accessed 01 September 2023.

Al-Amer, R., Ramjan, L., Glew, P., Darwish, M., & Salamonson, Y. (2016). Language translation challenges with Arabic speakers participating in qualitative research studies. *International Journal of Nursing Studies, 54*, 150–157.

Al-Barwani, T., Chapman, D. W., & Ameen, H. (2009). Strategic brain drain: Implications for higher education in Oman. *Higher Education Policy, 22*, 415–432.

Al-Buainain, H. A., Hassan, F. K., & Madani, A. (2011). Use of English by graduates of Qatar University in the workplace: A quantitative analysis. *Arab World English Journal, 2*(1), 140–185.

Aldashev, A., Gernandt, J., & Thomsen, S. L. (2009). Language usage, participation, employment and earnings: Evidence for foreigners in West Germany with multiple sources of selection. *Labour Economics, 16*(3), 330–341.

Aldosary, A. (2006). The nationalisation of the labor force in Saudi Arabia: logistical consideration and practical strategies. *King Fahd University for Petroleum and Mineral Journal, 1* (2).

Ali, Y., Nusair, M. M., Alani, F., Khan, F. R., & Al Badi, L. (2017). Employment in the private sector in Oman: Sector-based approach for localisation. *Humanities & Social Sciences Reviews, 5*(1), 01–20.

Alih, N. A. C., Yusof, M. A. M., Raof, A. H. A., Zakaria, W. Z. W., Shamsudin, S., Omar, N. A. M., & Jobil, J. (2018). Industries' requirements and expectations of graduates: A revisit. *LSP International Journal, 5* (1).

Al-Jahwari, N. S., Al Khalidi, M., Albalushi, S., Alghanami, M., AlOmairi, Z., & Khan, F. R. (2020). Job-search behaviour of Omani Graduates: Sohar University—a case study. *International Journal of Research in Entrepreneurship & Business Studies, 1*(2), 24–32.

Al-Mahrooqi, R., & Tuzlukova, V. (2014). English communication skills and employability in the Arabian Gulf: the case of Oman. *Pertanika Journal of Social Sciences & Humanities, 22* (2).

Arkoudis, S., Hawthorne, L., Baik, C., Hawthorne, G., O'Loughlin, K., Leach, D., & Bexley, E. (2009). *The impact of English language proficiency and workplace readiness on employment outcomes and performance of tertiary international students* (p. 45). Centre for the Study of Higher Education, University of Melbourne.

Belwal, R., Priyadarshi, P., & Al Fazari, M. H. (2017). Graduate attributes and employability skills: Graduates' perspectives on employers' expectations in Oman. *International Journal of Educational Management, 31*(6), 814–827.

Braun, V., & Clarke, V. (2012). *Thematic analysis*. American Psychological Association.

Cai, Y. (2013). Graduate employability: A conceptual framework for understanding employers' perceptions. *Higher Education, 65*, 457–469.

Cheong, K. C., Hill, C., Fernandez-Chung, R., & Leong, Y. C. (2016). Employing the 'unemployable': Employer perceptions of Malaysian graduates. *Studies in Higher Education, 41*(12), 2253–2270.

Clokie, T. L., & Fourie, E. (2016). Graduate employability and communication competence: Are undergraduates taught relevant skills? *Business and Professional Communication Quarterly, 79*(4), 442–463.

Crotty, M. J. (1998). *The foundations of social research: meaning and perspective in the research process*,, pp. 1–256. London: Routledge

Dakhli, M., & El-Zohairy, D. (2013). Emerging trends in higher education in the GCC: a critical assessment. *Innovation in business education in emerging markets*, 43–63.

Edwin, M. M. D., & Sabura, F. M. (2019). Critical analysis on employment of graduates in Oman. *Semantic Scholar*. https://api.semanticscholar.org/CorpusID:221134568

Fetters, M. D., Curry, L. A., & Creswell, J. W. (2013). Achieving integration in mixed methods designs—principles and practices. *Health services research, 48* (6pt2), 2134–2156.

General Secretariat for Development Planning (2008) Qatar National Vision 2030. Available at https://www.gco.gov.qa/en/about-qatar/national-vision2030/ (Accessed: August 2023).

Hart-Rawung, P., & Li, L. (2008). Globalisation and business communication: English communication skills for Thai automotive engineers. *World Academy of Science, Engineering and Technology, 48*, 320–330.

Joshi, D. H. (2013). Importance of English language for students of engineering. *VSRD International Journal of Technical and Non-Technical Research, 4*(3), 51–52.

Karoly, L. A. (2010). The role of education in preparing graduates for the labor market in the GCC countries. RAND Corporation.

Khan, R., & Chaudhury, T. A. (2012). The Bangladeshi employment sector: Employer perspectives concerning English proficiency. *Indonesian Journal of Applied Linguistics, 2*(1), 116–129.

Kroon, D. P., Cornelissen, J. P., & Vaara, E. (2015). Explaining employees' reactions towards a cross-border merger: The role of English language fluency. *Management International Review, 55*, 775–800.

Leveson, L. (2000). Disparities in perceptions of generic skills: Academics and employers. *Industry and Higher Education, 14*(3), 157–164.

Merriam, S. B., & Tisdell, E. J. (2016). *Basic qualitative research. Qualitative research: A guide to design and implementation* (4th ed.). Jossey-Bass.

Mustafawi, E., Shaaban, K., Khwaileh, T., & Ata, K. (2022). Perceptions and attitudes of Qatar University students regarding the utility of Arabic and English in communication and education in Qatar. *Language Policy, 21*(1), 75–119.

Nour, S. (2014). Education, training and skill development policies in Arab Gulf countries: Macro-micro overview.

Patil, S., & Karekatti, T. (2012). A study of language learning strategies used by engineering students. *English for Specific Purposes World, 35*(1), 1–19.

Planning and Statistics Authority, Qatar Second National Development Strategy 2018–2022, Hukoomi—Qatar e-Government, Doha, Qatar 2018, https://www.psa.gov.qa/en/nds1/pages/default.aspx (Accessed: August 2023).

Rajprasit, K., Pratoomrat, P., Wang, T., Kulsiri, S., & Hemchua, S. (2014). Use of the English language prior to and during employment: Experiences and needs of Thai novice engineers. *Global Journal of Engineering Education, 16*(1), 27–33.

Robson, C., & McCartan, K. (2016). *Real-world research.* John Wiley & Sons Ltd.

Sarudin, Isarji, Ainol Madziah Zubairi, Mohamad Sahari Nordin, and Mohd Azmi Omar. (2008). The English language proficiency of Malaysian public university students. *Enhancing the quality of higher education through research: Shaping future policy* (2008): 40–65.

Senanayake, P. C. (2019). Impact of Rajarata University, management graduates' English language proficiency on employability. *International Journal of Scientific and Research Publications, 9*(8), 556–565.

Singh, M. K. M., & Choo, J. C. S. (2012). Manufacturing industry employers' perception of graduates' English language skills proficiency. *International Journal of Applied Linguistics and English Literature, 1*(4), 114–124.

Stake, R. E. (1995). *The art of case study research.* Sage Publications.

Thomas, A., Piquette, C., & McMaster, D. (2016). English communication skills for employability: The perspectives of employers in Bahrain. *Learning and Teaching in Higher Education: Gulf Perspectives, 13*(1), 36–52.

Ting, S. H., Marzuki, E., Chuah, K. M., Misieng, J., & Jerome, C. (2017). Employers' views on the importance of English proficiency and communication skill for employability in Malaysia. *Indonesian Journal of Applied Linguistics, 7*(2), 315–327.

Webb, G. (2010). Enhancing the employability of language graduates. *Assessment, Teaching & Learning Journal, pp. 10*, 30–32.

Wu, Y., Xu, L., & Philbin, S. P. (2023). Evaluating the role of the communication skills of engineering students on employability according to the outcome-based education (OBE) theory. *Sustainability, 15*(12), 9711.

Zaharim, A., Yusoff, Y. M., Mohamed, A., Omar, M. Z., Muhamad, N., & Mustapha, R. (2010). Practical framework of employability skills for engineering graduate in Malaysia. In *IEEE EDUCON 2010 CONFERENCE* (pp. 921–927). IEEE.

Chapter 25
Conclusion

Evren Tok, Logan Cochrane, Esmat Zaidan, and Asmaa Al-Fadala

Abstract In the concluding chapter of this interdisciplinary exploration into the transformative journey of education in Qatar, we delve into a landscape characterized by significant ambition, challenge, and a commitment to a future that prioritizes culture, identity, and sustainability. These aims are outlined in the Qatar National Vision, which education is a key enabler and contributor to (via the human development pillar). The narrative and the analysis presented throughout this book underscore Qatar's strategic pivot from an oil and gas-dependent economy toward a diversified, knowledge-based economy (KBE).

25.1 Introduction

In the concluding chapter of this interdisciplinary exploration into the transformative journey of education in Qatar, we delve into a landscape characterized by significant ambition, challenge, and a commitment to a future that prioritizes culture, identity, and sustainability. These aims are outlined in the Qatar National Vision, which education is a key enabler and contributor to (via the human development pillar). The narratives and the analyses presented throughout this book underscore Qatar's strategic pivot from a hydrocarbon-dependent economy toward a diversified, knowledge-based economy (KBE). This transition, emblematic of the broader development trajectory of the Gulf Cooperation Council (GCC) nations, necessitates a profound reimagining of educational paradigms to foster innovation, entrepreneurship, and a workforce that is not only equipped for the job market of today but is adaptable to the challenges and opportunities of tomorrow (Tok et al., 2016). Yet, this transition is not easy, as many of the chapters in this collection highlight.

The challenges highlighted in this book in transforming the education system are mirrored in the challenges of transforming the economy and society. Thus, these challenges are not only interconnected and synergistic, they require systemic and

E. Tok · L. Cochrane · E. Zaidan (✉) · A. Al-Fadala
College of Public Policy, Hamad Bin Khalifa University, Doha, Qatar
e-mail: ezaidan@hbku.edu.qa

E. Zaidan et al. (eds.), *The Future of Education Policy in the State of Qatar*, Gulf Studies 21, https://doi.org/10.1007/978-981-97-9667-0_25

multi-sectoral change. Coordination between sectors is required, lest transformations in the sectors diverge and create policy incoherence. The Qatar National Vision 2030 has played a broad, strategic role in this orientation, while the specific National Development Strategies have guided the sectoral processes in working toward the 2030 goal (Zaidan et al., 2024). Balancing the economy and environment is just as critical as balancing the pursuit of global excellence while cherishing local languages and indigenous knowledge. The balance between global engagement and preserving local identity within this knowledge-based economy has been influenced by various systemic factors and requires on-going attention, adjustment and adaptation in the education system. Now, the future of education policy in Qatar is expected to better take these factors into consideration, aiming for education that targets multiple moving objectives. The future of education policy needs to adeptly respond to both endogenous and exogenous factors, underscoring the importance of Education for Development, Diversification, and Sustainability. This approach ensures that Qatar's educational evolution aligns with its broader economic, societal, and environmental ambitions, thereby preparing a generation that can navigate and contribute to a world defined by both local imperatives and global excellence (Tok et al., 2016).

25.2 Qatar's Educational Evolution: Strategic Perspectives for 2024–2030

The State of Qatar has achieved extraordinary growth in its K-12 education sector and this achievement has been presented in the Qatar National Development Strategy (NDS). This remarkable progress is highlighted by its leap from the 61st position to being ranked among the top 50 worldwide by 2022. However, the journey is not devoid of hurdles; the nation grapples with challenges like subpar performance in the OECD's Programme for International Student Assessment (PISA), a low tertiary enrollment rate among Qatari males, and a limited share of graduates in STEM disciplines. In this study we shed light on how these challenges do not merely reflect on educational achievements but also have wider repercussions on the integration of Qataris into the private sector.

The Qatar National Development Strategy (NDS) for 2024–2030 delineates an ambitious blueprint to transform the nation into a leading hub for higher education. This goal is anchored in capitalizing on prior investments and the multitude of institutions that have made Qatar their home. The strategy emphasizes the need to leverage these assets to nurture graduates of international caliber across diverse economic sectors. At the heart of this vision lies the objective to refine the higher education ecosystem, ensuring the production of graduates who are intellectually proficient and prepared to navigate the complexities of a rapidly evolving global economy.

A cornerstone of the NDS is the drive to attract private sector investments to diversify the economy and accommodate an expanding student demographic. This endeavor is critical for creating an environment where academic institutions and the private sector collaborate for the needs of today and those of tomorrow. Furthermore, escalating the output of STEM graduates is identified as vital for fostering innovation and propelling Qatar toward its goal of economic diversification. The strategy aims for a minimum of 18% of students to graduate in STEM fields by 2030, a strategic move to close the existing skill gap and synchronize Qatar's workforce with the imperatives of a knowledge-driven economy.

In sync with its vision to transition into a knowledge-centric economy, the NDS articulates bold objectives for workforce enhancement. These include a commitment to ensure that over 46% of the workforce occupies skilled and highly skilled roles, alongside a surge in Qatari participation in the private and semi-private sectors to over 20%. Realizing these goals requires a holistic strategy, integrating educational excellence, vocational prowess, and ongoing professional advancement.

Moreover, the strategy ambitiously targets an average PISA score of 483 by 2030. This benchmark transcends a mere numerical goal, embodying a pledge to elevate educational quality and equip Qatari students with the critical thinking and problem-solving aptitudes crucial in today's world. Various chapters in this study provide an academic and policy related background on how this target is critical for Qatar and the historical background.

The Qatar National Development Strategy 2024–2030 sets a foundation for the transformation of the nation's educational landscape and, consequently, its society and workforce. Through initiatives aimed at bolstering higher education, nurturing STEM competencies, engaging the private sector, and establishing stringent educational benchmarks, Qatar has the potential to elevate its educational stature globally. These strategic endeavors are pivotal in preparing Qatari citizens to actively engage in and contribute to a diversified, knowledge-based economy. The fruition of this ambitious strategy hinges on the synergistic efforts of governmental bodies, educational institutions, the private sector, and society, a new era in Qatar's educational and economic progression which have been analyzed in this book.

25.3 Education System at a Crossroads: Western and Non-Western Debates

The nexus between the aspirations of a nation and the evolution of its educational system is a recurring theme in the discourse about balancing global aspirations and retaining local language, culture and values. The educational sector has undergone transformations over the last three decades, primarily through a top-down approach, including shifts of system structure, language of instruction and direction of curriculum (Zaidan et al., 2023; Ghamrawi, 2023). With each change, lessons have been learned and further adaptations undertaken. Multiple chapters in this book highlighted the need to move away from standardized and imported curricula, training, and approaches and tailor learning to the specific context within which it occurs. Context is central for relevance, which is not necessarily provided by externally developed content. Learning may relate only in the abstract, but not in the lived experience, if content is developed for a very different socio-cultural and environmental setting. Beyond relevance, these forms of engagement may also alienate learners from their surrounding community, if they are not engaging with and being inspired by their surroundings. While localization resonates with many authors in this book, the implementation of that localization requires careful consideration, an area that requires much more interrogation. For example, in calling for non-Western approaches and content, at times this recommendation remains vague and/or debated regarding what is called for exactly. Does that mean, for example, a worldview, narrative, and orientation to tradition and culture which may be historical in focus? One chapter in this book called for a re-orientation of educational experiences from the globalized standard to localized ones, embedded within local epistemologies, ethics, and pedagogies. The authors in this book provoke us to think about change in the educational system, but in so doing we conclude this book by calling for additional work to specify exactly which new directions the education system should develop.

One area where K-12 and Higher Education experts appear to agree is the importance of improving Arabic language learning opportunities for all learners. This has pragmatic reasons, as the job market demands these skills, but also relates to the education system at a crossroads where many private K-12 schools and many higher educational institutions operate almost entirely in English (and/or have their primary language of instruction being English). This has not always been the case, and the State of Qatar has been at crossroads before—in the early 2000s English was adopted, and since the 2010s there has been a slow return to Arabic language instruction in private schools (with government schools returning to Arabic as a primary language of instruction). Multilingual learning settings are not unique, and many countries successful have such educational systems. At present, most K-12 and all Higher Educational Institutions are monolingual settings (in specific settings) with second language learning as a subject matter. As many authors have called for Arabic language expansion, the next step is learning from other countries on how this can be implemented (and likewise, learning the steps to avoid that have not been successful).

In this book, higher education emerges as a pivotal force in the shift toward a knowledge-based society, playing a crucial role in cementing Qatar's status as an education hub. This is through not only the training of workers in the knowledge economy but also teachers who are expected to convey all the transformations described in their classrooms in engaging and innovative ways. The training of future educators, as offered by higher education institutions in the country, is an area where scholars have called for reform. Scholars mentioned aspects for reform in the teacher training domain, such as novel approaches for teaching in multi-lingual environments and problem-based approaches, as well as ensuring teachers are appropriately trained in specialized subjects (e.g., being trained in sciences in English at the university and being expected to teach those subjects in Arabic). One chapter highlighted the use of futures thinking as a strategic planning approach students can engage with as an applied approach for teaching and learning that also embodies twentieth-century skills. The chapters also noted the broader role and consideration of emerging issues within learning objectives of teacher training programs, such as sustainable development, entrepreneurship and utilizing technological innovations as well as other twenty-first-century skills. Given the static nature of teacher training degree and the dynamic nature of changing teaching environments, ongoing professional development was highlighted in this collection as critical.

Attracting and retaining people to the teaching profession is not a challenge specific to Qatar, but one that it also faces. Additionally, it is a challenge that has some unique expressions within the country, such as the specific difficulty of attracting male nationals to the teaching profession. Chapters in this book have suggested improving the pathways into education, benefits, and ongoing support via professional development and governance strengthening. The use of technology was promoted as an opportunity to alleviate burdens, such as the potential for individualized tutoring using chatbots powered by artificial intelligence. While we are supportive of this potential, it also requires significant safeguards and monitoring by the government, to ensure the rights of students are protected (e.g., privacy) as well as ensuring the content aligns with the learning objectives of the curriculum. The latter point resonates with localization, not only of content, but also of technological solutions. And, once localized, to ensure that the professional development supports are in place to train teachers to effectively use them. Universities are already utilizing emerging technologies, as outlined by the partnership between higher education institutions in Oman and Qatar in this collection.

25.4 Disparities and Bridging Gaps

The expansion of privatization has created opportunities, but it has also created financial barriers. Access to the highest quality experiences has often been associated with the highest fees, which can be out of reach for the majority of families. This process runs the risk of turning a public good that aims to serve the public, to a service provided to customers who can afford it. Indeed, it is the responsibility of the government, as outlined in the Sustainable Development Goals and the 2030 Agenda, to ensure all people have access to quality and inclusive education. The pursuit of excellence, as is envisioned to occur via market mechanisms of competition, can have unintended impacts of exclusion. This is not only the case for financial exclusion, but it is also the case for inclusive education that caters to people of all abilities, which is discussed in the following sub-section in more detail. While gains have been made in some aspects of the education system, the transformation has not produced the envisioned equity within the educational system. The division between government and private schools in terms of primary language of instruction, (perceptions of) educational quality, socio-economic status, and implicit cultural values has created unintended gaps, limiting opportunities and instilling divergent worldviews. The call for localization, noted throughout the book, draws attention to these tensions in different ways.

Gender has emerged as another divisive factor, further complicating the socio-cultural landscape and necessitating future reforms aimed at bridging these gaps to promote a unified sense of identity and shared values, particularly among citizens (Zaidan et al., 2023, p. 21). This includes gendered differences throughout the educational ecosystem, from individual student experiences to employment opportunities as well as societal role models. The chapters in this book have highlighted gendered trends in the macro, but also identified specific issues, such as barriers for female learners as parents, challenges in the recruitment of male national teachers, and gendered barriers in employment particularly for mothers due to the lack of flexible work options. In making these points, the chapters also note that socio-cultural norms ought not to be imposed; the agency and personal choice of people was expressed in research studies, and not being employed while being highly educated is not necessarily a problem. We ought to avoid the instrumentalization of education, and recognize that for some the pursuit of knowledge for the sake of personal advancement alongside the choice to not work is not a failure, indeed it might also be celebrated. This is a second call for localization, as the metrics and norms set afar may not reflect the realities of people in all places, at all times.

How can these gaps be bridged? The chapters suggest that data-driven approaches combined with the integration of innovative uses of technology could play a positive role. There are also calls for new approaches to teaching and learning. Integrating twentieth-century skills, for example, has the potential to shift from content to skills-based learning and experiences. Several authors in this book noted the need to deepen education for sustainable development, while others called for entrepreneurship education, and another for financial education. These are not mutually exclusive recommendations, however, it leaves us with the challenge of how all of these calls for holistic approaches to multiple topics are feasible to deliver on. It may be that these aims need to look beyond the formal educational system.

The learning ecosystem approach provides an answer by complementing formal with informal educational experiences, wherein these various skills and knowledges can be engaged with. Indeed, the learning ecosystem takes these opportunities one step further than more out-of-school offerings. The SDGs envision access to quality and inclusive (formal) education as a goal, while the ecosystem approach suggests out-of-school (informal) experiences should also be entitlements every learner has. Furthermore, the ecosystem approach advocates, that these experiences should be validated so they can be communicated alongside the formal educational qualifications. Qatar has a wide range of experiences with programs of this type, as outlined in the chapter on the Qatar Reads program. Scholars in this book have provided ideas and examples for expansions, such as the implementation of entrepreneurship incubators as a support for entrepreneurship education, which may work in tandem with existing efforts such as the Qatar Scientific Club (among many others). Effectively implementing this ecosystem approach requires not only more informed choices, as one chapter called for support for school advisors and counselors, but that these guides would need to be aware of a much broader set of opportunities.

The disparities and gaps noted in this book, some of which have been summarized here, are substantial challenges. Strategic change management will be required, with strong monitoring and evaluation mechanisms to support organizational learning. The leader of this change will need to be the Ministry of Education and Higher Education, however, it cannot bear the burden alone, with all the actors in the educational ecosystem engaging, enabling, and supporting the transformation.

25.5 Educational Equity and Inclusion

As Qatar and its GCC counterparts stand at a critical juncture, the dual challenge of preserving cultural identity while integrating into the global knowledge economy becomes apparent. This complex scenario demands a multifaceted approach to education that considers job readiness, civic engagement, leading-edge research, and the socio-technological realities of the twenty-first century (Qatar Foundation, 2024). The concept of "Education for Resilience," as highlighted during the Teaching and Learning Forum organized by the Education Development Institute of Qatar Foundation, emphasizes adapting to these challenges through innovation, inclusivity, and global engagement. The integration of glocalized curriculums, which align global standards with local relevance, is a critical strategy for ensuring vibrant and meaningful education in Qatar and the Gulf region. By linking educational content to the United Nations Sustainable Development Goals and providing students with tools to navigate and shape a rapidly changing world, the education system fosters a sense of global citizenship and interconnectedness (UNESCO, 2021). However, the pursuit of educational excellence must also address the imperatives of equity and inclusion to overcome current disparities and exclusions.

The State of Qatar has made notable progress in providing access to quality education to people of diverse abilities. One chapter outlined the history, and rather rapid transformation that has occurred over the most recent decade in delivering on this aspiration. However, ensuring all people have access to quality and inclusive education remains an area that requires significant attention. The work ahead lies less in institution building and more in governance of the institutions, filling critical capacity gaps, while also raising broader public awareness and providing appropriate parental support. Relatively little has been written about this subject within Qatar, and this book made a significant contribution in this regard.

25.6 Looking Toward a Resilient Future and Role of Context in Policymaking

Each of the challenges noted in this book is an opportunity. Learning about areas of potential improvement is a chance for us to better serve the learners throughout the nation. In conclusion, the future of education in Qatar and the Gulf region presents a realm of opportunity characterized by innovation, inclusivity, and international collaboration. Qatar's educational reforms and policy changes offer valuable lessons for similar nations striving to enhance their educational systems while maintaining internal cohesion, values, and culture. The complex interplay between global trends and local traditions underscores the quest for a balanced, resilient, and knowledge-driven future. The evolution of Qatar's education system is a testament to the nation's dedication to building a future that honors its past while embracing the possibilities of the knowledge age. This journey, though filled with challenges, sets a benchmark for educational reform that is both locally grounded and globally resonant, positioning Qatar and its GCC peers as leaders in creating a sustainable, knowledge-based future. It is imperative that we continue to conduct research on balancing student expectations with the level of indigenization, and the feasibility of importing and preserving Western-style epistemology and pedagogy in a context like Qatar (Romanowski et al, 2023, p.12) to ensure we continue to adapt and advance as the needs of learners, and the world they engage with, change.

Bibliography

Amin, H., & Cochrane, L. (2023). The development of the education system in Qatar: assessing the intended and unintended impacts of privatization policy shifts. *British Journal of Middle Eastern Studies*, 1–21.

Ghamrawi, N. (2023). Toward agenda 2030 in education: policies and practices for effective school leadership. *Educational Research for Policy and Practice*, 1–23.

Huda, A. K. (2022). Centralize or decentralize? The question currently facing schools in Qatar. *International Journal of Learning, Teaching and Educational Research, 21*(2), 397–418.

Mihalache, S. (2024). Education Trends in GCC Countries: Over 1000 New Schools, a 1.5% Technology-Driven Growth by 2027. Accessed February 17, 2024. https://www.kinderpedia.co/education-trends-in-gcc-countries-by-2027.html.

Qatar Foundation (2024). Ideas for the Future of Schools Explored at QF's Teaching & Learning Forum. Accessed February 17, 2024. https://www.qf.org.qa/stories/ideas-for-the-future-of-schools-explored-at-qfs-teaching-and-learning-forum.

Romanowski, M. H., Tok, E., Amatullah, T., Amin, H., & Sellami, A. (2023). Globalisation, policy transferring and indigenisation in higher education: the case of Qatar's education city. *Journal of Higher Education Policy and Management*, 1–15.

Sellami, A., Sawalhi, R., & Al-Fadala, A. E. (2023). K-12 Educational Leadership and Administration in Qatar: Lessons Learned. In *Demystifying Educational Leadership and Administration in the Middle East and North Africa* (pp. 187–208). Routledge.

The Peninsula (2022). "Focus on Quality Education in Ministry's 2023–30 Plan," December 28, 2022. https://thepeninsulaqatar.com/article/28/12/2022/focus-on-quality-education-in-ministrys-2023-30-plan.

The Peninsula (2023). "Education Ministry Rolls out New Evaluation Policy," May 8, 2023. https://thepeninsulaqatar.com/article/08/05/2023/education-ministry-rolls-out-new-evaluation-policy.

Tok, M. Evren, Lolwah R. M. Alkhater, and Leslie A. Pal. (2016) "Policy-Making in a Transformative State: The Case of Qatar." In *Policy-Making in a Transformative State: The Case of Qatar*, edited by M. Evren Tok, Lolwah R. M. Alkhater, and Leslie A. Pal, 1–35. London: Palgrave Macmillan UK.

UNESCO (2021). *Reimagining Our Futures Together: A New Social Contract for Education.* UNESCO, 2021. https://doi.org/10.54675/ASRB4722.

Zaidan, E., Belkhiria, E., Wazen, C. (2023). Universities of the future as catalysts for change: Using the sustainable development goals to reframe sustainability—Qatar University as a case study. In: Al-Maadeed, M.A.S.A., Bouras, A., Al-Salem, M., Younan, N. (eds) *The Sustainable University of the Future*. Springer.

Zaidan, E., Momani, R., & Al-Saidi, M. (2024). Entrepreneurial universities and integrated sustainability for the knowledge-based economy: Self-perception and some structural challenges in the Gulf region. *Humanities and Social Sciences Communications, 11*(1), 1–14.

GPSR Compliance

The European Union's (EU) General Product Safety Regulation (GPSR) is a set of rules that requires consumer products to be safe and our obligations to ensure this.

If you have any concerns about our products, you can contact us on ProductSafety@springernature.com

In case Publisher is established outside the EU, the EU authorized representative is:

Springer Nature Customer Service Center GmbH
Europaplatz 3
69115 Heidelberg, Germany

Batch number: 08329220

Printed by Printforce, the Netherlands